BIOMECHANICS OF THE
MUSCULO-SKELETAL SYSTEM

BIOMECHANICS OF THE MUSCULO-SKELETAL SYSTEM

B.M. NIGG

W. HERZOG

Editors

University of Calgary
Calgary, Alberta, CANADA

JOHN WILEY & SONS

Chichester · New York · Brisbane · Toronto · Singapore

Reprinted with corrections May 1995

Other Wiley Editorial Offices

John Wiley & Sons, Inc., 605 Third Avenue,
New York, NY 10158-0012, USA

Jacaranda Wiley Ltd, 33 Park Road, Milton,
Queensland 4064, Australia

John Wiley & Sons (Canada) Ltd, 22 Worcester Road,
Rexdale, Ontario M9W 1L1, Canada

John Wiley & Sons (SEA) Pte Ltd, 37 Jalan Pemimpin #05-04,
Block B, Union Industrial Building, Singapore 2057

Library of Congress Cataloging-in-Publication Data

Nigg, Benno Maurus.
    Biomechanics of the musculo-skeletal system / B.M. Nigg, W.
Herzog.
        p.   cm.
    Includes bibliographical references and index.
    ISBN 0 471 94444 0
        1. Biomechanics.   2. Musculoskeletal system.   I. Herzog, Walter.
II. Title.
QP303.N478   1994
612.7′6—dc20   94-1827
                                                                      CIP

Library Cataloguing in Publication Data

A catalogue record for this book is available from the British Library

ISBN 0 471 94444 0

Produced from camera-ready copy supplied by the authors.
Printed and bound in Great Britain by Bookcraft (Bath) Ltd, Midsomer
Norton, Avon

TABLE OF CONTENTS

CONTRIBUTORS

Authors

Marcelo Epstein

Mechanical Engineering
The University of Calgary
2500 University Drive NW
Calgary, Alberta, CANADA
T2N 1N4

Cy B. Frank

Department of Surgery
The University of Calgary
3330 Hospital Drive NW
Calgary, Alberta, CANADA
T2N 4N1

Walter Herzog

Human Performance Laboratory
The University of Calgary
2500 University Drive NW
Calgary, Alberta, CANADA
T2N 1N4

Benno M. Nigg

Human Performance Laboratory
The University of Calgary
2500 University Drive NW
Calgary, Alberta, CANADA
T2N 1N4

Nigel G. Shrive

Civil Engineering
The University of Calgary
2500 University Drive NW
Calgary, Alberta, CANADA
T2N 1N4

Co-authors

Paul Binding

Department of Mathematics and Statistics
The University of Calgary
2500 University Drive NW
Calgary, Alberta, CANADA
T2N 1N4

Gerald K. Cole

Human Performance Laboratory
The University of Calgary
2500 University Drive NW
Calgary, Alberta, CANADA
T2N 1N4

Susan K. Grimston

Human Performance Laboratory
The University of Calgary
2500 University Drive NW
Calgary, Alberta, CANADA
T2N 1N4

Antonio C.S. Guimaraes

Human Performance Laboratory
The University of Calgary
2500 University Drive NW
Calgary, Alberta, CANADA
T2N 1N4

Barbara Loitz

Joint Injury and Arthritis Research Group
The University of Calgary
3330 Hospital Drive NW
Calgary, Alberta, CANADA
T2N 4N1

Ton J. van den Bogert

Human Performance Laboratory
The University of Calgary
2500 University Drive NW
Calgary, Alberta, CANADA
T2N 1N4

Yuanting Zhang

Human Performance Laboratory
The University of Calgary
2500 University Drive NW
Calgary, Alberta, CANADA
T2N 1N4

For further information contact the Human Performance Laboratory,
University of Calgary, Alberta, CANADA

Tel: (403) 220-3436
Fax: (403) 284-3553

PREFACE

Biomechanics of the Musculo-skeletal System is intended for students and researchers in biomechanics. In writing this book we have assumed that the reader has a basic background in mathematics and mechanics. Through this book, the authors hope to stimulate growth and development in the exciting field of biomechanics.

At present, thousands of researchers work in areas that are included under the umbrella of the term biomechanics. The research interests among biomechanists differ greatly. This book focuses on the biomechanics of the musculo-skeletal system and it contains chapters on specifically selected areas reflecting the interests of the authors:

> Definition and historical highlights
> Biomaterials
> Measuring techniques
> Modelling

Chapter 1 contains a definition of biomechanics followed by a section on selected historical highlights. Ideas, thoughts, and concepts about the human body and its movement existed in ancient Greek and Roman times, and were at the centre of interest for science and art through the centuries. Important contributions and historical developments are discussed within the socio-cultural environment, since the historical developments took place within, and were determined by the framework of the cultural conditions of that time.

Chapter 1 also includes a summary of basic principles in mechanics to the musculo-skeletal system. This section begins with the general equations of three-dimensional motion, the variables in these equations are explained, and a summary on how these variables can be determined experimentally or theoretically is given. This chapter does, however, not pretend to provide the information typical in an advanced textbook in mechanics.

Chapter 2, Biomaterials, discusses the morphology, histology, and mechanics of bone, cartilage, ligament, tendon, and muscle as they relate to biomechanics. Where appropriate, biomechanical properties are discussed as a function of age, gender, and activity. Selected injury and injury-repair mechanisms of the different biomaterials are described and compared. Chapter 2 emphasizes the *mechanical* aspects of the biomaterials.

Chapter 3, Measuring Techniques, describes selected methods for quantifying biomechanical data experimentally. This chapter includes sections on force measurements, accelerometry, measurement of motion with optical methods, electromyography, strain measurements, and discusses methods for determining inertial parameters of the human or animal body. The primary goals of this chapter are to explain the principles involved in the experimental techniques, and to compare the different techniques to each other. Little emphasis is given to technical details, since they are subject to rapid change.

Chapter 4, Modelling, starts with a philosophical discussion of the importance and the place of mathematical modelling in biomechanics. Then the concept of a free body diagram is introduced and discussed. The sections in this chapter are extensive because of the difficulty of finding publications which introduce modelling in biomechanics. The section energy considerations is an introduction to thermomechanical systems, with some considerations given to the energetics of contraction of skeletal muscle. In the last section of chapter 4 simulation as a tool of biomechanical research is discussed.

During the preparation of this book all the authors worked at the University of Calgary, Canada. In addition to collaborating on this publication, they were involved in joint research projects, providing their particular expertise in specific areas of biomechanics. Many examples in the book are taken from the authors' own work. The attempt was made, however, to distinguish between the general principles and the specific examples.

The various drafts of the book chapters or sections have been given to many colleagues for review in the process of writing. We gratefully acknowledge the helpful and valuable contributions of the following principal reviewers:

Andrews, J.,	The University of Iowa, USA
Binding, P.,	The University of Calgary, Canada
Bobbert, M.F.,	Free University Amsterdam, Netherlands
Denoth, J.,	ETH Zürich, Switzerland
Falck, J.,	The University of Calgary, Canada
Gautschi, G.,	KISTLER Instruments, Winterthur, Switzerland
Hay, J.G.,	The University of Iowa, USA
Lieber, R.,	The University of California, USA
Nicol, K.,	University of Münster, Germany
Norman, R.,	The University of Waterloo, Canada
Yeadon, M.R.,	Loughborough University, UK
Zernicke, R.,	The University of Calgary, Canada
Zhang, Y.,	The University of Calgary, Canada

Many co-workers of the different Calgary biomechanics research groups at the University of Calgary have contributed intellectually to the development of the final product. They provided critical comments, discussions, and were involved in the internal review process. We are indebted to them for their help in shaping the current form of the book. They include:

Allinger, Todd	Guimaraes, Antonio	Prilutzky, Boris
Anton, Michael	Hamilton, Gordon	Reinschmidt, Christoph
Bergman, Joseph	Loitz, Barbara	Ronsky, Janet
Cole, Gerald	Leonard, Timothy	Stefanyshyn, Darren
Damson, Erich	Mattar, Johnny	Stergiou, Pro
de Koning, Jos	Morlock, Michael	Vaz, Marco
Forcinito, Mario	Naumann, Andre	Wilson, Jackie
Gal, Julianna	Pitzel, Barb	

Additionally, we are indebted to many co-workers at the University of Calgary for their help in preparing different aspects of the book: Dale Oldham for the preparation of the figures; Karen Turnbull and Lindsay Campbell for the scientific editing of the text; Trish Jordan for reference checks; Geoff Elliott for proofreading; Glenda McNeil and Byron Tory for their help in computer related aspects; Colleen Robertson and Veronica Fisher for their help in many administrative capacities.

First editions of textbooks usually contain (often stupid) errors and inconsistencies. The authors welcome communications from students and colleagues about concerns or errors found in this edition of *Biomechanics of the Musculo-skeletal System,* or any other information that may make a possible second edition a better product.

Calgary, December 1993,   Benno M. Nigg
                                                                 Walter Herzog

1 INTRODUCTION

Biomechanics is a discipline of science, newly developed and in the process of becoming established. Two developments are often seen in such a situation: first, scientists attempt to define the new discipline, and second, scientists attempt to understand the roots and the historic development of their new discipline. This introductory chapter provides information on these two aspects. A proposed definition of biomechanics is introduced and selected historical facts are discussed describing their possible relevance to the development of modern biomechanics. Since biomechanics is a young discipline of science, historical highlights are not well-defined, allowing for many avenues to be opened for discussion and speculation. The reader may want to digest this chapter with care. Finally, a short comment about mechanics is presented, emphasizing how the different variables in the equations of motion can be determined.

1.1 DEFINITION OF BIOMECHANICS

NIGG, B.M.

Our undying curiosity concerning our own and other species' biological functioning dates back as far as any other scientific attempt to probe specific aspects of life. Initial studies included attempts at understanding the function of the lung, heart, peripheral and central nervous systems, muscles, and joints. One discipline borne of this early exploration was the study of the mechanics of living systems, a science we today call *biomechanics*. The definition of biomechanics has not yet achieved a consensus. Some authors propose broad definitions, others rather limited ones. For this book, an adaptation of Hay's (1973) definition of biomechanics is used:

> **Biomechanics is the science that examines forces acting upon and within a biological structure and effects produced by such forces.**

External forces acting upon a system are quantified using sophisticated measuring devices. Internal forces, which result from muscle activity and/or external forces, are assessed using implanted measuring devices or estimations from model calculations. Possible results of external and internal forces are:

* Movements of segments of interest, and
* Biological changes in the tissue(s) on which they act.

Consequently, biomechanical research, studies and/or quantifies:

* Movement of different body segments and factors that influence movement (e.g., body alignment, weight distribution, equipment), and
* Biological effects of locally acting forces on living tissue - effects such as growth and development or overload and injuries.

The definition suggested by Hay is analogous to the one proposed by Hatze (1971, p. 33). Hatze stated: "Biomechanics is the science which studies structures and functions of biological systems using the knowledge and methods of mechanics" (my rather liberal translation from German).

Biomechanical research addresses several different areas of human and animal movement. It includes studies on (a) the functioning of muscles, tendons, ligaments, cartilage, and bone, (b) load and overload of specific structures of living systems, and (c) factors influencing performance. Subjects of study for human biomechanics include the disabled and the non-disabled, athletes and non-athletes. Subjects range in age from children through to the elderly.

1.2   SELECTED HISTORICAL HIGHLIGHTS

NIGG, B.M.

This chapter discusses factors which are considered important in the development of biomechanics over the centuries. It was attempted to embed the facts of scientific, mechanical and/or biological development into the socio-cultural frame in which they emerged. The facts have been researched as carefully as possible. The interpretation, however, is a personal view, and some readers may not agree with some of the speculations and/or conclusions which have been made.

The historical highlights have been arbitrarily divided into periods which, for the development of biomechanics, have specific importance in the view of the author. The periods and their names used in this chapter are:

Antiquity	650 B.C. - 200 A.D.
Middle Ages	200 B.C. - 1450 A.D.
Italian Renaissance	1450 A.D. - 1600 A.D.
Scientific Revolution	1600 A.D. - 1730 A.D.
Enlightenment	1730 A.D. - 1800 A.D.
The Gait Century	1800 A.D. - 1900 A.D.
The 20th Century	1900 A.D. - . . . .

These periods do not have a well-defined beginning and end. The years indicated, therefore, are just an approximate guideline, enabling the reader to locate the different periods of the centuries.

1.2.1 THE SCIENTIFIC LEGACY OF ANTIQUITY

Ancient cultures such as the Maya, Egyptians, Mesopotamians, and Phoenicians had well developed knowledge in many disciplines, including astronomy and mathematics. Their attempts to understand nature combined knowledge and myth. The ancient Greeks were the first to attempt to separate knowledge from myth, developing what we would call today "true scientific inquiry". The emergence and development of Greek science from the 6th to the 4th century B.C. was influenced by the Greeks' relative freedom from political and religious restrictions. Their coastal harbors and inland caravan routes, which provided access to trade with Egyptians, Phoenicians, and the whole of Asia also provided the freedom necessary to inspire their ideas. The politics, commerce, and climate of Greece created leisure time, which the Greeks used to pursue mental and physical excellence, enhancing the opportunity for scientific discovery. Harmony of mind and body required athletic activity to complement the pursuit of knowledge. Kinematic representations of Greek athletics dominated artistic media, demonstrating an interest in sport and human movement. Greek artists developed increasingly sophisticated methods and media for depicting motion. Greek sculpture illustrated the dynamics of movement and a working knowledge of surface anatomy.

Scientific development as initiated by the Greeks can be characterized by four stages:

- The emergence of "natural philosophy",
- The "golden age" of Greek science,
- The Hellenistic age, and
- The conquest by the Roman Empire.

THE EMERGENCE OF "NATURAL PHILOSOPHY"

The "natural philosophy" movement was founded by Thales (624-545 B.C.). It was based on the study of natural science detached from religion. During this period, Greek science developed the basic elements of mathematics, astronomy, mechanics, physics, geography, and medicine. Two personalities are of particular interest in the context of this book, Pythagoras and Hippocrates.

Pythagoras (about 582 B.C.), born in a time of awakening, formed a brotherhood based on his philosophical and mathematical ideas which included both men and women living together in a communal atmosphere. His disciples were divided into two hierarchial groups: the scientists and the listeners. The scientists were permitted to ask the master questions and to postulate their own ideas, while the listeners were confined primarily to study.

According to Pythagoras "...all things have form, all things are form, and all forms can be defined by numbers" (from Koestler, 1968, p. 30). His arithmetic was based on dots that could be drawn in the sand, or with pebbles to depict geometrical shapes. Numbers like 3, 7, and 10 formed triangles, while 4, 9, and 16 formed squares. Some numbers, such as 10, were given mythical qualities. Experimentation with these shapes and numbers may have resulted in Pythagoras' famous theorem for rectangles and triangles: $a^2 + b^2 = c^2$. His definitions of the universe and the human body were based on his mathematical analysis of music. He regarded the universe and the human body as musical instruments, whose strings required balance and tension to produce harmony. Pythagoras' geocentric analysis of the universe was based on spherical planets, whose path around the earth created a musical hum. The hum for each planet was unique, producing an eternal music box.

Pythagoras attempted to synthesize the concepts of the Egyptians and Babylonians to create a foundation for future investigations. His lifelong devotion to mathematics set the tone and direction of science in the Western world. 2500 years after his death, he continues to shape our lives. His mathematics are used by school children, and his philosophies permeate our language. The word philosophy itself, along with harmony, tone, and tonic, have a Pythagorean heritage. Pythagoras believed that mathematical relations held the secrets of the universe.

Hippocrates (460-370 B.C.) travelled extensively throughout Greece and taught in Cos. The exceptional fame he enjoyed during his lifetime continued long after his death. Hippocrates' rational scientific approach to diagnosis freed medicine from supernatural constraints. For him, observation was based entirely on sense perceptions, and diagnostic errors were admitted and analysed. Hippocrates was well aware of the limitations of his era, expecting future generations to question and improve upon his work. His belief in the principle of causality "...that chance does not exist, for everything that occurs will be found to do so for a reason" (from Sarton, 1953, p. 135) confirms his commitment to forming a rational science. Hippocrates' emphasis on observation and experience of the senses pioneered the use of rational scientific thought in the practice of medicine.

THE "GOLDEN AGE" OF GREEK SCIENCE

The 4th century B.C. represented the "golden age" of Greek science, with the two main exponents, Plato and Aristotle. In this time, politics and ethics displaced the natural sciences as a means of attaining truth and knowledge. The disintegration of Greek civilization during the 4th century B.C. set the tone for both Plato and Aristotle's teachings. A century of constant war and political corruption demoralized the population. Materialism twisted the philosophy of personal excellence in mind and body, as a population of spectators developed to watch gladiators duel to the bitter end. In this environment of instability and violence, a great emphasis was placed on the higher morality of contemplation.

Plato (427-347 B.C.) believed the world of the senses to be an illusory shadow of reality. Ideas were the only reality, and true knowledge could not be obtained through the study of nature. The pursuit of truth required contemplation, not action. Plato discouraged artisans, engineers, and scientists from mechanical trades. His universe was subject to the powers of its divine creator, whereby, the stars and planets were spiritual entities whose movements and nature were derived from these powers.

Aristotle (384-322 B.C.). Aristotle's father, Nichomachos, served as physician to Amyntas II of Macedon, the grandfather of Alexander the Great. As a boy, Aristotle was influenced both by his life at court and medicine. For 20 years, Aristotle studied in Athens, spending several years at the Academy as Plato's pupil. His intellectual curiosity and independent mind eventually led him to question Platonic philosophy. Aristotle's garden or Lyceum, provided a forum for his teachings, eventually becoming the site of his famous Peripatetic school. One of Aristotle's famous pupils at that time was the young Alexander the Great, who was greatly influenced by Aristotle's ethics and politics. Aristotle was a universal research scientist as well as a social commentator. His interests lay in mechanics, physics, mathematics, zoology, physiology, chemistry, botany, and psychology. He believed the aim of science was to explain nature, and that mathematics provided a good model for a well-organized science.

Aristotle believed the senses revealed reality, and ideas were merely abstractions towards mental concepts (Keele, 1983). True knowledge of nature could only be gained through careful observation. Aristotle believed that contemplation was superior to mechanical labor. As a result, his methods for observation did not include verification or experimentation. Aristotle's universe consisted of the four observable elements; fire, air, water, and earth. The elements arranged themselves in concentric spheres, with the earth being the central sphere. Spheres of water, air, and fire extended outward from the centre. Beyond these were the spheres of the planets, stars, and then nothingness. The moon divided this universe into terrestrial and celestial zones, where movements were differentiated. Natural movements, such as a falling stone, were always in straight lines in the terrestrial zone, while movements in the celestial zone were always circular, eternal, and beyond terrestrial laws of nature.

Aristotle believed life was capable of mechanical expression. According to Aristotle, every motion presupposed a mover. His peculiar definition of motion required that all that is moved, be moved by something else. The motor must either be present within the mobile, or be in direct contact with it. Action at a distance was inconceivable. The motions of falling bodies and projectiles fascinated Aristotle. He assumed the average velocity of a falling body over a given distance, to be proportional to the weight of the falling body, and inversely proportional to the density of the medium. He believed that the nature of the objects, not externalities such as distance, determined the speed at which they fell.

Aristotle's theory of the four elements; fire, air, water, and earth, and their four characteristic qualities; heat, cold, humidity, and dryness, combined to produce the four main humors of the body; blood, phlegm, yellow bile, and black bile. Aristotle believed that the heart was the source of human intelligence. Muscular movements were the result of pneuma or breath which passed through the heart to the rest of the body. The concepts of movement and change dominated Aristotle's study of nature. His treatise *About the Movements of Animals*, which was based on observation, described movement and locomotion for the first time. The text provided the first scientific analysis of gait, and the first geometrical analysis of muscular action. Mechanical comparisons illustrated a deeper understanding of the functions of bones and muscles, and Aristotle explained ground reaction forces: "...for just as the pusher pushes, so the pusher is pushed" (from Cavanagh, 1990, p. 5).

THE HELLENISTIC AGE

The conquests of Alexander the Great gave rise to the Hellenistic age. In the 3rd century B.C., Alexandria became the centre for scientific specialization. The Museum of Alexandria, which contained an observatory and laboratories for physiology and anatomy, provided a safe haven for the research scientist. The seclusion of the Museum, and the political freedoms of Alexandria, enhanced the development of both anatomy and physiology.

Herophilos (about 300 B.C.) initiated the foundation of modern anatomy by creating a systematic approach to dissection, identifying numerous organs for the first time. Herophilos was the first to distinguish between tendon and nerve, attributing sensibility to the latter. He also made a distinction between arteries and veins, stating that arteries were six times as thick as veins, and that arteries contained blood not air (Sarton, 1953). Rejecting Aristotle's assertion that the heart was the seat of intelligence, Herophilos revived Alcmaeon's (about 500 B.C.) belief that the brain was the locus of intelligence.

Erasistratis (about 280 B.C.) studied at the Museum of Alexandria as a young contemporary of Herophilos, possibly as his assistant. Erasistratis distinguished himself as more of a theorist than Herophilos by applying physical concepts to the understanding of anatomy. He was the first to describe the muscles as organs of contraction. Like Hippocrates, Erasistratis rejected supernatural causes in explaining nature.

Archimedes (287-212 B.C.) was the son of Phedias, an astronomer who supported his early interest in mathematics and astronomy. Archimedes' fame resulted from his inventions, which included mechanical weapons such as catapults, ingenious hooks, and concave mirrors used to attack and burn marauding Roman ships. Using his compound pulley and the lever, Archimedes claimed that he would be able to move the Earth - if he only had a place to stand in order to do so! Archimedes used geometrical methods to measure curves and the area and volume of solids. Without the use of calculus, Archimedes used a close approximation for π to measure volumes and areas.

In his treatise *Equilibrium of Planes*, Archimedes proved that "...two magnitudes, whether commensurable or not, balance at distances reciprocally proportional to them" (from Sarton, 1953, p. 77). The distances to be considered were the respective distances of their centre of gravity from the fulcrum. He also demonstrated how to find the centre of gravity of a parallelogram, triangle, parallel trapezium, and parabolic segment. Archimedes' treatise *Floating Bodies* described the principle of water displacement, which he is said to have discovered while bathing! Movement occupied Archimedes' investigations as he solved the problem of how to move a given weight by a given force.

Archimedes' application of Euclidian methodology to mechanics formed the basis of rational mechanics, and his mathematical investigation of Aristotle's mechanical theories established statics and hydrostatics. His analyses, proofs, and methodology dominated statics and hydrostatics until the times of Simon Stevin (1548-1620) and Galileo Galilei (1564-1642). Archimedes died during the sack of Syracuse in 212 B.C. He is said to have been absorbed in the contemplation of geometrical figures drawn in the sand and to have shouted "keep off!" to an approaching Roman soldier. Not having appreciated Archimedes' particular mathematical concern, the soldier killed him on the spot.

THE CONQUEST BY THE ROMAN EMPIRE

The emergence of the Roman Empire during the 1st century B.C. eroded the foundations of Greek science. During this period, emphasis was placed on ethics, and scientific activity was reduced to the development of fortification, military weapons, and sources of amusement.

Hero of Alexandria (about 62 A.D.) taught physics at the University of Alexandria. Hero produced treatises on mechanics, optics, and pneumatics, which contained the principles used in his inventions. He is credited with inventing the first rudimentary "steam engine", in which steam propelled a globe in circular revolutions. By using compressed gases and saturated water vapors, Hero devised machines which opened and closed curtains for puppeteers, operated doors, and created a miniature carousel with mechanical birds. He invented a magic jug that used the principle of the siphon to control the flow of water out of it. Hero's mechanical inventions amused and entertained but were never used to provide practical improvements.

Galen (129-201 A.D.) began medical training in Pergamos. Inspired by a dream, his father ensured that Galen studied with the best physicians of the time. Galen's voracious appetite for learning, exceptional brilliance, overriding pride and ambitions, and energetic and sometimes violent nature manifested themselves in his passion to impress the world. Appointed physician to the College of Gladiators at the age of 28, Galen became probably the first "sports physician" and "team doctor" in history. For 4 years he practised surgery and dietetics among the gladiators, gaining substantial knowledge of the human body and human motion. Recognizing Galen's immense talents, Marcus Aurelius appointed him physician to the Emperor, where he remained for 20 years. Galen devoted much of this time to research, publishing over 500 medical treatises. He also wrote numerous philosophical books such as *On the Passions of the Soul and its Errors*, which was largely autobiographical. In fact, our knowledge of the work of many of his predecessors came from Galen's writings. After fire destroyed the temple of Pax and the majority of Galen's manuscripts, he returned to Pergamos, where at the age of 70 he died.

Galen believed medicine was a comprehensive science that included both anatomy and physiology. Like Aristotle, Galen was a teleologist, believing that nature arranged everything according to a distinct plan, with the soul ruling the body; the senses and organs subordinate and subservient. *De Usu Partium* (On the Use of the Parts) gave Galen his transcendent influence on the medical sciences. The first text on physiology, it quickly became the uncontested authority on medicine, and remained so for 1300 years. It was first printed in 1473, and over 700 copies were in existence by 1600, emphasizing the importance of the understanding of structure and function in diagnosis and therapy. Not until Vesalius (1514-1564) and William Harvey (1578-1657) did anatomy experience comparable growth.

De Motu Musculorum (On the Movements of Muscles) embodied Galen's lifelong passion for the mechanism of movement. He made enormous advances in the understanding of muscles, establishing the science of myology. Galen's emphasis on structure revealed the "...innervation that makes muscle substance into a muscle proper" (from Bastholm, 1950, p. 77). The flesh of the muscle contained webs of nerve endings which Galen believed transmitted *spiritus animalius* from the brain into the muscle, stimulating movement. Galen's pneuma theory included the contradictory discoveries that the arteries contained blood, that nerve endings pervaded muscle flesh, and Aristotle's teleological cause of movement - pneuma.

Galen looked to the function of muscles to distinguish between skeletal muscles and muscle parts, such as the heart and the stomach. He believed each muscle had only one possibility for movement and muscular movement that required contraction and the ability to contract, resulted from a property existing throughout the muscle. Galen described tonus and distinguished between agonistic and antagonistic muscles, and motor and sensory nerves.

Obsessed with the elegance of mathematics and trained by the best physicians of his time, Galen sought to raise medicine to the level of an exact science. As an anatomist, his descriptions of muscles and nerves far surpassed those of his predecessors. His correlation of injuries to various parts of the spinal chord and the cranial and cervical nerves, which provided paralysis of certain organs, were and are considered among the most brilliant medical observations of all times. Galen's era discouraged human dissection, and his observations were based almost entirely on the dissections of dogs, pigs, and apes. However, Galen was aware of this deficiency, encouraging students to travel to Alexandria to view the only human skeletons available to science. Unlike Hippocrates, Galen uncritically applied his principles, even in light of experience that contradicted them. Galen was the indisputable authority in his own time, and the Middle Ages further entrenched his authority, discouraging any investigations that revealed contradictory findings.

RELEVANCE TO BIOMECHANICS

The relevance of "Antiquity" to biomechanics lies in four major aspects:

- Knowledge and myth were separated,
- Mechanical and mathematical paradigms were developed,
- Anatomical paradigms were developed, and
- First biomechanical analysis of the human body was performed.

Thales was the first to attempt to separate knowledge and myth. Additionally, some of the ancient Greeks attempted to include observations to develop theories. Pythagoras, Aristotle, and Archimedes laid the foundations for mechanical and mathematical analysis in general, and for mechanical and mathematical analysis of movement specifically.

Hippocrates, Herophilos, Erasistratis, and Galen developed the first anatomical and neurophysiological concepts of the human body. Aristotle wrote the first book about human movement *(About the Movements of Animals),* a first scientific analysis of human and animal movement, based on observation, and describing muscular action and movement.

1.2.2 THE MIDDLE AGES

During the Middle Ages (200 B.C.-1450 A.D.) scientific development decreased, as religious and spiritual development increased. Arab scholars saved the scientific investigations of antiquity from disappearing completely by translating the works from Greek to Arabic. Neoplatonic doctrines, initially established during the 3rd century B.C., represented the main link between antiquity and medieval Europe. Plato's *Timaeus* provided the doctrine of macrocosm and microcosm, which compared the nature and structure of the universe to the nature and structure of man. Analogies were drawn combining the four elements; humors, organs, seasons, compass points, and astronomical events to describe nature, motion, and change. St. Augustine (354-430 A.D.) incorporated Neoplatonic doctrines into Christian beliefs. Discouraging scientific inquiry, St. Augustine explained that "...the only type of knowledge to be desired was the knowledge of God and the Soul, and that no profit was to be had from investigating the realm of nature" (from Koestler, 1968, p. 88). During the 13th century, Aristotle's writings were revived from Arabic translations and St. Thomas Aquinas (1277-74) integrated Aristotelian philosophy into Christian beliefs. Universities run by the Dominicans and Franciscans embraced Aristotle's doctrines.

The Middle Ages contributed little to the development of the field of biomechanics. The developments of the scientific age of antiquity remained relatively dormant throughout the period. Even the revival of Aristotle's writings did not create a renewed interest in investigating and understanding nature. Medicine existed, but any interest in the human body and human locomotion was discouraged. Universities became strongholds of Aristotelian thought, but did not conduct critical research.

RELEVANCE TO BIOMECHANICS

The relevance of the Middle Ages to the development of biomechanics is minimal. Scientific development, in general, was discouraged in this period and consequently, the previously created interest in human and animal anatomy, physiology, and locomotion was put to sleep for more than 1200 years. However, movement depiction flourished in Greek and Roman art, and it would be the artist, rather than the scientist, who would later revive the study of human movement.

1.2.3 THE ITALIAN RENAISSANCE

The Italian Renaissance, which flourished from 1450 until the sack of Rome in 1527, was characterized by freedom of thought, as it saw the revival of ancient Greek philosophy, literature, and art. The political chaos of 15th century Italy provided the backdrop for the eruption of moral and intellectual freedoms. The authority of the ancients replaced the authority of the Church. Man became the "measure of all things", as individual genius flourished in a time unparalleled since the "golden age of anatomy" in Alexandria. Encouraged by such liberal patrons as the Medici in Florence, small numbers of scholars and artists including Michelangelo, Leonardo da Vinci, and Machiavelli emerged as Renaissance men. With few exceptions, science played a limited role during the Renaissance, as an organized scientific community did not exist. The Renaissance, however, created the foundations for the scientific revolution in the 17th century. A revival of Platonic and Aristotelian philosophies broke down the rigid scholastic system of the Middle Ages, introducing the element of choice in intellectual pursuits. Most importantly, intellectual activity became a social adventure no longer subjugated by a predetermined orthodoxy and isolated by meditative study.

Leonardo da Vinci (1452-1519) was the illegitimate son of the notary Ser Pierro da Vinci and a peasant woman. Raised by his father's family in Vinci, he roamed the countryside, eagerly absorbing all that his senses could provide. Early education in Vinci provided only a rudimentary knowledge of reading, writing, and arithmetic. Never a "man of letters", da Vinci was proud of his self-taught education, often criticizing his colleagues for relying too heavily on the words of others rather than ideas of their own. Realizing da Vinci's immense artistic talent, Ser Pierro sent him to Florence as an apprentice to Verrochio, an artist. With his contribution to Verrochio's painting *The Baptism of Christ*, da Vinci made his mark as an artist. Verrochio, however, vowed never to touch another brush as his young pupil had so far surpassed him. Best known today as a brilliant artist, da Vinci was primarily a military and civil engineer. Constant poverty throughout his life forced him to expand his talents, creating festival productions and designing aqueducts and fortifications. The immense versatility of his mind, talents, and interests were reflected in the abundance of his inventions, which included distillation apparatus, water skis, a helicopter, a tank, a parachute, a steam cannon, and a hang glider. His personality was as well-documented as were his inventions. Strong, well-built, and handsome his "...charm, his brilliance and generosity were not less than the beauty of his appearance" (from Keele, 1983, p. 19).

Da Vinci contributed substantially to the understanding of mechanics in his time. He described the parallelogram of forces, divided forces into simple and compound forces, studied friction, and questioned Aristotle's relationship between force, weight, and velocity. Da Vinci prepared Newton's 3rd law in his analysis of the flight of birds "...an object offers as much resistance to the air as the air does to the object" (from Keele, 1983, p. 99). Da Vinci's mechanics, however, never reached the concepts of acceleration, inertia, or mass as opposed to weight. By dissecting the movements of man, not merely a motionless dead cadaver, da Vinci's achievements in mechanics flourished with his application of mechanical principles to the anatomy of man. His unique ability to communicate dynamic movements in visual form shed light on the mechanics of human beings that is seldom as bright in modern anatomical illustration.

Da Vinci's mechanical analysis of human movement included joints, muscles, bones, ligaments, tendons, and cartilage. Examining the structures of movement, da Vinci discarded the theory of pneuma. Spiritual force replaced it, acting as a form of energy through the nerves or fleshy muscle fibres, "transmuting" their shape, "enlarging" or broadening them, and shortening their tendons.

Da Vinci's anatomical studies fused art and science, stressing the importance of perspective in producing pictures of reality. Ball and socket joints such as the shoulder and hip joint, figured prominently in da Vinci's anatomical analysis. The "polo universale" allowed circular movement in many directions, with "polo" describing the fulcrum of balance for these rotating movements. Da Vinci illustrated the correct shape of the human pelvis for the first time (Fig. 1.2.1), describing the hip joint as the "polo dell' omo" (the pole of man).

Da Vinci depicted individual muscles as "threads", by demonstrating the origin, insertion, relative position, and interaction of each muscle. Mechanical action, dependent upon the muscle's shape, was represented in his drawings with forces acting along the line of muscle filaments (Keele, 1983). The accuracy of da Vinci's study of human musculature varied considerably. He was most successful illustrating the anatomy of the arm, elbow, and hand; slightly less so with his anatomy of the foot and lower leg; and least successful with the trunk and the thigh.

Da Vinci's ultimate goal was to integrate experience through geometrical analysis. As the first modern dissector and illustrator of the human body, he often realized contradictions and errors in Galen's comparative anatomy. With keen perception and amazing aptitude for communicating movements pictorially, da Vinci presented a comprehensive mechanical analysis of the human body. Unlike Galen, verification and experimentation figured prominently in da Vinci's studies. Currently, da Vinci's significance as a scientist remains contested. The first publication of da Vinci's works occurred 200 to 300 years after his death. The eclectic nature of his drawings and notes, which were disorganized, often having several seemingly unrelated subjects on one page, discouraged some scholars from qualifying his contributions as entirely scientific. However, even if his notes had been organized, their influence would have been limited in his time by the lack of circulation. Furthermore, the absence of any formal scientific community during the Renaissance, prevented da Vinci's ideas from circulating among colleagues with similar interests.

Vesalius (1514-1564) was born during the final years of da Vinci's illustrious career. His career in anatomy contrasted sharply with that of his predecessor. Unlike da Vinci, Vesalius received formal training in medicine, taught his anatomical theories, and published his research and methodology. Born into a medical family, Vesalius trained as a physician at the University of Padua and graduated *magna cum laude* in 1537. By 1538 he had produced a dissection manual for his students. Originally a proponent of Galen, Vesalius began to notice contradictions in his dissections and those done by Galen. In 1539 Judge Marcantonio Contarini became interested in Vesalius' anatomical studies, encouraging him in further detailed investigation by making available the bodies of executed criminals. For the first time in history, enough cadavers were available to make, repeat, and compare detailed dissections. Vesalius became increasingly convinced that Galen's anatomy was actually an account of the anatomy of animals and that it was wrongly applied to the human body. In 1539, Vesalius distinguished himself by publicly challenging the authority of Galen.

Two revolutionary views of nature were published in the year 1543. Copernicus challenged the notion that the universe was heliocentric in his *De Revolutionibus Orbium Coelestium,* and Vesalius' *De Humani Corporis Fabrica Libri Septem* revolutionized human anatomy by challenging a view that had dominated for 1300 years. Vesalius boldly declared that human anatomy could only be learned from the dissection and observation of the human body (Fig. 1.2.2). Numerous human dissections provided Vesalius with the opportunity to re-evaluate the anatomy of the muscles. Testing Galen's theories, Vesalius demonstrated that muscle shortened and became thicker during contraction. Contraction itself he attributed to a property of the muscle, as it contracted when both ends were sev-

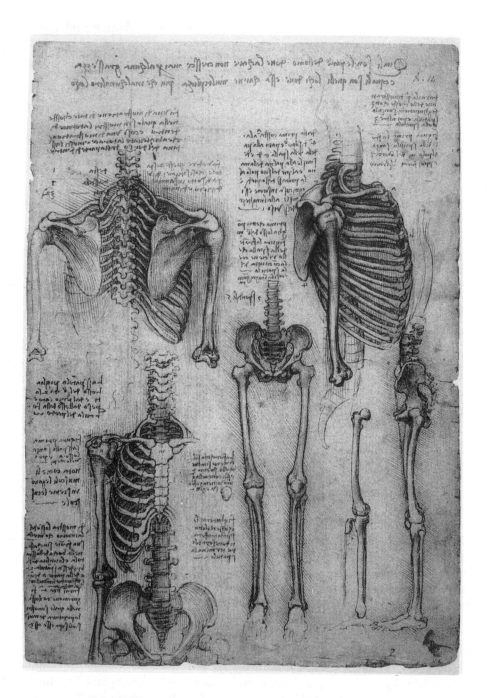

Figure 1.2.1   **Da Vinci's study of the human skeleton demonstrated the "polo dell' omo" and the shoulder joint (from Clayton, 1992, with permission).**

ered. Vesalius' experimentation stimulated scientific debate concerning the relationship between the nerves and muscles. Vesalius believed muscle to be "...composed of the sub-

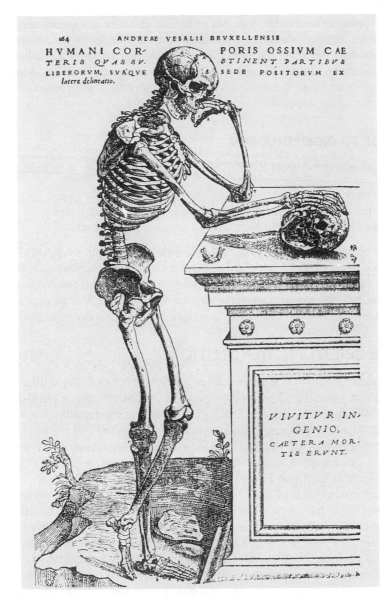

Figure 1.2.2   Human skeleton from Vesalius' book *De Humani Corporis Fabrica Libri Septem* (from Singer, 1959, with permission of Oxford University Press).

stance of the ligament or tendon divided into a great number of fibres and of flesh containing and embracing these fibres" (from Needham, 1971, p. 118), which received branches of arteries, veins, and nerves. Vesalius' observations inspired the investigations of Fallopius (1523-1562), who was keenly interested in the fibrous tissue in connection with movement. Fallopius stated that "...motion requires a fibrous nature in the actual body that is moved; since whatever moves itself does so by contraction or extension" (from Needham, 1971, p. 118).

Vesalius' demands on the development of human anatomy, his questioning of Galen's authority, and his detailed, descriptive anatomy laid the foundation for modern anatomy. Vesalius established the scientific principle of performing human dissections to understand human anatomy. Despite his criticism of Galen's anatomy, Vesalius adopted the theory of *spiritus animalis* as the cause of muscular contraction and carrier of the motor/ spiritual function of the brain.

RELEVANCE TO BIOMECHANICS

The relevance of the "Italian Renaissance" to biomechanics lies in three major aspects:

- Scientific work was revived,
- The foundations for modern anatomy and physiology were laid, and
- Movement and muscle action were studied as connected entities.

Scientific activities (the attempt to study the unknown), which remained dormant during the Middle Ages, experienced a revival. During the Italian Renaissance, da Vinci and Vesalius laid the foundations for modern anatomy by basing their teachings on observation and dissection of human cadavers. Parallel to the dissection of cadavers, human movements were "discussed" and factors contributing to human movement were studied.

1.2.4 THE SCIENTIFIC REVOLUTION

The environment in which the scientific revolution developed was similar to the environment in which the Italian Renaissance took place. Men of science were supported by private and political institutions such as Kings, Counts, wealthy families, Universities, and the Vatican in Rome. Intellectual freedom was highly respected and the interest for "new ideas and findings" was substantial. In addition, scientific societies began to emerge, encouraging the exchange of ideas and speculations, and intensive contact between scientists from different countries in Europe developed.

The scientific revolution of the 17th century, with which such great names as Galileo Galilei (1564-1642), Johannes Kepler (1571-1630), Rene Descartes (1596-1650), and Isaac Newton (1642-1727) are associated, saw a change in the understanding of nature and the way of doing scientific analysis. Led primarily by Galileo, the new scientific methodology revealed a picture of the universe that required updating interpretations of the Bible, and abandoning Aristotelian mechanics. Scientists turned to new methods of research. Experimentation became the cornerstone of the new scientific method. Descartes, Galileo, and Newton advocated a methodology that questioned certain truths by experimentation.

Galileo Galilei (1564-1642) received his early education in Florence around the end of the Italian Renaissance. Matriculating from the University of Pisa in 1581, he began systematically investigating Aristotle's doctrine of falling bodies. By 1591, he had discovered the impossibility of maintaining that the rate of a fall was a function of the falling object's weight. One year later, Galileo was appointed the Professor of Mathematics at the University of Padua.

Galileo's application of mechanical theory went beyond the inorganic world, as he proposed to publish a treatise on the mechanical analysis of animal movement. Had it been published, this treatise, *De Animaliam Motibus* (The Movement of Animals), would have preceded Borelli's famous treatise *The Movement of Animals* (1680-81) by 40 years. Galileo's topics included the biomechanics of the human jump, an analysis of the gait of horses and insects, and a determination of the conditions that allowed a motionless human body to float. Galileo's *Discourses on Two New Sciences* (1638) contained some of his biomechanical observations. His analysis of the strength of materials; beam strength, discussion of the strength of hollow solids, and particularly his analysis of resistance of solids to fracture and the cause of their cohesion, had application in the dynamics of the structure of bone. Galileo also compared the effects of changing the structures of biomaterials. One such comparison demonstrated the effect of increasing a normal femur to accommodate supporting an animal weighing three times the size (Fig. 1.2.3). The following quote illus-

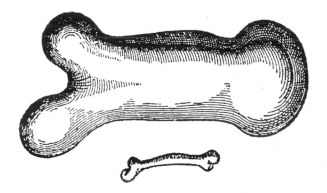

Figure 1.2.3   **Galileo's comparison of a normal femur with the femur needed to support an animal three times the size (from Singer, 1959, with permission of Oxford University Press).**

trates Galileo's theory: "You can't increase the size of structures indefinitely in art or nature. It would be impossible so to build the structures of men or animals as to hold together and function, for increase can be effected only by using a stronger material or by enlarging the bones and thus changing their shape to a monstrosity" (from Singer, 1959, p. 232).

Galileo applied this analysis to the predicament of enormous floating animals such as whales, in which he proposed that their weight was counterpoised by equivalent dispersal of water weight, altering the limit to their size. Ascenzi (1993) argues that Galileo's study on the mechanics of structure, function, and animal movements distinguishes him as the founder (father) of biomechanics.

Galileo, was a short, stocky man with fiery red hair and a forceful character. His clash with the Church which led to the condemnation of the Copernican theory in 1616, and to his permanent house arrest in 1633, appeared in great detail in historical literature. Of interest in this brief account was the immense role Galileo's personality played in the ultimate conflict with the Church. With the publication of *Letters on Sunspots* (1613) in which Galileo initially gave public commitment to the Copernican theory, not only did he not receive any opposition from the Church, but Cardinals Boroeo and Barberini (the future Pope Urban VIII) wrote letters to Galileo expressing their sincere admiration. In 1624, under the restraint of the 1616 decree, Galileo met with Pope Urban VIII to obtain permission to publish his analysis of the Copernican system. An admirer of Galileo, the Pope suggested a way to publish an analysis which avoided theological arguments, acknowledged the supremacy of God, and the inability of the human mind to understand all of God's mysteries. Galileo's arrogance sealed his fate as he placed the Pope's arguments in the mouth of a complete simpleton in his *Dialogue* (1631).

Galileo shaped the path of science for centuries to come. His theory of uniform motion, theory of projectiles, theory of the inclined plane, and definition of momentum provided foundations for Newton's three laws. Newton himself admitted: "If I have been able to see farther it was because I stood on the shoulders of giants" (from Koestler, 1959, p. 358). The giants Newton referred to were Galileo, Kepler, and Descartes. Galileo's primary contribution was the applicability of his theories to visible, tangible, organic, and inorganic objects.

Santorio Santorio (1561-1636) was a colleague of Galileo's and taught theoretical medicine at the University of Padua. Greatly influenced by Galileo, Santorio was the first scientist to apply mechanics to medicine, and the first to apply quantifiable methods to medicine. For 30 years, Santorio spent much of his time suspended from a steelyard weighing himself and his solid and liquid inputs and outputs (Fig. 1.2.4). As he tried to deduce the amount of "insensible respiration" lost through his lungs and skin, his experiments were laying the foundations of metabolism.

William Harvey (1578-1657), the eldest son of 7, was born into a family of yeoman farmers. He received his early education at Cambridge and completed his studies as a physician at the University of Padua. Moving back to London in 1602, Harvey was soon successful, becoming physician to James I and later Charles I. Harvey's main interest was medical research. Aristotle's theory of the primacy of the heart greatly influenced Harvey's discovery of circulation. By 1615 he had a clear conception of circulation, not publishing his results until 1628 in the slim, poorly produced book, *De Motu Cordis*, which was considered a scientific classic.

Vivisection of dogs allowed Harvey to see the motion of the heart. The active motion according to Harvey's observations was the heart's contraction - the systole - not the diastole as Galen had claimed. By measuring the capacity of the dissected heart, and estimating the number of heart beats every half hour, Harvey demonstrated that the heart pumped more blood into the arteries in half an hour than the body contained. Therefore, the blood had to be returning to the heart by another route. Acting without the aid of a microscope, Harvey assumed that a system of capillaries connected the arterial and venous systems. He had discovered the circulation of blood. This discovery altered physiological thought and inspired generations to utilize his experimental methods. He applied mechanistic theory to the action of the heart, describing its function as a pump and recognizing the mechanical nature of the vascular system. Harvey was the first "cardiac biomechanist". He died of a stroke at the age of 79.

Rene Descartes (1596-1650) was a philosopher and one of the dominant thinkers of the 17th century. Educated by the Jesuits, Descartes was exposed to the astronomical and mechanical observations of Galileo and rigorous mathematical training. Inheriting a modest wealth, Descartes spent his life travelling, serving as a soldier in Holland, Bohemia, and Hungary. In 1629 he settled in Holland, where he remained until persuaded to become the tutor to Queen Christina of Sweden in 1649. Descartes had the habit of sleeping late, claiming that his best thinking was done in the comfort of a warm bed. To his dismay, Queen Christina did not share his love for sleeping in, preferring to begin tutorials in philosophy at 5:00 a.m. in a freezing cold library. However, Descartes did not suffer too long

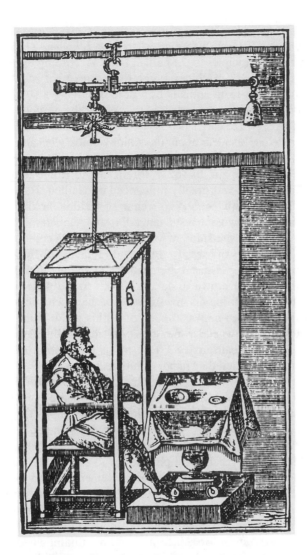

Figure 1.2.4   **Santorio in his balance measuring his solid and liquid input and output (Millar et al., 1989, p. 343, Chambers Concise Dictionary of Scientists, with permission of Chambers Harrap Publishers Ltd., Scotland, UK).**

from these untimely tutorials. Within five months of beginning to tutor the queen, Descartes died of lung disease. According to legend, Descartes invented the Cartesian coordinate system while lying in bed and observing the habits of a fly zooming around the room. As the fly flew into a corner of the room, Descartes realized the possibility of representing movement with a "Cartesian coordinate system".

Descartes became one of the originators of the mechanical philosophy that stated: changes observed in the natural world should be explained only in terms of motion and rearrangements of the parts of matter. He represented the first wave of mathematical analysis of mechanics, which shaped the development of mechanics during the 18th century. Descartes' mechanical philosophy influenced physiology so that by 1670 all major physiologists had developed a mechanical approach to their investigations (e.g., William Harvey, Santorio Santorio, and Borelli). Descartes' treatise *L'homme* (1664), written before 1637, applied mechanical principles to the human body. According to his mechanical philosophy, animals were organic machines running on auto pilot, while humans, also a machine, were distinguished from animals by virtue of having a soul. The first modern book devoted to physiology, *L'homme* stressed the role of the nervous system in co-ordinating movement. However, Descartes had no extensive knowledge of physiology. Often grotesquely wrong in detail, his ideas about physiology quickly became obsolete.

Giovanni Borelli (1608-1679) was born in Naples. His family was under constant suspicion of conspiracy against Spain. In 1624, Borelli fled to Rome to escape persecution under such suspicions. Borelli was greatly influenced by Galileo, Harvey, Kepler, Santorio, and Descartes. He was a student and colleague of Galileo, holding degrees in medicine and mathematics. His own interests ranged from geometry, physics, mechanics, physiology, and astronomy, to the study of volcanoes. Borelli served as the Professor of Mathematics at the Universities of Messina, Pisa, and Florence. In 1624, Queen Christina, the eager pupil who may have hastened the death of Descartes with her enthusiasm for early mornings, renounced her crown to devote herself to the pleasures of culture. She became Borelli's patron, supporting his investigation of the mechanics of the human body with annual endowments. When he was robbed of all his possessions, Borelli was compelled to dwell in a building owned by the Society of the Scholae Piae of San Pantaleo. There he remained until his death in 1680. Due to his publication *De Motu Animalium* (Fig. 1.2.5), Borelli is often called the "father of biomechanics".

Borelli's fundamental aim was to integrate physiology and physical science. *De Motu Animalium* (1680-81) demonstrated Borelli's geometrical method of describing complex human movements such as jumping, running, flying, and swimming (Fig. 1.2.5 and 1.2.6). His investigation of human movement included gait analysis and an analysis of the muscles. His work on the human gait and more complex movements was rather successful. However, he relied heavily on mechanical philosophy to explain the physiological nature of muscles. Borelli discussed several aspects of movement related to externally visible motion. He described muscle function, formulated mechanical lemmas to explain the movements produced by muscles, explained muscle function for the motion of the knee joint, and discussed the influence of the direction of muscle fibres for force production of the particular muscle in question. Borelli formulated his findings and hypotheses in the form of *propositions*. Examples from his propositions on "jumping" included:

> Proposition CLXXVIII: In jumping at an inclination to the horizon, the trajectory of the jump is parabolic.
>
> Proposition CLXXIX: Why a jump during running is longer and higher.

Borelli also discussed the mechanics of muscle contraction, the reason for fatigue, and offered some thoughts about pain. Some of his propositions for "muscle contraction" included:

> Proposition II: Muscle contraction does not consist of simple tension of the fibres similar to that exerted on a rope raising a weight.

Proposition XIV: Muscles do not contract by condensing the length of their fibres and bringing closer together their extremities but their hardness and tightening result from swelling.

Figure 1.2.5   **Cover of Borelli's book De Motu Animalium (from Borelli, 1680, translated 1989, with permission).**

Borelli's mechanical propositions were probably stronger than his propositions for muscle action and they had a significant impact on the subsequent development of biomechanics.

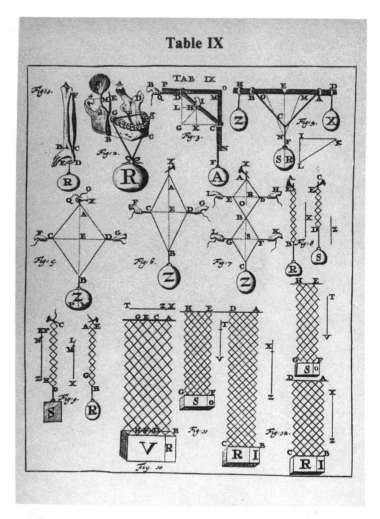

Figure 1.2.6   **Borelli's Table IX - The Movement of Animals - illustrating the models of muscular construction (from Borelli, 1680, translated 1989, with permission).**

Isaac Newton (1642-1727). On December 25th, 11 months after Galileo's death, Isaac Newton was born. A descendant of farmers, with no record of any notable ancestors, the young Newton quickly proved to be a very poor farmer, often absent-minded and lazy.

Preparing for University, Newton read the works of Kepler, Galileo, and Descartes. Although he was admitted to Trinity College, Cambridge, Newton returned to Woolsthorpe in 1666 to escape the plague. Conducting experiments in optics and roaming about the apple orchards of his home, Newton's fascination with mechanics peaked. As he explained: "I was in the prime of my age for invention and minded mathematics and philosophy more than at any other time" (from Singer, 1959, p. 248). Newton discovered the theory of gravitation during this time, but he "misplaced" it for 20 years when his interests turned to alchemy. He was a difficult man, remote and austere, often isolating himself from others in the pursuit of his science as well as secret metaphysical analysis. Newton gained recognition in his own time for his accomplishments. As the Master of the Royal Mint in 1696 and 1698, Newton reformed the currency and in 1705 was knighted for his effort. He became President of the Royal Society in 1703, a position he held until his death. However, Newton was ridiculed and satirized towards the end of his life in England as a scientist who concentrated on unimportant details.

Newton had several contributions which were important for science in general, and biomechanics specifically. However, his major contribution was the *synthesis* of many different pieces of the puzzle "mechanics". Koestler (1959) described Newton as the conductor of an orchestra of individual players, each absorbed in tuning his or her instrument; a conductor who pulled the orchestra together and made the caterwauling of the single instruments into a beautiful sound of harmony.

The pieces of the puzzle that Newton found were:

- Kepler's laws of the motion of heavenly bodies.
- Galileo's law of falling bodies and projectiles.
- Descartes law of inertia which required straight motion if no external force acted.

However, the pieces did not fit together. For example:

- Kepler's forces responsible for the movement of planets did not apply to movement on earth.
- Galileo's laws of movement on earth did not provide an apparent possibility to explain the movement of the planets.
- Kepler's planets moved in elliptic pathways.
- Galileo's planets moved in circular pathways.
- Kepler's driving forces for the planets were "spokes" from the sun.
- Galileo did not need a driving force since he proposed that circular motion was self-perpetuating.
- Descartes' inertia of bodies required planets to move in a straight line. However, the planets certainly did not move in a straight line.
- There was disagreement on the forces responsible for the movement of planets.

Newton's contribution was to *synthesize* these puzzle pieces into the law of gravity and the laws of motion. It was not completely clear how Newton found the solution. He claimed later, that he found the law of gravity by counterbalancing the moon's centrifugal force with the gravitational force. However, it was suggested that this explanation was not terribly convincing (Koestler, 1968). Some initial ideas (especially relating to the law of gravity) may have been present in 1666 when he was 24, and that the whole concept developed under the influence of discussions with his colleagues at the Royal Society (e.g.,

Hooke, Halley, Wren), and from continental Europe (e.g., Huygens in Holland). This path of development can probably never be reconstructed. One interesting aspect of this development was described by Koestler (1968, p. 20): "With true sleepwalker's assurance, Newton avoided the booby-traps strewn over the field: magnetism, circular inertia, Galileo's tides, Kepler's sweeping-brooms, Descartes' vortices - and at the same time knowingly walked into what looked like the deadliest trap of all: action-at-a-distance, ubiquitous, pervading the entire universe like the presence of the Holy Ghost".

With the publication of the four basic laws:

- The law of inertia,
- The law of acceleration due to an acting force,
- The law of action and reaction, and
- The law of gravity.

in Newton's *Philosophiae Naturalis Principia Mathematica* in 1686, all motion in the universe could be described and/or predicted, as long as the movement was with relative speeds which were small compared to the speed of light. In our days, more than 250 years after Newton's death, our view of the mechanical world is by and large still Newtonian.

Newton's second law of motion provided the fundamental tool for kinetic and kinematic analysis of movement. Additionally, Newton was credited with the first general statement of the parallelogram of force based on his observation that, a moving body affected by two independent forces acting simultaneously, moved along a diagonal equal to the vector sum of the forces (Rasch, 1958). When applied to muscular action, Newton's observation provided insight "...as two or more muscles may pull on a common point of insertion, each at different angles with a different force" (from Rasch, 1958, p. 573).

The improvement of the understanding of mechanics during the period of the "scientific revolution" was a quantum leap of scientific development. However, besides these exciting mechanical developments, other disciplines also made significant developments. Emerging biomechanical and physiological experiments benefited greatly from the invention of the microscope and the advent of experimental method. In 1663, the experiments of Jan Swammerdam (1637-1680) on frogs demonstrated the constancy of muscle volume during contraction. Nerve section experiments by William Croone (1633-1684) in 1664, illustrated that the brain must send a signal to the muscles to cause contraction. Between 1664 and 1667, Niels Stensen (1638-1686) provided precise descriptions of muscular structures. The theory of irritability, the muscles' ability to react to stimuli, was proposed by Francis Glisson (1597-1677). In 1691, Clopton Havers (1655-1702) used the microscope to conduct the first complete, systematic study of bone. He discovered that bone was composed of both inorganic and organic strings and plates arranged around cavities in tubular form.

RELEVANCE TO BIOMECHANICS

The relevance of the "scientific revolution" to biomechanics lies in two major aspects:

- Experiment and theory were introduced as complementary elements in scientific investigation, and
- The Newtonian mechanics were established, providing a complete theory for mechanical analysis.

Experimentation became an accepted approach to understanding nature. Instruments improved the scope of the experiments. As the telescope revolutionized the science of mechanics, the microscope revolutionized physiology. Separate fields of study began to emerge, and would come to fruition during the 18th century. The foundations for biomechanics were created during the 17th century, as scientists such as Galileo, Borelli, and Harvey used experimentation to understand the human body and its movements.

Newton's synthesis established a new way of studying motion. The emergence of the mechanical philosophy and Newton's laws provided the impetus for the study of human movement, and the tools to understand it. The human body became the focus of mechanical investigation. The emergence of the mechanical philosophy in the 17th century created a focus on movement and motion that persisted well into the 18th century.

1.2.5 THE ENLIGHTENMENT

The scientific revolution resulted in a "quantum leap" for science, specifically for the natural sciences. Galileo's, Kepler's, Descartes', and especially Newton's contributions revolutionized science, and the former "natural philosophy" was replaced by a new foundation, the comprehensive mechanical paradigms. A new group of scientists developed; the mechanical philosophers. As typical in such situations, some concepts were not yet understood. Much work was still needed, requiring the brilliant mathematical brains of the 18th century to clarify many unsolved parts of the puzzle, parts that would not change the basic foundation, but parts that would prove important for further development of the mechanical understanding of movement.

One such difference between the leading thinkers of this time was the theory concerning the causes of motion. The mechanical philosophers disagreed whether matter was moved by external, internal, or no force at all. The concept of force was not clear at this time. Descartes argued that there were no forces or power in matter. Newton argued that matter consisted of inertia particles, and that forces acted between every pair of particles. Leibnitz proposed that force was internal to matter. Typically, these positions were influenced by religious beliefs.

This period of increased understanding was called "le siècle des lumières" in France, "die Aufklärung" in Germany, and is known today as the "Enlightenment". It was coined by mathematicians who were convinced that mathematics was the most important revolutionizing force of the scientific revolution. Mathematical analysis was proposed as the solution to the ills of society. In 1699, Fontanelle had argued that the new "geometric spirit" could improve political, moral, and literary works. D'Alembert believed that the Spanish Inquisition could be undermined by smuggling mathematical thinking into Spain (Hankins, 1985). It was in this atmosphere that natural philosophers were made into heroes, and in France, Newton was the greatest hero of them all. He was, in the language of modern times, a star. By 1784, there were 40 books written on Newton in English, 17 in French, and 3 in German.

18th century science was not organized the way it is known today. The subdivision into the current disciplines began to develop during this time, as a reflection of the changing understanding of nature and the way its study progressed. Three mathematicians contributed significantly to the development of natural science during the Enlightenment as it related to biomechanics: Euler, d'Alembert, and Lagrange.

Newton's laws described the movement of mass points and could appropriately be applied to the movement of celestial bodies. However, they could not describe the motion of rigid bodies, the motion of fluids, or the vibrations of a stretched spring. It was a group of mathematicians from Basel, Switzerland who concentrated on the solution of these questions, the brothers Jakob and Johann Bernoulli, their nephew Daniel Bernoulli, and Johann's pupil Leonhard Euler (1707-1783). Euler was considered as one of the ablest, most brilliant, and most productive mathematician and scientist of the 18th century, if not of all times. Euler developed mathematical theories to describe the motion of vibrating bodies, and the buckling of beams and columns, to mention just a few. One of the major contributions of Euler was the expansion of Newton's laws to the applications of rigid and fluid bodies on earth. Furthermore, Euler established on a solid mathematical background, the concept of conservation of energy, a concept which was not recognized by Newton but started to develop during the Enlightenment as "vis viva".

As a baby, Jean le Rond d'Alembert (1717-1783) was found on the steps of the St. Jean le Rond Church. As an adolescent, he began studying law and was to appear before the bar in 1741, but his interests strayed to medicine and finally to mathematics. In 1743, 2 years after his acceptance at the Academie des Sciences, d'Alembert published the *Traité de Dynamique*. Contained within was d'Alembert's principle stating "...that Newton's third law of motion holds not only for fixed bodies but also for those free to move" (from Millar, 1989, p. 103). The implications of d'Alembert's work for the development of biomechanics included the application of his principles to kinetics.

Joseph Louis Lagrange (1736-1815) demonstrated a keen interest in mathematics at an early age. He studied mathematics at the Royal Artillery School, and in 1766 replaced Euler as the director of the Berlin Academy of Sciences. Moving to Paris in 1797, Lagrange became the Professor of Mathematics at the Ecole Polytechnique. A victim of periods of severe depression, Lagrange had virtually abandoned mathematics by his late 40s. He began work on his *Méchanique Analytique* (1788) at the age of 19, but did not complete it until the age of 52. Lagrange generally treated mechanical problems with the use of differential calculus. His treatise did not contain any diagrams, nor geometrical methods as did Newton's *Principia*. Lagrange's equations expressed Newton's second law in terms of kinetic and potential energy.

Physiologists of the 18th century adopted mechanical philosophy to explain the structures and functions of the human body. However, advances in chemistry began to provide a new approach to physiology. The study of the body's pumps, pulleys, and levers gave way to the investigation of growth, regeneration, nutrition, and the chemical functioning of the human body. Vitalistic theory began to challenge mechanistic theories. Marie Francois Bichat's (1771-1802) philosophy that "form follows function" began to influence physiology. Research in muscle physiology had been facilitated by the microscope in the 17th century. In the 18th century discovery of electricity increased interest and understanding of the nature of muscles.

During the mid 18th century, Albrecht von Haller (1708-1777) became one of the leading physiologists, and a proponent of vitalism. Drawing upon Hermann Boerhaave's (1668-1738) theory that fibres were the basic structural element of the body, von Haller focused on the structure and function of muscles. He expanded Glisson's theory of irritability or contraction, suggesting that contractility was an innate property of muscle. Experiments demonstrated that the muscle retained its ability to contract even after death, and that contraction of the muscles could be provoked by mechanical, thermal, chemical, or electrical stimuli. He believed that nerves played an important role in contraction. Electricity had been found to produce muscle contraction, and physiologists suggested that electricity in the body took a fluid form, flowing through the nerves carrying sense stimuli and motor commands (from Hankins, 1985, p. 121). However, von Haller urged caution in interpreting the electrical signal as the mysterious *spiritus animalius* that controlled movements. He believed the experimental technique of the 18th century was not capable of revealing the secrets of electro-chemical nerve impulses. This resistance to accepting chemical theory as an all encompassing explanation resulted from the failure of purely mechanistic interpretations to explain everything.

In the 18th century, muscle physiology began with the observations of Baglivi (1688-1706) who differentiated between the structures and functions of smooth and striated muscle in 1700. James Keill (1674-1719), a leading proponent of the mechanical approach to physiology, calculated the number of fibres present in certain muscles, and calculated the amounts of tension per fibre required to lift a given weight. In 1728, Daniel Bernoulli (1700-1728) developed a mechanical theory of muscular contraction, replacing Borelli's fermentation hypothesis. He also studied the mechanics of breathing, and the mechanical work of the heart. Charles Dufay (1698-1739) believed that all living bodies had electrical properties. Nicholas Andre (1658-1742) coined the term "orthopaedics" in 1741 and believed that muscular imbalances created skeletal deformities. In 1750, Jallabert was the first to re-educate paralyzed muscles with electricity, writing the first book on electrotherapy. The first demonstrations of reflex action in the spinal chord were done by Robert Whytt (1714-1766) in 1751. He also localized the sites of single reflexes. Between 1776 and 1793, John Hunter (1728-1793) produced a descriptive analysis of muscle functions stating that "...muscle was fitted for self motion and was the only part of the body so fitted" (from Rasch, 1958 p. 574). Hunter believed that muscle function should be studied with live subjects, not cadavers.

RELEVANCE TO BIOMECHANICS

The relevance of the "Enlightenment" to biomechanics lies in four major aspects:

- The concept of force became more clearly understood,
- The concepts of conservation of momentum and energy started to develop,
- A mathematical consolidation of the different mechanical laws took place, and
- Muscle contraction and action became an event influenced by mechanical, bio-chemical, and electrical forces.

The development of Newtonian mechanics and a Newtonian world view during the 18th century, stimulated discussion and debate regarding what force was exactly and the effects of such force. The relationship between force and movement became important. From this debate, the laws of conservation of energy and momentum developed, forming the mechanical foundation of biomechanics.

Mathematical analysis during the 18th century advanced the study of mechanics. Lagrange and d'Alembert developed methods of analysis, based on Newton's mechanics, that facilitated the study of dynamic human movements. As the tools for studying human movements were becoming more sophisticated, physiology adapted a more holistic approach to understanding the human body. The structures and functions of biomaterial became clearer in terms of a combination of chemical and mechanical processes. A greater understanding of the mechanisms of movements, improved analysis of their movements in future biomechanical investigations.

1.2.6 THE GAIT CENTURY

The development of science, as it related to human movement in the 19th century, was influenced substantially by three events in the second half of the 18th century:

- Jean Jacques Rousseau's novel *Emile* in 1762,
- The invention of the steam engine by James Watt in 1777, and
- The storming of the Bastille in 1789.

Jean Jacques Rousseau's (1712-1778) novel *Emile* (1762), revived the ancient idea of a complementary development of body and intellect, by encouraging a return to nature and physical activity. Rousseau depicted movement and sport as an ideal form of human activity and fulfillment. The invention of the steam engine in 1777 by James Watt (1736-1819), heralded the start of the industrial revolution, and the development and need for leisure time and recreation. The storming of the Bastille in 1789 signalled the beginning of the French revolution, and an end of the monopoly on sport and leisure by the upper class. Developments in physiology paralleled this increased interest in physical activity, as Nicholas Andre proposed that exercise during childhood could prevent musculo-skeletal deformities in adulthood.

The development of sport and leisure during the late 18th century created a renewed scientific interest in human locomotion. The 19th century was characterized by the development of instruments and experimental methods to increase the understanding of how we move. The analysis of human gait occupied physiologists, engineers, mathematicians, and adventurers. The study of locomotion began as an observational science, and by the end of the 19th century, photography had revolutionized and quantified the study of human and animal movement.

In 1836, 150 years after Borelli published his mechanical analysis of the movement of animals, the two brothers Eduard (1795-1881) and Wilhelm Weber (1804-1891) published their treatise *Die Mechanik der menschlichen Gehwerkzeuge* (On the Mechanics of the Human Gait Tools). The treatise contained almost 150 hypotheses about human gait which were derived by the Weber brothers from observations and/or theoretical considerations. As in Borelli's time, a lack of instrumentation prevented a quantified analysis of movement, forcing scientists to rely on their senses and intuition. Many of the Webers' hypotheses were incorrect, some were correct, and others were to be verified. However, the primary importance of these hypotheses lay - as proposed by Cavanagh (1990) - not in the accuracy and appropriateness of their statements, but rather in establishing an agenda for further research of the human gait.

Etienne Jules Marey (1838-1904) transformed the study of locomotion from an observational science to one based on quantification. Marey's numerous inventions were "...designed entirely for providing a quantified and unbiased description of movement" (from Bouisset, 1992, p. 85). With the assistance of the French government, Marey developed the most extensive facility ever devoted exclusively to biomechanics (Cavanagh, 1990). Built on what is presently the Roland Garros Tennis Courts in the Parc des Princes, the Station Physiologique included a 500 m circular track equipped with monitoring equipment. Here he analysed the movements of adults and children during sport and work, as well as the movements of horses, birds, fish, insects, and even a jellyfish. Marey's study of locomotion was boundless, his analysis comprehensive.

Marey is often given more recognition as a pioneer of cinematography than of biomechanics. Yet his inventions correlated ground reaction forces with movement, using pneumatic devices such as shoes (Fig. 1.2.7), hoof covers, and the dynamometric table which

Figure 1.2.7   **Marey's pneumatic analysis of human locomotion (from Centre National d'Art et de Culture Georges Pompidou, 1977, with permission of Ville de Beaune, Conservation des Musées).**

was the first serious force plate. He quickly embraced the potential of the photographic plate "...to overcome the defectiveness of our senses and the insufficiency of our traditional language" (from Tosi, 1992, p. 52). Marey developed the existing technology to record sequential motion at comparatively high speeds. His photographic rifle worked on a similar principle as the Kodak disk camera. In 1889 Marey also developed the "Chronophotographe a pellicule" or modern cinecamera. Marey preferred frame-by-frame analysis of movement (Fig. 1.2.8), arguing that the screen portrayed images he could not see with his own eyes. Marey's data collection techniques quickly became in high demand in such fields as cardiology, microscopy, mechanics, music, civil engineering, and hydrodynamics.

Marey greatly influenced the development of biomechanics, providing the ability to quantify movements and the rigorous scientific nature of his investigations. He was the first to combine and synchronize kinematic and force measurement, and inspired comprehensive analysis of locomotion. Marey's research also provided insight regarding the storage and reutilization of elastic energy, variations in ground reaction forces and centre of gravity motion, and the dependency of physiological cost on movement characteristics. The variety and abundance of his data collection methods inspired others to adapt and create devices for the quantification of motion.

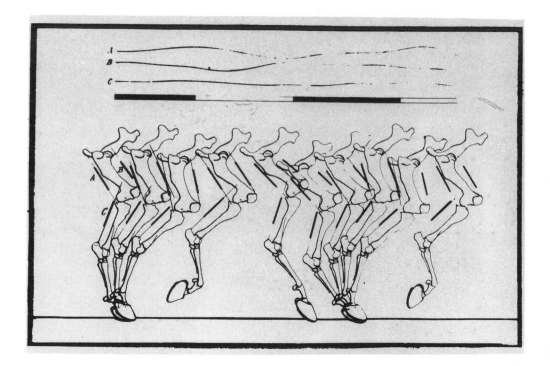

Figure 1.2.8   **Marey illustrated the movements he analysed from film recordings with scientific drawings such as this one (from Tosi, 1992, with permission).**

Edweard Muybridge (1830-1904) began his career in the study of locomotion at the initiation of Mr. Leland Stanford. Stanford was an avid horse racing fan, and while owning and training many of his own horses he became interested in their anatomy and movements. Stanford believed that during a trot, there was a period in which all four hooves were off the ground. Marey had proven this with his pneumonic device at slower gaits but was unable to do so at faster speeds. Stanford commissioned Muybridge to photograph his horse Occident at a trot, to verify his belief in unsupported transit. Muybridge was successful and subsequently began a lifetime devotion to documenting the sequential motion of humans and animals. Initially, he continued his collaboration with Stanford, documenting the gaits of Stanford's horses on what is today the site of Stanford University. However, after some controversy over the rights of the publication of the research in *Horses in Motion*, Stanford and Muybridge ended their partnership.

Muybridge's contribution to biomechanics was the sheer quantity of pictures he produced to document movement. Between 1884 and 1885, Muybridge produced 781 plates for a total of 20,000 images. These images were eventually published in *Animal Locomotion*, *Animals in Locomotion,* and *The Human Figure in Motion*. However, recent analysis by Marta Braun indicates that "...the relationship in some 40% of the plates is not what Muybridge states it to be" (from Braun, 1993, p. 1). Muybridge's publications were plagued with several inaccuracies, but the richness of his pictures were testimony to the importance of photography as a new language of science. Marey's acknowledgment of the Stanford-Muybridge study accorded it the status of scientific research. Muybridge and Marey shared a fascination for locomotion and a passion for the new photographic language, but unlike Marey, Muybridge lacked a scientific methodology. Muybridge and Marey collaborated in Paris to document the flight of birds. Marey was interested in Muybridge's methods and adopted photography as a tool for the study of locomotion.

In 1891, Wilhelm Braune and Otto Fischer made precise mathematical analysis possible by conducting the first tri-dimensional analysis of the human gait. To complete a mathematical study, the centre of gravity and moments of inertia of the body and all its parts were required. The centres of gravity were determined experimentally with the use of frozen cadavers. Two cadavers were nailed to a wall with long steel spits, allowing dissection with a saw to intersect longitudinal, sagittal, and frontal planes of the centre of gravity. The points of intersection of the centre of gravity were then recorded on lifesize drawings, and compared photographically with those of over 100 soldiers. A subject with the same measurements was then dressed in a black suit with thin light tubes and passed by 4 cameras (Maquet, 1992). The light tubes left imprints on the sensitive plates with each discharge of a Rhumkorff coil. A network of coordinates was later photographed over the picture of the subject passing through the cameras. These experiments were conducted at night and took 10-12 hours of constant effort, the preparation of the subject alone took up to 8 hours.

Braune and Fischer's *Der Gang des Menschen* (1895-1904) contained the mathematical analysis of "...3 transits of the human gait including 2 on free walking and 1 walking with army knapsack, 3 full cartridge pouches and 88 rifle in the shoulder arms position" (from Cavanagh, 1990, p. 20). The data analysed was gathered in a single night of experimentation. Braune died soon after the experiments, leaving the arduous task of analysis to Fischer. Today, the calculations that took Fischer several years could be done by computer in a matter of hours or even minutes. Their methods essentially are the same as those used by today's investigators in the biomechanics of gait.

In the 19th century biology became diversified into specialized fields of study. Darwin's theories of evolution coincided with the discovery and increased awareness of the development of the structures and their functions over time. As seen in the development of bone physiology, contributors from different fields of science such as engineering began to influence developments in biology.

The foundations of electromyography were laid by the pioneers Du Bois Reymond and Duchenne during the 19th century. Galvinism and the regeneration of animal electricity sparked the imagination of Mary Shelly to create Frankenstein in 1816. The popularity of electricity filtered into the 19th century as many charlatans sold it as a "cure all" to desperate aristocrats. It was in this environment that Du Bois Reymond (1818-1922) refined the methods for measuring currents in 1841, and traced electricity in contracting muscle to its independent fibres. His experimental methods dominated electromyography for a century and laid the foundations for future researchers. In 1866 Duchenne (1806-1875) published *Physiologie des Mouvements* which described the muscle action of every important superficial muscle. By developing electrodes for the surface of the skin, he was able to observe the actions of the superficial muscles on hundreds of normal and abnormal subjects. He was the first in this respect to use abnormal muscle function to analyse normal muscle function. Unfortunately this work was not translated into English until 1949, but was believed to be "...one of the greatest books of all times" (from Rasch, 1958, p. 641).

Bone physiology experienced dramatic development in the 19th century. An emphasis on the importance of proper alignment of the skeleton for good health, prompted increased awareness and interest in bone growth and function. Andrew Still (1828-1917) developed osteopathy as a separate school of medicine during this time. Mechanical forces were studied to increase the understanding of bone physiology. In 1855, Breithaupt (1791-1873) described stress fractures in military recruits of a Prussian unit, inspiring further analysis of mechanical forces and bone physiology. Volkmann described the effects of pressure on bone growth in 1862, indicating an inverse relationship between increasing pressure and bone growth. In 1867, van Meyer described the relationship between the architecture of bone and its function. Engineering principles were introduced to bone physiology by Culmann (1821-1881) in 1867, when he observed that the "...interior architecture of the femur coincided with the graphostatic determination of the lines of maximum internal stress in the Forborne Crane" (from Rasch, 1958, p. 643). Wolff synthesized many of the prevalent ideas about bone physiology in 1870 with the formulation of Wolff's law. Recognizing the interdependence between form and function he proposed that physical laws strictly dictated bone growth.

RELEVANCE TO BIOMECHANICS

The relevance of the "gait century" to biomechanics lies in three major aspects:

- Measuring methods were developed to quantify kinematics and kinetics of movement, and were applied extensively to human gait analysis,
- Measuring methods were developed to quantify electrical current during muscular activity, and
- Engineering principles were applied in biological and biomechanical analysis.

The transformation of biomechanics from an observational and intuitive science, to one based on quantification and mathematical analysis, created an exciting new perspective for the analysis of human movement and locomotion. Instrumentation developed by Marey, Muybridge, and Braune and Fischer, allowed the quantification of movement for the first time. The developments in photography illustrated the subtle stages of motion that our eyes could not see. Additionally, muscle action could be quantified with EMG measurements, opening a new avenue of the understanding of muscle function. Biomaterials were understood in greater detail, combining an interest in their mechanical functions and their growth.

1.2.7 THE 20TH CENTURY

The 20th century was characterized by several factors which, in turn, effected the development of biomechanics: the mechanical and technological developments resulting from two World Wars, the increased popularity, social and financial recognition of sport in society, and the explosion of financial support for medical and health care research. In this environment, the discipline *biomechanics* developed both with respect to knowledge and understanding, and with respect to the number of researchers and research centres in biomechanics. The following sections will discuss some milestone developments in biomechanics, and some external indications for the development of the field.

In 1920 Jules Amar published in his book *The Human Motor* (1920), an analysis of physical and physiological components of work, taking into consideration the worker's environment and individual movements. A product of the post World War I industrialization, Amar's study focused on the efficiency of human movements.

Nicholas Bernstein (1896-1966) descended from a family of physicians. After a year of service in a war hospital during 1914, Bernstein chose to pursue a medical career. Upon completion of his studies he served on the Siberian front during the Russian Civil War. By 1921, Soviet science began to develop a new direction. The scientific organization of work psychophysiology dominated Soviet science from the early 1920's to 1940's, with Bernstein leading the way (Jansons, 1992). Bernstein developed a method for measuring movement based on mathematical analysis. His analysis of the human gait as "...a whole, unified complex of biological symptoms" (from Jansons, 1992, p. 155), was even more comprehensive than the one done by Braune and Fischer (1895-1904). Bernstein's "biodynamic" studies also included an analysis of the proper use of tools such as the hammer and the saw, which saw the redesign of the driver's cab in Moscow's trams; an analysis of the movements of working women; and bridge dynamics. Bernstein's analysis of the coordination and regulation of movement in both children and adults, provided the basis for his theories of motor control and coordination. He also established that adults ran far more economically than did children. Isolated from the west, Bernstein's work was not published in North America until 1967.

A.V. Hill (1886-1977) began his career in 1915 at Cambridge as a mathematician, but after 2 years of studies in mathematics he switched to physiology. In 1923, Hill received the Nobel prize for physiology and medicine. His main research activities focused on the explanation of the mechanical and structural function of human muscle. Based on experimental evidence from experiments using isolated frog sartorius muscles (Hill, 1970), Hill developed theories for mechanical and structural muscle action. Hill's classical paper (Gasser and Hill, 1924) discussed the dynamics of muscular contraction, signifying the beginning of an applicable understanding of muscle physiology to muscle functioning. Hill and his co-workers also studied human locomotion, contributing to the understanding of the efficiency of running.

Interest in human movement flourished during the 20th century and was not confined to mechanical analysis. Although he was not typically considered a contributor to the field of biomechanics, Rudolph Laban (1879-1958) developed a method of representing a series of complex human movements, which is still used in dance choreography today. Breaking down movements into symbols, Laban gained substantial knowledge regarding the use of movement as an expression of the individual. Escaping the Nazis and arriving in England in 1938, Laban found his knowledge of human movement could be used in industry. Laban increased the efficiency of individual factory workers through planned movement sessions with groups of workers. Teaching them to find the rhythm of their individual movements, Laban trained them to balance movements such that "...strong movements are compensated for light, and narrow ranging by wide..." (from Hodgson and Preston-Dunlop, 1990, p. 54). Productivity increased as the worker's movements became balanced and efficient. Laban's techniques were also used by the Air Ministry to create a more efficient way of parachute jumping.

In 1938 and 1939, Elftman quantified internal forces in muscles and joints. In order to perform these estimations, he developed a force plate to quantify ground reaction forces and the centre of pressure under the foot during gait. Elftman concluded that muscles act, regulating energy exchange, by using the strategies of transmission, absorption, release, and dissipation (Elftman, 1939).

A.F. Huxley (1924-) studied physics at Cambridge, worked on radar in World War II, and eventually applied his knowledge of physics to muscle physiology. Huxley also worked at the Massachusetts Institute of Technology, at Cambridge. Beginning in the 1950's, Huxley gained recognition for his work with the sliding filament model of muscle contraction, and the development of X-ray diffraction and electron microscopy. Huxley made a significant breakthrough in the study of muscle action when he proposed his sliding filament theory to explain muscle shortening in 1953. Huxley later expanded his work by proposing connecting mechanisms between both actin and myosin filaments, constituting his Cross-bridge Theory.

The development and growth of biomechanics in the 20th century may be illustrated by various political and professional activities, and by the numerical expansion of the biomechanical community. The first breakthrough of biomechanics into the curricula of Universities was in sport related disciplines. In the early 20th century, some Universities started to teach "biomechanics" in Faculties of Physical Education. At that time, there was discussion as to whether the field should be called kinesiology or biomechanics, a question which is solved today, kinesiology being the science dealing with various aspects of movement (mechanical, physiological, neurological, etc.).

From August 21-23 of 1967, the First International Seminar on Biomechanics was held in Zürich, Switzerland. Initiated by E. Jokl, organized by J. Wartenweiler, and sponsored by the International Council of Sport and Physical Education of UNESCO, the congress attracted approximately 200 participants from Europe, Asia, and America discussing the following major topics:

- Technique of motion studies.
- Telemetry.
- Principles of human motion studies: general aspects of coordination.
- Applied biomechanics in work.
- Applied biomechanics in sports.
- Clinical aspects.

This first international seminar was repeated biannually. During the 1973 conference in Penn State, the International Society of Biomechanics, ISB, was founded and J. Wartenweiler became the society's first president. For the 1975 Congress in Jyväskylä, Finland, the name of the Congress was changed to "International Congress of Biomechanics", a name which is still used today. The major trust of the International Society of Biomechanics and its congresses was and is locomotion, musculo-skeletal mechanics, ergonomics, sport biomechanics, and clinical biomechanics. The biannual congresses are ongoing, attracting between 500 and 1000 participants in the early 1990's.

In the 1980's a movement started to emerge, allowing biomechanists from all sub-disciplines of biomechanics (e.g., locomotion, orthopaedic, sport, muscle, material, tissue, dental, cardiac, etc.) to meet periodically as one group to exchange ideas and findings. In 1989, under the chairmanship of Y.C. Fung, a group of biomechanists from all the sub-disciplines, in cooperation with most of the international, regional, and national biomechanic organizations, organized the First World Congress of Biomechanics in San Diego, USA. The Congress attracted approximately 1200 participants and included all areas of biomechanics. A "Second World Congress of Biomechanics" with the same goals and interests is presently being organized in Amsterdam for 1994.

Overall, there are now thousands of biomechanists working on all continents. Biomechanics is a recognized discipline with many Universities offering courses and graduate programs. Results from biomechanical research directly influences medicine, work, and sport equipment development as well as many other aspects of human life. Recently, biomechanical research became integrated into multi-disciplinary research projects, contributing the mechanical aspect of understanding the biological systems.

RELEVANCE TO BIOMECHANICS

The relevance of the "20th Century" to biomechanics lies in three major aspects:

- Biomechanics developed as a discipline at Universities with graduate programs and faculty positions,
- Biomechanical research results were increasingly used in practical, medical, and industry applications, and
- Biomechanics became a player in a multi-disciplinary attempt to understand human and animal movement and effects of movement on the musculo-skeletal system.

Today, most major Universities have faculty members with the title "Professor of Biomechanics". Hundreds of students are enrolled in graduate programs, studying for a masters or a Ph.D in biomechanics. Many biomechanical research centres have contracts with various industries, helping them to improve their products regarding performance and/or safety. Many hospitals have gait laboratories to quantify medical treatment or to help in the decision making process for difficult cases. Many biomechanical research centres have moved into a multi-disciplinary co-operation with other disciplines to answer complex questions.

1.2.8 FINAL COMMENTS

The section entitled "selected historical highlights" included a multitude of facts. One may risk losing sight of the "forest through the trees" with such a lengthy exchange of information. This section attempts to provide a short overview and to synthesize the different facts into one picture. In order to do so, a graph has been developed, showing the intensity of the development of areas such as mechanics, mathematics, anatomy, muscle, and locomotion research (Fig. 1.2.9). The graph also illustrates the different phases of development.

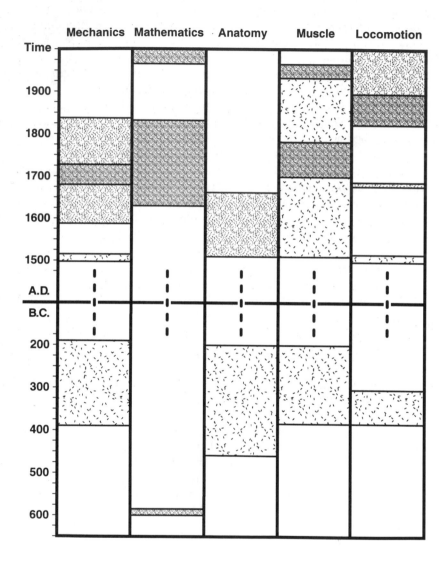

Figure 1.2.9   Schematic illustration of the time periods with significant developments in the areas of mechanics, mathematics, anatomy, muscle, and locomotion research as they relate to biomechanics. Dark shades indicate the greatest development, medium shades indicate medium development, and light shades indicate light development. Note: greatest, medium, and light have been set arbitrarily.

An initial phase (650 B.C.-200 A.D.) was characterized by an awakening of western scientific activity with initial developments in mechanics, mathematics, anatomy, muscle, and locomotion research. This initial phase was followed by a phase of very little development (200 A.D.-1450 A.D.). The Renaissance (1450 A.D.-1600 A.D.) changed the scientific picture, opening doors for further scientific development. Interest in the human and animal body awakened and the first locomotion studies were published.

The most decisive step for further biomechanical advancement occurred during the scientific revolution (1600 A.D.-1730 A.D.). This period was characterized by a new understanding of scientific research and a questioning of old concepts. It culminated in Newton's three laws of motion and a highly mechanistic world view. Some sources describe this period as the birthdate of biomechanics, with the publishing of the work *De Motu Animalium* by Borelli, as the initiator of many future studies.

The mechanical understanding developed during the scientific revolution was ideal for celestial mechanics. However, when applied to terrestrial problems, there were still many unsolved problems. The concept of force was not clearly understood, and the question of conservation of energy or momentum was debated intensively with no consensus being found. It was during the Enlightenment (about 1730 A.D.-1800 A.D.) when mathematicians such as d'Alembert, Lagrange, Leibnitz, Euler, and others worked intensively on these questions, providing the framework for mechanics which we use today. Conceptually, this was the period that changed mechanical thinking most dramatically.

The 19th century, the "gait century", yielded the development of a variety of rather sophisticated techniques to analyse human and animal movement experimentally. The 20th century provided an explosion of sophisticated experimental methods and, with the development of the computer, a wealth of numerical mathematical methods which could be applied to biomechanical research. Furthermore, it was characterized by an increased complex understanding of bone, cartilage, tendon, ligament, and especially muscle. Biomechanics developed into a discipline with University courses, departments, and graduate students. There are many thousands of biomechanical researchers today working in Universities or in various industries. Results from biomechanical research contributed to the broadening of the understanding of the human body as well as to many practical applications in medicine, ergonomy, sport, and equipment.

1.3 MECHANICS

NIGG, B.M.

1.3.1 DEFINITIONS AND COMMENTS

Dynamics:

- Direct dynamics: Mechanical analysis of a system that starts with force and determines movement.

- Inverse dynamics: Mechanical analysis of a system that starts with movement and determines force.

Equation of motion: Mathematical set of equations combining variables that describe the forces and movement of a system.

Particle: Matter that is assumed to occupy a single point in space. In practical terms, this means that the volume of the body of interest is small compared to the space in which its behaviour is of interest.

Reference frame (RF):   A system of coordinate axes.

- Inertial (IRF):   A system of coordinate axes that have a constant orientation with respect to the fixed stars, and whose origin moves with constant velocity with respect to the fixed stars. An inertial reference frame, IRF, is a reference frame in which Newton's laws are valid.

Rigid body: Matter that is assumed to occupy a finite volume in space and that does not deform if subjected to forces. A rigid body consists of a number of mass particles, with the distance between any two particles being constant. The rigid body concept may be valid where the deformations of the body are insignificant relative to the motion of the whole body.

Vector: Throughout this text, vectors are depicted with **bold** symbols.

1.3.2 SELECTED HISTORICAL HIGHLIGHTS

1686 Newton   An English philosopher, physicist, and mathematician published the three laws of motion in his *Philosophiae Naturalis Principia Mathematica* (The Mathematical Principles of Natural Sciences).

1743 d'Alembert   A French mathematician published an alternate form of Newton's second law of motion in his *Traité de Dynamique*.

1750's Euler	A Swiss mathematician expanded Newton's laws to the application of rigid and fluid bodies on earth.
1788 Lagrange	A French mathematician published *Méchanique Analytique*, which included the Lagrangian equations, allowing the formulation of Newton's second law in terms of kinetic and potential energy.
1835 Hamilton	An Irish mathematician and astronomer. His text *On a General Method in Dynamics* provided a different expression of Newton's second law by using coordinate position and energy.
1905 Einstein	Worked as a "technical expert (class III)" at the Swiss patent office in Bern where he developed the theory of relativity which revolutionized the mechanical and philosophical thinking of humankind.

1.3.3 NEWTON'S LAWS OF MOTION

Newtonian mechanics was based on the three laws of motion, which were first formulated by Sir Isaac Newton in *Philosophiae Naturalis Principia Mathematica* in 1686. At least three other scientists, however, contributed significantly to the puzzle that Newton had to solve: Kepler, Galileo, and Descartes (Koestler, 1968). Kepler (who died about 30 years before Newton's attempt to synthesize the knowledge), contributed the laws of motion of the heavenly bodies. Galileo (who died about 20 years before Newton's synthesis), contributed the laws of motion of bodies on earth. Descartes contributed the idea of inertia, which forced bodies into straight motion if no external force was acting. Before Newton, these ideas did not seem to fit together. Some were even contradictory (e.g., planetary motion in ellipses or circles, and the inertia concept). Newton's contribution was to *synthesize* these fragments into one concept, the three laws of motion, providing a framework for explaining them all.

Newton's laws were formulated for single particles and were valid in an inertial frame of reference, a frame at rest or moving with uniform velocity relative to the distant, fixed stars. The three laws were:

First law:	**A particle will remain in a state of rest or move in a straight line with constant velocity, if there are no forces acting upon the particle.**
	If $\mathbf{F} = 0$ then $\mathbf{v} = $ constant
Second law:	**A particle acted upon by an external force moves such that the force is equal to the time rate of change of the linear momentum.**
	where:

$$\mathbf{F} = k\frac{d(m\mathbf{v})}{dt}$$

F   =   resultant force acting on the particle
p   =   linear momentum of the particle
m   =   mass of the particle
v   =   velocity of the particle
t   =   time
k   =   constant of proportionality where the con-
          stant of proportionality, k, is equal to unity
          for dynamically consistent unit systems (e.g.,
          SI).

If the mass of a particle is assumed to be constant in time, Newton's second law reduces to:

$$\mathbf{F} = m\mathbf{a} = m\frac{d\mathbf{v}}{dt} = m\frac{d^2\mathbf{r}}{dt^2}$$

where:

a   =   acceleration of the particle
r   =   position of the particle

Note: velocity and linear momentum are measured relative to an inertial reference frame, IRF.

Third law: **When two particles exert force upon one another, the forces act along the line joining the particles and the two force vectors are equal in magnitude and opposite in direction.**

$$\mathbf{F}_{12} = -\mathbf{F}_{21}$$

where:

$\mathbf{F}_{ik}$ =   force acting on particle i from particle k

Because of their importance for classical mechanics, Newton's laws are reproduced in their original Latin version (Cajori, 1960):

Lex I: **Corpus omne preservare in statu suo quiescendi vel movendi uniformiter in directum, nisi quatenus illud a viribus impressis cogitur statum suum mutare.**

Lex II: **Mutationem motis proportionalem esse vi motrice impressae, et fieri secundum lineam rectam qua vis illa imprimitur.**

Lex III: **Actioni contrariam semper et aequalem esse reactionem: sive corporum duorum actiones in se mutuo semper esse aequales et in partes contrarias dirigi.**

1.3.4 EQUATIONS OF MOTION FOR A RIGID BODY

This chapter attempts to summarize aspects of Newtonian mechanics as they relate to biomechanical applications. Derivation of formulas and detailed discussions of mechanics can be found in mechanics textbooks (e.g., Meirovitch, 1970), or in selected biomechanics textbooks (e.g., Fung, 1990). The reader is advised to consult these sources for further details. Specifically, the derivation of the equations from particle to rigid body mechanics is well documented in textbooks.

In this section, 3-D descriptions of translation and rotation is presented. Then, assumptions are made to reduce these general descriptions to less general ones and/or 2-D descriptions. In all cases, an outline of how the variables in the equations can be determined is provided.

GENERAL 3-D CASE FOR TRANSLATION

The equation of motion for the centre of mass of a rigid body is:

$$\mathbf{F} = \frac{d\mathbf{p}_{CM}}{dt}$$

or for a rigid body with constant mass:

$$\mathbf{F} = m\mathbf{a}_{CM} - m\frac{d^2\mathbf{r}_{cm}}{dt^2} = m\ddot{\mathbf{r}}_{CM}$$

where:

$\mathbf{F}$ = resultant force acting on the rigid body

$\mathbf{r}_{CM}$ = position of the centre of mass of the rigid body

m – mass of the rigid body

$\mathbf{a}_{CM}$ = acceleration of the centre of mass of the rigid body

$\mathbf{p}_{CM}$ = linear momentum of the centre of mass of the rigid body

or in matrix form:

$$\begin{bmatrix} F_x \\ F_y \\ F_z \end{bmatrix} = m \begin{bmatrix} a_{CMx} \\ a_{CMy} \\ a_{CMz} \end{bmatrix} \tag{1.3.1}$$

Determination of variables in the equation

Mass: The mass of a rigid body as applied to biomechanical problems can be determined using methods described in chapter 3.7. In general, experimental mass determination is appropriate for the required accuracies in biomechanical projects.

Acceleration: Accelerations can be measured with accelerometers or from film or video pictures using the second time derivative of position. Both approaches have methodological problems, as described in chapters 3.3 and 3.4. However, often both methods provide accelerations with acceptable errors for movements with non-excessive accelerations (e.g., impacts).

SIMPLIFIED 2-D CASE FOR TRANSLATION

Often it is sufficient and appropriate to discuss planar movement. In this special case, the accelerations in the third dimension are assumed to be zero. Defining the plane of interest as the x-y-plane, the set of translational equations of motion is:

$$F_x = ma_{CMx}$$

$$F_y = ma_{CMy}$$

GENERAL 3-D CASE FOR ROTATION

For the subsequent discussion the following definitions and assumptions are used:

- X, Y, Z is an inertial reference frame with origin O.
- x, y, z is a body fixed reference frame with origin at the centre of mass, CM.

In the inertial reference frame, X, Y, Z, the rotational equation of motion is:

$$\mathbf{M}^{XYZ} = \dot{\mathbf{H}}^{XYZ} \tag{1.3.2}$$

where:

$\mathbf{H}^{XYZ}$ = the angular momentum of the body with respect to the inertial frame X, Y, Z, with:

$$\mathbf{H}^{XYZ} = \int_V (\mathbf{r} \times \dot{\mathbf{r}})\, dm$$

where:

$\mathbf{r}$ = vector from origin to a point in the body
$\mathbf{M}^{XYZ}$ = resultant moment of the rigid body with respect to the origin of the inertial frame, X, Y, Z

Equation (1.3.2) may be split into two parts:

$$\mathbf{M}^{XYZ} = \mathbf{M}_O^{XYZ} + \mathbf{M}_{CM}^{xyz} = \dot{\mathbf{H}}_O^{XYZ} + \dot{\mathbf{H}}_{CM}^{xyz} = \dot{\mathbf{H}}^{XYZ} \tag{1.3.3}$$

with:

$$M_{CM}^{xyz} = \dot{H}_{CM}^{xyz} \tag{1.3.4}$$

$$M_{O}^{XYZ} = \dot{H}_{O}^{XYZ} \tag{1.3.5}$$

where:

H_{O}^{XYZ} = angular momentum due to the centre of mass revolving about the origin of the X, Y, Z inertial

H_{CM}^{xyz} = angular momentum due to the centre of mass rotating about the origin of the body fixed coordinate system, x, y, z with:

$$H_{CM}^{xyz} = \int_{V} (R \times \dot{R})\, dm$$

where:

R = vector from CM to a point in the body

M_{O}^{XYZ} = moment about the origin of the X, Y, Z inertial frame

M_{CM}^{xyz} = moment about the origin of the body fixed coordinate system, x, y, z

The expansion of H_{CM}^{xyz} in equation (1.3.4) yields:

$$\{M_{CM}^{xyz}\} = [I]\{\alpha\} + \{\omega\} \times [I]\{\omega\}$$

where:

$\{\}$ = vector symbol

$[]$ = matrix symbol

$\{M_{CM}^{xyz}\}$ = resultant moment vector about the CM expressed in components of x, y, z

$[I]$ = inertia tensor

$\{\alpha\}$ = angular acceleration vector with respect to the body fixed reference frame x, y, z

$\{\omega\}$ = angular velocity vector

or in an expanded version:

$$
\begin{bmatrix} M_{CMx} \\ M_{CMy} \\ M_{CMz} \end{bmatrix} = \begin{bmatrix} I_{xx} & -I_{xy} & -I_{xz} \\ -I_{yx} & I_{yy} & -I_{yz} \\ -I_{zx} & -I_{zy} & I_{zz} \end{bmatrix} \begin{bmatrix} \alpha_x \\ \alpha_y \\ \alpha_z \end{bmatrix} + \begin{bmatrix} 0 & -\omega_z & \omega_y \\ \omega_z & 0 & -\omega_x \\ -\omega_y & \omega_x & 0 \end{bmatrix} \begin{bmatrix} I_{xx} & -I_{xy} & -I_{xz} \\ -I_{yx} & I_{yy} & -I_{yz} \\ -I_{zx} & -I_{zy} & I_{zz} \end{bmatrix} \begin{bmatrix} \omega_x \\ \omega_y \\ \omega_z \end{bmatrix}
$$

where:

ω_i =   angular velocity component relative to the X, Y, Z system expressed in components of x, y, z

α_i =   angular acceleration component relative to the X, Y, Z system expressed in components of x, y, z

I_{ij} = $\begin{cases} i=j & \text{moment of inertia with respect to the principal axes} \\ i \neq j & \text{products of inertia} \end{cases}$

where for the moments of inertia:

$$I_{xx} = \int_V (y^2 + z^2)\, dm$$

$$I_{yy} = \int_V (x^2 + z^2)\, dm$$

$$I_{zz} = \int_V (x^2 + y^2)\, dm$$

and for the products of inertia:

$$I_{xy} = \int_V xy\, dm$$

$$I_{xz} = \int_V xz\, dm$$

$$I_{yz} = \int_V yz\, dm$$

Determination of variables in the equation

Moments of inertia:

Experimental methods for determining the moments of inertia, I_{xx}, I_{yy}, and I_{zz} are discussed in chapter 3. Several methods are available, and the moments of inertia can usually be determined with sufficient accuracy.

Products of inertia:

The same experimental methods can be used to determine the products and moments of inertia. In most biomechanical applications, the body of interest is symmetrical enough to allow the principal axes, x, y, z, to be chosen such that the products of inertia are approximately zero.

Angular velocity:

Experimental methods for determining the angular velocity of a rigid body are described in chapter 3. For typical biomechanical applications these methodologies provide sufficiently accurate results.

Angular acceleration:

Experimental methods for determining the angular acceleration of a rigid body are described in chapter 3. For most biomechanical applications, these methodologies provide results that suffer from the typical problems of double differentiation.

SIMPLIFIED 3-D CASE FOR ROTATION

If the body's fixed axes are aligned along the principal axes, the products of inertia are zero and the inertia tensor becomes diagonal. For this case, the rotational equations of motion are:

$$M_{CMx} = I_{xx}\alpha_x + (I_{zz} - I_{yy})\omega_y\omega_z$$

$$M_{CMy} = I_{yy}\alpha_y + (I_{xx} - I_{zz})\omega_x\omega_z$$

$$M_{CMz} = I_{zz}\alpha_z + (I_{yy} - I_{xx})\omega_x\omega_y$$

For the special case where the body fixed axis system is aligned along the principal axes, these equations are called the modified Euler equations. These equations are commonly used for 3-D biomechanical analysis.

SIMPLIFIED 2-D CASE FOR ROTATION

If the motion is planar (e.g., in the x-y-plane), the rotational equations of motion take a further simplified form. The angular velocity vector has only one component (in this example the z-component):

$$\omega_x = 0$$
$$\omega_y = 0$$
$$\alpha_x = 0$$
$$\alpha_y = 0$$

and the resulting rotational equation of motion is:

$$M_{CMz} = I_{zz}\alpha_z$$

This equation is used for most simplified rotational applications with typical forms such as:

$$I_{zz}\alpha_z = |\mathbf{r}_1 \times \mathbf{F}_1| + |\mathbf{r}_2 \times \mathbf{F}_2| + \ldots + |\mathbf{r}_n \times \mathbf{F}_n|$$

1.3.5 GENERAL COMMENTS

Mechanical analysis may concentrate on forces (kinetics) and/or on movement (kinematics). Forces are the cause; movement is the result. Mechanical analysis can proceed from forces to movement or from movement to forces.

If the analysis starts with the cause (the force), the process is called direct or forward dynamics. A unique movement is the result of a defined set of forces. Consequently, the direct or forward dynamics approach has one solution and the approach is deterministic.

If the analysis starts with the result (the movement), the process is called inverse dynamics. A specific movement, however, can be the result of an infinite number of combinations of individual forces acting on a system. The inverse dynamics approach, therefore, has an infinite number of possible solutions and is not deterministic.

Newton's second law is essential for explaining and describing motion. Many researchers since Newton have centered their work on this law, and some of them have attempted to propose other (hopefully more elegant) mathematical procedures for applying Newton's second law of motion. The most popular ones are the approaches developed by Lagrange, d'Alembert, and Hamilton. This text concentrated primarily on Newtonian mechanics. The approaches of Lagrange, d'Alembert, and Hamilton are described in detail in advanced mechanics textbooks, and the reader should reference these sources for more specifics.

1.3.6 REFERENCES

Amar, J. (1920) *The Human Motor: Or the Scientific Foundations of Labour and Industry.* E.P. Dutton, New York.

Ascenzi, A. (1993) Biomechanics and Galileo Galilei. *Journal of Biomechanics.* **26 (2)**, pp. 95-100.

Asmussen, E. (1976) Movement of Man and Study of Man in Motion: A Scanning Review of the Development of Biomechanics. *Biomechanics V-A* (ed. Komi, P.V.). University Park Press, Baltimore. pp. 23-40.

Basmajian, J. and De Luca, C. (1985) *Muscles Alive: Their Functions Revealed by Electromyography* (5th Ed.). Williams & Wilkins, London.

Bastholm, E. (1950) *The History of Muscle Physiology, from the Natural Philosophers to Albrecht von Haller.* Kopenhagen.

Bernstein, N. (1923) *Studies of the Biomechanics of the Stroke by Means of Photoregistration.* Res. Central Institute of Work, Moscow. **N1**, pp. 19-79. (In Russian).

Bernstein, N.A. (1967) *The Coordination and Regulation of Movements.* Pergamon Press, Oxford.

Borelli, G.A. (1680) *De Motu Animalium. Pars Prima* (trans. Bernabò, A., 1989). Springer Verlag, Berlin.

Borelli, G.A. (1681) *De Motu Animalium. Opus Posthumun. Pars Altera* (trans. Maquet, P., 1989). Springer Verlag, Berlin.

Borelli, G.A. (1734) *De Motu Animalium.* G. Bernoulli. *De Motu Musculorum et de Effervescentia et Fermentatione.* F. Mosca Publ., Naples, Italy.

Bouisset, S. (1992) *Etienne-Jules Marey. On When Motion Biomechanics Emerged as a Science.* ISB Series, pp. 71-87.

Braun, G.L. (1941) Kin. from Aristotle to 20th Century. *Research Quarterly.* **12 (2)**, pp. 163-173.

Braun, M. (1993) *The Moment as it Flies.* University of Chicago Press, Chicago.

Braune, W. and Fischer, O. (1987) *The Human Gait* (trans. Maquet, P. and Furlong, R.). Springer Verlag, Berlin.

Braune, W. and Fischer, O. (1900) *Der Gang des Menschen* (trans. Hirzel, S.). Leipzig.

Cajori, F. (1960) *Sir Isaac Newton's MATHEMATICAL PRINCIPLES of Natural Philosophy and his System of the World.* University of California Press, Berkley, CA.

Cappozzo, A. and Marchetti, M. (1992) Borelli's Heritage (eds. Capozzo, A., Marchetti, M., and Tosi, V.). *Biolocomotion: A Century of Research Using Moving Pictures*. Rome, Italy. ISB Series - Volume I, Promograph, pp. 33-47.

Cavanagh, P.R. (1990) The Mechanics of Distance Running: A Historical Perspective (ed. Cavanagh, P.R.). *Biomechanics of Distance Running*. Human Kinetics Publishers, Champaign, IL. **1**, pp. 1-34.

Centre National d'Art et de Culture Georges Pompidou (1977) Museé National d'Art Moderne, Paris. *E.J. Marey. 1830/1904*. p. 22.

Clarys, P. and Lewille, L. (1992) *Clinical and Kin. Electromyography by Le Dr. Duchenne (de Boulogne)*. ISB Series, pp. 90-112.

Clayton, M. (1992) *Leonardo da Vinci; Anatomy of Man*. Museum of Fine Arts, Houston.

Cooper, J.M. The Historical Development of Kinesiology with Emphasis on Concepts and People. *Kinesiology: A Natural Conference on Teaching* (eds. Dillman, C. and Sears, R.). University of Illinois Press, Urbana-Champaign. pp. 3-15.

d'Alembert, M. (1743) *Traité de Dynamique*. Chez David l'Aîné, Paris.

Dijksterhuis, E.J. (1961) *The Mechanization of the World Picture: Pythagoras to Newton*. Oxford University Press, Princeton.

Elftman, H. (1938) The Force Exerted by the Ground in Walking. *Arbeitsphysiologie*. **10**, pp. 485-491.

Elftman, H. (1939) Force and Energy Changes in the Leg During Walking. *Am. J. of Physiology*. **125 (2)**, pp. 339-366.

Foster, M. (1901) *Lectures on the History of Physiology During the Sixteenth, Seventeenth, and Eighteenth Centuries*. Cambridge University Press, Cambridge.

Fung, Y.C. (1981) *Biomechanics*. Springer Verlag, New York.

Fung, Y.C. (1990) *Biomechanics: Motion, Flow, Stress, and Growth*. Springer Verlag, New York.

Furusawa, K., Hill, A.V., and Parkinson, J.L. (1927) The Energy Used in Sprint Running. *Proceedings of the Royal Society*. **102 (B)**, pp. 43-50.

Gasser, H.S. and Hill, A.V. (1924) The Dynamics of Muscular Contraction. *Proc. Roy. Soc.* **96 (B)**, pp. 398-437.

Gillespie, C.C. (1973) *Dictionary of Scientific Biography*. Charles Scribner's Sons, New York.

Gray, Sir J. (1953) *How Animals Move*. Cambridge University Press, Cambridge.

Gray, Sir J. (1968) *Animal Locomotion*. Norton, New York.

Haas, R.B. (1976) *Muybridge: Man in Motion*. University of California Press, Berkeley.

Hamilton, W.R. (1834) On a General Method in Dynamics. *Phil. Trans. Royal Soc.* pp. 247-308.

Hankins, T. (1985) *Science and Enlightenment*. Cambridge University Press, New York.

Hatze, H. (1971) Was ist Biomechanik. *Leibesübungen Leibeserziehung*. **25**, pp. 33-34.

Hay, J.G. (1973) *Biomechanics of Sports Techniques*. Prentice Hall, Inc., Englewood Cliffs, NJ.

Hill, A.V. (1970) *First and Last Experiments in Muscle Mechanics*. Cambridge University Press, Cambridge.

Hirt, S. (1955) What is Kinesiology?: A Historical Review. *Physical Therapy Reviews*. **35 (18)**, pp. 419-426.

Hodgson, J. and Preston-Dunlop, V.M. (1990) *Rudolf Laban: An Introduction to his Work and Influence*. Northcote House, Plymouth, England.

Huxley, H.E. (1953) Electron Microscope Studies of the Organization of the Filaments in Striated Muscle. *Biochem. Biophys. Acta*. **12**, pp. 387-394.

Jansons, H. (1992) Bernstein: The Microscopy of Movement (eds. Capozzo, A., Marchetti, M., and Tosi, V). *Biolocomotion: A Century of Research Using Moving Pictures*. Rome, Italy. ISB Series - Volume I, Promograph, pp. 137-174.

Kardel, T. (1990) Neils Stensen's Geometric Theory of Muscle Contraction: 1667: A Reappraisal. *Journal of Biomechanics*. **23 (10)**, pp. 953-965.

Keele, K.D. (1983) *Leonardo da Vinci's Elements of the Science of Man*. Academic Press, New York.

Koestler, A. (1968) *The Sleepwalkers*. Hutchinson, London.

Lagrange, J.L. (1788) *Méchanique Analytique*. Veuve Desaint, Paris.

Lorini, G., Bossi, D., and Specchia, N. (1992) The Concept of Movement Prior to Giovanny Alphonso Borelli (eds. Capozzo, A., Marchetti, M., and Tosi, V.). *Biolocomotion: A Century of Research Using Moving Pictures*. Rome, Italy. ISB Series - Volume I, Promograph, pp. 23-32.

Maquet, P. (1992) *The Human Gait by Braune and Fischer*. ISB Series, pp. 115-125.

Marey, E.J. (1873) *La Machine Animale*. Librairie Germer Baillière, Paris.

Marey, E.J. (1878) *La Méthode Graphique*. G. Masson Editeur, Paris.

Marey, E.J. (1885) *La Méthode Graphique* (2nd Ed.). G. Masson Editeur, Paris.

Marey, E.J. (1972) *Movement*. Arno, New York. (Original 1895).

Marey, E.J. (1977) *La Photographie Mouvement*. Conservation des Musée de Beaune, France.

Meirovitch, L. (1970) *Methods of Analytical Dynamics*. McGraw-Hill, Toronto.

Millar, D., Millar, I., Millar, J., and Millar, M. (1989) *Chambers Concise Dictionary of Scientists*. W. & R. Chambers Ltd., Edinburgh.

Muybridge, E. (1887) *Animal Locomotion (1-11)*. University of Pennsylvania, Philadelphia.

Muybridge, E. (1955) *The Human Figure in Motion*. New York, Dover. pp. 1-3. (Original 1887).

Needham, D. (1971) *Machina Carnes*. Cambridge University Press, Cambridge.

Nelson, R.C. (1976) *Contribution of Biomechanics to Improved Human Performance*. The Academy Pages. **10,** pp. 61-65.

Newton, J.S. (1686) *Philosophiae Naturalis Principia Mathematica*. Jussu Societatis Regiae ac Typis Josephi Streater, London.

Nussbaum, M.C. (1978) *Aristotle's De Moto Animalium*. Princeton University Press, Princeton, NJ.

Rasch, P.J. (1958) Notes Toward a History of Kinesiology (Parts 1, 2, and 3). *Journal of American Osteopathic Ass*. pp. 572-574, 641-644, 713-715.

Sarton, G. (1953) *A History of Science: Ancient Science Through the Golden Age of Greece (1-2)*. W.W. Norton & Company Inc., New York.

Singer, C.J. (1959) *A Short History of Scientific Ideas to 1900*. Clarendon Press, New York.

Symposia of the Society for Experimental Biology (1980) *Symposia XXXIV. The Mechanical Properties of Biologist Materials*. Cambridge University Press, Cambridge.

Tosi, V. (1992) Marey and Muybridge: How Modern Biolocomotion Analysis Started (eds. Capozzo, A., Marchetti, M., and Tosi, V.). *Biolocomotion: A Century of Research Using Moving Pictures*. Rome, Italy. ISB Series - Volume I, Promograph, pp. 51-69.

Wartenweiler, J. (1972) Zur Geschichte der Biomechanik. *Jugend und Sport*. **12,** pp. 323-324.

Weber, W. and Weber, E. (1836) *Mechanik der menschlichen Gehwerkzeuge*. Dietrichsche Buchhandlung, Göttingen.

Weisheipl, J. (1985) *Nature and Motion in the Middle Ages* (ed. Carroll, W.). The Catholic University of America Press, Washington, D.C.

Westfall, R.S. (1971) *The Construction of Modern Science*. Cambridge University Press, New York.

Wolf, J. (1986) *The Law of Bone Remodelling* (trans. Maquet, P.). Springer Verlag, Berlin. (Original 1892).

Zernicke, R.F. (1981) Emergence of Human Biomechanics. *Perspectives on the Academic Disciplines of Physical Education* (ed. Brooks, G.A.). Human Kinetics Publications, Champaign, IL. pp. 124-136.

2 BIOMATERIALS

The human body is constructed of bone, cartilage, ligament, tendon, muscle, and other connective tissues. The anatomical components of the human body can be divided into *active* and *passive* structures. Active structures produce force; passive structures do not. Muscles are active, whereas bones, cartilage, ligaments, tendons, and the remaining soft tissues are passive structures.

The musculo-skeletal system can be used to influence the environment. Its actions can be interpreted mechanically (e.g., translation or rotation) or psychologically (e.g., expression of joy or pain by facial movements). The purpose of this chapter is to provide insight into the mechanical aspects of the construction and function of bone, cartilage, tendon, ligament, and muscle. The different biomaterials function together due to the joints which allow the transfer of forces from one segment to the next as well as relative movement between segments.

2.1 BONE

NIGG, B.M.
GRIMSTON, S.K.

2.1.1 DEFINITIONS AND COMMENTS

Anatomy

Bone modelling:
: Bone modelling is the process by which bone mass is increased.

Bone remodelling:
: Bone remodelling is the process by which bone mass is maintained or decreased.

Cancellous bone:
: Bone of reticular or spongy structure formed by thin spicules (trabeculae).
Comment: *cancellous* bone is also called *spongy* bone.

Cortex:
: An external layer of material.

Cortical bone:
: A solid external layer of material comprising the walls of diaphyses and external surfaces of bone.
Comment: *cortical* bone is also called *compact* bone.

Diaphysis:
: The elongated cylindric portion of a long bone between the ends (epiphyses).

Epiphysis:
: The end of a long bone.

Mineral content:
: The ratio of the unit weight of the mineral phase of bone to the unit weight of dry bone.

Osteoblast:
: A cell that arises from a fibroblast and that, as it matures, is associated with the formation and mineralization of bone.

Osteoclast:
: A large multinuclear cell associated with the absorption of bone.

Periosteum:
: Connective tissue covering the outer surface of bone.

Primary lamellar bone:
: Bone material which is arranged in circular form around the inner (endosteal) and outer (periosteal) circumference of a bony structure.
Comment: cancellous bone (at the end of long bones) belongs to the group of primary lamellar bone.

Primary osteon:
: A set of concentric lamellae together with associated bone cells (osteocytes) and a central vascular channel.

Trabecula:
: Little beam.

Trabecular bone:	Bony spicules forming a meshwork of intercommunicating spaces that are filled with bone marrow.
	Comment: the terms trabecular and cancellous bone are used interchangeably.
Water content:	The ratio of extracted water divided by the volume of the specimen (bone).

Mechanics

Density:	Mass per unit volume.
Elastic modulus:	The ratio of stress divided by strain.

$$E \; = \; \frac{\sigma}{\varepsilon}$$

where:

σ	=	stress
ε	=	strain
E	=	elastic modulus

with the units:

$[E]$	=	N/m^2	=	Pa
$[\sigma]$	=	N/m^2	=	Pa
$[\varepsilon]$	=	%		

Note: the unit "percent" is obtained by multiplying the measured relative length change with the number 100.

Load:	Sum of all the forces and moments.
Properties:	
• Material:	Properties of a material which describe its general behaviour without including any information about its size and shape.
	Comment: material properties include stress and strain, etc.
• Physical:	Properties of a material which relate to its physics.
	Comment: physical properties include density, specific weight, etc.
• Structural:	Properties of a specific sample of a material which describe the behaviour of that sample including effects of its size and shape.
	Comment: structural properties include force to failure, deformation, etc.
Strain:	Relative change of length.

$$\varepsilon = \frac{\Delta L}{L_o}$$

where:

ε = strain
ΔL = change in length = $L - L_o$
L_o = original length

Strain rate: Change of strain over time.

$$\dot{\varepsilon} = \frac{d\varepsilon}{dt}$$

with the unit:

$$[\dot{\varepsilon}] = \frac{1}{s}$$

Strength: Maximal force a material can sustain before failure.

- **Compressive strength:**   Maximal force in compression a material can sustain before failure (= ultimate compressive strength).

- **Tensile strength:**   Maximal force in tension material can sustain before failure (= ultimate tensile strength).

Stress: Force per unit area.

$$\sigma = \frac{F}{A}$$

where:

σ = stress
F = force
A = area

- **Compressive stress:**   Stress perpendicular to the surface that acts to compress an object.

- **Shear stress:**   Stress parallel to the surface of an object.

- **Tensile stress:**   Stress perpendicular to the surface that acts to elongate an object.

- **Ultimate stress:**   The highest stress experienced by the tissue before complete failure.

2.1.2 SELECTED HISTORICAL HIGHLIGHTS

The following historical highlights have been selected from several sources (Steindler, 1964; Bouvier, 1989; Martin and Burr, 1989). Fossil records of bone, in the form of a dermal armour surrounding the heads of fish, have been found that date from the Paleozoic era, about 500 million years ago. Further fossil records suggest that about 50 to 100 million years later, bone had an established place in the evolutionary development of fish and land vertebrates. Structure and function of bone has been discussed for several centuries. Significant progress in the mechanical and morphological understanding of the construction, functioning, and growth of bone did not, however, occur until the second half of the 19th century.

1674	van Leeuwenhoeck	Published microscopic observations in the Philosophical Transactions of the Royal Society of London.
1678	van Leeuwenhoeck	Reported observations of an extensive canal system in bone.
1691	Havers	Proposed that bone is composed of organic and inorganic strings and plates, respectively which are arranged around cavities in tubular form. He found pores between these plates.
1739	du Hamel	Demonstrated the lamellar nature of bone.
1742	Lieutaud	Suggested that bone is composed of laminae and compact fibres.
1754	Albinus	Recognized that the pores were used as vascular channels.
1776	Monro	Understood that bone resorption and formation occurs throughout life.
1816	Howship	Found that interstitial bone could be removed by "absorption".
1841	Burns	Reincarnated van Leeuwenhoeck's observation of osteons in bone.
1855	Breithaupt	Described stress fractures in military recruits of a Prussian military unit.
1856	Fick	Stated that bone is a passive structure and that the surrounding muscles determine the form of bone.
1856	Virchow	Stated that bone plays an active role in developing its form and structure.
1862	Volkmann	Suggested that pressure inhibits bone growth and release of pressure promotes it.
1867	Culman	Stated that there is a similarity between the trabecular arrangement in bone and the structural elements of a crane. In both cases the principles of highest efficiency and economy are used.
1867	van Meyer	Suggested that there is a relationship between architecture and function of bone.
1870	Wolff	Summarized various suggestions and statements about bone by stating that there is an interdependence between form and function of bone. Physical laws have strict control over bone growth.

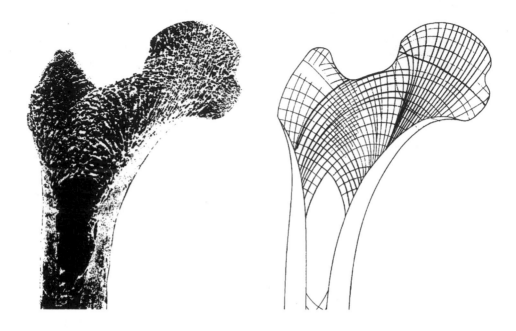

Figure 2.1.1   Cross-section of the upper end of a femur of a 31 year old male (left) and schematic representation of the same picture (right), from Wolff (1870) in the English translation by Maquet and Furlong, 1986, with permission.

1883	Roux	Proposed that the orientation of the trabecular system corresponds to the direction of tension and compression stresses and is developed using the principle of maximum economy of use of material (same as Wolff). Architecture of bone follows good engineering principles.
1897	Stechow	Made first radiographic verification of stress fractures.
1920	Jores	Proposed that bone cells act as sensors for structural alterations.
1931	Greig	Suggested that microscopic local damage may stimulate bone remodelling.
1942	Maj	Showed experimentally that cortical tissue of bone becomes weaker with advancing age.
1984	Lanyon & Rubin	Demonstrated the influence of bone loading on bone remodelling.

1986	Frost	Proposed a controlling mechanism (mechanostat) for the structural and functional adaptations of bone.

The schematic construction of bone has often been depicted in ways that illustrate the similarity between theoretical considerations of three-dimensional trajectorial systems and the actual arrangement of the trabeculae in the human bone. One of the earliest examples of such an illustration, by Wolff (1870), is shown in Fig. 2.1.1.

2.1.3 MORPHOLOGY AND HISTOLOGY

MORPHOLOGY AND FUNCTION

Bone is the hard part of the connective tissue which constitutes the majority of the skeleton of most vertebrates. It consists of an organic component (the cells and the matrix) and an inorganic or mineral component (Dorland's Illustrated Medical Dictionary). Bone performs several mechanical and physiological functions:

Mechanical functions:

- To provide *support* for the body against external forces (e.g., gravity).
- To act as a *lever* system to transfer forces (e.g., muscular forces).
- To supply *protection* for vital internal organs (e.g., the brain).

Physiological functions:

- To form blood cells (*hematopoiesis*).
- To store calcium (*mineral homeostasis*).

Support

Both cortical and trabecular bone provide support for the soft tissues and for the skeletal construction of the body. The different positions of human and animal postures would not be possible without the bony structures in them.

Lever system

Bones provide points of attachment for skeletal muscle-tendon units. Through muscular attachment and the articulation of bones at joints, bones and muscles function together to transfer forces in lever systems.

Protection

The flat bones of the skeleton, composed of a layer of trabecular bone wedged between two cortical plates, are largely responsible for protecting vital structures such as the heart, lung, and brain. The brain, for instance, is protected by the bones of the skull, the bladder and internal reproductive organs by the pelvis, and the heart and lungs by the rib cage.

Hematopoiesis

Hematopoiesis, the process of blood cell formation, occurs in red bone marrow. In adults red bone marrow is found in bone regions composed largely of trabecular bone, e.g., the vertebrae, proximal femur, and iliac crest. In these regions erythrocytes, leukocytes, and thrombocytes are formed.

Mineral homeostasis

Bone is the body's largest reservoir of calcium, with 99% of total body calcium stored in the skeleton. Other critical minerals, such as phosphorus, sodium, potassium, zinc, and magnesium, are also stored in bone. Since calcium is critical for a number of vital metabolic processes, maintaining serum calcium homeostasis always takes priority over the calcium requirements of bone. The hormones that regulate serum calcium balance include para thyroid hormone (PTH), calcitonin (CT), cholecalciferol (vitamin D), reproductive hormones, and growth hormones.

Table 2.1.1   Summary table of the bones in the human body.

NAME	NUMBER OF BONES
VERTEBRAL COLUMN, SACRUM, AND COCCYX	26
CRANIUM	8
FACE	14
AUDITORY OSSICLES	6
HYOID BONE, STERNUM, AND RIBS	26
UPPER EXTREMITIES	64
LOWER EXTREMITIES	62
TOTAL	206

Bone is a dynamic tissue that is constantly remodelling and may undergo modelling given the appropriate stimuli. The morphology of any bone can undergo substantial changes during its lifetime. This section discusses human adult bone morphology at the organ and tissue levels.

The adult human skeleton consists of 206 distinct bones (Table 2.1.1). Although the bones vary considerably in size and shape, they are similar in structure and development. Bone may be classified into a few basic but overlapping shapes. The different bones types, selected examples, and their functions are given in Table 2.1.2.

Despite the variety of external forms in the skeleton, the morphology of bone at the organ, tissue, and cellular levels is relatively consistent. It is assumed that the reader is familiar with basic anatomy, so no detailed discussion of skeletal anatomy will be presented beyond the definitions of the structural components of a long bone (Fig. 2.1.2).

Parts of long bones can be described based on their centres of ossification. The *epiphyses* develop from secondary ossification centres and are found at the ends of long bones. The epiphyses articulate with other bones and are protected by a layer of hyaline cartilage referred to as articular cartilage. Between the two epiphyses is the shaft of the long bone, the *diaphysis*, which develops from the primary ossification centre. The diaphysis is a hollow structure surrounding the *medullary cavity*. The medullary cavity, which is used as a fat storage site, is lined by a thin, largely cellular connective tissue membrane, the *endosteum*. There is no medullary cavity in a flat bone. A portion of the flared ends of long bones are called the *metaphyses* and are the growth zone between the epiphyses and the diaphysis during development.

Table 2.1.2   **Different bone types, classified according to shape.**

SHAPE	EXAMPLES	FUNCTION
LONG	femur tibia radius	to act as levers
SHORT	carpal bone tarsal bone	to provide strength
FLAT	sternum ribs skull bones ilium scapula	to provide protection and points of attachment for tendons and ligaments
IRREGULAR	ischium pubis bone vertebrae	various functions
SESAMOID	patella	improved lever situation

Surrounding and attached to the bone (except for those areas covered by cartilage) is a tough, vascular, fibrous tissue called the *periosteum*. The outer layer of the periosteum is well supplied with blood vessels and nerves, some of which enter the bone. The inner layer is anchored to the bone by collagenous bundles called *Sharpey's fibres*, which penetrate the bone. Some of the periosteum fibres are intertwined with fibres of tendons, which provide attachment for muscles. Projections of bone, called processes, act as sites of attachment for ligaments and tendons, while a depression in a bone allows for articulation with a process of another bone.

At the gross level all bones in the adult skeleton have two basic structural components: the cortical and the cancellous bone. *Cortical* (or *compact*) bone, is the solid, dense material comprising the walls of diaphyses and external surfaces of bones. This type of bone is solid, strong, and resistant to bending. The thickness of cortical bone varies between and within bones as a function of the mechanical requirements of the bone. *Cancellous* (or *spongy*) bone is named after the thin bony spicules that form it, called *trabeculae*. These trabeculae have been observed to orient themselves in the direction of the forces applied to

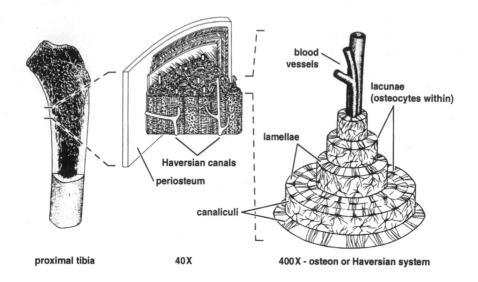

blood vessels

lacunae (osteocytes within)

lamellae

Haversian canals

periosteum

canaliculi

proximal tibia   40X   400X - osteon or Haversian system

Figure 2.1.2   **Illustration of the gross and microscopic structure of bone (from White, 1991, with permission).**

the bone. They are found in processes, in the vertebral bodies, in the epiphyses of long bones, in short bones, and sandwiched between the two layers of compact bone that make up the flat bones (called *diploe*). Irregular interconnecting spaces occur between the trabeculae, reducing the weight of the bone. Trabecular bones are highly resistant to compressive loads.

Red bone marrow, a hemopoietic tissue that produces red and white blood cells and platelets, is found in areas of trabecular bone. The yellow bone marrow found in the medullary canal of long bones consists of fat cells. Bones, therefore, provide an active site of hematopoiesis.

Bone is a composite of two materials. The *organic matrix* is approximately 95% collagen fibres, greatly strengthened by deposits of calcium and phosphate salts in the form of *hydroxyapatite*. The calcium and phosphate deposits give bone its strength, hardness, and rigidity, while the collagen fibres give bone its limited flexibility.

HISTOLOGY

Histology examines the structure of cells, tissues, and organs in relation to their function. Histological examinations are usually conducted at the microscopic level.

Microscopic structures of bone types

In general, there are three broad categories of bone microstructure. These are determined by differences in organization and/or control. Here, each is introduced, with the greatest emphasis placed on primary and secondary bone.

Woven bone:

Unlike the regularly oriented collagen fibres of lamellar bone, the collagen fibres in woven bone are randomly oriented. The result is a less dense bone, although there is generally no mineralization deficit. Woven bone further differs from other forms of bone in that it can be deposited *de novo*, i.e., without any previous hard tissue or cartilage model. The most characteristic form of woven bone is formed during embryonic life and at the growth plate during endochondral ossification. The most common form of woven bone in the adult skeleton is *callus* formation, where bone fractures heal as a result of damage to or tension on the periosteum. Woven bone provides a quickly developing source of mechanical strength and the framework for the slower development of lamellar bone.

Primary bone:

There are several types of primary bone, which differ in how they develop. Primary bone cannot be deposited *de novo*. It requires a pre-existing substrate such as a cartilaginous model. The three main categories of primary bone differ not only morphologically, but also in their mechanical and physiological properties.

(i) Primary lamellar bone:

Primary lamellar bone, or circumferential lamellar bone is arranged in circular rings around the endosteal and periosteal circumference of a bone. The trabeculae in the epiphyses of long bones are also primarily lamellar bone, and they are closely associated with marrow and vascular tissues. This proximity allows for rapid exchange of calcium between bone and serum, and explains in part why regions of cancellous bone are the first to exhibit osteopenia (reduced bone mass). Primary lamellar bone is mechanically competent. Where there is a large surface area adjacent to marrow and blood, however, the requirements of hematopoiesis and mineral metabolism may override mechanical requirements.

(ii) Plexiform bone:

Like primary lamellar bone, plexiform bone must be deposited on pre-existing surfaces. Like woven bone, it forms rapidly, but it has better mechanical properties. Structurally, plexiform bone appears as highly oriented cancellous bone with the trabecular plates thickening due to endosteal and/or periosteal surface apposition. This type of bone is most commonly seen in rapidly growing large animals, such as the cow, whose rapid growth demands mechanical competence of the skeleton (Martin and Burr, 1989). However, it has also been observed in children during the growth spurt (Amprino, 1947).

(iii) Primary osteons:

A set of lamellae arranged in concentric rings around a vascular channel (versus around the entire bone cortex) is called a primary osteon. Until recently, primary osteons had not been distinguished from secondary osteons (see below); however, developmentally, morphologically, and possibly mechanically, they are quite different (Currey, 1984). The greatest distinction is that primary osteons do not have cement lines (reversal lines) because they are not developed through bone remodelling. As well, they appear to have smaller vascular channels and fewer lamellae than secondary osteons, so primary osteons may be stronger than secondary osteons (Martin and Burr, 1989). Primary osteons develop through the sequential filling in of vascular channels with layers of lamellar bone and are found in well organized primary lamellar bone.

Secondary bone:

When bone is the product of resorption of previously existing bone tissue and the deposition of new bone to replace it, it is referred to as secondary bone. In cortical bone the result of bone resorption by osteoclasts, that is followed by apposition of bone by osteoblasts, is the *secondary osteon*. A secondary osteon has a central vascular channel larger than that found in primary osteons and referred to as a *Haversian Canal*. This is surrounded by concentric lamellae arranged in a circular fashion containing osteocytes. The two distinguishing morphological features of the secondary osteon are the presence of cement or reversal lines which separate the osteon from the extraosteonal bone matrix, and the organization of lamellae around the Haversian Canal.

Each "trunk" in the cross-section of compact lamellar bone is referred to as a *Haversian System*. Haversian Systems in healthy adult humans measure approximately 300 microns in diameter and are approximately 3-5 mm in length, with their long axes parallel to those of the long bone. Blood, lymph, and nerve fibres pass through the Haversian Canal. Smaller *Volkmann's Canals* pierce the bone tissue obliquely and perpendicular to the periosteal and endosteal surfaces, linking the Haversian Canals and creating a network for blood, nerve, and lymph supply to bone cells. Small spaces within the lamellae called *lacunae* contain mature bone cells - osteocytes - which receive nutrients via minute fluid-filled channels called *canaliculi*, which radiate out from the Haversian Canal.

Three primary mature bone cell types are responsible for forming, resorbing, and maintaining bone. *Osteoclasts* are the cells responsible for resorbing bone, thereby releasing calcium into the serum. The bone forming cells are *osteoblasts*, which synthesize the collagen matrix (*osteoid*) and later deposit bone mineral (*hydroxyapatite*) within that matrix, producing mineralized bone. Once it has surrounded itself with mineralized bone tissue, the osteoblast is referred to as an *osteocyte*, the mature bone cell believed to be responsible for maintaining bone tissue. The processes of bone resorption followed by bone formation occur throughout life and comprise the process of bone *remodelling*. Remodelling can maintain existing bone mass or result in a reduction of bone mass. For detailed discussion of the bone remodelling process and its ramifications the reader is referred to more detailed publications.

Skeletal processes

There are essentially 4 skeletal processes that occur at various stages of human life: growth (endochondral and intramembranous), modelling, remodelling, and repair. These processes are fundamentally distinct, operating under different controls, in different locations, and at different ages in the human, yet the same types of bone cells are involved in all. Each will be discussed briefly, and the reader is referred to books on anatomy, physiology, and the publications of H.M. Frost for greater detail.

Bone growth:

(i) Intramembranous ossification:

Intramembranous ossification refers to the growth of bones such as the frontal and parietal bones of the skull. It occurs through the apposition of bone on tissue within an embryonic tissue membrane. During their development, membrane-like layers of connective tissue are supplied with dense networks of blood vessels, which attract the connective tissue cells. These cells enlarge and differentiate into osteoblasts, which deposit osteoid and then mineralize it to produce a matrix of trabecular bone. The cells of the primitive connective tissue that lie outside the developing trabecular matrix give rise to the periosteum. Osteoblasts beneath the periosteum generate a layer of compact bone to cover the surface of newly formed trabecular bone.

(ii) Endochondral ossification:

The majority of bones in the skeleton grow through the process of endochondral ossification, in which bone is preceded by cartilage. Ribs, vertebrae, the cranial base, and bones of the extremities begin as cartilage models *in utero*, an environment in which the function of support is not necessary. Ossification occurs within the cartilage model at the centre of the diaphysis, as it is penetrated by blood vessels (nutrient foramen). This region is called the primary ossification centre, and bone radiates from it towards the ends of the cartilaginous model. Like the periosteum on bone, a thin membrane called the perichondrium surrounds the cartilage model. Osteoblastic progenitor cells in this region produce osteoblasts that deposit a thin layer of compact bone around the primary ossification centre. The resultant periosteum then continues to deposit bone layer upon layer, increasing the diameter of the diaphysis. With this increase, osteoclasts on the endosteal surface resorb bone while osteoblasts on the periosteal surface deposit bone. Thus, appositional growth allows diaphyseal diameter to increase and the medullary canal to develop. The compact bone of an adult limb bone is, therefore, periosteal in origin.

Increases in bone length arise through the formation of secondary ossification centres in the epiphyses of long bones. The metaphysis and epiphysis in a growing bone are separated by the *epiphyseal plate* (growth plate). As trabecular bone develops in all directions from the secondary and primary ossification centres, the cartilaginous centre of the plate is replaced by bone on the diaphyseal side. As the bone grows, the epiphyseal plate is pushed further from the bone's primary ossification centre thereby lengthening the bone. Ossification and growth of the bone come to a halt when cells at the growth plate stop dividing and the epiphysis fuses with the metaphysis of the shaft. Most long bones develop two secondary centres in addition to the primary centres of ossification. A few long bones develop only one secondary centre, and the majority of carpal and tarsal bones develop completely from the primary ossification centre.

The growth of bone in a maturing skeleton is the only process that continually creates new trabeculae. There is no known counterpart in the adult skeleton. Noticeable changes in the size and shape of adult bones occur through the process of modelling.

Bone modelling and remodelling

Bone modelling is defined as the process by which bone mass is increased.

Bone remodelling is defined as the process by which bone mass is maintained or decreased.

Bone modelling can add but not reduce bone mass. Modelling can occur through modelling drifts, which thicken cortices and trabeculae in children, or through minimodelling drifts, which can thicken trabeculae in children and adults. Modelling becomes inefficient in cortical bone of adult mammals, including humans.

Bone remodelling in both cortical and trabecular bone occurs at Basic Multicellular Units (BMU). Bone remodelling at BMUs occurs on all bone surfaces throughout life, at varying rates. Where bone is in contact with marrow, resorption exceeds formation, resulting in gradual enlargement of the marrow cavity and decreases in trabecular bone volume.

The distinction between modelling and remodelling is critical to bone biology. Modelling differs from remodelling in a number of fundamental ways. The most profound differences are with respect to the temporal and spatial relationship of osteoblasts and osteoclasts. In modelling, resorption and formation occur on different bone surfaces. They have no spatial relationship, and no fixed temporal "coupling" exists between them. Osteoclastic and osteoblastic activity appear to be subject to independent regulation. Modelling involves the activities of osteoblasts and osteoclasts at their respective surfaces for extended periods of time with no apparent interruption. In remodelling, osteoclastic and osteoblastic activities are coupled and occur in cycles of resorption, formation, and quiescence. In remodelling, the advancement of the "formation front" through tissue space is slower than the advancement of the "resorption front", while in modelling the endosteal resorption front progression is slightly slower than that of the periosteal formation front. Thus, in modelling the cortex of bone increases in both thickness and external diameter. Finally, during the growth period, modelling processes involve almost the entire periosteal and endosteal surface at all times. In remodelling only a fraction of the surface is active at any one time.

Bone repair

Bone repair occurs at two levels. The first is the constant repair of the microdamage that results from exposure to physiological loads, by the remodelling process. This naturally occurring process may not produce woven bone as a result. If this repair mechanism at the tissue level is inhibited, for example by the introduction of some pharmacological agent that suppresses osteoclastic activity, microdamage will accumulate and fatigue fracture may result. In contrast, repair of a fracture at the organ level, for example in response to abnormal and/or excessive loads, will first result in the formation of woven bone, often referred to as a callus.

The process of repair is initiated with blood flow into the fracture region that normally coagulates to form a hematoma. The fracture ruptures the periosteum, stimulating the rapid formation of the callus or woven bone. It provides temporary strength and support for the fractured bone. Mineralization of the final callus takes approximately 6 weeks in the human adult. It is then gradually remodelled to produce lamellar bone. The final orientation of the broken bone ends will depend in part on its similarity to the original and on the orientation of loads applied during the healing process.

2.1.4 PHYSICAL PROPERTIES

Bone's material properties can be quantified from a mechanical point of view. The structural and mechanical properties of bone, however, change with the forces acting on the bone. The first analyses of the implications of this fact were performed in the second part of the 19th century. They were summarized by Wolff (1870) in what is known today as *Wolff's laws of functional adaptation*.

WOLFF'S LAW OF FUNCTIONAL ADAPTATION

In his classic publication in 1870 Wolff wrote:

> **"The shape of bone is determined only by the static stressing..."**

> **"Only static usefulness and necessity or static superfluity determine the existence and location of every bony element and, consequently of the overall shape of the bone".**

Several comments seem appropriate in the context of these statements. First, stress has different effects on bone. Stress may effect growth as seen in the healing process of a fracture in a child, it may effect modelling as in an overload situation, and/or remodelling as in an under load situation. Second, growth of bone is, among other factors, influenced by heredity. If a person has an inherited bone deformity, stress may not necessarily correct the inherited form. Third, it has been shown (Lanyon and Rubin, 1984) that the type of loading influences bone formation and/or remodelling processes. Dynamic loading seems to be more important for bone modelling than static loading. Considerations of this kind suggest that Wolff's law of functional adaptation of bone should be restated in a more general way. Specific forms of stress are not the only factor that determines bone growth, but they are an important one. Adapted, Wolff's law may be worded thus:

> **Physical laws are a major factor influencing bone modelling and remodelling.**

SELECTED PHYSICAL PROPERTIES FOR BONE

Physical properties of biomaterials can be divided into two categories; structural and material properties. Structural properties describe properties to a specific specimen, for instance a bone or a ligament. A typical example would be a breaking force for a specific tibia. Material properties are independent of a specific specimen and describe the material itself. A typical example for a material property would be a stress for which the bone breaks under compression. In this chapter, both properties are discussed, however, more emphasis is given to the material properties.

Selected physical properties of bone (and, for comparison, selected other materials) are summarized in Table 2.1.3.

The elastic modulus, E, represents the stress needed to double the length of an object. The elastic modulus is a constant which describes the material characteristic. It does not suggest that the particular material can in fact be stretched to double its length. For bone the typical values (order of magnitude) for E are:

- Trabecular (spongy) bone 10^9 Pa   =   1 GPa
- Cortical (compact) bone $2 \cdot 10^{10}$ Pa =   20 GPa
- Metals 10^{11} Pa   =   100 GPa

For bone, the elongation required to cause a fracture is only a small fraction of this doubling length. As a rule of thumb one can write:

$$F_{fracture} \approx (1 : 200) \cdot F_{double}$$

Table 2.1.3   **Selected physical properties of bone based on data from Yamada (1970); Burstein et al. (1976); Noyes et al. (1984); and Ascenzi and Bonucci (from Martin and Burr, 1989, with permission of Williams & Wilkins, Baltimore, Maryland).**

VARIABLE	COMMENT	MAGNITUDE	UNIT
DENSITY	cortical bone	1700-2000	kg/m^3
	lumbar vertebra	600-1000	kg/m^3
	water	1000	kg/m^3
MINERAL CONTENT	bone	60-70	%
WATER CONTENT	bone	150-200	kg/m^3
ELASTIC MODULUS (TENSION)	femur (cortical)	5-28	GPa
TENSILE STRENGTH	femur (cortical)	80-150	MPa
	tibia (cortical)	95-140	MPa
	fibula (cortical)	93	MPa
COMPR. STRENGTH	femur (cortical)	131-224	MPa
	tibia (cortical)	106-200	MPa
	wood (oak)	40-80	MPa
	limestone	80-180	MPa
	granite	160-300	MPa
	steel	370	MPa

EXAMPLE 1

Question:

Estimate the ultimate tensile force to break a trabecular bone. Calculate the ultimate tensile force for trabecular bone in general and for the femur in particular.

Symbols:

E	=	elastic modulus
A	=	cross-sectional area
L_o	=	unloaded length
ΔL	=	change in length from unloaded to loaded situation
A_{tib}	=	cross-sectional area of the tibia

Assumptions:

(1) E   = 10^9 Pa
(2) A   =   1 mm² = 10^{-6} m²
(3) $\Delta L/L_o$ =   1:200
(4) A_{tib} =   800 mm²= $8 \cdot 10^{-4}$ m²
(5) The question can be treated as a one-dimensional problem.

Solution:

In the first step, the force required to break a bone sample with the diameter of 1 mm² is determined:

$$\varepsilon \;=\; \frac{1}{E} \cdot \sigma \;=\; \frac{1}{E} \cdot \frac{F}{A} \;=\; \frac{\Delta L}{L_o}$$

$$F \;=\; \frac{1}{L_o} \cdot \Delta L \cdot E \cdot A$$

$$F \;=\; 0.005 \cdot 10^9 \cdot 10^{-6} \, \text{N}$$

$$F \;=\; 5 \, \text{N}$$

A force of 5 N is needed to break a trabecular bone sample of 1 mm² cross-sectional area in tension. For a specific bone of 800 mm² cross-section (e.g., at the epiphysis level) a force of about 4000 N is needed to break it in tension.

The ultimate stresses for tension and compression are different. For cortical bone, the ultimate stress for compression, σ_{comp} is about 30 to 50% higher than the ultimate stress for tension σ_{tens} (Reilly and Burstein, 1975; Steindler, 1977):

$$\sigma_{comp} \;\approx\; 1.4 \cdot \sigma_{tens} \quad \text{(cortical bone)}$$

The ultimate stress for compression and tension of cancellous bone is at least one order of magnitude smaller than that of compact bone. The order of magnitude of averaged data from the literature is illustrated in Fig. 2.1.3.

There is only a small region of elongation where bone follows a linear law between force and elongation (Hooke's law). Beyond this limit, deformation and force are no longer arithmetically proportional and when force is removed, bone does not return to its original form but retains some deformity.

2.1.5 BONE MECHANICS AS RELATED TO FUNCTION

BONE ANISOTROPY

Bone is a non-homogeneous and anisotropic material. It's mechanical properties change as a function of the specific location in the bone and the direction in which force is applied. Stiffness with respect to tension is maximal for axial forces and minimal for perpendicular forces. Stiffness for axial forces is about twice the magnitude of stiffness for perpendicular forces. Additionally, the ultimate strain for bone (cortical bone of human femur) loaded in the axial direction is about twice the ultimate strain for bone loaded in perpendicular direction (Frankel and Burstein, 1970).

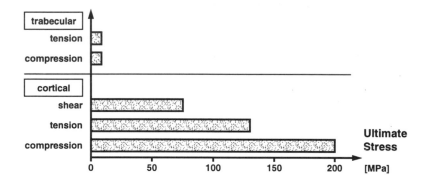

Figure 2.1.3   **Order of magnitude of ultimate stress for cortical and trabecular bone based on data from Yamada, 1970; Steindler, 1977; Reilly and Burstein, 1975; Martin and Burr, 1989; with permission of Williams & Wilkins, Baltimore, Maryland.**

Bone strain and/or stress capacity is assumed to be related to the osteonal strain and/or stress capacity, a concept reminiscent of the relationship between the whole muscle and the sarcomere. The differences in mechanical properties for different types of osteons has been summarized (Martin and Burr, 1989) from data produced by Ascenzi and Bonucci over several years. They distinguished three types of bone specimens based on collagen fibre orientation:

L = longitudinal fibre orientation
A = alternating fibre orientation
T = circumferential fibre orientation

The results include ultimate stress, elastic modulus, and ultimate strain (Table 2.1.4).

The simple version of Hooke's law does not describe the anisotropic nature of bone. For this purpose a more generalized version of Hooke's law is applicable (Martin and Burr, 1989). It uses the assumption that the specific stress components are proportional to the corresponding strain components. This generalized version of Hooke's law is:

$$\sigma_1 = c_{11}\varepsilon_1 + c_{12}\varepsilon_2 + c_{13}\varepsilon_3 + c_{14}\varepsilon_4 + c_{15}\varepsilon_5 + c_{16}\varepsilon_6$$

$$\sigma_2 = c_{21}\varepsilon_1 + c_{22}\varepsilon_2 + c_{23}\varepsilon_3 + c_{24}\varepsilon_4 + c_{25}\varepsilon_5 + c_{26}\varepsilon_6$$

$$\sigma_3 = c_{31}\varepsilon_1 + c_{32}\varepsilon_2 + c_{33}\varepsilon_3 + c_{34}\varepsilon_4 + c_{35}\varepsilon_5 + c_{36}\varepsilon_6$$

$$\sigma_4 = c_{41}\varepsilon_1 + c_{42}\varepsilon_2 + c_{43}\varepsilon_3 + c_{44}\varepsilon_4 + c_{45}\varepsilon_5 + c_{46}\varepsilon_6$$

$$\sigma_5 = c_{51}\varepsilon_1 + c_{52}\varepsilon_2 + c_{53}\varepsilon_3 + c_{54}\varepsilon_4 + c_{55}\varepsilon_5 + c_{56}\varepsilon_6$$

$$\sigma_6 = c_{61}\varepsilon_1 + c_{62}\varepsilon_2 + c_{63}\varepsilon_3 + c_{64}\varepsilon_4 + c_{65}\varepsilon_5 + c_{66}\varepsilon_6$$

where:

c_{ik} = elastic constants

with:

$c_{ik} = c_{ki}$

Often, materials have characteristics that reduce the number of elastic constants. *Isotropic* material (identical material behaviour in all directions) has only two independent elastic constants, the elastic modulus, E, and the shear modulus, G.

Table 2.1.4   **Mechanical properties of different types of osteons (based on data from Ascenzi and Bonucci). From Martin and Burr, 1989, with permission of Raven Press, Ltd., New York.**

LOADING MODE	TYPE OF OSTEONS	ULTIMATE STRESS [MPa]	ELASTIC MODULUS [GPa]	ULTIMATE STRAIN [%]
TENSION	L	114 (17)	11.7 (5.8)	6.8 (2.9)
	A	94 (15)	5.5 (2.6)	10.3 (4.0)
COMPR.	L	110 (10)	6.3 (1.8)	2.5 (0.4)
	A	134 (9)	7.4 (1.6)	2.1 (0.5)
	T	164 (12)	9.3 (1.6)	1.9 (0.3)
SHEAR	L	46 (7)	3.3 (0.5)	4.9 (1.1)
	A	55 (3)	4.1 (0.4)	4.6 (0.6)
	T	57 (6)	4.2 (0.4)	4.6 (0.6)

Anisotropic material, such as bone and other biological structures, often has internal structures of symmetry. Often such material can be approximated as *orthotropic* (mechanical properties are different in 3 orthogonal directions). In long bone such axes may be the longitudinal and 2 orthogonal axes. In this case, the matrix of elastic constants has 9 independent values and is:

$$
\begin{bmatrix}
c_{11} & c_{12} & c_{13} & 0 & 0 & 0 \\
c_{12} & c_{22} & c_{23} & 0 & 0 & 0 \\
c_{13} & c_{23} & c_{33} & 0 & 0 & 0 \\
0 & 0 & 0 & c_{44} & 0 & 0 \\
0 & 0 & 0 & 0 & c_{55} & 0 \\
0 & 0 & 0 & 0 & 0 & c_{66}
\end{bmatrix}
$$

In discussing the mechanical properties of bone different aspects may be of interest. One may be interested in the local strain and/or stress properties. The outlined procedures are applicable in this case. The anisotropic mechanical behaviour of the entire bone, however, is currently not covered by a comprehensive theory. For further, more complete discussion of anisotropy and elastic theory of bone, the reader is referred to *Theory of an Anisotropic Elastic Body* (Lekhnitskii, 1963).

MAXIMAL STRESS

Maximal stress at the surface of bone for compression, tension, or torsion can be estimated for bone with circular shape by using the formulas listed in Table 2.1.5.

Table 2.1.5   Formulas to estimate maximal stress on bone surfaces with a circular cross-section.

CROSS-SECTION	MAXIMAL STRESS DUE TO BENDING σ_{be}	MAXIMAL STRESS DUE TO TORSION σ_{tor}
GENERAL	$\dfrac{M_{be} \cdot y}{I_{xx}}$	$\dfrac{M_{be} \cdot R}{I_{zz}}$
FULL CIRCLE	$\dfrac{4 \cdot M_{be}}{\pi R^3}$	$\dfrac{2 M_{tor}}{\pi R^3}$
HOLLOW CIRCLE	$\dfrac{4 \cdot M_{be} \cdot R}{\pi (R^4 - r^4)}$	$\dfrac{2 M_{tor} R}{\pi (R^4 - r^4)}$

where:

M_{be} = bending moment responsible for compression and tension
M_{tor} = moment responsible for torsion
R = outer radius of the structure (bone) of interest
r = inner radius of the hollow structure (bone) of interest
y = (maximal) distance from the neutral axis
I_{xx} = moment of inertia about the neutral axis
I_{zz} = polar moment of inertia about the axis of rotation

with the units:

$[M]$ = $N \cdot m$
$[R]$ = m
$[r]$ = m
$[\sigma]$ = Pa
$[y]$ = m
$[I_{xx}]$ = $kg \cdot m^2$
$[I_{zz}]$ = $kg \cdot m^2$

EXAMPLE 2

Question:

An idealized bone has the shape which is illustrated in Fig. 2.1.4. The cross-section is assumed to be circular. The radius of the bone is $R = 1$ cm and a constant force, F, of 4000 N is acting as illustrated with a distance of 2 R from the long axis of the bone. Estimate the maximal tensile and compressive stress at the cross-section, S: if the force, F, acts as illustrated in the sketch, the bone bends to the right. Consequently, the right bone surface will show compression and the left one tension.

Assumptions:
(1) The weight of the bone can be neglected.
(2) Bone is isotropic (same material properties over the whole bone).
(3) The problem can be treated two-dimensionally.

Solution:

The compressive stress, σ_{comp}, acting on the right outside of the structure (the bending side) is a result of the axial stress, σ_{ax}, and the stress due to bending, σ_{be}:

$$\sigma_{comp} = \sigma_{ax} + \sigma_{be}$$

Similarly, the tensile stress, σ_{tens}, is a result of the axial and the bending stress. However, in this case the bending stress is in the direction opposite to the axial stress:

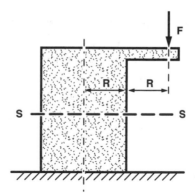

Figure 2.1.4   **Schematic illustration of the loading of a "bony structure" with isotropic material and a circular cross-section.**

$$\sigma_{tens} \ = \ \sigma_{ax} - \sigma_{be}$$

$$\sigma_{ax} \ = \ \frac{F}{\pi R^2}$$

$$\sigma_{be} \ = \ \frac{4M_{be}}{\pi R^3}$$

$$\sigma_{be} \ = \ \frac{4F \cdot 2R}{\pi R^3}$$

$$\sigma_{be} \ = \ 8 \cdot \frac{F}{\pi R^2}$$

consequently:

$$\sigma_{comp} \ = \ 9 \cdot \frac{F}{\pi R^2}$$

$$\sigma_{tens} \ = \ -7 \cdot \frac{F}{\pi R^2}$$

or numerically for the maximal compression or tension (rounded results):

$$\sigma_{comp} \ = \ 115 \ \text{MPa} \ = \ 115 \ \text{N/mm}^2$$

$$\sigma_{tens} \ = \ 89 \ \text{MPa} \ = \ 89 \ \text{N/mm}^2$$

The stress distribution in a cross-section of the bone structure presented is illustrated in Fig. 2.1.5. The maximal compression stress is greater than the maximal tension stress. The result is in the same direction as the previously mentioned result, which indicates that the ultimate compression stress is about 30% higher than the ultimate tension stress. It may be that the higher ultimate stress for compression is a result of an adaptation.

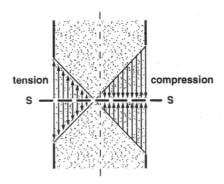

Figure 2.1.5   **Stress distribution due to a bending force, F, in a cross-section, S, for the assumptions used in Example 2.**

The stresses estimated due to bending in the above example are about one order of magnitude larger than the axial stresses. This illustrates that the geometry of the acting forces is extremely important. If the resultant force is not acting along the axis of a bone, the total stress on the surface of the bone increases and can easily reach multiples of the stresses produced by the axial forces.

The results of this example have at least 4 practical implications:

(1)   Joint forces (bone-to-bone contact forces in joints) that do not act along a bone axis are often compensated for by muscular forces that reduce the maximal stresses on the surface of the bone.

(2)   Malalignment of the skeleton may necessitate increased muscular compensation to reduce the maximal stresses on the bone surfaces (e.g., at the diaphysis of a long bone). Muscular atrophy as a consequence of injury or aging may disturb the muscular balance and place excessive stresses on the bone surfaces, which may lead to fractures.

(3)   Movements in which external forces do not act along the bone axes in the human body (e.g., forces on the foot in different shoes) may produce high internal stresses on the bone surfaces.

(4)   Commonly, bone is loaded in different modes (tension, compression, torsion, and shear). An example for the complex loading of bone will be discussed in the section "Applications".

EXAMPLE 3

Question:

An idealized bone has a shape like that in Fig. 2.1.4 but the structure is hollow. The outer radius is R = 1 cm and the inner radius is r = 0.5 cm. A force of 4000 N acts at a distance of 2 R = 2 cm from the axis of the structure. Determine the maximal compression and tension stresses for the hollow cylindrical structure, and compare the results with the results for the full structure.

Assumptions:

The same assumptions as in the previous example are used.

Solution:

The same procedure is used as in the previous example:

$$\sigma_{comp} = \sigma_{ax} + \sigma_{be}$$

$$\sigma_{tens} = \sigma_{ax} - \sigma_{be}$$

The numerical values used for this example are:

$$R = 0.01 \text{ m}$$
$$r = 0.005 \text{ m}$$
$$F = 4000 \text{ N}$$

The rounded maximal stresses on the surface of the hollow structure are:

$$\sigma_{ax} = 17 \text{ MPa}$$

$$\sigma_{be} = 109 \text{ MPa}$$

Therefore, the maximal compressive and tensile stresses are:

$$\sigma_{comp} = 126 \text{ MPa} \quad \text{(for solid structure:} \quad \sigma_{comp} = 115 \text{ MPa)}$$

$$\sigma_{tens} = 92 \text{ MPa} \quad \text{(for solid structure:} \quad \sigma_{tens} = 89 \text{ MPa)}$$

The maximal compressive stress on the surface of the hollow structure is only marginally higher than the maximal compressive stress for the solid (filled) structure. The difference for this example is less than 10%.

The hollow structure has less mass and will, therefore, have a smaller inertia. This is advantageous for bone segments involved in frequent, quick movements such as those in the extremities. In fact, most extremities in human and animal bodies have hollow bones - bones such as femur, tibia, metatarsals, radius, and others. As the examples illustrate, the cost of the advantages of less mass and less inertia is only a small increase in the maximal stress.

This result is illustrated in Fig. 2.1.6 in a general form. The two graphs compare the stress per unit force for a solid structure $(R_1/R_2 = 0)$ and an arbitrarily selected hollow

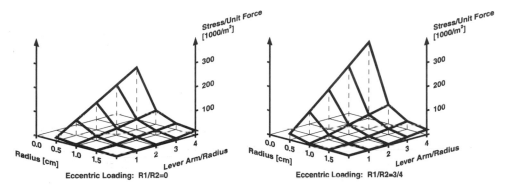

Figure 2.1.6   Illustration of the effect of the diameter of bone and the moment arm of the acting force on the maximal compression stress for a solid (filled) and a hollow structure (from Nigg, 1985, with permission).

structure ($R_1/R_2 = 0.75$). The comparison is made for changes in the moment arm of the acting force and changes of the actual radius of the force. The graphs indicate that, in both cases, the compressive stress is high for:

- A small diameter of the (bony) structure, and/or
- A large moment arm of the acting (resultant) force.

A small diameter of the bony structure is a given boundary condition and cannot be changed in short time spans. A large moment arm of the resultant force is determined by the components contributing to the resultant force. In the hip joint example, the contributing structures include the bone-to-bone contact force in the joint as well as the contribution from the muscles crossing the joint (e.g., gluteus maximus, sartorius, rectus femoris, etc.). If these muscles do not contribute appropriately, the stress on the femur will increase. Fractures in elderly people, in some cases, may be influenced by an inadequate muscular contribution to an axial resultant force.

The cylindrical construction of the bones of the extremities has been discussed not only by biomechanists but also by biologists. A hollow cylinder proves efficient when its mass-strength ratio is examined. Efficiency depends on the ratio of column diameter to wall thickness (Currey, 1984). For pure bending, a maximal ratio is most efficient, while for eccentric loading and failure due to combined compression and tension a minimal ratio is most efficient. Different animal types have different mass-strength ratios, depending on the function of the bone (Oxnard, 1993).

MECHANICAL ADAPTATION

While a fractured bone heals, the bone parts sometimes do not lie in their original configuration. Physicians have observed, however, that bone, as it heals, tends to straighten itself. As bone straightens, material must be removed from the convex side (where tensile strain and stress predominate), and bone apposition must occur at the concave side (where compressive strain and stress predominate).

In order to explain this observation, Frost (1964) proposed that remodelling is influenced by the tangential bone wall strain and the effect of the applied load on the curvature of the bone surface (which, of course, influences the tangential bone wall strain). He suggested that, as a general rule:

Increase in surface convexity   →   osteoclastic activity
Decrease in surface convexity   →   osteoblastic activity

It is difficult to distinguish between stress and strain as causative factors for bone adaptation and formation. Currently, bone strain is assumed to be the important mechanical variable in bone modelling and remodelling. Strain has many different aspects which may be important. They include:

* Strain mode.
* Strain direction.
* Strain rate.
* Strain frequency.
* Strain distribution.
* Strain energy.

Strain mode

There is evidence that both strain modes, compression and tension, provide stimuli for bone formation (Lanyon, 1974; Pauwels, 1980). The relative importance of these two strain modes is not completely understood. However, the greatest bone density often appears in the compression cortex of bone, an indication that compression strain may be the dominant influence.

Strain direction

There is evidence that osteons are better oriented in the principal loading direction (Lanyon and Bourn, 1979) since osteons are weak in longitudinal shear loading modes.

Strain rate

Strain rate seems to be an important variable in bone's response to load and strain. Strain can be produced with static or dynamic load. If only the magnitude and direction of strain were important for bone modelling and remodelling, static and dynamic loading would produce the same effects on bone. However, it is well documented (e.g., Liskova, 1965) that bone responds differently to static and dynamic load. Bone formation occurs as a result of intermittent load (and consequently, strain), and the loading rate is important. Continuous static loading leads to bone resorption in cortical and increased porosity in cancellous bone (Lanyon and Rubin, 1984). However, some studies suggest that constant static combined with dynamic load may promote increased net bone formation (Storey and Feik, 1982).

Strain frequency

Repetitive bone strain promotes net bone formation. It has not yet been determined, though, whether or not there is an optimal frequency range for maximal net bone formation. There is evidence that impact loading may be advantageous for net bone formation. It has been shown that juvenile gymnasts have higher bone mineral density than matched juvenile swimmers (Grimston et al., 1993). One major difference between the two activities is the frequency content of the external forces. Gymnasts experience more impact forces with higher frequencies than the forces typical for swimming. Higher loading rates for the external forces may result in higher strain rates for the loaded bones, which, in turn, may be responsible for the observed increase in bone mineral density.

Strain distribution

The strain stimulus required to effect net bone formation depends on the pattern of loading. The necessary strain stimulus is lower for load patterns that differ from the usual pattern of loading (Lanyon, 1987).

Strain energy

Strain energy is equal to the product of stress and strain. It is a scalar quantity and is always positive, which means that strain energy includes less information than strain itself. The influence of strain energy on bone formation is not well understood. However, strain energy has been proposed recently to be the controlling variable for bone density (Carter et al., 1987; Whalen et al., 1987).

2.1.6 FAILURE OF BONE

FRACTURE OF BONE

Fracture of bone may occur due to:

- Excessive stress, and/or
- Weak material.

Stress may be excessive for different reasons:

- The acting external force is excessive,
- The dimensions of the structure are small,
- The geometry of the acting forces is unfavorable, and/or
- Excessive frequency of load application.

Excessive external forces may occur during a skiing accident when a skier crashes into a tree. In such a case, the actual force acting on the bone is in excess of the ultimate force limits. Except in such accidents, typically during sport activities, external forces are rarely of a magnitude which results in bone fracture.

Small bone dimensions have been shown to be the cause of bone "fatigue fractures". Milgrom et al. (1989) showed that the area moment of inertia of the tibia had a statistically significant correlation with the incidence of tibial, femoral, and total stress fractures in military recruits. The actual forces acting on the bone were within the physiological range for bone, but due to the small dimensions of the structures, the stresses were beyond their limits.

Forces acting on a structure have a line of action and a direction. They produce moments with respect to joints. The magnitude of these moments determines the internal loading. If forces act with a moment arm that is large in relation to the relevant joints, the local stresses in bone may be excessive. An example of the influence of the geometry of the acting forces is shown in Fig. 2.1.4. One can easily replace the schematic Fig. 2.1.4 with the head and shaft of the femur, realizing quickly how, for instance, a change in the geometry of the head of the femur affects the loading of the shaft of the femur.

Loading and overloading of bone always has two aspects: first, the local stresses during one cycle (e.g., in running), and second, the number of repetitions of these stresses. Bone needs stress or strain stimuli for its development. However, the stimuli must be in an optimal range. It is difficult to define this optimal range. In a recent review article, a model describing the development of stress fractures was proposed (Grimston and Zernicke, 1993). In this model, a series of factors were identified as important in the development of stress fractures. Included were biomechanical factors, muscle fatigue, and hormone perturbations. With respect to optimal strain levels a lower value of about 150 microstrain and an upper level of about 2500 microstrain has been proposed.

2.1.7 APPLICATIONS

MECHANICAL PROPERTIES AS A FUNCTION OF AGE AND GENDER

Age

Changes of mechanical properties of bone with *age* have been quantified by several authors. The variable used for this discussion is the aging rate of a mechanical property of interest (Yamada, 1970). The aging rate of a mechanical property describes relative changes (the aging rate of a mechanical property is defined as the percentage change from the standard). The mechanical properties of the age group from 20 to 29 years of age are used as the standard.

The aging rates for cortical bone, as summarized from the literature (Yamada, 1970; Burstein et al., 1976), show reductions of (Fig. 2.1.7):

- Ultimate compressive stress $\approx 15\%$
- Ultimate tensile stress $\approx 30\%$
- Ultimate compressive strain $\approx 5\%$
- Ultimate tensile strain $\approx 30\%$
- Elastic modulus in compression $\approx 15\%$
- Elastic modulus in tension $\approx 10\%$

Mechanical properties of bone depend in part, therefore, on age. They are reduced between 20 and 80 years of age by about 10 to 40% as long as no other influences improve or worsen the situation. Possible improvements may result from physical activity. Possible negative effects may result from hormonal changes, e.g., due to menopause, or from sedentary activities.

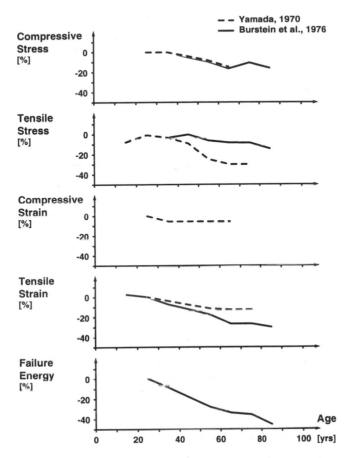

Figure 2.1.7   **Summarized results of the effect of age on mechanical properties of bone, based on data from Yamada (1970) and Burstein et al. (1976), with permission of Williams & Wilkins, Baltimore, Maryland.**

New data on cortical bone from 235 specimens ranging in age from 20 to 102 years (McCalden et al., 1993) show an average deterioration of cortical bone material properties as follows:

- Ultimate stress   5% per decade,
- Ultimate strain   9% per decade, and
- Energy absorption   12% per decade.

Additionally, the porosity of bone was reported to increase significantly with age, while mineral content of bone was not affected by age. Cortical width measured from bone biopsies taken from the iliac crest of 92 subjects ranging in age from 17 to 79 years, was reported to decrease with age for women but not for male bone samples (Christiansen et al., 1993).

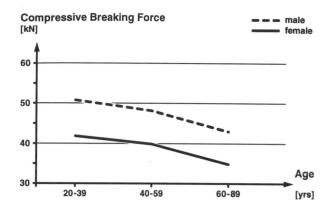

Figure 2.1.8   Compressive breaking forces in the middle portion of wet femoral shafts in longitudinal direction (based on data from Yamada, 1970, with permission of Williams & Wilkins, Baltimore, Maryland).

Gender

The influence of *gender* on the mechanical properties of bone has been studied extensively. Bone mass decreases with increasing age equally in the female and male (Atkinson, 1965; Atkinson and Weatherell, 1967; Yamada, 1970; Burstein et al., 1976; Martin and Atkinson, 1977) (Fig. 2.1.8). However, the mechanical properties of female bone after menopause are less than those of male bones of the same age. This phenomenon has been explained as showing that although the reduction in material strength is compensated for by changes in the overall structure of the long bone, this compensatory mechanism is more efficient in males than in females (Martin et al., 1989). Age related reductions in material strength were compensated by radial expansion of the cortex in men but not in women (Martin and Atkinson, 1977).

The compressive breaking forces of long bone decrease similarly for female and male bones (Yamada, 1970). The compressive breaking force decreases for bone from both genders by about 15 to 20% between the younger age group (20 to 39) and the older age group (60 to 89). However, the relative difference for the compressive breaking forces between males and females is constant for the two age groups at about 17%. Consequently, female bones fail at smaller forces compared to male bones (Fig. 2.1.8). This difference is one reason why bone fractures occur more often in elderly women than in elderly men.

2.1.8 REFERENCES

American Academy of Orthopaedic Surgeons (1988) *Injury & Repair of the Musculo-skeletal Soft Tissues* (eds. Woo, S.L.Y. and Buckwalter, J.A.). Amer. Academy of Ortho. Surg., Park Ridge, IL.

Astrand, P. and Rodahl, K. (1977) Physiological Bases of Exercise. *Textbook of Work Physiology* (2nd Ed.). McGraw-Hill, New York.

Atkinson, P.J. (1965) Changes in Resorption Spaces in Femoral Cortical Bone with Age. *J. Pathology and Bacteriology.* **89,** pp. 173-178.

Atkinson, P.J. and Weatherell, J.A. (1967) Variations in the Density of the Femoral Diaphysis with Age. *J. Bone Jt. Surg.* **49 (B),** pp. 481-488.

Bouvier, M. (1989) The Biology and Composite of Bone. *Bone Mechanics* (ed. Cowin, S.C.). CTC Press, Boca Raton, FL. pp. 1-13.

Burstein, A.H., Reilly, D.T., and Martens, M. (1976) Aging of Bone Tissue: Mechanical Properties. *J. Bone Jt. Surg.* **58 (A-1),** pp. 82-86.

Carter, D.R., Orr, T.E., Fyhrie, D.P., Whalen, R.T., and Schurman, D.J. (1987) Mechanical Stress and Skeletal Morphogenesis, Maintenance, and Degeneration. *Transactions of the Orthopaedic Research Society.* Adept Printing, Chicago, IL. **12,** p. 462.

Christiansen, P., Steiniche, T., Brockstedt, H., Mosekilde, L., Hessov, I., and Melsen, F. (1993) Primary Hyperparathyroidism: Iliac Crest Cortical Thickness, Structure, and Remodelling Evaluated by Histomorphometric Methods. *Bone.* **14,** pp. 755-762.

Currey, J.D. (1984) *The Mechanical Adaptations of Bones.* Princeton University Press, Princeton.

Dorland's Illustrated Medical Dictionary (26th Ed). W.B. Saunders Company, Philadelphia, PA.

Frankel, V.H. and Burstein, A.H. (1970) *Orthopaedic Biomechanics.* Lea & Febiger, Philadelphia, PA.

Frost, H.M. (1964) *The Laws of Bone Structure.* Charles C. Thomas, Springfield, IL.

Frost, H.M. (1986) *Intermediary Organization of the Skeleton.* Vols. I and II. CTC Press, Boca Raton, FL.

Frost, H.M. (1973) *Bone Remodelling and its Relationship to Metabolic Bone Disease.* Charles C. Thomas, Springfield, IL.

Frost, H.M. (1973) *Bone Modelling and Skeletal Modelling Errors.* Charles C. Thomas, Springfield, IL.

Frost, H.M. (1989) *Mechanical Usage, Bone Mass, Bone Fragility: A Brief Overview. Clinical Disorders in Bone and Mineral Metabolism* (eds. Kleerekoper, M. and Krane, S.). Mary Ann Liebow, New York. pp. 15-49.

Grimston, S.K., Willows, N.D., and Hanley, D.A. (1993) Mechanical Loading Regime and its Relationship to Bone Mineral Density in Children. *Med. Sc. Sports Exercise.* (Submitted).

Grimston, S.K. and Zernicke, R.F. (1993) Exercise-related Stress Responses in Bone. *J. Appl. Biomechanics.* **9 (1),** pp. 2-14.

Guralnik, D.B. (ed.) (1979) Webster's New World Dictionary. William Collins Publishers, Cleveland.

Hole, J.W. (1987) *Human Anatomy and Physiology* (4th Ed.). Wm. C. Brown, Dubuque.

Lanyon, L.E. (1974) Experimental Support for the Trajectorial Theory of Bone Structure. *J. Bone Jt. Surg.* **56 (B),** pp. 160-166.

Lanyon, L.E. and Boiurn, S. (1979) The Influence of Mechanical Function on the Development and Remodeling of the Tibia. An Experimental Study in Sheep. *J. Bone Jt. Surg.* **61 (2),** pp. 263-273.

Lanyon, L.E. (1987) *Strain and Remodelling.* Hard Tissue Workshop, Sun Valley, ID.

Lanyon, L.E. and Rubin, C.T. (1984) Static vs Dynamic Loads as an Influence on Bone Remodelling. *J. Biomechanics.* **17,** pp. 897-905.

Lekhnitskii, S.G. (1963) *Theory of Elasticity of an Anisotropic Elastic Body.* Holden-Day, Inc., San Francisco, CA.

Liskova, M. (1965) The Thickness Changes of the Long Bone after Experimental Stressing During Growth. *Plzensky Lekarsky Sbornik.* **25,** pp. 95-104.

Martin, R.B. and Atkinson, P.J. (1977) Age and Sex Related Changes in the Structure and Strength of the Human Femoral Shaft. *J. Biomechanics.* **10 (4),** pp. 223-231.

Martin, R.B. and Burr, D.B. (1989) *Structure, Function, and Adaptation of Compact Bone.* Raven Press, New York.

McCalden, R.W., McGeough, J.A., Barker, M.B., and Court-Brown, C.M. (1993) Age-related Changes in the Tensile Properties of Cortical Bone. *Journal of Bone and Jt. Surg.* **75 (A-8),** pp. 1193-1199.

Milgrom, C., Giladi, M., Simkin, A., Rand, N., Kedem, R., Kashtan, H., Stein, M., and Gomor, M. (1989) The Area Moment of Inertia of the Tibia: A Risk Factor for Stress Fractures. *J. Biomechanics.* **22 (11-12),** pp. 1243-1248.

Nicholas, J.A. (1986) *The Lower Extremity and Spine in Sports Medicine.* C.V. Mosby Company, St. Louis.

Nigg, B.M. (1985) Biomechanics, Load Analysis, and Sports Injuries in the Lower Extremities. *Sports Medicine.* **2,** pp. 367-379.

Oxnard, C.E. (1993) Bone and Bones, Architecture and Stress, Fossils and Osteoporosis. *J. Biomechanics.* **26 (1),** pp. 63-79.

Pauwels, F. (1980) *Biomechanics of the Locomotor Apparatus: Contributions on the Functional Anatomy of the Locomotor Apparatus* (trans. Maquet, P. and Furlong, R.). Springer Verlag, Berlin.

Recker, R.R. (1983) *Bone Histomorphometry: Techniques and Interpretation.* CTC Press, Boca Raton, FL.

Reilly, D.T. and Burstein, A. (1975) The Elastic and Ultimate Properties of Compact Bone Tissue. *J. Biomechanics.* **8 (6),** pp. 393-406.

Spence, A.P. (1987) *Human Anatomy and Physiology* (3rd Ed.). Benjamin Cummings, Menlo Park, CA.

Steindler, A. (1964) *Kinesiology of the Human Body Under Normal and Pathological Conditions.* Charles C. Thomas, Springfield, IL.

Storey, E. and Feik, S.A. (1982) Remodelling of Bone and Bones: Effects of Altered Mechanical Stress on Analages. *British Journal of Experimental Pathology.* **6,** pp. 184-193.

van Leeuwenhoeck, A. (1678) Microscopical Observations on the Structure of Teeth and Other Bones. *Phil. Trans. Royal Soc.* London. **12,** pp. 1002-1003.

Weineck, J. (1986) *Functional Anatomy in Sports.* Year Book Medical, Chicago, IL.

Whalen, R.T., Carter, D.R., and Steele, C.R. (1987) The Relationship Between Physical Activity and Bone Density. *Trans. Orthopaedic Research Society.* Adept Printing, Chicago, IL. **12,** p. 463.

White, T.D. (1991) *Human Osteology.* New York, Academic Press.

Wolff, J. (1986) *The Law of Bone Remodelling* (trans. Maquet, P. and Furlong, R.). Springer Verlag, Berlin. (Original 1870).

Yamada, H. (1970) *Strength of Biological Materials* (eds. Evans, F. and Gaynor, F.). Williams & Wilkins, Baltimore, Maryland. p. 49.

2.2   ARTICULAR CARTILAGE

SHRIVE, N.G.
FRANK, C.B.

2.2.1 DEFINITIONS AND COMMENTS

Anatomy

Articular cartilage:	A thin layer of fibrous connective tissue on the articular surface of bones in synovial joints.
Alymphatic:	Not supplied with lymphoid tissue.
Aneural:	Not supplied with nerves.
Avascular:	Not supplied with blood vessels.
Cell:	Basic unit of life within a tissue.
Chondrocyte:	A mature cartilage cell embedded in a small cavity (lacuna) in the cartilage matrix.
Collagen:	A protein substance of the matrix.
Isotropic:	Having physical properties (e.g., wave conductivity, elasticity) that are the same in all directions.
• Anisotropic:	Having physical properties that are not the same in all directions.
Homogeneous:	Of a uniform quality throughout.
Lacuna:	Small cavity.
Matrix:	The intercellular substance of a tissue from which the structure develops.
• Interterritorial matrix:	Matrix that lies outside the territorial matrix.
• Pericellular matrix:	Matrix completely surrounding the chondrocytes (cells), which contains small amounts of collagen, proteoglycan, and other proteins.
• Territorial matrix:	Matrix that surrounds the pericellular matrix. It contains fibrillar collagen, which adheres to the pericellular matrix.
Proteoglycans:	Group of glycoproteins found in connective tissue. Formed of subunits of disaccharides linked together and joined to a protein core.
Synovial fluid:	Transparent, alkaline, viscous fluid secreted by the synovial membrane contained in joint cavities.
Tropocollagen:	The molecular unit of collagen fibrils, about 1.4 nm wide and about 280 nm long. Tropocollagen is a heli-

cal structure consisting of three polypeptide chains. Each chain is composed of about a 1000 amino acids coiled around each other to form a spiral.

Zone:

Region or area with specific characteristics.

- Deep zone:

Layer of articular cartilage between the bone and the transitional zone.

- Superficial zone:

Thin part of articular cartilage situated near the surface.

- Transitional zone:

Layer of articular cartilage between the transitional and the deep zone.

Mechanics

Density:

Mass per unit volume.

Elastic modulus:

The ratio of stress divided by strain.

$$E \; = \; \frac{\sigma}{\varepsilon}$$

where:

σ = stress
ε = strain
E = elastic modulus

with the units:

$[E]$ = N/m^2 = Pa
$[\sigma]$ = N/m^2 = Pa
$[\varepsilon]$ = %

Note: the unit "percent" is obtained by multiplying the measured relative length change with the number 100.

Load:

Sum of all the forces and moments.

Properties:

- Material:

Properties of a material which describe its general behaviour without including any information about its size and shape.
Comment: material properties include stress and strain, etc.

- Physical:

Properties of a material which relate to its physics.
Comment: physical properties include density, specific weight, etc.

- Structural:

Properties of a specific sample of a material which describe the behaviour of that sample including

effects of its size and shape.
Comment: structural properties include force to failure, deformation, etc.

Strain:

Relative change of length.

$$\varepsilon \quad = \quad \frac{\Delta L}{L_o}$$

where:

$\varepsilon \quad = \quad$ strain
$\Delta L \quad = \quad$ change in length $\quad = \quad L - L_o$
$L_o \quad = \quad$ original length

Strain rate:

Change of strain over time.

$$\dot{\varepsilon} \quad = \quad \frac{d\varepsilon}{dt}$$

with the unit:

$$[\dot{\varepsilon}] \quad = \quad \frac{1}{s}$$

Strength:

Maximal force a material can sustain before failure.

- Compressive strength:

Maximal force in compression a material can sustain before failure (= ultimate compressive strength).

- Tensile strength:

Maximal force in tension material can sustain before failure (= ultimate tensile strength).

Stress:

Force per unit area.

$$\sigma \quad = \quad \frac{F}{A}$$

where:

$\sigma \quad = \quad$ stress
$F \quad = \quad$ force
$A \quad = \quad$ area

- Compressive stress:

Stress perpendicular to the surface that acts to compress an object.

- Shear stress:

Stress parallel to the surface of an object.

- Tensile stress:

Stress perpendicular to the surface that acts to elongate an object.

- Ultimate stress:

The highest stress experienced by the tissue before complete failure.

2.2.2 SELECTED HISTORICAL HIGHLIGHTS

1742   Hunter

First to associate articular cartilage with mechanical concepts. Stated of cartilages that "by their elasticity, the violence of any shock, which may happen in running, jumping et cetera is broken and gradually spent... which must have been extremely pernicious, if the hard surfaces of bones had been immediately contiguous. As the course of the cartilaginous fibres appears calculated chiefly for this last advantage, to illustrate it, we need only reflect upon the soft undulatory motion of coaches, which mechanics want to procure by springs". Hunter also described a fibrous infrastructure radiating perpendicularly from the bone surface and deforming upon compression.

1898   Hultkrantz

Illustrated a surface fibre pattern by pricking the cartilage surface with a sharp tool and noting the direction of the "splits" produced (Fig. 2.2.1).

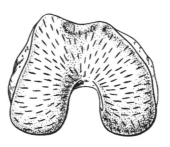

human femoral condyles

Figure 2.2.1   **Schematic diagram of Hultkrantz' "split lines" on the surface of articular cartilage. The split lines illustrate a surface fibre pattern (from Hultkrantz, 1898, with permission).**

1925   Benninghoff

Confirmed and expanded Hultkrantz's "split line" theory and investigated the deeper substance of the cartilage body. Introduced a detailed "arcade" theory, which confirmed Hunter's ideas and which described fibres radiating out from the bone and curving obliquely until parallel to the articular surface (Fig. 2.2.2).

1920's and 1930's

Scientists attempted to measure elastic properties of articular cartilage (Bar, 1926; Göcke, 1927; Muller, 1939; Policard, 1936) to try to understand arthritic diseases.

1960's

Ultrastructure of articular cartilage was described in great detail with the help of the newly developed polarized light microscope and the scanning electron microscope (Goodfellow and Bullough, 1968; Hunter and Finlay, 1973).

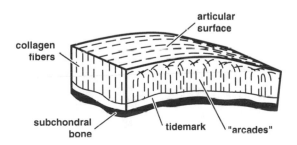

Figure 2.2.2   Schematic diagram of Benninghoff's "arcades" (from Benninghoff, 1925, with permission).

2.2.3 MORPHOLOGY AND HISTOLOGY

MORPHOLOGY AND FUNCTION

Articular cartilage is a thin layer of fibrous connective tissue on the articular surfaces of bones in synovial joints. Articular cartilage consists of cells ($\approx$5%) and an intercellular matrix ($\approx$95%), which is substantially water (65-80%). Cartilage's mechanical behaviour is viscoelastic, and in conjunction with synovial fluid, cartilage provides an extremely low co-efficient of friction ($\approx$0.0025) to joints. The major functions of articular cartilage include:

- *Transferring forces* between articulating bones,
- *Distributing forces* in joints, and
- Allowing *joint movement* with minimal friction.

Intercellular matrix

A matrix is the intercellular substance of a tissue from which the structure develops. With respect to the chondrocytes, one may identify three types of matrices: pericellular, territorial, and interterritorial. These regions differ in collagen content, collagen diameter, and proteoglycan content (Fig. 2.2.3).

The *pericellular matrix* completely surrounds the chondrocytes and contains small amounts of collagen, proteoglycan, and other proteins. The *territorial matrix* surrounds the pericellular matrix and contains fibrillar collagen which adheres to the pericellular matrix in the form of nests that help to protect the chondrocytes during cartilage deformation. The *interterritorial matrix* lies outside the territorial matrix. It consists of widely spaced, thick, mostly parallel collagen fibrils.

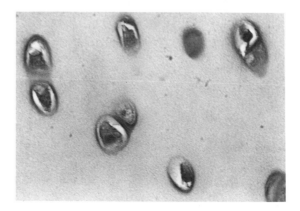

Figure 2.2.3   Photographic illustration of matrix territories in relation to chondrocytes. The pale grey area is the interterritorial zone while the darker grey and almost black areas are the territorial and pericellular territories respectively.

Structure

Articular cartilage is structurally heterogeneous. It changes with depth from the joint surface. The constitutive changes are continuous but may be divided into four zones (Fig. 2.2.4):

- Superficial zone.
- Transitional zone.
- Deep (or radial) zone.
- Calcified zone.

Superficial zone

The superficial zone is the thinnest, most superficial region of articular cartilage. It has:

- A surface layer (*lamina splendens*), and a
- Deeper layer.

The *surface layer* is approximately 2 μm thick and consists of random flat bundles of collagen fibrils.

The *deeper layer* of the superficial zone consists of dense collagen fibres lying parallel to the plane of the joint surface and following the "split line" pattern described by Hultkrantz (Fig. 2.2.1), which follows the directions of normal joint movement. The superficial zone seems to be structured to resist the shear stress that develops during joint motion.

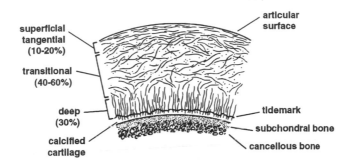

superficial
tangential
(10-20%)

transitional
(40-60%)

deep
(30%)

calcified
cartilage

articular
surface

tidemark

subchondral bone

cancellous bone

Figure 2.2.4   Collagen fibre and chondrocyte arrangement between the heterogeneous zones of
articular cartilage (from Mow and Proctor, 1989, with permission of Lea &
Febiger, Baltimore, Maryland).

The strength and stiffness of collagen fibres is greater in directions parallel to the "split lines" (which coincide with the direction of joint movement) than perpendicular to it (Kempson et al., 1973).

The surface of cartilage shows *collagen crimp* (Fig. 2.2.4). Crimped (or wavy) collagen is relaxed, much like an elastic band before it is stretched. The crimp appears as a vast array of ridges and dips between 2 and 6 μm deep (Dowson, 1990). In a microscopic sense therefore, cartilage does not have an entirely smooth surface.

This deeper layer of the superficial zone also contains elongated *chondrocytes*, whose long axes lie parallel to the joint surface. The presence of only a few mitochondria, golgi bodies, and endoplasmic reticula suggests that these cells are metabolically relatively inactive. Very little proteoglycan is found in this area.

Of all the zones of articular cartilage, the superficial zone also contains the highest concentration of *water*, approximately 80%. Water concentration decreases in a nearly linear manner with increasing depth until it is approximately 65% in the deep zone (Lipshitz et al., 1976; Maroudas, 1979; Torzilli et al., 1982; Torzilli, 1985).

Transitional zone

The transitional zone consists of collagen fibrils with larger diameters than those in the superficial zone. These fibrils mostly lie parallel to the plane of joint motion, but they are less parallel than those in the superficial zone. The chondrocytes are spherical (Fig. 2.2.4) and contain endoplasmic reticula, golgi bodies, and mitochondria, a characteristic which suggests that the cells in the transitional zone play a greater role in matrix synthesis than those in the superficial zone.

Deep zone

The deep zone features large numbers of big collagen bundles running perpendicular to the plane of joint motion. Proteoglycan content is highest in the deep zone. The lack of homogeneity in proteoglycan content among the zones produces a non-homogeneous swelling phenomenon in the cartilage body. The functional significance of this swelling, as will be seen, is great. Water content is relatively low in this zone. The chondrocytes are round and stacked on top of each other in a column perpendicular to the joint surface (Fig. 2.2.4). The fact that the cells contain intracytoplasmic filaments, glycogen granules, endoplasmic reticula, and golgi bodies shows that much protein synthesis occurs in the deep zone.

Calcified zone

The calcified cartilage zone marks the transition from soft articular cartilage to stiffer subchondral bone. It is characterized by hydroxyapatite, an inorganic constituent of bone matrix and teeth which provides rigidity. The calcified zone is separated from the deep zone by the "tidemark", an undulating line 2-5 μm thick. The collagen fibres from the deep zone anchor the cartilage to the bone by fixing themselves into the subchondral bone. The heterogeneous composition and structural framework of articular cartilage result in a highly anisotropic material.

HISTOLOGY

Articular cartilage consists mostly of *matrix* (approximately 95%) and a sparse population of *cells* (approximately 5%). It is avascular, aneural, and alymphatic.

Cells

The cells of articular cartilage account for less than 10% of the tissue's volume (Stockwell, 1979; Muir, 1980 and 1983) and are called *chondrocytes*. They vary in size, shape, and density according to their location. Being highly metabolically active, they are responsible for the synthesis and degradation of the matrix. Chondrocytes are isolated, lie in lacunae, and receive their nourishment via the diffusion of substrates through the extracellular matrix. The mechanical stresses and strains on cartilage force fluid in and out of the cartilage body. The fluid takes nutrients and wastes with it. Cell processes extend a short distance into the matrix but do not touch other cells. The plasma membrane is therefore rough, reflecting the constant, active cell secretion process necessary for the metabolic role these cells play. Chondrocytes in different cartilages of the body have different synthetic activities. In articular cartilage they specialize in producing Type II collagen and proteoglycan. Chondrocytes may contain:

- *Nucleus*   - to give the cell its genetic identity.
- *Mitochondria*   - to generate energy.
- *Endoplasmic reticulum*   - to produce proteins.
- *Glycogen granules*   - to store sugar.
- *Intracytoplasmic filaments*   - to support the cell structurally.
- *Golgi bodies*   - to sort and secrete proteins to the matrix.
- *Lysosomes*   - to degrade protein.

These cells are designed for metabolic purposes. Chondrocytes are similar to the osteocytes of bone in that they play an active metabolic role.

Matrix

The intercellular matrix consists of structural macromolecules and tissue fluid.

Tissue fluid comprises 60-80% of the wet weight of cartilage, while the structural macromolecules comprise 20-40%. These substances, which are produced by the chondrocytes, interact to produce the fluid and solid mechanical behaviour associated with articular cartilage. The interactions of tissue fluid with structural macromolecules produce mechanical entanglement, electrostatic bonding, and excluded volume effects.

The structural macromolecules consist of *collagen*, proteoglycan, and other proteins. Collagen in articular cartilage is the same insoluble fibrous protein found in bone in a slightly different form. It retains the characteristic stiff helical structure, but the triple helix of articular cartilage contains three identical alpha chains, where bone collagen has two identical and one different. Bone collagen is called Type I, and the collagen of articular cartilage is called Type II. Each chain in articular cartilage has a high hydroxylysine content and covalently attached carbohydrates, characteristics that make it associate readily with and adhere to proteoglycans. Types I and II collagen are fibrillar, so they do not appear vastly different under gross examination.

Collagen accounts for much of the structural framework of cartilage and gives it much of its tensile stiffness and strength. The tensile strength of individual collagen fibres has not yet been tested, but tendon, of which collagen comprises 80% (dry weight), has a tensile strength of about 70 MPa. Its tensile strength is comparable to that of nylon (80 MPa) and pure aluminium (70 MPa) (van B. Cochran, 1982). Like rope, however, collagen offers little resistance to compression or shear, due to the natural propensity of a thin column to buckle in compression (Fig. 2.2.5). It is laid out such that advantage can be taken of its tensile strength.

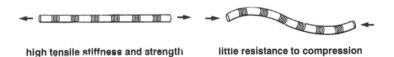

high tensile stiffness and strength   little resistance to compression

Figure 2.2.5   Schematic diagram of collagen fibres in (left) tension and (right) compression (from Myers and Mow, 1983, with permission).

On a molecular level, inter- and intra-molecular cross-links exist between collagen fibrils. These cross-links may further increase tensile strength, stiffness, and integrity. As well, molecular chains from the glycosaminoglycan and polysaccharide found in proteoglycan play a role in collagen binding. The collagen fibres and fibrils in articular cartilage are not only strong individually but, being entwined and bonded with other structural macromolecules and their own molecules, they are stiff and relatively immobile.

Proteoglycans are a group of glycoproteins formed of subunits of disaccharides linked together and joined to a protein core. They are present in connective tissue. Proteoglycans are found in relatively large portions in articular cartilage ($\approx 10\%$). They are much less common in bone ($\approx 1\%$).

Proteoglycans are enmeshed in the network of collagen fibrils and are virtually immobile. They are an important part of the structural framework of articular cartilage. Proteoglycan molecules are widely spaced and branching. Proteoglycans, because of their molecular structure, are ideally suited to resisting compressive forces, which is why more proteoglycans are found in articular cartilage than in the other soft tissues of the synovial joint.

The *glycosaminoglycan* part of the proteoglycan (95%) contains sugars on a protein core which have a high negative electrostatic charge and are attracted to water. The negatively charged molecules repel each other and attempt to become as far apart as possible. The result is the "bottle brush" effect (Fig. 2.2.6). Proteoglycans' attraction to water means that the aggregates are soluble. *In vitro* proteoglycans can hold up to 50 times their weight in water. *In vivo* however, proteoglycans are restrained by the collagen mesh and exist in as little as 20% of the volume they would have if free: they are compressed (Hascall, 1977; Muir, 1983). The result is a stiff viscoelastic "gel" surrounding the collagen fibres.

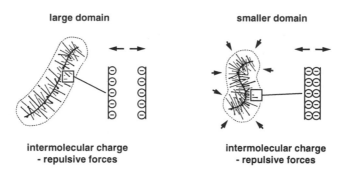

Figure 2.2.6   Proteoglycan aggregate domain. Left: aggregate in solution; the glycosaminoglycan chains' fixed negative charge groups repel each other (at right). Right: compressive stress reduces the volume (domain) of the aggregate solution, which increases the charge density and the repulsive forces (from Mow and Rosenwasser, 1987, with permission).

Proteoglycans are kept from passing into the solution by the tension in the surrounding collagen network. The negative charges are forced closer together, which increases the repulsive forces, swelling pressure, and frictional drag of fluid. Cartilage, therefore, is pretensioned. When subjected to external loads, the proteoglycans are further compressed. The molecules are forced yet closer together, and an even higher repulsive force develops.

Tropocollagen, the molecular unit of collagen fibrils, is about 1.4 nm wide and about 280 nm long. Tropocollagen is a helical structure consisting of three polypeptide chains. Each chain is composed of about a 1000 amino acids coiled around each other to form a spiral. Tropocollagen is rich in glycine, proline, hydroxyproline, and hydroxylysine, all types of collagen being formed by interaction of these units. Tropocollagen is tightly packed and, therefore, better suited than proteoglycans to resist tensional forces.

Tissue fluid

Tissue fluid, the main constituent of articular cartilage, contains mostly water and is found in the form of a viscous gel along with the structural macromolecules. The vast majority of the tissue fluid is free to move both inside and outside of the cartilage body. It is, therefore, closely associated with the synovial fluid of the joint capsule and with joint lubrication.

Tissue fluid within cartilage is closely associated with proteoglycans which attempt to restrain its movement. The combination of the proteoglycans and the collagen framework makes cartilage sponge-like. The pores of cartilage, from which the water can escape, are small. They have been calculated to be in the order of 2.5 to 6.5 nm (McCutchen, 1962; Maroudas, 1975 and 1979; Mow et al., 1984).

Permeability measures how easily a fluid can flow through a porous material. Tissue permeability is inversely proportional to any frictional drag exerted by the fluid as it flows through the material. In other words, permeability is proportional to the force needed to force the fluid to flow at a given speed through the porous tissue. The mathematical relationship is:

$$k = \frac{\beta^2}{\kappa}$$

where:

k = apparent permeability
κ = diffusive drag coefficient
β = partitioning coefficient

In cartilage, collagen and proteoglycan are dispersed in the fluid, making cartilage a microporous material with low permeability. Low permeability is particularly significant for compression and joint lubrication since it means that any substantial exchange of fluid between the cartilage body, and any lubricating film that may exist, must take place over a period of time that is substantial (e.g., minutes), compared to the physiological loading cycle which is typically one second. Consequently, cartilage maintains its stiffness under compression; its low permeability inhibits any rapid escape of fluid when the cartilage is compressed.

Synovial fluid, the lubricant of synovial joints, is a transparent, alkaline, viscous fluid secreted by the synovial membrane contained in joint cavities. It is a dialysate of blood with a protein polysaccharide. The polysaccharide is hyaluronic acid, which makes up a significant proportion of cartilage proteoglycan. Synovial fluid is viscous (this is dictated by the hyaluronic acid) but its viscosity is more akin to that of water than silicone oil for example. Another glycoprotein "lubricin" is also found in synovial fluid. This substance seems to be found on the lubricating surfaces but not in the cartilage layer. The estimated thickness of the film of synovial fluid between two articulating surfaces is in the range of 0.5 to 1.0 μm.

Synovial fluid exerts a strong influence on the mechanical behaviour of cartilage and permits the diffusion of gases, nutrients, and waste products back and forth between the chondrocytes and the body, making it essential to the health of the tissue.

2.2.4 PHYSICAL PROPERTIES

GENERAL PHYSICAL PROPERTIES

Selected general physical properties are summarized in Table 2.2.1.

Table 2.2.1   Selected physical properties for cartilage tissue.

PROPERTY	UNIT	VALUE
DENSITY	$kg \cdot m^{-3}$	1300
DENSITY	$g \cdot cm^{-3}$	1.3
WATER CONTENT (WET)	%	75
ORGANIC CONTENT (WET)	%	20
ORGANIC CONTENT (DRY)	%	90
ULTIMATE COMPRESSIVE STRESS*	MPa	5
ULTIMATE COMPRESSIVE STRESS*	$N \cdot cm^{-2}$	500
COEFFICIENT OF FRICTION (DRY)		0.0025
COEFFICIENT OF FRICTION OF HEALTHY JOINT**		≤ 0.02
THICKNESS OF FLUID FILM IN JOINT***	μm	0.5

* Approximate values based on data from Yamada (1973). ** From Dowson (1992). *** From Dowson (1990).

TENSILE PROPERTIES

The behaviour of articular cartilage under tension varies from one structural zone to another, primarily due to the different orientations of their collagen fibrils. The tensile properties of articular cartilage are, therefore, highly anisotropic.

Cartilage shows differences in tensile strength when tested parallel to the predominant fibre direction (Fig. 2.2.7a) and perpendicular to it (Fig. 2.2.7b). Tensile strength is consis-

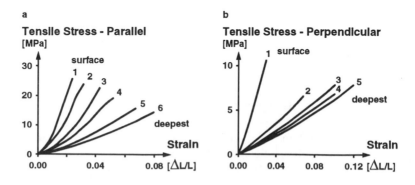

Figure 2.2.7   Variation in tensile stress behaviour with depth in articular cartilage (a) parallel to "split line", and (b) perpendicular to "split line" (from Kempson, 1972, with permission).

tently higher parallel to the split line than in the corresponding perpendicular direction. The shapes of the curves also show a more non-linear "toe" region for specimens tested parallel to the split line, which may indicate that the collagen fibrils become realigned and that their crimp is straightened out as they take up load.

Furthermore, the tensile modulus decreases with increasing depth from the cartilage surface because the orientation of the collagen fibres of the deeper layers is rather random compared to the parallel uniformity at the surface. The implication of this pattern is that the surface fibres seem designed to resist more tensile stress parallel to the split lines than those of the deeper zones.

The properties of cartilage with respect to tensile loads are far less than the equivalent ones of tendons and ligaments. This is because cartilage contains less collagen and also because of its non-homogeneous structure compared to the strictly parallel arrangement found in tendons.

COMPRESSIVE PROPERTIES

Compressive properties vary with the zone tested and are related to proteoglycan concentration. When proteoglycan content increases so does the tissue's compressive stiffness (Fig. 2.2.8 and Fig. 2.2.9). This pattern is consistent with the distribution of proteoglycans described earlier, in that the cartilage is stiffer in the deeper zones where proteoglycan concentration is greatest. The increase in compressive stiffness is due to the fact that having more proteoglycans in the matrix increases the net negative charge in a given volume of matrix, which means that a greater externally applied load has to be applied in order to overcome the repulsive charges in the matrix.

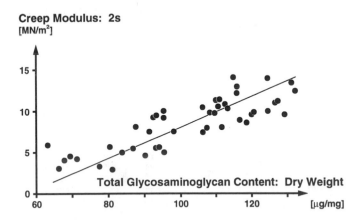

Figure 2.2.8   Relationship between glycosaminoglycan content and compressive stiffness in articular cartilage (from Kempson et al., 1970, with permission).

Thus, the compressive stiffness of articular cartilage is least at the cartilage surface and most in the middle zones, a characteristic that suggests that the surface of articular cartilage cannot play a major role in resisting compression. The deeper zones, with their high proteoglycan concentration, may.

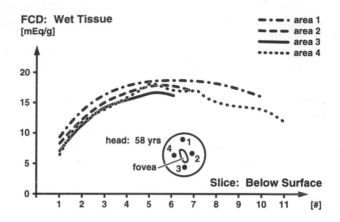

Figure 2.2.9 Variation in glycosaminoglycan content (expressed as fixed charge density) with depth in articular cartilage (from Maroudas, 1979, with permission).

SHEAR PROPERTIES

Shear properties have not yet been measured with respect to depth within the tissue. It is understood, however, that the interactions of the solid components of the matrix are responsible for its shear properties. Fig. 2.2.10 shows a cube of cartilage subject to infinitesimal shear deformation. Under these conditions, the fluid remains in the tissue while the matrix deforms. This may put collagen fibrils under tension and move the proteoglycans in such a way as to increase the overall resistance to deformation.

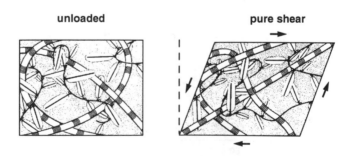

Figure 2.2.10 Schematic diagram of articular cartilage in pure shear (from Mow and Proctor, 1989, with permission of Lea & Febiger, Baltimore, Maryland).

The synovial fluid is an highly non-Newtonian fluid. At low shear rates (around 0.1 s), its viscosity is in the order of a few 10s of pascals, but at higher rates (1000 s), it is only a 1000th of this value.

VISCOELASTICITY

The viscoelastic properties of articular cartilage are mainly associated with the movement of water in the tissue. The matrix material itself, however, may also behave in a viscoelastic manner. Fluid flow is proportional to pressure gradients in the pore water and is described by the coefficient of hydraulic permeability. The inverse of this coefficient gives a measure of the material's resistance to fluid flow and is known as the diffusive drag coefficient. The higher the pressures and compressive strains, the less permeable cartilage becomes (Fig. 2.2.11). The areas where proteoglycans are most concentrated are logically the areas of highest frictional drag to fluids. Permeability is thus greatest at the tissue's surface and least in the deeper zones.

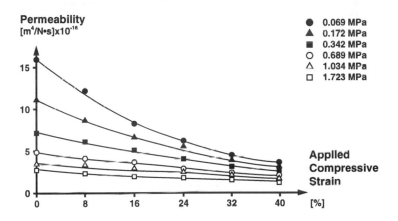

Figure 2.2.11   Experimental curves for articular cartilage permeability show a strong dependence on compressive strain and applied pressure (from Mow and Rosenwasser, 1987, with permission).

If a material is subjected to constant force or deformation and its response varies over time, its mechanical behaviour is said to be viscoelastic.

Viscoelasticity may be revealed through creep or stress relaxation. Creep is the increase in strain over time with constant stress (Fig. 2.2.12). When the stress σ_a is applied at time $t = t_0$, an initial strain ε_i occurs. The initial modulus is:

$$E_i = \frac{\sigma_a}{\varepsilon_i}$$

This modulus is usually dependent on how fast the force is applied. The faster the application of force, the smaller ε_i and the greater E_i. The strain increases with time. The increase in strain is called creep, and at time, t, the total strain is made up of the initial and creep strains:

$$\varepsilon_{tot} = \varepsilon_i + \varepsilon_{cr}$$

Creep strain is often modelled mathematically by an exponential function or a set of such functions, as there is usually a limiting value of creep as time tends to infinity.

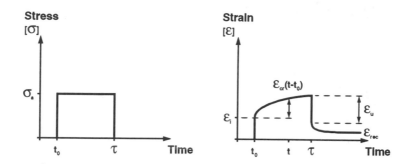

Figure 2.2.12   Schematic of stress application (left) and removal in time and resulting strains for a viscoelastic material (right).

If the stress is removed, there is an immediate reduction in strain, ε_u. The strain continues to diminish and is thus "recoverable". Residual strain at the time infinity would be deemed to be irrecoverable strain.

Stress-relaxation, on the other hand, occurs under constant strain (Fig. 2.2.13). A strain is applied at $t = t_o$, causing an initial stress σ_i. The initial modulus is now:

$$E_i \;=\; \frac{\sigma_i}{\varepsilon_a}$$

and is again typically dependent on how fast the strain is applied. The stress then diminishes over time. The amount the stress diminishes (relaxes) is the stress relaxation. Removal of the applied strain can cause a stress reversal that decays with time. This is not always observed; for example, a ligament might buckle after a stress reversal from an applied tension, and only a minimal compressive load would be recovered.

There are many mathematical formulations for modelling viscoelastic behaviour, both creep and stress relaxation. With biological tissues it is important to remember that these phenomena are relatively more rapid than for many other materials. It is thought that much of this rapid viscous behaviour is due to fluid motion through the tissue.

Most loading in the body is not constant. For example, we do not usually stand still for long periods of time and apply creep loading to the joints of the legs. Most loading varies with time, and, in activities like walking and running, the loading is cyclic. Under cyclic loading, most soft tissues quickly settle into a "steady-state" response. Usually in less than 30 cycles the "steady-state" is achieved. There is sometimes some slow secondary creep of the steady state cycle, but this is typically negligible compared to the differences in response over the first few cycles of loading. The schematic response shown in Fig. 2.2.14 would be for one loading cycle of cartilage subject to compression or of tendons or ligaments subject to tension. The important feature is the inelastic response: the stress-strain relationship on unloading is not the same as on loading (=hysteresis).

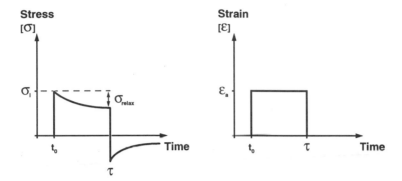

Figure 2.2.13   Schematic of the phenomenon of stress-relaxation (left), and for an applied constant strain (right).

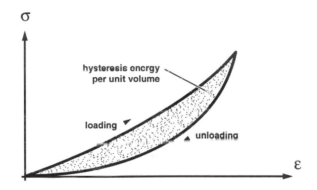

Figure 2.2.14   Schematic of steady state response to cyclic loading.

The area between the two curves is the difference between the energy stored in the tissue on loading and that returned to the loading system on unloading. The area between the curves is the energy absorbed by the tissue on each loading cycle and is called the hysteresis energy. The lower this energy is on each cycle, the more elastic the response of the tissue. The hysteresis energy may dissipate as heat or in the drag of fluid being exuded and imbibed during different stages of the cycle. Indeed it is thought that in steady state loading, the volume of fluid exuded on each cycle equals that imbibed and that this movement of fluid may be associated with tissue health. Fluid imbibed brings nutrients in, while fluid exuded can take waste out.

The relationship between the tissue's porous permeability (fluid flow) and its viscoelasticity may be seen in the creep and stress-relaxation graphs in Fig. 2.2.15. For creep, the graph (top) shows that initially the fluid exudes rapidly, as evidenced by the early rate of increased deformation. The increases in deformation gradually diminish until the flow ceases and any solid matrix creep has finished. At this point the compressive stress developed within the solid matrix must balance the applied load. The rate of creep is, therefore,

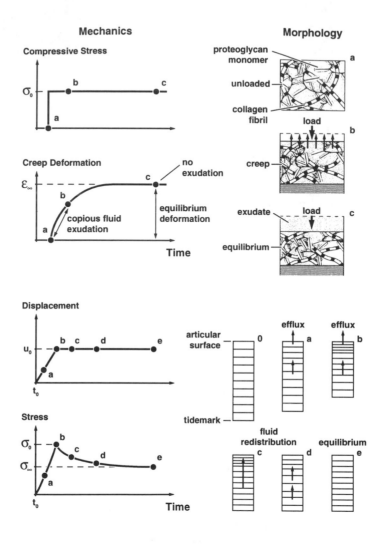

Figure 2.2.15   Creep and stress-relaxation response of articular cartilage. Creep (top) is accompanied by exudation of fluid. The rate of exudation decreases over time from point a to b to c (from Armstrong and Mow, 1980, with permission of Lea & Febiger, Baltimore, Maryland). The stress-relaxation response (bottom) shows a characteristic stress increase during the compressive phase and then the stress decreases during the relaxation phase until equilibrium is reached (point e). Fluid flow and solid matrix deformation during the compressive process give rise to the stress-relaxation phenomena (from Mow and Proctor, 1989, with permission of Lea & Febiger, Baltimore, Maryland).

dependent on the permeability of the tissue. For human and bovine articular cartilages, which are relatively thick (in the order of 4 mm), it can take between 4 and 16 hours to reach equilibrium, whereas rabbit articular cartilage, which is about 1 mm thick, takes only around 1 hour. The stress-relaxation graph (bottom) shows that stress rapidly increases when the load is initially applied and gradually becomes less. The initial high stress is induced by the confinement of fluid in the tissue. As the fluid flows away from the loaded region, the stress diminishes. The viscoelasticity of articular cartilage (and many other tissues) means that its properties are time-dependent. As such, rapidly applied loads compress cartilage less than slowly applied loads of the same magnitude. Thus, at very low loading rates, cartilage is more compliant. The same is true for synovial fluid.

2.2.5 ARTICULAR CARTILAGE MECHANICS AS RELATED TO FUNCTION

The bearing material in synovial joints is a porous, permeable, viscoelastic material whose surface topography is determined by its underlying collagenous framework. It is supported by a rigid bone backing and consists of a solid matrix and water. The surface layer may be distinguished from the deeper layers by its richness in collagen and water, and its paucity in proteoglycan. The water is closely associated with the synovial fluid of the joint cavity, a highly non-Newtonian viscous fluid.

COMPRESSION

Compressive loads are experienced when the two opposing cartilaginous surfaces come into "contact" with each other. Load is transferred from one cartilage layer to the other, either through the direct contact of asperities, through a thin film of pressurized fluid between the layers, or through a combination of the two. With static compression, fluid from the transitional and deep zones does not flow towards the contact surface but flows laterally away from the contact zone. Compression forces proteoglycans closer together, increasing the negative charge in the aggregations and reducing the permeability of the surface layer or even sealing it. Thus, the fluid can only flow laterally. At the edges of the contact zone, the fluid will begin to flow up and out of the cartilage layer.

Upon initial contact, the fluid in both layers is pressurized and carries most of the force down to the bone through hydrostatic pressure (Fig. 2.2.16). This is an efficient way to transmit compression across a tissue that is mainly fluid. In time, the tissue fluid flows away from the contact zone, the cartilage deforms more and more (creep), the contact area increases and then decreases as the fluid moves almost completely away from the contact area, and the matrix becomes more heavily stressed (Fig. 2.2.16a, b, and c). In normal activity (walking and running), the forces come on and off so rapidly that there is little fluid flow to the sides of the contact zone, and the forces are transmitted mainly by the fluid in each cartilage layer.

The flow of fluid through the matrix is accompanied by drag forces, the interaction of the fluid and the matrix. These forces must be resisted by the matrix. First then, consider the flow of fluid laterally away from the contact zone. The drag of the fluid pulls the matrix away from the contact zone in a lateral direction, causing lateral tension in the matrix (Fig. 2.2.17). As the flow subsides over time, so does the tension in the matrix. This tension, in the superficial zone, is resisted by collagen aligned parallel to the surface. At the edges of the contact zone, there are two effects to consider. The cartilage "bulges", first, where the load at the contact area irons out the ridges of the crimped collagen and increases the surface contact area, and second, where the fluid moves into the volume by moving up and

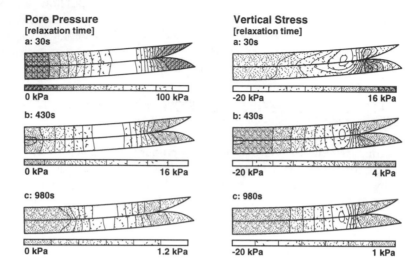

Figure 2.2.16   Pore pressures in a poroelastic finite element model of articular cartilage during stress relaxation. Left: excess pore pressure in MPa at time intervals of (a) 30s, (b) 430s, and (c) 980s. Right: vertical stress in MPa at time intervals of (a) 30s, (b) 430s, and (c) 980s. As contact is made between the two asperities, the internal fluid is forced to flow sideways then upwards (from van der Voet et al., 1992, with permission).

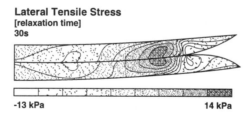

Figure 2.2.17   A poroelastic finite element model showing zones of lateral tensile stress in the superficial layer of articular cartilage (from van der Voet, 1992, with permission).

out of the cartilage layer (Fig. 2.2.18). The surface area is stretched by the bulging caused by the fluid pushing into this previously unpressurized area from the pressurized zone. The stretching causes tension parallel to the surface which is the orientation of the collagen. In addition, drag from the fluid flow causes tension in the matrix as it tries to rise up off the subchondral bone, hence the need for vertical and angled collagen in the deeper zones to resist this drag (Fig. 2.2.19). McConaill's comment is apt: "As iron filings are to a magnetic field so are collagen fibres to a tension field" (MacConaill, 1951, p. 253).

As well as providing a restraining network for proteoglycans, collagen must meet static load transfer requirements. The proteoglycans swell, since they are hydrophilic, and

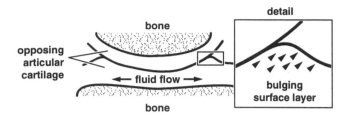

Figure 2.2.18   Schematic diagram of fluid flow in the superficial layer of articular cartilage under a compressive load.

the swelling is constrained by the collagen network acting like ropes over a pressurized tent. The network also prevents the proteoglycans from floating away.

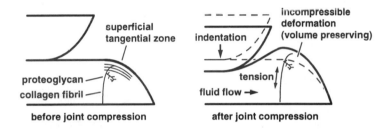

Figure 2.2.19   Role of collagen in the middle and deep zones of articular cartilage (from van der Voet, 1992, with permission).

The regular flow of fluid in and out of the tissue also provides nutrients and removes waste. Physical activity is, therefore, necessary for the maintenance of a healthy tissue.

SHEAR

The magnitude of frictional shear stress due to the motion of one bone relative to another is usually ignored because the coefficient of friction is so small. However, the shear stress is not negligible.

When accounted for, the intersurface frictional shear is initially carried in the superficial zone as a tensile stress. This stress may be transmitted down to the subchondral bone as shear through the matrix, and through tensing the angled collagen fibres in the middle and deeper zones of the cartilage layer. The need to resist joint frictional shear is one reason for the fibres of the superficial zone to be oriented in the direction of normal joint motion (Fig. 2.2.20).

Synovial joints experience micro-elastohydrodynamic lubrication. The pressure generated within the fluid (which is related to its viscosity), combined with elastic deformation of the contact surface, produces an average coefficient of friction for a healthy joint of about 0.02 or less (Dowson, 1992). The undulations on the rough surface of the cartilage (2-6 µm (Dowson, 1990)) become effectively "smoothed out" by self-generated displace-

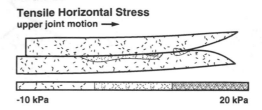

Figure 2.2.20   **Distribution of tensile horizontal stress in articular cartilage during joint motion (from van der Voet et al., 1992, with permission).**

ments derived from hydrodynamic pressures generated during articulation (Fig. 2.2.21). As well, fluid rises to the surface immediately adjacent to the contact zone, where it becomes the thin film required for dynamic lubrication. However, although unlikely in the dynamic situation, despite the 0.5 μm estimated thickness of the fluid film (Dowson, 1990), direct contact of some asperities is not necessarily prevented, as the fluid film needs to be pressurized and/or the asperities touch in order to transmit compressive loads from one layer to another.

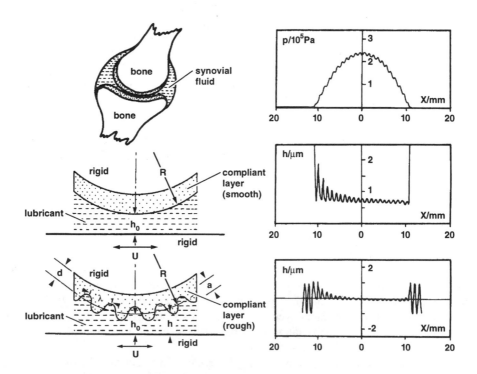

Figure 2.2.21   **Micro-elastohydrodynamic lubrication in synovial joints (from Dowson, 1992, reprinted by permission of the Council of the Institution of Mechanical Engineers).**

EXAMPLE 1

Question:

Estimate the superficial tensile stress generated by friction between two moving cartilage surfaces (Fig. 2.2.22).

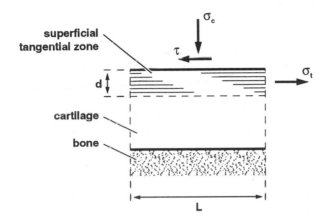

Figure 2.2.22   **Schematic illustration of joint contact. The sliding friction stress is assumed to be carried as tension in the superficial zone entirely, although some will be transmitted to the underlying bone through shear in the proteoglycans and tension in the angled collagen fibres arcading down to the bone.**

Symbols:

σ_c	=	normal contact stress	=	1.5 MPa
L	=	length of contact zone	=	25 mm
μ	=	dynamic friction coefficient	=	0.002
d	=	thickness of superficial zone	=	200 µm
F	=	shear force per unit width		
σ_t	=	tensile stress for superficial zone		

Solution:

A dynamic coefficient of friction produces a shear force per unit width of the contact force of:

$$F = \mu \cdot L \cdot \sigma_c$$

For a superficial zone thickness of 200 µm the tensile stress would, therefore, be:

$$\sigma_t = \frac{F}{d} = 0.375 \text{ MPa}$$

2.2.6 FAILURE OF CARTILAGE

The incongruency of articular surfaces in synovial joints is crucial to the health of articular cartilage and the smooth lubrication of joint movement. Pooling of synovial fluid in areas of incongruency permits nutrients to accumulate and provides enough fluid to create the elastohydrodynamic fluid film. Mechanical loading and unloading keep the tissue healthy by causing the influx of nutrients and the efflux of waste as well as lubrication. Disuse has been associated with degeneration. Greenwald and O'Connor (1971), backing up the work of Goodfellow and Bullough (1968), showed that in the hip joint, areas of habitual non-contact matched areas of known degeneration. Older patients are less likely to experience the high load conditions required to produce direct contact of opposing surfaces. Reduced mechanical stimulation results in a lack of nourishment, so the cartilage degenerates.

When cartilage is damaged there is a remodelling response. The remodelling of articular cartilage originates in the chondrocytes, which are responsible for the ongoing synthesis and degradation of the matrix. Investigators have concluded, however, that cartilage has a *very limited ability to remodel* itself when damaged. The normal pattern of activity sees the clustering of chondrocytes around the site of damage. Their synthetic activity is high, but in general, they fail to restore the matrix to normal, even if the defect is very small. From a mechanical perspective, wear of cartilage and its subsequent failure may be caused by acute and/or chronic loading.

Wear/failure	Acute	• **Active Loading**
		• **Impact Loading**
	Chronic	• **Interfacial**
		• **Fatigue**

Acute failure of cartilage results when the local stress exceeds the ultimate strength in that stress state. Failure usually occurs where there is a combination of high external forces and a small contact area between two adjacent bones. The external forces may be active or impact forces. Active forces that can produce excessive local stress in a joint include forces from heavy lifting. Impact forces that can produce excessive stress in a joint include forces from collisions (e.g., with the boards in ice-hockey or car accidents). It is speculated that acute cartilage failure is most frequently associated with excessive impact loading.

Chronic failure of cartilage may develop because of interfacial problems and/or fatigue phenomena. The source of *interfacial wear* is a lack of lubrication at the bearing surface. It can be adhesive, whereby particles from the bearing surfaces adhere to each other and tear away from their surfaces, or abrasive, whereby a soft particle becomes scraped and damaged by a harder one. Interfacial wear should not occur in a normal joint with efficient lubricating mechanisms, but in a faulty or degenerative joint with impaired lubrication, interfacial wear may easily occur. Of course, with increasing abrasion or adhesion, interfacial wear becomes worse.

Fatigue wear occurs in articular cartilage tissue when the proteoglycan-collagen matrix is damaged by cyclic stressing. It has been suggested that this kind of stressing causes "proteoglycan washout". The washing out results from repeated, massive exudation and imbibition of the tissue fluid. Fatigue wear may result from the application of high active or impact forces over long periods of time (e.g., from running). The intensity, duration, frequency and magnitude of stresses can cause structural changes in the properties of normal cartilage that affect its ability to resist mechanical wear. Such changes start to take place at the molecular level, where macromolecules, cells, and/or the collagen fibrillar network are lost or damaged.

2.2.7 APPLICATIONS

The destructive process of *arthritic diseases* (diseases that involve inflammation of the joint) are well documented. Osteoarthrosis likely originates in cartilage. It involves the fibrillation and softening (chondromalacia) of the tissue body in the areas of greatest pressure and movement. Splitting occurs when the fibres in the superficial zone cannot resist the lateral tension of the fluid drag. Its cause is not known, although it is thought to be age- and wear-related. The damage caused by osteoarthrosis increases with wear.

Two pathological effects that do not originate in articular cartilage are chronic hemarthrosis and rheumatoid arthritis. Chronic hemarthrosis, the repeated hemorrhaging of blood into the joint (as seen in hemophilia for example) causes fibrillation and erosion, which eventually exposes the bone. This and the mechanical effect of fibrous adhesions on the articular surface destroy the joint. Rheumatoid arthritis is a joint inflammation that starts in the synovium. A sheet of fibrogranular tissue with protrusions and adhesions attaches to the superficial cartilage, causing interfacial wear. Rheumatoid arthritis is a chronic, insidious disease that is thought to begin mainly in the immune system. It is highly destructive to the whole joint.

Injury and *surgery* may produce cracks, fissures, and lacerations in the cartilage. Depending upon the severity of the insult, fibrillation, fragmentation, and ulceration may result. Generally, such damage is degenerative because of the limited remodelling ability of cartilage. Investigators who have undertaken long-term studies have described cartilage's overall inability to repair itself, despite its high concentration of chondrocytes, as "striking".

Aging effects are minimal in cartilage as it is a highly durable material able to withstand normal use for many years without negative effect. Upon gross examination, aging cartilage may be more pitted and opaque than younger cartilage and may have minor but non-progressive erosive features on its surface. Morphologically, chondrocyte density decreases slightly with age (which may affect the tissue's capability for repair) and there may be a slight increase in the number of filaments and the quantity of lipids (fats). Overall, the synthesis and degradation of the matrix remain about the same.

2.2.8 REFERENCES

Armstrong, C.G. and Mow, V.C. (1980) Friction, Lubrication, and Wear of Synovial Joints. *Scientific Foundations of Orthopaedics and Traumatology* (eds. Owen, R., Goodfellow, J., and Bullough, P.). pp. 223-232.

Bar, E. (1926) Elastizitätsprüfungen der Gelenkknorpel. *Arch. F. Entwicklungsmech. D. Organ.* **108**, p. 739.

Benninghoff, A. (1925) Form und Bau der Gelenknorpel in ihren Beziehungen zur Funktion. Zeitsch. Zellforsch. *Mikrosk. Anat.* **2**, p. 814.

Buckwalter, J., Rosenberg, L., Coutts, R., Hunziker, E., Hari Reddi, A., and Mow, V. (1987) Articular Cartilage Injury and Repair. *Injury and Repair of the Musculo-skeletal Soft Tissues* (eds. Woo, S.L.Y. and Buckwalter, L.). American Academy of Orthopedic Surgeons, Park Ridge, IL. pp. 465-482.

Bullough, P.G. and Goodfellow, J. (1968) The Significance of the Fine Structure of Articular Cartilage. *J. Bone Jt Surg.* **50 (B)**, pp. 852-857.

Dowson, D. (1990) Bio-tribology of Natural and Replacement Synovial Joints. *Biomechanics of Diarthrodial Joints*, Volume II (eds. Mow, V. C., Ratcliffe, A., and Woo, S. L.Y.). Springer Verlag, New York. pp. 305-345.

Dowson, D. (1992) Engineering at the Interface. *Proceedings of the Institution of Mechanical Engineers. Part C: Mechanical Engineering Science.* Mechanical Engineering Pubs. Limited, Suffolk. **206** (3), pp. 149-165.

Göcke (1927) Elastizitätsstudien am jungen und alten Gelenkknorpel. *Verhandl. D. Deutsch. Orthop. Gesellsch.* pp. 130-147.

Goodfellow, J. and Bullough, P.G. (1968) Studies on Age Changes in the Human Hip Joint. *J. Bone Jt. Surg.* **50 (B)**, p. 222.

Greenwald, A.S. and O'Connor, J.J. (1971) The Transmission of Load Through the Human Hip Joint. *J. Biomechanics.* **4**, pp. 507-528.

Hascall, V.C. (1977) Interactions of Cartilage Proteoglycans with Hyaluronic Acid. *J. Supramol. Struct.* **7**, pp. 101-120.

Hultkrantz, W. (1898) Über die Spaltrichtungen der Gelenkknorpel. *Verh. Anat. Ges.* **12**, pp. 248-256.

Hunter, J.A. and Finlay, B. (1973) Scanning Electron Microscopy of Connective Tissues. *Int. Rev. Connect. Tissue Res.* **6**, p. 218.

Hunter, W. (1742) Of the Structure and Diseases of Articulating Cartilages. *Phil. Trans.* **42**, pp. 513-521.

Kempson, G.E., Muir, H., Freeman, M.A.R., and Swanson, S.A.V. (1970) Correlations Between Stiffness and the Chemical Constituents of Cartilage on the Human Femoral Head. *Biochimica et Biophysica Acta.* Amsterdam. **215**, pp. 70-77.

Kempson, G.E. (1972) The Tensile Properties of Articular Cartilage and Their Relevance to the Development of Osteoarthrosis. *Orthopaedic Surgery and Traumatology. Proceedings of the 12th International Society of Orthopaedic Surgery and Traumatology.* Tel Aviv. Exerpta Medica, Amsterdam. pp. 44-58.

Kempson, G.E., Muir, I.H.M., Pollard, C., and Tuke, M. (1973) The Tensile Properties of the Cartilage of Human Femoral Condyles Related to the Content of Collagen and Glycosaminoglycans. *Biochim. Biophys. Acta.* Elsevier Science Publishers BV. **297**, pp. 456-472.

Kempson, G.E. (1979) Mechanical Properties of Articular Cartilage. *Adult Articular Cartilage* (ed. Freeman, M.A.R.). Elsevier Science Publishers BV. pp. 333-414.

Lipshitz, H., Etheredge III, R., and Glimcher, M.J. (1976) Changes in the Hexosamine Content and Swelling Ratio of Articular Cartilage as Functions of Depth from the Surface. *J. Bone Jt. Surg.* **58 (A-8)**, pp. 1149-1153.

MacConaill, M.A. (1951) The Movements of Bones and Joints - 4. The Mechanical Structure of Articulating Cartilage. *J. Bone Jt. Surg.* **33 (B-2)**, pp. 251-257.

Maroudas, A. (1975). Biophysical Chemistry of Cartilaginous Tissues with Special Reference to Solute and Fluid Transport. *Biorheology.* **12**, pp. 233-248.

Maroudas, A. (1979) Physiochemical Properties of Articular Cartilage. *Adult Articular Cartilage* (ed. Freeman, M.A.R.). Pitman, London. pp. 215 - 290.

McCutchen, C.W. (1962) The Frictional Properties of Animal Joints. *Wear.* **5**, p. 1.

Mow, V.C., Holmes, M.H., and Lai, W.M. (1984) Fluid Transport and Mechanical Properties of Articular Cartilage: A Review. *J. Biomechanics.* **17 (5)**, pp. 377-394.

Mow, V.C. and Proctor, C.S. (1989) Biomechanics of Articular Cartilage. *Basic Biomechanics of the Musculo-skeletal System* (eds. Nordin, M. and Frankel, V.H.). Lea & Febiger, Baltimore, Maryland. pp. 31-58.

Mow, V.C. and Rosenwasser, M. (1987) Articular Cartilage: Biomechanics. *Injury and Repair of the Musculo-skeletal Soft Tissues* (eds. Woo, S.L.Y, and Buckwalter, L.). American Academy of Orthopaedic Surgeons, Park Ridge, IL. pp. 427-463.

Muir, I.H.M. (1978) The Chemistry of the Ground Substance of Joint Cartilage. *The Joints and Synovial Fluid* (ed. Sokoloff, L.). Academic Press, New York. pp. 27-94.

Muir, I.H.M. (1983) Proteoglycans as Organisers of the Intercellular Matrix. *Biochem. Soc. Trans.* **11 (6),** pp. 613-622.

Müller, W. (1939) *Biologie der Gelenke.* Leipzig.

Myers, R. and Mow, V.C. (1983) Biomechanics of Cartilage and its Response to Biomechanical Stimuli. *Cartilage. Volume I. Structure, Function and Biochemistry.* (ed. Hall, B.K.). Academic Press, New York, NY. pp. 313-341.

Policard, A. (1936) *Physiologie Générale des Articulations a l'État Normal et Pathologique.* Masson et Cie, Paris.

Stockwell, R.S. (1979) *Biology of Cartilage Cells.* Cambridge University Press, Cambridge.

Torzilli, P.A. (1985) Influence of Cartilage Conformation on its Equilibrium Water Partition. *J. Orthop. Res.* **3,** p. 473.

Torzilli, P.A., Rose, D.E., and Dethmers, D.A. (1982) Equilibrium Water Partition in Articular Cartilage. *Biorheology.* **19,** pp. 519-537.

van B. Cochran, G. (1982) Biomechanics of Orthopaedic Materials. *A Primer of Orthopaedic Biomechanics* (ed. van B. Cochran, G.), Churchill Livingstone, New York. pp. 71-142.

van der Voet, A.F. (1992) *Finite Element Modelling of Load Transfer Through Articular Cartilage*, Ph.D. Thesis, University of Calgary. Nat'l Library of Canada, Ottawa.

van der Voet, A.F., Shrive, N.G., and Schachar, N.S. (1993) Numerical Modelling of Articular Cartilage in Synovial Joints: Poroelasticity and Boundary Conditions. *Proceedings of the International Conference on Computer Methods in Biomechanics and Biomedical Engineering* (eds. Middleton, J. and Pande, G.N.). Books and Journals International, Swansea. pp. 200-209.

2.3   LIGAMENT

FRANK, C.B.
SHRIVE, N.G.

2.3.1 DEFINITIONS AND COMMENTS

Anatomy

ACL:	Anterior Cruciate Ligament of the knee.
Elastin:	Elastic fibres present in various quantities in ligaments. Comment: elastin is found in small amounts ($\approx 1.5\%$) in extremity ligaments. In elastic ligaments (e.g., lig. flavum), however, elastin fibres are about twice as common as collagen fibres.
Epiligament:	Surface of a ligament. Loose envelope that encloses the ligament.
Fibroblasts:	Ligament cells, usually ovoid or spindle-like and oriented longitudinally along the length of the ligament.
Fibronectin:	A protein composed of two 220-kD subunits joined by a single disulfide bridge and containing about 5% carbohydrate.
Insertion:	The area at which a ligament inserts into bone.
MCL:	Medial collateral ligament of the knee.
Matrix:	Body of the ligament; consists of water, collagen, proteoglycans, fibronectin, elastin, actin, and other glycoproteins.
Midsubstance:	Central part of the ligament, halfway between insertions.
Non-axial fibres:	Fibres not running parallel to the long axis of the ligament.
Proteoglycans:	Group of glycoproteins present in connective tissue. They are composed of subunits of disaccharides linked together and joined to a protein core.

Mechanics

Creep:	Increase in deformation over time under a constant force or a force reached repetitively in a cyclic fashion.
Density:	Mass per unit volume.

Elastic modulus:

The ratio of stress divided by strain.

$$E \ = \ \frac{\sigma}{\varepsilon}$$

where:

σ = stress
ε = strain
E = elastic modulus

with the units:

$[E]$ = N/m^2 = Pa
$[\sigma]$ = N/m^2 = Pa
$[\varepsilon]$ = %

Note: the unit "percent" is obtained by multiplying the measured relative length change with the number 100.

Force relaxation:

Decrease in force when a ligament is pulled to a particular deformation, either once or in cyclic succession.

Load:

Sum of all the forces and moments.

Properties:

• Material:

Properties of a material which describe its general behaviour without including any information about its size and shape.
Comment: material properties include stress and strain, etc.

• Physical:

Properties of a material which relate to its physics.
Comment: physical properties include density, specific weight, etc.

• Structural:

Properties of a specific sample of a material which describe the behaviour of that sample including effects of its size and shape.
Comment: structural properties include force to failure, deformation, etc.

Strain:

Relative change of length.

$$\varepsilon \ = \ \frac{\Delta L}{L_o}$$

where:

ε = strain
ΔL = change in length = $L - L_o$
L_o = original length

Strain rate: Change of strain over time.

$$\dot{\varepsilon} = \frac{d\varepsilon}{dt}$$

with the unit:

$$[\dot{\varepsilon}] = \frac{1}{s}$$

Strength: Maximal force a material can sustain before failure.

- Compressive strength: Maximal force in compression a material can sustain before failure (= ultimate compressive strength).

- Tensile strength: Maximal force in tension material can sustain before failure (= ultimate tensile strength).

Stress: Force per unit area.

$$\sigma = \frac{F}{A}$$

where:

σ = stress
F = force
A = area

- Compressive stress: Stress perpendicular to the surface that acts to compress an object.

- Shear stress: Stress parallel to the surface of an object.

- Tensile stress: Stress perpendicular to the surface that acts to elongate an object.

- Ultimate stress: The highest stress experienced by the tissue before complete failure.

Most ligaments are identified by the points where they attach to bone (e.g., talo-fibular), their shape (e.g., deltoid), their gross functions (e.g., capsular), their relationships to a joint (e.g., collateral), or their relationships to each other (e.g., cruciate). Some of these ligaments have functional subdivisions as well, so there are probably at least twice as many functionally discrete ligaments as have so far been named. Only a few of these ligaments have been studied scientifically.

2.3.2 SELECTED HISTORICAL HIGHLIGHTS

3000 B.C. Smith Papyrus   Described joint sprains.

400 B.C. Hippocrates   Described treatments for ligament injuries.

300 B.C. Herophilus   First person known to have performed extensive anatomical dissections of ligaments.

100 B.C. Hegator   Provided the first anatomical definition of a ligament.

129	Galen	Provided further anatomical definitions of ligaments. Distinguished ligaments from tendons.
1514	Vesalius	Detailed anatomical definitions of joint tissues.
1830	Schleiden & Schwann	Discovered cells and long fibres in dense connective tissues.
1850s	Rudinger & Hilton	Discovered nerves in joint tissues and postulated the existence of ligament-muscle feedback systems that mediate movement.
1911	Fick	Published the first biomechanical review of ligaments.
1980	Noyes	Provided the first definition of the complex structure-function relationships of ligaments.
1981	Woo	Reported on the viscoelastic properties of ligaments.

2.3.3 MORPHOLOGY AND HISTOLOGY

MORPHOLOGY AND FUNCTION

The word "ligament" is derived from the Latin word "ligare" which means to bind. Ligaments consist of elastin and collagen fibres and attach one articulating bone to another across a joint. The major functions of ligaments are as follows:

- To *attach* articulating bones to one another across a joint,
- To *guide* joint movement,
- To maintain joint *congruency*, and
- Possibly to act as a positional bend or strain *sensor* for the joint.

Collagen is the main protein present in ligaments. It is found chiefly in fibrillar form, oriented between insertions such that it will resist tensile forces. The hierarchical structure of the collagen in the ligament midsubstance includes fibres, fibrils, subfibrils, microfibrils, and tropocollagen (Fig. 2.3.1). Tightly packed tropocollagen molecules, approximately 1.5 nm in diameter, aggregate into groups of 5, becoming microfibrils of approximately 3.5 nm in diameter. The microfibrils group into subfibrils, which, in turn, aggregate to form fibrils. Fibrils are approximately 50-500 nm in diameter with a periodicity (spacing) of 64 nm. Fibres are an aggregation of fibrils and are 50-300 μ in diameter. They are the smallest unit of the collagen hierarchy that can be seen under a light microscope. Fibres have an undulating crimp with a distance between amplitudes of about 50 μ. The crimp period can vary quite dramatically in different locations within the ligament. In the rabbit MCL, fibroblasts tend to align in rows between fibre bundles and are elongated along the long axis in the direction of normal tensile stress (see Fig. 2.3.2). Fibres and fibre bundles may or may not aggregate into "fascicles" (Fig. 2.3.3). In the MCL, fascicles are not as obvious as they are in the ACL.

The surface of a ligament comprises a loose envelope known as the epiligament. Its collagen fibrils are of a smaller diameter than those of the midsubstance and are oriented in a multitude of directions. The epiligament contains a variety of cell types. These cells appear to have a greater proliferative ability than those of the midsubstance.

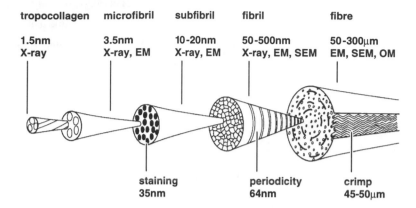

Figure 2.3.1   Schematic illustration depicting the hierarchical structure of collagen in ligament midsubstance. EM = electron microscope; SEM = scanning EM; OM = optical microscope (from Kastelic et al., 1978, with permission of Gordon and Breach, Science Publishers Ltd.).

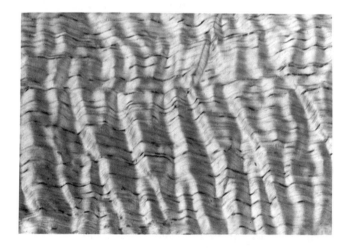

Figure 2.3.2   Photograph illustrating crimped pattern of collagen in ligament. Fibroblasts may be seen interspersed between the collagen fibres.

Unlike the midsubstance, the epiligament encloses nerves and blood vessels (which occasionally branch into the midsubstance). The function of the epiligament appears to be to protect the midsubstance of the ligament from abrasion, to support the neurovasculature, to control the water and metabolite flux, and possibly to act as a source for matrix, cells, and vasculature during maturation and healing.

The structure of ligaments where they insert into bone is different from the midsubstance. Insertions anchor the rest of the ligament into the rigid, non-compliant bone. Insertions are of two general types, *direct* and *indirect*:

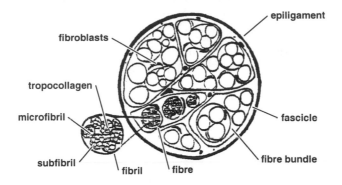

Figure 2.3.3   Schematic diagram of a ligament in cross-section.

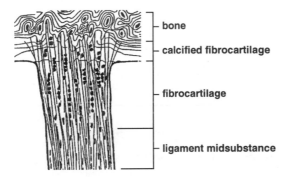

Figure 2.3.4   Schematic diagram of a "zonal" ligament insertion into bone. The bone is at the top
of the diagram, the ligament at the bottom (from Matyas, 1985, with permission).
The tidemark would separate the bone from the calcified fibrocartilage.

(1)   Direct insertions (e.g., the femoral insertion of the MCL) occur where the ligament inserts directly into the bone. Direct insertions contain four different cellular
zones al! of which occur within approximately 1 mm (Fig. 2.3.4) of each other.
The first zone is normal ligament midsubstance, with organized parallel collagen
bundles, some elastin, and elongated fibroblasts. The second zone consists of
non-mineralized fibrocartilage in which the cell numbers increase, become more
ovoid, increase in size, and lie in rows. The collagen fibrils continue to extend
into this region. The third zone may be characterized as mineralized cartilage and
is clearly distinguishable from the previous zone by the "tidemark", an undulating dark line. The cell morphology remains the same as in the second zone, but
mineral crystals appear, aggregated into masses. The fourth zone is where liga-

ment collagen blends directly with bone collagen. The thickness of these zones is tissue and age specific (Matyas et al., 1990).

(2)   Indirect insertions occur where the ligament, temporarily during growth and development, inserts into the periosteum, which is, in turn, connected to bone. A typical example is the tibial insertion of the rabbit MCL (Matyas, 1990).

It is assumed that the morphology and mechanical properties of insertions change gradually. The insertion is mechanically stiffer near bone than in the ligament substance. Such stiffening could lessen stress concentrations and reduce the risk of tearing due to shearing of the tissue at the interface.

Nerves

Recent years have seen much research into the role of the neural components of ligaments. As yet, the exact function of nerves in ligaments is not clear. It is speculated that they permit the sensing of joint position, the monitoring of ligament tension and integrity, and that they initiate protective reflexes. Some investigators have produced muscle activity by stimulating the nerves in ligaments (Sojka et al., 1989; Barrack and Skinner, 1990; Kraupse et al., 1992). For the knee joint, two groups of nerves have been identified, one anterior and one posterior. These nerves respond to active and passive motion throughout the joint's range of motion. Their greatest responses occur at the extremes of knee motion, possibly warning of impending injury. Experimental stimulation of these receptors has been shown to cause reflex activation in the surrounding muscles (Adams, 1977; Kennedy et al., 1982).

Blood vessels

In the epiligament of the MCL lies a fine network of blood vessels. A few vascular channels penetrate from the epiligament into the ligament substance and then course longitudinally between collagen fascicles. Ligament insertions, however, are poorly supplied with blood. The blood supply appears to nourish the fibroblasts, enabling them to remain metabolically active.

HISTOLOGY

The healthy ligament looks like a simple white band of homogeneous fibrous tissue, but it is actually highly complex and dynamic. It is composed of a few cells in a largely collagenous matrix.

Cells

Fibroblasts:

Ligament cells are called fibroblasts. Fibroblasts are not homogenous in ligament tissue and vary in size, shape, orientation, and number. Fibroblasts are usually ovoid or spindle-like and are generally oriented longitudinally along the length of the ligament body. Both metabolic and histological experiments suggest that there may be fibroblast subtypes within the ligament, but these remain to be defined (Frank and Hart, 1990). Fibroblasts are responsible for synthesizing and degrading the ligament matrix in response to various stimuli. Presumably, fibroblasts prevent or repair on-going microscopic damage. Fibroblasts therefore, despite their relative scarcity, are crucial to maintaining the *status quo* of ligaments.

Matrix

The matrix comprises virtually the entire body of the ligament. It consists of water, collagen, proteoglycans, fibronectin, elastin, actin, and a few other glycoproteins.

Water:

Water makes up approximately two thirds of the wet weight of a ligament. Water can be associated with other ligament components in a variety of ways. It can be structurally bound to other matrix components. It can be bound to polar side chains, be so-called "transitional water" (loosely bound), or be freely associated with the interfibrillar gel. Most water in ligaments is freely bound or transitional. Although the exact function of water in ligaments is unknown, it appears to be crucial for at least three main reasons. First, its interaction with the ground substance and particularly the proteoglycans influences the tissue's viscoelastic behaviour (Amiel et al., 1990; Bray et al., 1991). Second, it seems to provide lubrication and facilitate inter-fascicular sliding (Amiel et al., 1990; Bray et al., 1991). Third, it carries nutrients to the fibroblasts and takes waste substances away.

Collagen:

Collagen comprises approximately 70-80% of the dry weight of ligament (Amiel et al., 1983; Frank et al., 1983a; Frank et al., 1983b). Ligament collagen is of several types:

Type I:	The majority of ligament collagen is fibrillar, Type I collagen.
Type III:	The second most common type is Type III, which is also fibrillar.
Types V, VI, X, and XII:	Most ligaments also have small quantities of Types V, VI, X, and XII collagen.

For more information about these collagen types, the reader should consult Miller and Gay (1992).

Collagen fibres within ligaments vary in size (diameter) from 10-1500 nm, a range that appears to be age, tissue, and species specific (Parry et al., 1978a; Parry et al., 1978b; Frank et al., 1989; Yahia and Drouin, 1989). Fibre size may affect the strength of the material. Ligaments with larger bimodal collagen distributions tend to be stronger and able to sustain higher stresses. Those with smaller unimodal diameters are more suited to resisting lower stresses.

Collagen is enormously strong due to a combination of biochemical bonds known as molecular cross-links.

Proteoglycans:

Ligament proteoglycans (mainly so-called "small dermatan sulphate proteoglycans") comprise less than 1% of the dry weight of ligament, more than is found in tendon but considerably less than in cartilage (3-10%). The molecular structure of proteoglycans is described in section 2.4.3. The role of proteoglycans in ligaments remains to be determined. However, ligaments (and tendons), because they experience primarily tensile forces, do not need the "cushioning" effect that proteoglycans give cartilage. Proteoglycans, with their pronounced hydrophilic properties, may instead be involved in regulating the amount and movement of water within the tissue. Proteoglycans would, therefore, mainly influence the viscoelastic behaviour of the ligament.

Fibronectin:

Fibronectin is a protein composed of two 220-kD subunits joined by a single disulfide bridge. This protein contains about 5% carbohydrate. In ligament, as in tendon (see section 2.4.3), fibronectin is found in small quantities in the matrix, usually in association with several other matrix components and blood vessels. Fibronection also interacts with portions of the cell surface that are known to attach to intracellular elements, possibly forming part of an important matrix-cell feedback mechanism.

Elastin:

Elastin is an elastic substance that is found in very small amounts in most skeletal ligaments (approximately 1.5%) in fibular form. In elastic ligaments (e.g., ligamentum flavum), however, elastin fibres are about twice as common as collagen fibres. When unstressed, the insoluble globular proteinaceous elastin molecule takes on a complex, coiled arrangement, probably maintained in part by lysine derived cross-links. Elastin stretches into a more ordered configuration when it is stressed, reverting to its globular form when unstressed again. This behaviour probably accounts for part of the tensile resistance in ligament tissue and some of its elastic recoverability. The role of elastin is probably related to recovering ligament length after stress is removed. Elastin probably "protects" collagen, at least at low strains.

Interactions of these components:

Interactions between the various components of the extracellular matrix have, until recently, received little attention. A recent study (Bray et al., 1990) revealed, through a combination of cationic stains and enzymatic digestion, a network of electron-dense "seams", which connect cells and subdivide the matrix into compartments. The seams contain microfilaments of Type VI collagen (Bray et al., 1990; Bray et al., 1993), microfibrils, and chondroitin sulphate-containing proteoglycan granules (Fig. 2.3.5). The granules are suspended on the microfilaments and spiralled through the seams. The discovery of seams in the extracellular matrix of ligaments has led to the speculation that seams influence the viscoelastic behaviour of ligaments. The proteoglycans attached to the microfilament network may provide functional divisions within the matrix. As well, because of the hydrophilic nature of proteoglycans, water bound within the seams may act as a lubricant, facilitating sliding between adjacent collagen fascicles. The seams would, therefore, be an integral part of a ligament's viscous element.

2.3.4 PHYSICAL PROPERTIES

The physical properties of all soft connective tissues can be classified into two general categories, structural and material. The structural properties of a ligament are derived from the behaviour of a bone-ligament-bone complex and thus involve the midsubstance

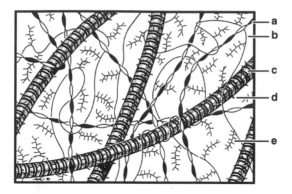

Figure 2.3.5   Schematic diagram showing the proposed interrelationship between the elements of rabbit and human extracellular matrix. a = Type VI collagen; b = Non-beaded microfilaments; c = Banded collagen fibrils; d = Dermatan sulphate proteoglycan; e = Chondroitin sulphate proteoglycan (from Bray et al., 1993, with permission).

of the ligament, the insertions, and the bone local to the insertions. The material properties describe the material irrespective of geometry. They are usually measured in the midsubstance of the ligament. In this section, both types of properties are discussed briefly as they pertain to the ligament model with which we have greatest experience, the rabbit MCL. The reader is encouraged to read the analogous section in chapter 2.4 on tendon, as tendinous properties can be characterized similarly and are, at least in a gross sense, somewhat similar.

STRUCTURAL PROPERTIES

The non-linear force-deformation curve

Fig. 2.3.6 shows a typical force-deformation curve for ligaments. The stiffness (rate of change of force with deformation) of ligaments varies non-linearly with force. This non-linear behaviour allows ligaments to permit initial joint deformations with minimal resistance. The area under the curve in region I of Fig. 2.3.6 is small compared, for example, to a straight line relationship. Together with other ligaments, bone geometry and active muscles, ligaments work within their low-force range to guide bones through normal movement. At higher forces, ligaments become stiffer, providing more resistance to increasing deformations. It is assumed that such stiffening protects the joint.

The non-linear force-deformation behaviour of ligaments occurs for at least two reasons:

Flattening out of collagen crimp:

As described in more detail in section 2.2.3, collagen (the main tensile-resisting substance in the ligament) is crimped. As with tendon, this crimping is thought to allow some extensibility of the ligament along its length under low forces. As tension is added, the crimp gradually flattens out. Once no more crimp can be removed, stiffness increases and an ever-increasing amount of force is required for further displacement. The toe region of

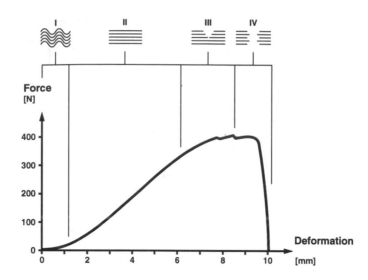

Figure 2.3.6   A typical force-deformation curve for ligament for monotonic forcing. I = Toe region; II = Linear region; III = Region of microfailure; IV = Failure region. At top are schematic representations of fibres going from crimped (I) through recruitment (II) to progressive failure (III and IV).

the force-deformation curve (area I in Fig. 2.3.6) is thought to correspond to the stretching out of crimp. With the crimp gone and the whole matrix under tension (start of area II in Fig. 2.3.6), a region with more constant linear stiffness (the linear region) begins.

Heterogeneous distribution of fibres:

The fibres that run across ligaments are heterogeneous. Collagen crimp is not homogeneously distributed. When the ligament ends are distracted as bones are displaced, the ligament fibres become straight at different displacements and are recruited progressively into force-bearing. As noted above, when all the fibres have been recruited, the stiffness behaviour becomes more linear until some fibres (presumably those first recruited) fail. At this point the net stiffness of the structure begins to drop, as shown in the fracture region of the force-deformation curve (area III in Fig. 2.3.6). As some fibres fail, the force is redistributed onto the remaining fibres, increasing the force on them and the likelihood of their failure. It then takes little additional deformation to produce gross structural failure of the ligament through all the remaining fibres (area IV of Fig. 2.3.6).

Microscopic examination shows some fibres crossing between parallel fibres in the ligament substance, some running perpendicular to the long axis, and some at every angle in-between. While there are not many *non-axial fibres* (Liu et al., 1991) and they tend to be smaller than their longitudinally oriented partners, they should nonetheless contribute to non-linear stiffness behaviour as the ligament is forced. Depending on how and where the non-axial fibres are connected, they could serve as "tethers" for longitudinal fibres. Alternatively, the crossing fibres could be connected in a separate network from end to end in some oblique fashion. At this point in time, this microarchitecture of even the relatively simple MCL is not known. What is known, however, is that gross midsubstance strains from around 8% (in up to 5 mm gauge lengths) are sufficient to cause failure of that area of the rabbit MCL (at least under certain *in vitro* boundary conditions (Lam, 1988)).

Force relaxation and creep

Like other connective tissues, ligaments exhibit force relaxation and creep. When a ligament is pulled to a particular deformation, either once, or repeatedly in cyclic succession, the force in a ligament decreases in a predictable way (Fig. 2.3.7). This predictable

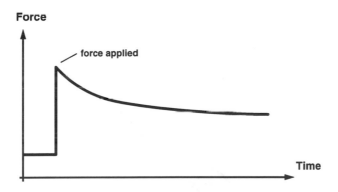

Figure 2.3.7   Schematic force-relaxation curve for ligament.

decrease in force is *force relaxation*. Force relaxation is a result of the viscous component of the ligament's response to force, and the most rapid part of the force decay occurs immediately after loading. Decay continues non-linearly until a steady-state value of force is achieved: this is the elastic component of the force. Then there is a non-linear decay to a

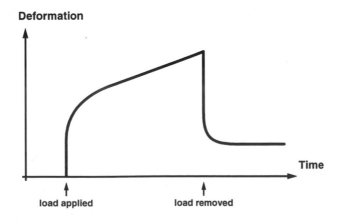

Figure 2.3.8   Schematic creep curve for ligament.

steady-state value. This latter value reflects the elastic component of behaviour. The relative proportions of the viscous and elastic components in any one force depend on the rate of forces and the previous forced history, as well as in a variety of test-specific parameters, e.g., temperature and solution in which the test is carried out. The faster a force is applied, the less time there is for the viscous component to dissipate. A ligament will appear stronger and slightly stiffer under rapid force than under slow force. *Creep* is the analogous behaviour of a ligament under a fixed force when the force is either held or reached repetitively in a cyclic fashion. Creep is the increase in length over time under a constant force. With creep, as with force relaxation, manifestation of the viscous component through time-dependent force or strain changes eventually ceases (Fig. 2.3.8).

During cyclic force however, some of the viscous component can be recovered in each cycle (Fig. 2.3.9). When the ligament is unforced, the viscous component, while never

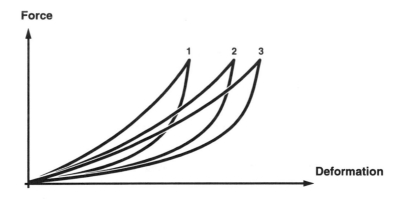

Figure 2.3.9   Schematic force-deformation graph showing three successive cycles of forcing and unforcing, illustrating the viscoelastic creep effect of cycling upon a ligament.

recovering completely (at least during *in vitro* tests), can recover to over 90% of its original state after many hours in a relaxed condition. One can speculate about what is being recovered, but this recovery probably involves some combination of water influx, returning collagen crimp, elastin tensile force, and increasing collagenous disorganization under unforced conditions.

Material Properties

Non-linear behaviour:

The material behaviour of ligaments (i.e., stress-strain behaviour) is also non-linear (Woo et al., 1982b; Lam, 1988), both under monotonic forces and under conditions in which viscoelastic behaviour is demonstrated. With increasing strain, increasing stress in the ligament occurs, as shown in Fig. 2.3.10. In the "toe" region (area I) there is little stress, relative to the strain applied. As described above, this area is thought to correspond to the straightening out of the collagen fibres. There follows a more linear region (area II in Fig. 2.3.10), presumably as the collagen fibres take up force. A measurement of the tangent in the linear region is the tangent modulus which often but erroneously is called the "elastic modulus". Finally, in region III of Fig. 2.3.10, the curve flattens out, eventually

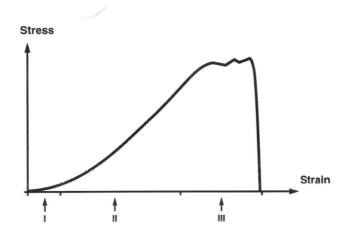

Figure 2.3.10   Schematic stress-strain curve for ligament (I = Toe region; II = Linear region; III = Failure region).

dropping dramatically towards the strain axis. The flattening is presumably related to increasingly rapid microfailure, followed by catastrophic failure.

The reasons for the non-linear behaviour at the stress-strain level are not as easy to explain as that of the structural behaviour, since less is known about the actual molecular nature of the ligament components or their interactions. Certainly there must be some overlap with structural explanations. Collagen fibres themselves may display non-linear characteristics as they are tensed, due to a molecular rearrangement or some internal reordering of their relationships with other elements (elastin, fibronectin, etc.). It is beyond the scope of this work to speculate on such mechanisms.

Water:

Ligaments are two-thirds water, a fact that suggests that water must play important functional roles in the ligament. While these functions may be mainly "biological" (e.g., supply or removal of nutrients, etc.), water is known to contribute to the non-linear viscoelastic behaviour of the ligament in some significant way (Chimich et al., 1992). The interfibrillar amount of water may determine collagen stiffness, or water may determine the distances at which fibres can interact in a physical or biochemical sense. Further, because of its attraction to the highly negatively charged proteoglycan molecules, water may induce pressure gradients in the tissue that are further manipulated by movement.

Problems in measuring physical properties of ligaments:

Measuring the physical properties of ligaments is not as simple as one might naively expect. Selected problems with these procedures are discussed in the following paragraphs.

Source of the tissue sample:

The source of the tissue sample may have a considerable effect on the mechanical properties of the sample. All the properties of ligaments noted above are to some extent "boundary condition" specific. The "boundary conditions" include:

- Species,
- Type of ligament,
- Gender,
- Age,
- Activity,
- Drugs, and
- Diet.

Immature, mature, and aging tissue specimens, for example, have different properties. As tissue matures, it strengthens due to increases in the size of collagen fibres and the number of molecular cross-links. Aging tissues, by contrast, experience a gradual, limited reversal of the maturation process. The *in vivo* force history of a tissue is another factor that vastly influences its properties. Inactivity results in a weaker tissue, while mobility and exercise increase tissue strength.

The fact that so many "boundary conditions" may influence the mechanical properties of ligaments is one reason why consensus about the precise biomechanical behaviour of ligaments is lacking, and why we prefer not to publish either structural or material properties in this text which could be misinterpreted as being "absolute".

Different aspect ratios and trouble securing the ends of ligaments:

Testing isolated ligaments is complicated by the different aspect ratios (length/width) of different ligaments and by difficulties in effectively securing the cut ends. Putting the free ends in clamps induces end-effects and often results in stress concentration at the grips, which damages the tissue and may contribute to premature failure or defective data. A bone-ligament-bone preparation provides more secure clamping but increases the difficulty in separating the properties of the ligament midsubstance itself from those of the insertion sites. Special devices such as buckle transducers and magnetic field (Hall effect) displacement transducers have been used to measure ligament forces during testing, but they unfortunately rely on direct contact with the tissue sample and may influence testing results. Measurements of ligament strains are also flawed. Optical analysers that do not require contact with tissues have been used to quantify strains. However, the reported accuracy levels are known to be boundary condition specific (Lam et al., 1993), and the dye lines marked on the specimen may affect the strain being measured.

Measuring the cross-sectional area of a ligament:

Measurements of stress have been compromised by the lack of an ideal method for measuring the cross-sectional area of a tissue sample. The irregular, complex shape and geometry of these tissues make direct measurements difficult and errors large. Rigid calliper measurements, which are frequently used, require approximations, as the callipers are unable to take irregularities into consideration. Flexible callipers, which are able to follow contours better, are more accurate but still require tissue contact. The only "non-contact" methods of measuring areas are optical, using either photography or laser refraction to quantify distances. The latter technique involves placing a specimen perpendicular to the path of a laser beam and rotating it by 180° or vice versa. The data of profile width and position are then recorded via a microprocessor. The centre of rotation and upper and lower boundaries are determined for each increment of rotation and an iterative procedure is used to reconstruct the cross-sectional shape, from which the area may be calculated. Unfortunately, this technique misses depressions in an irregular surface. Like other techniques, this method is also unable to determine the circumference of the ligament with complete accuracy.

A "zero" strain position:

The definition of a true "zero" strain position poses problems. It should be understood that the zero point on all the graphs used above is boundary condition specific. The zero point depends on environment, force, and force history. The zero point is critical when properties are compared in "absolute terms" (i.e., at "comparable forces and displacements", or "comparable stresses and strains"). A slight shift in the zero point of a test can drastically alter such comparisons (see Fig. 2.3.10). Differences in environmental temperature can cause similar inaccuracies.

The necessity for well-documented biomechanical test procedures has been demonstrated. All the factors mentioned above must be taken into consideration before conclusions about ligament behaviour are drawn.

2.3.5 APPLICATIONS

The knee joint is used to illustrate ligament biomechanics, because, like all weight-bearing joints, the knee is fundamentally concerned with stability over mobility. The four ligaments of the knee joint are illustrated in Fig. 2.3.11. Their main passive restraining functions are summarized in Table 2.3.1.

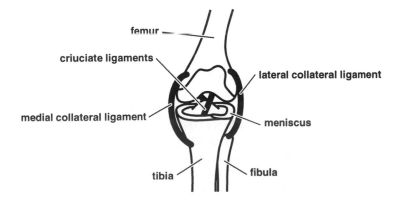

Figure 2.3.11   Schematic diagram of the human knee joint from an anterior view. The "cruciate" (shaped like a cross) ligaments may be seen at the centre of the joint. The anterior cruciate ligament (ACL) attaches superiorly to the posterior lateral femoral condyle and inferiorly to the lateral anterior tibial spine. The posterior cruciate ligament (PCL) attaches superiorly to the medial femoral condyle and inferiorly to the posterior tibia. The "collateral" (parallel) ligaments lie lateral or to the side of the joint. The medial collateral ligament (MCL) is attached to the medial femoral condyle and the medial tibia, while the lateral collateral ligament (LCL) runs from the lateral epicondyle of the femur to the lateral fibular head.

The anatomical terms used in Table 2.3.1 are defined as follows:

- Anterior: toward the front part.
- Posterior: toward the back part.
- Internal rotation: the frontal aspect of a body rotates toward the inside.
- External rotation: the frontal aspect of a body rotates toward the outside.
- Valgus: bent outward with respect to the proximal bone
                            (for the knee, valgus means X-shaped).
- Varus: bent inward with respect to the proximal bone
                            (for the knee, varus means O-shaped).

Table 2.3.1   Restraining functions of the four ligaments of the knee joint.

LIGAMENT	PRIMARY RESTRAINT	SECONDARY RESTRAINT	COMMENTS
ANTERIOR CRUCIATE LIGAMENT (ACL)	Anterior tibial displacement	Internal tibial rotation	No restraint to posterior tibial displacements
POSTERIOR CRUCIATE LIGAMENT (PCL)	Posterior tibial displacement	External tibial rotation	No resistance to varus/valgus angulation
MEDIAL COLLATERAL LIGAMENT (MCL)	Valgus angulation and external tibial rotation	Anterior tibial displacement	
LATERAL COLLATERAL LIGAMENT (LCL)	Varus angulation and internal tibial rotation	Anterior and posterior tibial displacement	

The knee joint has a particularly complex ligament structure. In principle, the tibia has six possible degrees of freedom in relation to the femur (Fig. 2.3.12): three translational (two sliding and one direct impingement/distraction) and three rotational. The ligaments, acting with the fibres of the joint capsule and the articular surfaces of the knee, restrict the range of motion, even when the muscles are completely relaxed.

The joint may be flexed through about 150°. The tibia can be rotated internally and externally relative to the femur through a range of about 35°. The range of varus/valgus movement is only about 5°. Only millimeters of translational movements parallel to the three axes are possible, these being restricted by bone surface geometry and the ligaments. In normal physiological movements it is, however, rare for only one degree of freedom to be utilized. For example, knee flexion is associated primarily with rotation in the sagittal plane, but the tibia also translates along the frontal plane and rotates around the transverse plane while flexion occurs. Motions such as these are referred to as "coupled motions". External forces seldom transmit only one component of force or moment across the joint. When examining the role of ligaments, however, we will simplify the possible combinations of the six degrees of motion.

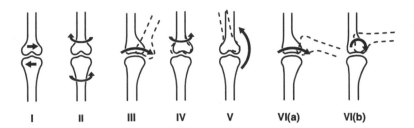

Figure 2.3.12   Schematic diagrams illustrating the six possible degrees of freedom for the knee joint where (I) shows medial-lateral-medial translation, (II) shows rotation, (III) shows anterior-posterior-anterior translation in the sagittal plane, (IV) shows tibial and femoral rotation, (V) shows varus-valgus rotation, and (VI) shows (a) the cruciate ligaments and (b) the collateral ligaments during flexion and extension in the sagittal plane.

Cruciate ligaments

Fig. 2.3.13 shows a "four bar linkage". AD and CB are, respectively, tibial and femoral links and are more or less parallel to the plateaux of the joints. The anterior cruciate AB and posterior cruciate CD are crossed (hence their name) and attach one condyle of one articulating bone to the opposite condyle of the apposing bone. Fig. 2.3.13 shows how the

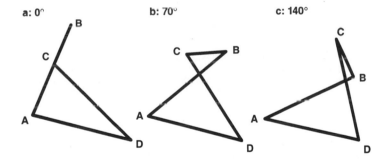

Figure 2.3.13   The cruciate "four bar linkage" ABCD at (a) full knee extension, (b) 70° of knee flexion and (c) 140° of knee flexion. Between (a) and (c) the femoral link CB rotates through 140° relative to the tibial link AD, and the cruciate ligaments AB and CD rotate through 40° about their tibial attachments A and D and through 100° about their femoral attachments B and C (from O'Connor et al., 1990, reprinted by permission of the Council of the Institution of Mechanical Engineers).

shape of the "four bar" cruciate linkage changes during one particular degree of motion, flexion, and extension. Fig. 2.3.13a shows the joint at full extension, (b) at 70° of flexion and (c) at 140° of flexion. Between (a) and (c), the femoral link CB rotates 140° relative to the tibial link AD, and the cruciate links AB and CD rotate through 40° about their tibial attachments A and D and through 100° about their femoral attachments B and C.

During this movement, the femoral condyles roll and slide on the tibia. The contact area moves backwards and outwards on the tibia in flexion and forwards and inwards in extension. The femoral condyles diverge posteriorly as well as having reduced curvature. Thus, they sit in the menisci in an essentially anterior-posterior orientation in extension, but in flexion, the diverging condyles force the menisci to assume a more medio-lateral curvature. It is through this mechanism that the condyles stay in contact with the menisci.

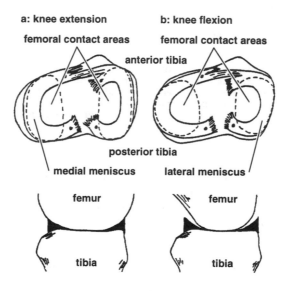

Figure 2.3.14   Schematic diagram of the knee joint showing distortion of the menisci during (a) extension and (b) flexion. When the menisci are viewed from above (top illustrations) the movement backwards and outwards from flexion to extension becomes apparent. The black dots show the points where each meniscus inserts firmly into the tibia eminence.

The menisci are actually major force-bearers in the knee, making the femur and tibia far more congruent than just a roller on a flat surface (Shrive et al., 1978). It takes about half a body weight to make the femoral cartilage actually contact the tibial cartilage. At lower forces, typically all the force is carried by the menisci.

The cruciate ligaments play essential roles in controlling motion.

The stresses that develop in the cruciate ligaments vary with joint angle. The twisted configuration of these ligaments and their wide insertion sites (Fig. 2.3.15) appear well designed to deal with the varying force patterns.

Uniform stress is seldom achieved. The fact that the fibres are twisted suggests that the ACL is forced in torsion as well as tension. The "screw home" mechanism as the knee comes into extension, for example, involves a torsional rotation of the femur relative to the tibia. As the knee undergoes flexion and extension (and similarly internal, external, varus, and valgus rotations), the length and orientation of the fibres change (Fig. 2.3.16). The broad attachments to the femur and the tibia, combined with inter-fascicular sliding, allow various portions of the ligament to be relatively taut while others are lax, depending upon the knee motion. For example, in the ACL, the anterior fascicles are taut in flexion

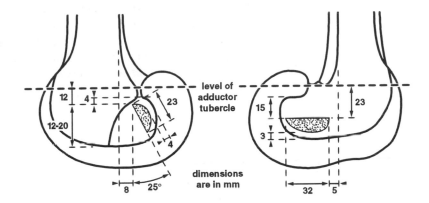

Figure 2.3.15   Average measurements and body relations of the femoral attachment of the ACL (left) and the PCL (right). Note that the ACL attachment site is about 23 mm in length and about half that wide. The attachment site for the PCL is about 32 mm in length and almost half that wide (from Girgis, 1975, with permission of J.P.

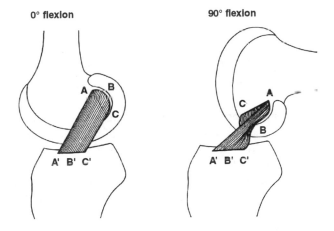

Figure 2.3.16   Schematic drawing representing changes in the shape and tension of the ACL in extension (left) and flexion (right). In flexion there is a lengthening of the anteromedial band (A-A') and a shortening of the posterolateral aspect of the ligament (C-C'). Between (A-A') and (C-C') is an intermediate band (B-B') in which fascicles experience varying degrees of tension (from Arnoczky and Warren, 1988, with permission of J.P. Lippincott Co., Philadelphia, PA).

but are relaxed to a certain degree in extension. The situation is the reverse for the posterior bundles, hence the advantage of separate "sliding" fascicular fibre bundles and the variations of collagen crimp. The tension and relaxation in fibre bundles is clearly evident in cadaver specimens missing gravity and muscle forces. The implication is that during normal activity the pattern of tension and relaxation is the same. Compressive forces, however, can affect inter-insertional distances. The muscle forces associated with normal

activity that initiate compressive forces in a joint may cause the bones to develop orientations to each other that are slightly different from those that occur in most cadaver tests.

In normal joint motion, together with the other ligaments and musculo-tendon units,
the cruciate ligaments cope with varied angles of stress. The anterior cruciate ligaments
are stronger in directions similar to their predominant fibre orientation than in directions
aligned with the tibial axis (Woo et al., 1990). The results are illustrated in Fig. 2.3.17,

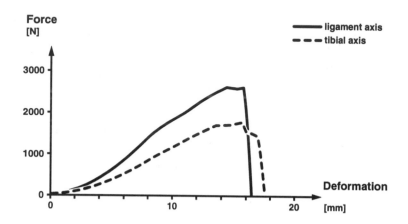

Figure 2.3.17   **Typical force-displacement curves for the femur-anterior cruciate ligament-tibia
complex tested along the ACL axis (unbroken line) and along the tibial axis
(broken line) (from Woo and Adams, 1990, with permission).**

with typical force-displacement curves obtained during tensile testing of ACLs from
young donors. The specimens forced along the axis of the ACL demonstrate a steeper linear stiffness and higher ultimate forces than those deformed along the tibial axis. These
results indicate that the cruciate ligaments are favorably oriented to take up force in flexion and extension and varus/valgus movements that approximate to the ligament's angle
of alignment. Restraint of these movements is considered to be the prime function of this
ligament complex in the knee.

Collateral ligaments

The collateral ligaments are found outside of the joint capsule (Fig. 2.3.18), and,
unlike the cruciate ligaments, they are not twisted.

The collateral ligaments are found on the medial and lateral sides of the knee joint.
Their main function is to restrain varus/valgus angulation and rotation. During varus
angulation the lateral collateral ligament takes up the force while the MCL relaxes. In valgus angulation the roles of the collateral ligaments are reversed (see Fig. 2.3.12(V)).
When medial/lateral tibial rotation occurs, the femoral condyles ride up the central eminence of the tibia, thus distracting the joint. The collateral ligaments resist the distraction
and consequently the rotation. During rotation in the transverse plane, the collateral ligaments work together with the cruciate ligaments. During medial rotation, the MCL takes
up a similar angulation to the anterior cruciate, and during lateral rotation, the lateral col-

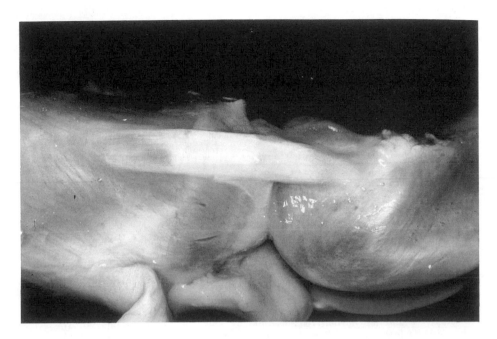

Figure 2.3.18   Gross appearance of the rabbit MCL. The relatively parallel alignment of collagen fibres gives the MCL a cord-like appearance (from Bray et al., 1991, with permission).

lateral takes up a similar angulation to the posterior cruciate (see Fig. 2.3.12(IV)). During flexion and extension, the collateral ligaments play a role secondary to the cruciate ligaments in restraining the anterior/posterior displacement of the femur and the tibia, because alongside an increase in the joint angle during flexion, the posterior bundles of the collaterals relax, while the anterior bundles retain their length and take up some strain (Fig. 2.3.19). This secondary role illustrates the advantage of "sliding" fascicles and the benefit of having different portions of ligaments able to take up stress at different angles of force.

2.3.6 LIGAMENT BIOMECHANICS AS RELATED TO FUNCTION

Some of the most important factors that affect the normal integrity of ligaments include exercise, immobilization, aging and maturing, pregnancy, and injuries.

EXERCISE

Studies of the effect of exercise on ligaments must be evaluated with care because a given exercise may not be increasing the stress in the specific ligament being investigated. Stress, however, is a positive stimulus for connective tissue and is necessary for growth. The following comments, must, therefore, be read carefully.

Studies with exercised dogs (Laros et al., 1971) did not show any *morphological* changes of the femoral and tibial insertions of the MCL. However, it has been shown that

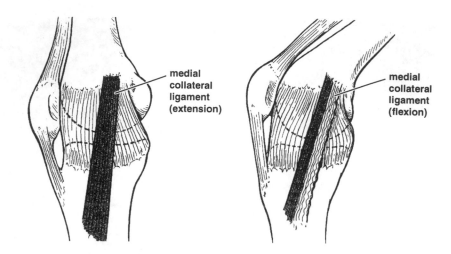

Figure 2.3.19   The MCL at knee extension (left) and knee flexion (right). When the knee is fully extended, the fibres of the MCL are equally aligned and fully resist vagus angulation and tibial rotation. When the knee is in the flexed position, the anterior portion of the MCL remains taut while the posterior portion is relaxed. The anterior portion remains able to resist valgus and rotational stresses (from Indelicatio, 1988, with permission of Churchill Livingstone, New York).

the ligaments of exercised rats have more small diameter collagen fibres than rats that have not been exercised (Tipton et al., 1970; Binkley and Peat, 1986).

Exercise seems to affect the *biochemical* composition of ligaments modestly. Increases in collagen concentration, glycosaminoglycan, collagen turnover, and collagen non-reducible cross-links have been reported (Tipton et al., 1970).

The *physical properties* of ligaments seem to change with exercise. Trained ligaments fail at somewhat greater forces and are modestly stiffer than untrained ligaments (Tipton et al., 1967; Viidik, 1967; Tipton et al., 1970; Tipton et al., 1974).

IMMOBILIZATION

To immobilize a ligament is to deprive it of stress. *Morphologically*, immobilization increases bone resorption and, consequently, the number of osteoclasts under the indirect ligament insertions (Laros et al., 1971; Noyes et al., 1974; Noyes, 1977; Woo et al., 1983). Furthermore, immobilization increases the number of large diameter fibrils and decreases the number of small diameter fibrils and the density of collagen fibrils (Frank et al., 1988).

Biochemically, immobilization decreases the glycosaminoglycan and the water content of ligaments. The rate of collagen synthesis and degradation increases, indicating that there are more "new" collagen fibres. The total amount of collagen mass decreases, and the number of reducible cross-links increases with immobilization (Akeson et al., 1977; Akeson et al., 1980; Akeson et al., 1987).

The *physical properties* of ligaments alter substantially with immobilization. Force to failure, stress to failure, energy absorbed to failure, and stiffness are reduced by immobilization. As well, immobilized ligaments tend to fail at the bony insertion rather than in mid-substance, a behaviour also characteristic of immature ligaments.

Changes induced by short term immobilization seem to be reversible. The physical properties of primate ACLs were studied after 8 weeks of immobilization and after 5 and 12 months of remobilization (Noyes et al., 1974). After 12 months the force to failure compared to control values was 91%, the stiffness 98%, and the energy absorbed to failure 92%. Similar work with rabbit MCLs (Woo et al., 1983) suggested that the longer the immobilization, the more harmful the effect on the ligament's physical properties.

LIGAMENT PROPERTIES AS FUNCTIONS OF AGE

Biochemically, aging increases the collagen concentration and reduces hexosamine and water. The number of non-reducible cross-links in ligament increases throughout the aging process. The number of reducible cross-links increases rapidly during maturation, and at some point in the aging process the production of reducible cross-links stops.

The *physical properties* of ligament also change with age, structural properties more than material ones. During growth and maturation the failure properties improve rapidly, levelling off in early adulthood. Then the failure properties decline. Ultimate tensile stress of ACLs was compared for young adults and for people about 65 years old (Noyes and Grood, 1976). The results showed a reduction of ultimate tensile stress of more than 60% in the older ligaments! Young ligaments are more viscous than older ones.

INJURY AND FAILURE

Accidental ligament ruptures in humans usually result from excessive forces. The rate of strain affects the type of ligament failure (Noyes et al., 1974). A study on primates revealed that at a slow stretch rate only 29% ligamentous (or midsubstance) failures occurred, compared to 57% failures by tibial avulsion (insertion site at tibia). However, when the strain-rate was increased a hundred-fold, the percentage of ligament midsubstance tears increased to 66% and the percentage of tibial avulsions declined to 28%.

Ligaments which do heal do so normally, first, by red blood cells and inflammatory white blood cells (disease preventing cells) entering the wound. Within days, a fragile fibrous scar composed mainly of blood-clot components and water appears in the wound. After about a week, metabolically active fibroblasts migrate into the wound to form an extracellular scar matrix. Eventually, these fibroblasts become the predominant cell type, as the fluid and blood gradually dissipate.

Morphologically, even 40 weeks after an injury, the collagen fibrils of injured ligaments are different than those of ligaments that have not been injured. Injured ligaments also appear more vascularized. The cross-sectional areas of scars are larger for injured ligaments than for uninjured ligaments, even 40 weeks after the injury. With respect to the physical properties of injured ligaments, changes in viscoelastic behaviour have been reported. The viscoelastic properties of healing rabbit MCLs are initially lower than those of normal rabbit MCLs but improve over time to within 10 to 20% of normal.

Remodelling occurs over many months or years. Scar "flaws" are removed, the number of collagen fibrils present increases, fibre alignment improves, the number of fibroblasts present declines, and vascularity decreases. There is, however, no documented return to "normal ligament".

2.3.7 REFERENCES

Adams, J.A. (1977) Feedback Theory of How Joint Receptors Regulate the Timing and Positioning of a Limb. *Psychol. Rev.* **84**, pp. 504-523.

Akeson, W.H., Amiel, D., Abel, M.F., Garfin, S.R., and Woo, S.L.Y. (1987) Effects of Immobilization on Joints. *Clin. Orthop. Rel. Res.* **219**, pp. 28-37.

Akeson, W.H., Amiel, D., Mechanic, G.L., Woo, S.L.Y., Harwood, F.L., and Hamer, M.L. (1977) Collagen Cross-linking Alterations in Joint Contractures: Changes in the Reducible Cross-links in Periarticular Connective Tissue Collagen After Nine Weeks of Immobilization. *Connect. Tiss. Res.* **5**, pp. 15-19. (Abstract).

Akeson, W.H., Amiel, D., and Woo, S.L.Y. (1980) Immobility Effects on Synovial Joints: The Pathomechanics of Joint Contracture. *Biorheology.* **17**, pp. 95-110.

Amiel, D., Billings, E., and Akeson, W.H. (1990) Ligament Structure, Chemistry, and Physiology. *Knee Ligaments: Structure, Function, Injury, and Repair* (eds. Daniel, D.D., Akeson, W.H., and O'Connor, J.J.). Raven Press, New York. pp. 77-91.

Amiel, D., Frank, C.B., Harwood, F.L., Fronek, J., and Akeson, W.H. (1983) Tendons and Ligaments: A Morphological and Biochemical Comparison. *J. Orthop. Res.* **1**, pp. 257-265.

Arnoczky, S.P. (1983) Anatomy of the Anterior Cruciate Ligament. *Clin. Orthop.* **172**, pp. 19-25.

Arnoczky, S.P. and Warren, R.F. (1988) Anatomy of the Cruciate Ligaments. *The Crucial Ligaments* (ed. Feagin Jr., J.A.). Churchill Livingstone, New York. pp. 179-195.

Barrack, R.L. and Skinner, H.B. (1990) The Sensory Function of Knee Ligaments. *Knee Ligaments: Structure, Function, Injury, and Repair* (eds. Daniel, D.D., Akeson, W.H., and O'Connor, J.J.). Raven Press, New York. pp. 95-114.

Binkley, J.M. and Peat, M. (1986) The Effects of Immobilization on the Ultrastructure and Mechanical Properties of the Medial Collateral Ligament of Rats. *Clin. Orthop. R. Res.* **203**, pp. 301-308.

Bray, D.F., Bray, R.C., and Frank, C.B. (1993) Ultrastructural Immunolocalization of Type VI Collagen and Chondroitin Sulphate in Ligament. *J. Orthop. Res.* **25 (10)**, pp. 1227-1231.

Bray, D.F., Frank, C.B., and Bray, R.C. (1990) Cytochemical Evidence for a Proteoglycan-associated Filamentous Network in Ligament Extracellular Matrix. *J. Orthop. Res.* **8 (1)**, pp. 1-11.

Bray, R.C., Frank, C.B., and Miniaci, A. (1991) Structure and Function of Diarthrodial Joints. *Operative Arthroscopy* (ed. McGinty, J.B.). Raven Press, New York. pp. 79-123.

Chimich, D.D., Shrive, N.G., Frank, C.B., Marchuk, L., and Bray, R.C. (1992) Water Content Alters Viscoelastic Behaviour of the Normal Adolescent Rabbit Medial Collateral Ligament. *J. Biomechanics.* **25 (18)**, pp. 831-837.

Diamant, J., Keller, A., Baer, E., Litt, M., and Arridge, R.G.C. (1972) Collagen: Ultrastructure and its Relation to Mechanical Properties as a Function of Aging. *Proc. R. Soc. Lond.* **180 (B)**, pp. 293-315.

Frank, C.B., Amiel, D., and Akeson, W.H. (1983a) Healing of the Medial Collateral Ligament of the Knee: A Morphological and Biochemical Assessment in Rabbits. *Acta. Orthop. Scand.* **54**, pp. 917-923.

Frank, C.B., Woo, S.L.Y., Andriacchi, T., Brand, R., Oakes, B., Dahners, L., DeHaven, K., Lewis, J., and Sabiston, P. (1988) Normal Ligament: Structure, Function, and Composition. *Injury and Repair of the Musculo-skeletal Soft Tissues* (eds. Woo, S.L.Y. and Buckwalter, J.A.). American Academy of Orthopaedic Surgeons, Rosemont, IL. pp. 45-101.

Frank, C.B., Bray, D.F., Rademaker, A., Chrusch, C., Sabiston, C.P., Bodie, D., and Rangayyan, R.M. (1989) Electron Microscopic Quantification of Collagen Fibril Diameters in the Rabbit Medial Collateral Ligament: A Baseline for Comparison. *Connect. Tissue Res.* **19**, pp. 11-25.

Frank, C.B. and Hart, D.A. (1990) The Biology of Tendons and Ligaments. *Biomechanics of Diathrodial Joints* (eds. Mow, V.C., Ratcliffe, A., and Woo, S.L.Y.). Springer Verlag, New York. pp. 39-62.

Frank, C.B., Woo, S. L.Y., Amiel, D., Harwood, F.L., Gomez, M.A., and Akeson, W.H. (1983b) Medial Collateral Ligament Healing: A Multi-disciplinary Assessment in Rabbits. *Am. J. Sports Med.* **11 (6)**, pp. 379-389.

Frank, C.B., Woo, S.L.Y., Andriacchi, T., Brand, R., Oakes, B., Dahners, L., De Haven, K., Lewis, J., and Sabiston, P. (1988) Normal Ligament: Structure, Function, and Composition. *Injury and Repair of the Musculo-skeletal Soft Tissues* (eds. Woo, S.L.Y. and Buckwalter, J.A.). American Academy of Orthopedic Surgeons, Park Ridge, IL. pp. 45-101.

Girgis, F.G., Marshall, J.L., and Al Monajem, A.R.S. (1975) The Cruciate Ligaments of the Knee Joint. Anatomical, Functional, and Experimental Analysis. *Clin. Orthop.* **106,** pp. 216-231.

Indelicatio, P.A. (1988) Injury to the Medial Capsuloligamentous Complex. *The Crucial Ligaments* (ed. Feagin Jr., J.A.). Churchill Livingstone, New York. pp. 197-216.

Kastelic J., Galeski, A., and Baer, E. (1978) The Multi-composite Structure of Tendon. *Connect. Tissue Res.* **6,** pp. 11-23.

Kennedy, J.C., Alexander, I.J., and Hayes, K.C. (1982) Nerve Supply of the Human Knee and its Functional Importance. *Am. J. Sports Med.* **10 (6),** pp. 329-335.

Kraupse, R., Schmidt, M.B., and Schaible, H.G. (1992) Sensory Innervation of the Anterior Cruciate Ligament. *J. Bone Jt. Surg.* **74 (A),** pp. 390-397.

Lam, T.C. (1988) *The Mechanical Properties of the Maturing Medial Collateral Ligament.* Ph.D. Thesis, University of Calgary. Nat'l Library of Canada, Ottawa.

Lam, T.C., Frank, C.B., and Shrive, N.G. (1993) Changes in the Cyclic and Static Relaxations of the Rabbit Medial Collateral Ligament Complex During Maturation. *J. Biomech.* **26 (1),** pp. 1-8. (In press).

Laros, G.S., Tipton, C.M., and Cooper, R.R. (1971) Influence of Physical Activity on Ligament Insertion in the Knees of Dogs. *J. Bone Jt. Surg.* **53 (A),** pp. 275-286.

Liu, Z.Q., Rangayyan, R.M., and Frank, C.B. (1991) Statistical Analysis of Collagen Alignment in Ligaments by Scale-space Analysis. *I.E.E.E. Trans. Biomed. Eng.* **38 (6),** pp. 580-588.

Matyas, J.R. (1985) *The Structure and Function of Tendon and Ligament Insertions into Bone.* M.Sc. Thesis, Cornell University, New York.

Matyas, J.R. (1990) *The Structure and Function of the Insertions of the Rabbit Medial Collateral Ligament.* Ph.D. Thesis, University of Calgary. Nat'l Library of Canada, Ottawa.

Matyas, J.R., Bodie, D., Andersen, M., and Frank, C.B. (1990) The Developmental Morphology of a "Periosteal" Ligament Insertion: Growth and Maturation of the Tibial Insertion of the Rabbit Medial Collateral Ligament. *J. Orthop. Res.* **8 (3),** pp. 412-424.

Miller, E.J. and Gay, S. (1992) Collagen Structure and Function. *Wound Healing: Biochemical and Clinical Aspects* (eds. Cohen, I.K., Diegelmann, R.F., and Lindblad, W.J.). W.B. Saunders, Philadelphia. pp. 130-151.

Noyes, F.R. (1977) Functional Properties of Knee Ligaments and Alterations Induced by Immobilization. *Clin. Orthop. Rel Res* **123,** pp. 210-312.

Noyes, F.R., DeLucas, J.L., and Torvik, P.J. (1974) Biomechanics of Anterior Cruciate Ligament Failure: An Analysis of Strain-rate Sensitivity and Mechanisms of Failure in Primates. *J. Bone Jt. Surg.* **56 (A-2),** pp. 236-253.

Noyes, F.R. and Grood, E.S. (1976) The Strength of the Anterior Cruciate Ligament in Humans and Rhesus Monkeys. *J. Bone Jt. Surg.* **58 (A-6),** pp. 1074-1082.

Noyes, F.R., Torvik, P.J., Hyde, W.B., and DeLucas, J.L. (1974) Biomechanics of Ligament Failure II: An Analysis of Immobilization, Exercise, and Reconditioning Effects in Primates. *J. Bone Jt. Surg.* **56 (A),** pp. 1406-1418.

O'Connor, J.J., Shercliff, T., Fitzpatrick, D., Bradley, J., Daniel, D.M., Biden, E., and Goodfellow, J. (1990) Geometry of the Knee. *Knee Ligaments: Structure, Function, Injury, and Repair* (eds. Daniel, D.M., Akeson, W.H., and O'Connor, J.J.). pp. 163-199.

Parry, D.A.D., Barnes, G.R.G., and Craig, A.S. (1978a) A Comparison of the Size Distribution of Collagen Fibrils in Connective Tissues as a Function of Age and a Possible Relation Between Fibril Size Distribution and Mechanical Properties. *Proc. R. Soc. Lond.* **203 (B),** pp. 305-321.

Parry, D.A.D., Craig, A.S., and Barnes, G.R.G. (1978b) Tendon and Ligament from the Horse: An Ultrastructural Study of Collagen Fibrils and Elastic Fibres as a Function of Age. *Proc. R. Soc. Lond.* **203 (B),** pp. 293-303.

Rundgren, Å. (1974) Physical Properties of Connective Tissue as Influenced by Single and Repeated Pregnancies in the Rat. *Acta. Physiol. Scand. Supplement.* **417,** pp. 1-138.

Shrive, N.G., O'Connor, J.J., and Goodfellow, J.W. (1978) Load-bearing in the Knee Joint. *Clin. Orthop.* **131,** pp. 279-287.

Sojka, P., Johansson, H., Sjolander, P., Lorentzon, R., and Djupsjobacka, M. (1989) Fusimotor Neurones can be Reflexly Influenced by Activity in Receptors from the Posterior Cruciate Ligament. *Brain Res.* **483 (1),** pp. 177-183.

Tipton, C.M., James, S.L., Mergner, W., and Tcheng, T.K. (1970) Influence of Exercise on Strength of the Medial Knee Ligaments of Dogs. *Am. J. Physiol.* **218,** pp. 894-902.

Tipton, C.M., Matthes, R.D., and Sandage, D.S. (1974) In Situ Measurement of Junction Strength and Ligament Elongation in Rats. *J. Appl. Physiol.* **37 (5),** pp. 758-761.

Tipton, C.M., Schild, R.J., and Tomanek, R.J. (1967) Influence of Physical Activity on the Strength of Knee Ligaments in Rats. *Am. J. Physiol.* **212,** pp. 783-787.

Viidik, A. (1967) The Effect of Training on the Tensile Strength of Isolated Rabbit Tendons. *Scand. J. Plast. Reconstr. Surg.* **1,** pp. 141-147.

Walsh, S. (1989) *Immobilization Affects Growing Ligaments.* M.Sc. Thesis, University of Calgary. Nat'l Library of Canada, Ottawa.

Walsh, S., Frank, C.B., and Hart, D.A. (1992) Immobilization Alters Cell Metabolism in an Immature Ligament. *Clin. Orthop.* **277,** pp. 277-288.

Woo, S.L.Y. and Adams, D.J. (1990) The Tensile Properties of Human Anterior Cruciate Ligament (ACL) and ACL Graft Tissues. *Knee Ligaments: Structure, Function, Injury, and Repair* (eds. Daniel, D.M., Akeson, W.H., and O'Connor, J.J.). Raven Press, New York. pp. 279-289.

Woo, S.L.Y., Gomez, M.A., Seguchi, Y., Endo, C.M., and Akeson, W.H. (1983) Measurement of Mechanical Properties of Ligament Substance from a Bone-ligament-bone Preparation. *J. Orthop. Res.* **1 (1),** pp. 22-29.

Woo, S.L.Y., Gomez, M.A., Woo, Y.K., and Akeson, W.H. (1982a) Mechanical Properties of Tendons and Ligaments II: The Relationships of Immobilization and Exercise on Tissue Remodelling. *Biorheology.* **19,** pp. 397-408.

Woo, S.L.Y., Gomez, M.A., Woo, Y.K., and Akeson, W.H. (1982b) Mechanical Properties of Tendons and Ligaments. I. Quasi-static and Non-linear Viscoelastic Properties. *Biorheology.* **19,** pp. 385-396

Yahia, L.H. and Drouin, G. (1989) Microscopical Investigation of Canine Anterior Cruciate Ligament and Patellar Tendon: Collagen Fascicle Morphology and Architecture. *J. Orthop. Res.* **7 (2),** pp. 243-251.

2.4 TENDON

HERZOG, W.
LOITZ, B

2.4.1 DEFINITIONS AND COMMENTS

Anatomy

Aponeurosis:	Tendinous expansion into which the muscle fibres insert, serving mainly to connect muscle and tendon.
Cell:	Basic unit of life within a tissue.
Chondrocyte:	A mature cartilage cell embedded in a small cavity (lacuna) within the cartilage matrix.
Collagen:	The protein substance of the white fibres (collagen fibres), composed of molecules of tropocollagen.
Matrix:	The intercellular substance of a tissue from which the structure develops.
Myotendinous junction:	Region where the muscle fibres join with the tendon.
Proteoglycans:	Group of glycoproteins present in connective tissue. Formed of subunits of disaccharides linked together and joined to a protein core.
Sheaths:	Tubular structure enclosing or surrounding some organ part.
• Endotenon:	Connective tissue separating the fascicles of a tendon.
• Epitenon:	Fibrous tissue sheath covering a tendon.
• Paratenon:	Fibrous tissue sheath covering a tendon.
• Peritenon:	Connective tissue investing larger tendons and extending as septa (partitions) between the fibres composing them.
Tendon:	A fibrous, cord-like structure of connective tissue by which a muscle is attached to bone.
Tropocollagen:	The molecular unit of collagen fibrils, about 1.4 nm wide and about 280 nm long. Tropocollagen is a helical structure consisting of three polypeptide chains. Each chain is composed of about a thousand amino acids coiled around each other to form a spiral.

Mechanics

Density:	Mass per unit volume.

Elastic modulus
(for a linearly elastic structure):   The ratio of stress divided by strain.

$$E = \frac{\sigma}{\varepsilon}$$

where:

σ = stress
ε = strain
E = elastic modulus

with the units:

$[E] = N/m^2 = Pa$
$[\sigma] = N/m^2 = Pa$
$[\varepsilon] = \%$

Note: the unit "percent" is obtained by multiplying the measured relative length change with the number 100.

Load: Sum of all the forces and moments.

Properties:

• Material: Properties of a material which describe its general behaviour without including any information about its size and shape.
Comment: material properties include stress and strain, etc.

• Physical: Properties of a material which relate to its physics.
Comment: physical properties include density, specific weight, etc.

• Structural: Properties of a specific sample of a material which describe the behaviour of that sample including effects of its size and shape.
Comment: structural properties include force to failure, deformation, etc.

Strain: Relative change of length.

$$\varepsilon = \frac{\Delta L}{L_o}$$

where:

ε = strain
ΔL = change in length = $L - L_o$
L_o = original length

| Strain rate: | Change of strain over time. |

$$\dot{\varepsilon} \;=\; \frac{d\varepsilon}{dt}$$

with the unit:

$$[\dot{\varepsilon}] \;=\; \frac{1}{s}$$

Strength:	Maximal force a material can sustain before failure.
• Compressive strength:	Maximal force in compression a material can sustain before failure (= ultimate compressive strength).
• Tensile strength:	Maximal force in tension material can sustain before failure (= ultimate tensile strength).
Stress:	Force per unit area.

$$\sigma \;=\; \frac{F}{A}$$

where:

σ = stress
F = force
A = area

• Compressive stress:	Stress perpendicular to the surface that acts to compress an object.
• Shear stress:	Stress parallel to the surface of an object.
• Tensile stress:	Stress perpendicular to the surface that acts to elongate an object.
• Ultimate stress:	The highest stress experienced by the tissue before complete failure.

2.4.2 SELECTED HISTORICAL HIGHLIGHTS

129-201	Galen	Stated that muscles consisted of fibres that were connected to tendons.
1500's	da Vinci	Stated that tendons were mechanical instruments playing a passive role in carrying out as much work as was put upon them.
1828	Bell	Observed that mechanical influences governed the growth and development of tendons.
1847	Wertheim	Noted that the stress-strain curve of tendon was non-Hookean.
1850	Kölliker	Stated that tendons were relatively avascular. In 1855, he noted that tendons contained "waves" (banding)

which disappeared when the tissue was tensioned. He attributed these waves to the influence of elastic fibres upon collagen.

| 1949 | Wickoff | Observed from electron microscopic examination that there was a helical (spiral) pattern within the collagen fibrils. |
| 1950 | Rollhäuser | Stated that the extensibility of tendons increases with age. |

2.4.3 MORPHOLOGY AND HISTOLOGY

MORPHOLOGY AND FUNCTION

Tendons are fibrous cords which attach the muscle to the bone. As such, the tendon includes not only the external fibrous bundles that are readily identifiable, but also a juxta-muscular component into which the muscle fibres insert. The major functions of tendons include:

- To *transfer forces* between muscle and bone, and
- To *store elastic energy.*

The internal portion of the tendon is called an *aponeurosis* (Zajac, 1989) (Fig. 2.4.1). The aponeurosis is often neglected when describing tendon morphology, but it plays a critical role in the biomechanical behaviour of the muscle-tendon unit.

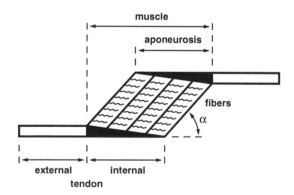

Figure 2.4.1   Schematic diagram of the relationship between muscle fibres, the portion of the tendon into which the muscle fibres insert, and the extramuscular tendon (from Zajac, 1989, with permission of CRC Press, Boca Raton, FL).

The form taken by a particular tendon depends on the muscle's contractile properties and its orientation with respect to the relevant joint. The proximal tendon of the human semimembranous muscle, for instance, is a flat, band-like tendon. It's shape reflects the shape of the short, angled fibres of the muscle belly that insert into the tendon as it courses distally (Fig. 2.4.2a). By contrast, the fibres of the human biceps brachii muscle converge and insert into a cord-like proximal tendon (Fig. 2.4.2b). Other factors that influence ten-

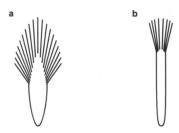

Figure 2.4.2   Muscle fibres may insert into a broad band-like tendon (a) or they may converge before inserting into a cord-like tendon (b).

don morphology are the availability of a bony surface into which the tendon can insert, and the need to dissipate force by increasing the insertional area. Some tendons are so short that the white glistening fibres are not discernible. This is the case in the broad proximal attachments of many muscles, such as the gluteal muscles of the buttock or the vasti muscles of the anterior (front) thigh. To a casual observer, these muscle fibres appear to insert directly into the bone, yet this is not the case.

Generally, tendon structure can be viewed hierarchically (Fig. 2.4.3). Tropocollagen is

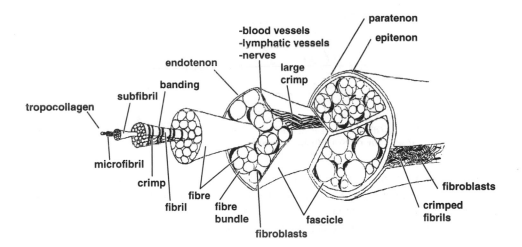

Figure 2.4.3   The structural hierarchy of a tendon, from the tropocollagen molecule to the entire tendon. Connective tissue layers or sheaths envelop the collagen fascicles (endotenon), bundles of fascicles (epitenon), and the entire tendon (paratenon). Note that blood and lymphatic vessels and nerves are cut in the cross-section within the endotenon (from Kastelic et al., 1978, with permission).

the fundamental unit of collagen. Five tropocollagen molecules group together to form a microfibril; microfibrils aggregate into subfibrils which in turn group together into fibrils. Fibrils are the basic load bearing units of tendon and because of the repeating, overlapping structure of the tropocollagen, appear banded under electron microscopic examination. Groups of fibrils, or fibres, bundle together and with a few individual fibrils, aggregate to form fascicles. Individual fascicles are not interconnected (Kastelic et al., 1978) but can slide past one another smoothly, apparently lubricated by the tissue fluid. Fibroblasts are distributed sparsely among the fibres and fibre bundles within the fascicles.

In addition to the parallel-fibred collagen fibrils and fascicles, collagen and (rarely) elastin fibrils also form connective tissue sheaths that surround the fibrils and fibres (Fig. 2.4.3). The sheaths are more cellular than the bundles of collagen fibres, with the fibroblasts flattened among the loosely organized collagen. The *endotenon* surrounds the collagen fascicle and contains a well-ordered criss-cross pattern of collagen fibrils and filament-sized proteoglycans and elastin (Rowe, 1985). Blood vessels, nerves, and lymphatics are also enclosed within the endotenon. Groups of fascicles are surrounded by the *peritenon* and the entire tendon is enclosed by the *epitenon* or *paratenon*. This outermost layer comprises small collagen fibrils widely separated by fibroblasts and possibly elastin (Rowe, 1985) and functions as an elastic sleeve to permit free movement of the tendon against surrounding tissues.

Tendons may be constrained as they cross joints with the purpose of maintaining their orientation during joint motion. These constraints may be formed by bony prominences around or through which the tendon passes (e.g., bicipital groove in the humerus) or by specialized connective tissue sheaths or retinacula. The sheaths are particularly important in the hands and feet (Fig. 2.4.4), as tendons passing distally (toward the fingers or toes)

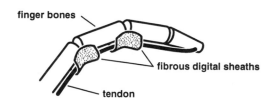

Figure 2.4.4   **Fibrous digital sheaths of the finger form a tunnel with the bone to keep the tendon close to the bone as the finger bends.**

over the numerous small joints would be susceptible to injury if allowed to be displaced during finger or toe motions. Also, the sheaths function as pulleys to improve (and maintain) the mechanical advantage of the tendons relative to the joint over which they cross. Some tendons of the hands and feet pass through digital sheaths enclosed within a double-

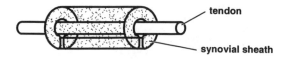

Figure 2.4.5   **A synovial sheath envelops a tendon. The potential space between the layers of the sheath is exaggerated in this diagram to illustrate that the layers of the sheath can slide upon one another as the tendon moves.**

layered membrane called a synovial sheath (Fig. 2.4.5). Between the layers, a viscous fluid (synovial fluid) is produced that is important in reducing the friction between the tendon and the digital sheath. This fluid allows the tendon to move freely without damage.

The region where the muscle fibres join with the tendon is a specialized structure called the *myotendinous junction*. Here, the contractile proteins or myofibrils of the muscle and the collagen of the tendon form a four-layered region of longitudinal infoldings. The membranous folds increase the surface area of the junction up to 22.2 times greater than a junction formed by the myofibrils and collagen fibres meeting end-to-end (Tidball, 1984). In addition, the collagen fibres of the tendon and the membranous layer of the muscle meet at an acute angle such that the junction sustains primarily shear loads (Fig. 2.4.7). This ori-

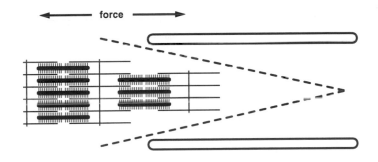

Figure 2.4.6   **Diagram of a myotendinous junction. Muscle force is applied parallel to the longitudinal axes of the myofilaments and the collagen fibres. The junctional membrane lies at an angle relative to the myofilaments (from Tidball, 1983, with permission).**

entation enhances junctional strength because cell membranes are known to be more resistant to shear stress than to a tensile stress that would develop if the tendon and muscle fibrils met more end-to-end (Fig. 2.4.7). The degree of infolding also reflects the rate of

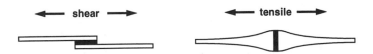

Figure 2.4.7   **The acute angle with which the muscle and collagen fibrils meet creates a shear stress between the fibrils. If the fibrils met end-to-end, the junction would be loaded in tension (from Tidball, 1983, with permission).**

tension development within the muscle, with slow twitch muscle fibres having a larger junctional area than fast twitch fibres (Trotter et al., 1985). It has been estimated that a junction of a fast twitch muscle fibre bears an average stress of 15 kPa during muscle contraction (Tidball, 1984).

Tendon morphology also changes gradually as the tendon approaches the bone to form the "osteotendinous junction" or "insertion". Two types of osteotendinous insertions are described (Woo et al., 1988). When tendons insert into epiphysial bone, four distinct zones of the insertion can be identified (Fig. 2.4.8a). The *fibrous* tissue of the tendon gradually

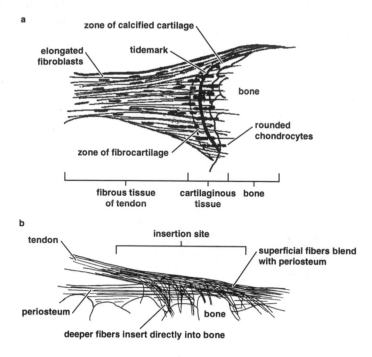

Figure 2.4.8   Schematic diagrams illustrating tendon insertion morphology. (a) The fibrous tissue of the tendon gradually changes to fibrocartilage then to calcified fibrocartilage before inserting into the bone. (b) shows tendon approaching bone at an acute angle, the superficial fibres blending directly with the periosteum while the deeper fibres insert into the bone (from Benjamin et al., 1986, with permission of Cambridge University Press).

changes to *fibrocartilage*, which is characterized by a more dense matrix rich in proteogly-cans. The cells gradually change from the spindle-shaped fibroblasts of the tendon to rounded chondrocytes. The chondrocytes and cartilage matrix lie among the collagen bundles that continue from the tendon proper. Blood vessels are generally absent from the zone of fibrocartilage (Benjamin et al., 1986). The amount of fibrocartilage present in an insertion varies widely among tendons and between regions of the same insertion. In tendons that attach close to the joint surface, the zones of fibrocartilage are continuous with the articular cartilage that covers the bony end (Benjamin et al., 1986). The third insertional area forms when the fibrocartilage calcifies and becomes the *calcified fibrocartilage* zone. With the light microscope, one or more basophilic lines, or tidemarks, mark the transition between the non-calcified and the calcified fibrocartilage. Chondrocytes are less numerous on the bony side of the tidemark. The calcified fibrocartilage zone is adjacent to the *bone*. The functional significance of this type of insertion is controversial, with some speculating that the gradual transition from fibrous tissue into bone helps to distribute forces over the attachment site, thereby minimizing local stress concentration (Cooper and Misol, 1970; Noyes et al., 1974; Woo et al., 1988). Benjamin et al. (1986) suggest that fibrocartilage is typical of epiphysial attachments because these tendons undergo a greater angular change during joint movement than do tendons inserting along the bony shafts.

The other type of insertion is one in which the tendon fibres approach the bone surface at an acute angle. The superficial tendon fibres blend with the periosteum that covers the bony surface while the deeper fibres of the tendon insert directly into the bone (Fig. 2.4.8b).

From the muscle belly to the osteotendinous junction, one may generalize that the tendon is loaded in tension. However, areas of tendons may withstand compressive loads as the tendon passes around a bony prominence or turns through a digital sheath. Within this area, a specialized fibrocartilage "button" develops that is similar to the fibrocartilage that develops pathologically when a tissue that usually withstands tension is subjected to compression (Scapinelli and Little, 1970). Merrilees and Flint (1980) examined the differences in the ultrastructure of a rabbit tendon that sustains tension along its superficial surface and compression on its deep surface. Collagen in the tensile surface was found to orient longitudinally, with typical elongated fibroblasts distributed among the collagen fibres. In contrast, in the zone that was compressed, the collagen assumed a basket-weave orientation and enclosed chondrocyte-like (cartilage) cells rather than fibroblasts. The authors concluded that the difference in morphology represented an adaptive response to the compressive loads sustained by the tendon. In a similar study, Gillard et al. (1979) used the same rabbit tendon model and altered the loading pattern of the tendon by surgically relocating the tendon, thereby eliminating the compressive load. Glycosaminoglycan content of the typically compressed region decreased 60% within 8 days of removing the compressive load. When the tendon was again compressed, glycosaminoglycan content returned towards its previous level, although the transition was much slower than that which occurred when the compression was removed. This data suggests that tendon may be quite sensitive to the local loading environment and apparently responds to changes in the loads by altering its biochemical makeup and collagen orientation.

The glistening white appearance suggests incorrectly that tendon is an avascular (lacking blood vessels) structure. Certainly, if compared to muscle or skin, tendon vascularity is minimal. The lack of blood supply reflects the minimal metabolic requirements of tendon. As a result, few vessels are needed. In the tendon midsubstance, longitudinal arterioles surround the collagen fascicles and are fed by similar vessels oriented transversely; arterioles are small blood vessels that bring oxygen- and nutrient-rich blood to the tissues from the heart and lungs. Veins accompany the arterioles, collecting blood from the capillary beds and returning it to the heart and lungs to be reoxygenated. At the myotendinous junction, vessels surrounding the muscle fibres continue around the actual junction into the endotenon. Vessels do not cross through the junction. Similarly, at the osteotendinous junction, vessels do not cross the actual junction, but are seen coursing around it, continuous between the periosteum and the endotenon. While the major role of the blood supply is to nourish the tendon fibroblasts, the cells may also derive nutrients from the synovial fluid (Lundborg and Rank, 1978; Lundborg et al., 1980; Manske and Lesker, 1982). This appears to be particularly important in areas of tendons that are enclosed within synovial sheaths.

Specialized nerve endings (golgi tendon organs) are embedded within the tendon close to the myotendinous junction and in series with the contractile proteins of the muscle. Each receptor connects to the tendinous fascicles associated with approximately 10 muscle fibres and sends a thick myelinated afferent fibre to the spinal cord. Afferent nerve fibres carry sensory information from the periphery to the spinal cord. The velocity with which the neural signals are conducted depends, in part, upon the fibre diameter (greater conduction velocity in large fibres) and whether the fibre is covered in a fatty-protein layer called myelin. A myelinated nerve fibre looks similar to an insulated electrical wire, with the myelin being analogous to the insulation. Conduction velocity of myelinated nerve fibres is markedly greater than that of unmyelinated fibres. The tendon organ is more sensitive to dynamic versus static tension in the muscle, as noted by the slightly decreased discharge rate if muscle stretch is maintained statically. Similarly, the tendon organ is more sensitive to tension produced within the muscle-tendon unit than to tension produced by pulling the muscle passively (Schmidt, 1978). The tendon organ, therefore, senses dynamic muscle-tendon tension, particularly when tension is generated by the muscle.

HISTOLOGY

Tendon has two components:

- Sparse slender cells, and
- The more abundant extracellular matrix.

Cells

Tendon cells, or *fibroblasts*, vary somewhat in size, shape, and distribution (Frank and Hart, 1990), although most are spindle-shaped and orient parallel to the tendon longitudinal axis.

Matrix

The extracellular matrix contains structural macromolecules and fluid. The fluid phase (primarily water) makes up 58 to 70% of a tendon's wet weight (Curwin and Stanish, 1984). The structural macromolecules of tendon include the proteins collagen, elastin, proteoglycan, and fibronectin. *Collagen* is the most abundant macromolecule, accounting for approximately 30% of the tendon by wet weight and roughly 80% by dry weight (Curwin and Stanish, 1984). The fibres are oriented primarily along the tendon longitudinal axis, which provides tensile strength during muscular contraction. Electron microscopic examination of human tendons has revealed collagen fibrils oriented transversely, apparently binding adjacent longitudinal fibres together (Jozsa et al., 1991). The presence of these transverse fibres suggests that tendons not only withstand uniaxial loads applied by muscle contraction but may also sustain rotational forces as the tendon reorients during joint motion.

Fibronectin is a large glycoprotein that plays an important role in the interaction between the fibroblasts and the extracellular matrix, particularly in cell-to-cell adhesion, fibroblast to collagen attachment, and cell migration. It may also play an indirect role in the control of matrix production (Viidik, 1990).

2.4.4 PHYSICAL PROPERTIES

The primary role of tendon is to transmit the force of its associated muscle to bone. As such, tendon needs to be relatively strong and stiff in tension. Tensile stiffnesses (load per unit deformation) and ultimate strengths have been measured for a variety of tendons from a variety of vertebrate species, using conventional engineering testing machines. Mechanical testing of tendons, however, has proven difficult. The main problem is that specimens tend to slip out of the standard testing machine grips. Serrated and roller grips (Fig. 2.4.9)

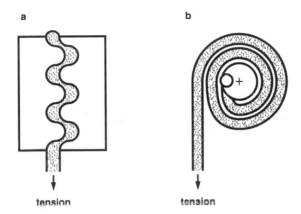

Figure 2.4.9   **Different grips for holding tendons. Serrated grips in (a) may have different shapes but rely on friction between serration and specimen to help hold the specimen in place. Roller grips (b) are based on reducing the tension in the specimen as it is wrapped around the roller, before the end is gripped.**

have been used, but stress concentration at the edge of the grips, particularly the serrated type, still causes failure of the specimen at the grip. Such tests do not give a true measure of the strength of the tissue, but relative, low level stress/strain properties can be obtained for the midsubstance if an accurate method of strain measurement is used, and the specimen has a large enough aspect ratio for end effects near the grips not to affect the midsubstance. Specimens can also be tested by gripping the bone at one end and the muscle or tendon at the other. If the material properties of the tendon proper are being sought, structural effects involving the insertion must then be separated from material effects. Recently, cryo-grips have been used: the specimen is frozen where it is gripped. Since frozen tissue is much stronger than "fresh" tissue, this technique eliminates the failure problem at the specimen-grip interface.

A typical stress-strain curve for tendon appears in Fig. 2.4.10. A region of considerable strain deformation with little accompanying resistive stress (toe region, between 0 and 10 MPa for the diagram), is followed by a linear region. The slope of this linear portion has been used to traditionally define the "elastic" or "tangent modulus" of the tendon. Loading beyond the linear region causes permanent deformation if the specimen is unloaded, or failure. Tendons show viscoelastic behaviour in that a proportion of the energy expended during tensile deformation is not recovered when the deforming force is removed (the hysteresis area in the diagram).

Typically, the "toe" region lies within 3% strain (Rigby et al., 1959). The "ultimate" or "failure" strain of tendon is about 8-10% (Rigby et al., 1959). The region of "reversible" or "linear" strain extends to about 4% (Wainwright et al., 1982), beyond which permanent deformation occurs.

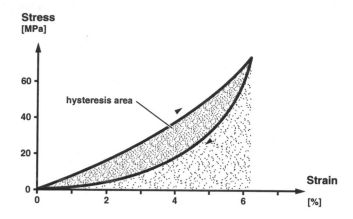

Figure 2.4.10   A typical stress strain curve for a tendon for both loading and unloading. Loading is the upper line, unloading the lower. The area under the loading curve is the single shaded area, while the hysteresis area is the double shaded area.

Although tendon is relatively inextensible compared to other structural proteins such as elastin, its orders of magnitude are stiffer and stronger. The elastic modulus for rat tail tendons is approximately 1.0 GPa (i.e., 1.0×10^9 Nm^{-2}, Rigby et al., 1959). The mean tangent moduli, derived from several aquatic and terrestrial mammalian species, ranges from 1.25-1.65 GPa, with a mean over all species of 1.5 GPa (Bennett et al., 1986). Most of these measurements were obtained in dynamic tests at 1.0-2.2 Hz, but little variation with frequency was found in the range of 0.2-11 Hz.

The strength of any material depends ultimately on the presence of flaws. Since the number and/or severity of flaws can vary between specimens, values for the "ultimate" or "failure" strength of vertebrate tendon can be equally variable (Wainwright et al., 1982). Elliot (1965) reported ranges of tendon strengths of 20-140 MPa. More recently, Bennett et al. (1986), measured the ultimate strengths of various mammalian digital and tail tendons, using cryo-grips. Mean values ranged from 90-107 MPa, with a global mean over all species tested of about 100 MPa.

Besides being relatively stiff and strong in tension, tendon is relatively resilient. Ker (1981) reported that the hysteresis area of sheep plantaris tendon was only about 6% of the area under the loading curve, and that this value was effectively independent of frequency in the range of about 0.2-11 Hz. Similarly, Bennett et al. (1986) showed that hysteresis measurements of several vertebrate tendons ranged from 6-11%, and were equally frequency independent. Thus, for a biological material, tendon shows relatively limited viscoelastic behaviour, and least within the likely range of physiologically relevant frequencies of deformation.

The low initial stiffness of tendon is thought to be due, in part, to the straightening of the collagen crimp. Rigby et al. (1959) observed that rat tail tendon lost its crimping pattern beyond strains of about 3%. Hooley and McCrum (1980), however, attributed the toe region to a shearing action between the collagen fibrils and the matrix of the tendon.

The high tensile stiffness and strength of tendon is attributed to its relatively high collagen content (70-80% dry weight, Elliot, 1965), and the hierarchical organization of the tendon. The elongated tropocollagen molecules are linearly assembled to form microfibrils, subfibrils, fibrils, and finally fibres. The helical interweaving of the units varies among the levels (e.g., tropocollagen has a left-hand helix while microfibrils have a right-hand one), in order to keep the material from unwinding when loading. Finally, because the majority of the collagen fibres lie parallel to the long axis of the tendon, tensile properties are further enhanced (Table 2.4.1).

The significance of the observed tensile properties can be appreciated by considering tendon function. Tendon must be sufficiently stiff and strong in order to transmit muscle forces to bone, without being substantially deformed in the process. Ker et al. (1988) studied the relative size of muscle and tendon dimensions. They argued that thin tendons require long-fibred muscles (that allow for great changes in length), in order to compensate for tendon deformation during muscle contraction, whereas, thick tendons deform less than thin tendons, and so, may not need extra long fibres in the corresponding muscle. Ker et al. (1988) calculated an optimal ratio for the cross-sectional area of muscles and their associated tendons, based upon the isometric properties of mammalian striated muscle, and minimizing muscle-tendon mass. They found that many mammalian species, including human, maintained their calculated optimal ratio (Ker et al., 1988; Cutts et al., 1991). Minimizing body mass is an important aspect of the functional design of living systems.

Table 2.4.1   Selected physical properties for tendon tissue.

PROPERTY	UNIT	VALUE
LINEAR STRAIN	%	2 to 8
ULTIMATE STRESS	MPa	80 to 120
ELASTIC MODULUS	GPa	0.8 to 2
RELATIVE HYSTERESIS AREA	%	2.5 to 20

Because of the low hysteresis of tendons, they are capable of storing and releasing considerable elastic strain energy. When muscle is activated, and the muscle-tendon unit is stretched, as on impact during locomotion, the energy associated with the deformation of the tendon is almost completely recovered when the deforming force is removed. The energetic consequences of elastic energy storage have been well established for many vertebrate species. During locomotion, the gravitational potential energy of the body increases and decreases as the centre of mass rises and falls. If the potential energy can be stored as elastic strain energy in deforming the tendon during impact, and subsequently can be recovered during take-off, the energy released in elastic recoil could be used to help propel the body through the subsequent stride. This mechanism seems to work particularly well for animals with long distal tendons that are attached to massive, very short-fibred proximal muscles (for example, Alexander et al., 1982; Alexander, 1983; Alexander, 1984; Ker et al., 1986; Alexander, 1988b). Muscle mass increases with animal mass, as does the area and the length of tendons. The result is that, per unit of animal mass, especially for tendons subjected to large stresses, the tendons of larger mammals have a greater capacity for storing elastic strain energy than those of smaller animals (Pollock, 1991). The role of tendon elasticity in human locomotion seems less certain. Cavagna et al. (1964) calculated the mechanical efficiency of running humans. They found that the efficiency of running was greater than that of contracting an isolated muscle in the absence of any prestretch. They argued that during impact and take-off, muscle-tendon complexes may have been stretched and relaxed, thus storing and releasing elastic strain energy.

They were unable, however, to differentiate between the potential for muscle strain energy and tendon strain energy. More recently, van Ingen Schenau (1984) calculated the potential for elastic strain energy storage during running in humans. His calculations were based at least in part, upon the tensile properties of tendons, and measurements of the ground reaction forces during running. van Ingen Schenau (1984) showed that compared to the total energy required to run, the amount of energy that could be saved by means of tendon elastic energy storage was insignificant. Thus, while the role of tendon elasticity seems well established for some vertebrate species, the debate continues as to the relevance of this mechanism during human locomotion.

Tendon elasticity has typically been associated with energy savings during cyclic movements. This motion has probably developed because tendons have been shown to "store" elastic energy (e.g., Alexander, 1984), and then "release" this elastic energy. The loss of energy during "release" is typically just a few percent of the "storage" energy (e.g., Ker, 1981), thus, the tendon system appears efficient. What sometimes gets forgotten in these considerations is the fact that the muscle consumes metabolic energy while its tendon is stretched, and also when it shortens. The question, thus arises, what is metabolically more efficient, a tendon that is stretched (while the muscle fibres remain at a constant length), or a muscle fibre that stretches (while its stiff tendon does not change length). Assuming that the force-time profile in both cases is the same, it is quite clear that the tendon stretch is metabolically more costly, because an isometric fibre contraction at a given force is associated with a higher metabolic cost than an eccentrically contracting fibre at the same force. During shortening, the roles are reversed; the isometrically contracting muscle fibre requires less energy to produce a given force than a contracting fibre. So, which is the more efficient way to perform a cyclic stretch-shortening movement; a muscle with a stiff tendon where all the work is done by the fibres, or a muscle with a compliant tendon, where much of the work (negative and positive) may be done by the tendon? The appropriate experimental evidence is outstanding.

2.4.5 FACTORS INFLUENCING METABOLIC BALANCE

Mature tendon is generally in a metabolic balance, with resorption and deposition of fibres occurring as ongoing processes. This balance can be altered, however, by a number of hormonal, mechanical, and drug-related factors.

Mechanically-induced changes in tendons have been investigated in many studies. If the Achilles tendon of a rabbit is cut, glycosaminoglycan content and fibroblast number increase, and the number of small collagen fibres increases (Flint, 1982). Interestingly, the area of the tendon closest to the muscle belly shows the greatest change, which suggests that tendon adaptation is region-specific. If the tendon is sutured so that some tensile load is transmitted, only minimal changes occur. Denervation, or the elimination of neural impulses to the muscle, eliminates the development of tension due to active muscle contraction and dramatically increases collagen turnover in the rat Achilles tendon (Klein et al., 1977) but produces no change in the relative amounts of the various tendon components. Tensile loads generated by the muscle, therefore, appear to exert an important influence over collagen turnover.

Changes in tendon attributed to increased loads have been equivocal, in part because the various exercise regimes that are typically used to increase tendon load are variable, and no doubt, result in different loads being applied to the tendon. Long-term running has been reported to increase the size and collagen content of swine extensor tendons but did not give a corresponding change in the flexor tendons (Woo et al., 1980 and 1982). This may reflect the different recruitment patterns or contractile properties of flexor and extensor muscles. In young mice, treadmill running increased the dry weight of the Achilles tendon (Kiiskinen, 1977), while in senescent mice, treadmill running or voluntary exercise appeared to slow the age-related loss of glycosaminoglycans (Viidik, 1979; Vailas et al., 1985). In growing chicks, strenuous exercise slowed the maturation of collagen by increasing the turnover of matrix and collagen (Curwin et al., 1988). The response of tendon to overload appears to be not only specific to the tendon but also to the relative maturation of the animal. Growth- and exercise-related stimuli may conflict when growing animals are exercised, which suggests that the stimuli responsible for adaptation may change during maturation.

The use of anabolic steroids to enhance athletic performance has become a common concern among sports medicine specialists. Laboratory investigations studying the effects of anabolic steroids on tendon biomechanics and biochemistry suggest that long-term steroid use, particularly in combination with exercise, decreases tendon stiffness (Wood et al., 1988) and ultimate strength by stimulating collagen degeneration (Michna and Stang-Voss, 1983) and the proliferation of small rather than large collagen fibrils. Furthermore, anabolic steroids impair the healing of tendon injuries (Kramhoft and Solgaard, 1986; Bach et al., 1987; Herrick and Herrick, 1987). Ironically, substances ingested to enhance performance may ultimately predispose individuals to problems that will force them out of competition.

2.4.6 FAILURE OF TENDON

The most common tendon injuries are lacerations, ruptures, and tendinitis. Lacerations and ruptures may be considered traumatic injuries, although a tendon may be predisposed to rupture by chronic tendinitis or repeated injury. Tendon healing can be divided into three basic phases: inflammation with cellular infiltration, proliferation of new matrix and collagen, and remodelling of the newly formed tissue. Maintaining the blood supply to the healing tendon is critical to prevent necrosis of the collagen. Another aspect of tendon healing that affects the long-term result, particularly in injuries to tendons of the hands, is the tendency for adhesions to form between the repair site and the tendon sheath that prevent the tendon from gliding easily during joint motion. Research suggests that suturing to maintain contact between the tendon ends, early protected passive motion, and the prevention of adhesions between the tendon and the overlying sheath is the combination of factors that provides the best healing result (Gelberman et al., 1988).

Tendinitis (tendon inflammation) is a more common tendon injury and can be quite disabling if it becomes chronic. Generally, tendinitis is described as an overuse syndrome, where the tendon is loaded repeatedly to the point that the normal repair of everyday microdamage cannot take place (Curwin and Stanish, 1984). When this happens, a typical healing response is initiated but the inflammatory stage persists. Chronic inflammation may ultimately lead to permanent scarring of the tendon. For the individual with tendinitis, pain is the most common symptom. Treatment consists of pain management with analgesics and anti-inflammatory agents to decrease the inflammation around the tendon. Other physical agents, such as ice or ultrasound, may be beneficial, particularly in the early stages. The insulting activity that initiated the process should also be curtailed until the symptoms have subsided. This may be difficult, particularly if the symptoms are due to a work-related, repetitive task. Tendinitis can have tremendous impact if typical daily and work-related activities must be limited to allow healing to occur.

2.4.7 MUSCLE-TENDON PROPERTIES

In section 2.4.4, the physical properties of isolated tendinous tissue were discussed, and in section 2.5.5, the physical properties of isolated muscle are presented. Muscle and tendon, however, are anatomically linked into an arrangement, whereby, the properties of tendon may influence the properties of muscle, and vice versa.

Two examples, on how tendon properties may influence the force producing properties of skeletal muscle, are discussed; the first example is aimed at illustrating the effects of tendon elasticity on the force-length property of muscle; and the second example provides a possible explanation for the well known phenomenon of force enhancement in muscle during a concentric contraction that is preceded by an eccentric contraction (e.g., Cavagna, 1970; Komi and Bosco, 1978).

Fig. 2.4.11a shows schematically the force-length property of a muscle-tendon complex. Length, on the horizontal axis, represents the distance from origin to insertion of the

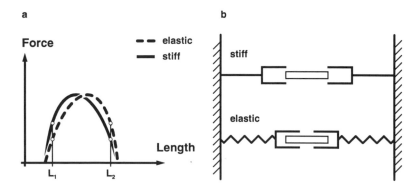

Figure 2.4.11   Schematic force-length relation of a muscle with a stiff tendon, and the same muscle with an elastic tendon (a), and a corresponding schematics of the contractile element and its associated (stiff and elastic) tendons (b). The length axis represents the muscle-tendon complex length (a).

muscle, thus, it includes the muscle belly length and the tendon length. The force-length relation shown with the solid line assumes a stiff (non-elastic) tendon; the force-length relation shown with the intermittent line assumes that the tendon is elastic.

For any given force, F > 0, the elastic tendon will be stretched in accordance with the stress-strain property of the tendon. Therefore, at a given muscle-tendon complex length (and F > 0), the contractile element of the muscle with the elastic tendon will always be shorter than the contractile element of the muscle with the stiff tendon. This result is schematically shown in Fig. 2.4.11b. A shorter contractile element length is associated with less force on the ascending limb (section 2.5.5) of the force-length relation (Fig. 2.4.11a, L_1), and is associated with a higher force on the descending limb (section 2.5.5) of the

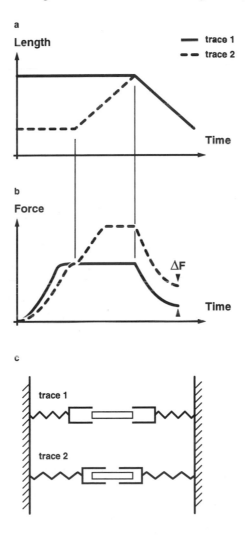

Figure 2.4.12   **Length-time (a), Force-time (b), and schematic contractile element and tendon lengths (c) of a muscle which contracts isometrically before shortening (trace 1), and the same muscle contracting eccentrically (stretching) before shortening (trace 2). Stimulation starts at time 0.**

force-length relation (Fig. 2.4.11a, L_2). Therefore, the force-length relation of the muscle with the elastic tendon, is shifted to the right on the length-axis relative to the corresponding relation of the muscle with the stiff tendon. This shift becomes increasingly more pronounced for a decreasing stiffness of the tendon in series with the muscle belly.

When a muscle-tendon complex is allowed to shorten after an initial isometric contraction (Fig. 2.4.12a, trace 1), it's force is known to decrease (Fig. 2.4.12b, trace 1) in accordance with the force-velocity property of skeletal muscle (Hill, 1938). If the same muscle is allowed to shorten in the same way as above, but the shortening phase is preceded by a stretch (rather than an isometric contraction, Fig. 2.4.12a, trace 2), the force just prior to shortening will be increased, and will remain higher during the shortening phase than the corresponding force of the shortening contraction that was preceded by an isometric contraction (Fig. 2.4.12b, trace 2). This increased force during muscle shortening has been associated with storage of elastic energy in the tendon during the stretching of the muscle (Cavagna et al., 1965; Asmussen and Bonde-Petersen, 1974), with a potentiation of the force on the cross-bridge level (Cavagna, 1977), and with a decreased velocity of shortening of the contractile elements during the shortening phase of a stretch-shorten cycle, compared to the velocity of shortening of the contractile elements during the shortening phase of an isometric shortening cycle (Hof et al., 1983; Alexander, 1988a). This last point is of particular interest when talking about muscle-tendon interactions.

Imagine that the muscle-tendon complex lengths of the isometrically contracting and the eccentrically (stretched) contracting muscle, are the same just prior to the shortening phase (Fig. 2.4.12a), however, the force in the stretched muscle is higher (Fig. 2.4.12b), which implies that the elongation of the tendon is more in the stretched, compared to the isometrically contracting muscle (Fig. 2.4.12c). During the shortening phase, the muscle-tendon shortening of both muscles is the same (Fig. 2.4.12a), but the decrease in force is typically much larger in the stretched, compared to the isometric muscle. Therefore, the shortening in the tendon will be more, and correspondingly, the average velocity of shortening in the contractile element will be smaller in the muscle that is stretched prior to shortening, compared to the muscle that is contracting isometrically before shortening. A smaller velocity of shortening, according to the force-velocity property (Hill, 1938), will be associated with larger forces, thus, the force enhancement, ΔF, shown in Fig. 2.4.12b may be partly explained.

2.4.8 REFERENCES

Alexander, R.McN., Maloiy, G.M.O., Ker, R.F., Jayes, A.S., and Warui, C.N. (1982) The Role of Tendon Elasticity in the Locomotion of the Camel (Camelus Dromedarius). *J. of Zool.* London. **198**, pp. 293-313.

Alexander, R.McN. (1983) *Animal Mechanics* (2nd Ed.). Oxford, Blackwell, UK.

Alexander, R.McN. (1984) Elastic Energy Stores in Running Vertebrates. *Amer. Zool.* **24**, pp. 85-94.

Alexander, R.McN. (1988a) The Spring in Your Step: The Role of Elastic Mechanisms in Human Running. *Biomechanics XI-A* (eds. De Groot, G., Hollander, A.P., Huijing, P.A., and van Ingen Schenau, G.J.). Free University Press, Amsterdam. pp. 17-25.

Alexander, R.McN. (1988b) *Elastic Mechanisms in Animal Movement.* Cambridge University Press, Cambridge, UK.

Asmussen, E. and Bonde-Petersen, F. (1974) Apparent Efficiency and Storage of Elastic Energy in Human Muscles During Exercise. *Acta. Physiol. Scand.* **92 (4)**, pp. 537-545.

Bach Jr., B.R., Warren, R.F., and Wickiewicz, T.L. (1987) Triceps Rupture: A Case Report and Literature Review. *Am. J. Sports Med.* **15 (3)**, pp. 285-289.

Bell, C. (1828) *Animal Mechanics.* Baldwin & Craddock, Edinburgh. pp. 64-72.

Benjamin, M., Evans, E.J., and Copp, L. (1986) The Histology of Tendon Attachments to Bone in Man. *J. Anat.* **149**, pp. 89-100.

Bennett, M.B., Ker, R.F., Dimery, N.J., and Alexander, R.McN. (1986) Mechanical Properties of Various Mammalian Tendons. *J. of Zool.* London. **209 (A)**, pp. 537-548.

Cavagna, G.A., Saibene, F.P., and Margaria, R. (1964) Mechanical Work in Running. *J. of Appl. Physiol.* **19**, pp. 249-256.

Cavagna, G.A., Saibene, F.P., and Margaria, R. (1965) Effect of Negative Work on the Amount of Positive Work Done by an Isolated Muscle. *J. Appl. Physiol.* **20 (1)**, pp. 157-158.

Cavagna, G.A. (1977) Storage and Utilization of Elastic Energy in Skeletal Muscle (ed. Hutton, R.S.). *Exercise and Sport Sci. Reviews.* **5**, pp. 89-129.

Cavagna, G.A. (1970) The Series Elastic Component of Frog Gastrocnemius. *J. Physiol.* **206**, pp. 257-262.

Cooper, R.R. and Misol, S. (1970) Tendon and Ligament Insertion: A Light and Electron Microscope Study. *J. Bone Jt. Surg.* **52 (A)**, pp. 1-20.

Curwin, S.L. and Stanish, W.D. (1984) *Tendinitis: Its Etiology and Treatment.* Collamore Press, Toronto.

Curwin, S.L., Vailas, A.C., and Wood, J. (1988) Immature Tendon Adaptation to Strenuous Exercise. *J. Appl. Physiol.* **65 (5)**, pp. 2297-2301.

Cutts, A., Alexander, R.McN., and Ker, R.F. (1991) Ratio of Cross-sectional Areas of Muscles and Their Tendons in a Healthy Human Forearm. *J. of Anat.* **176**, pp. 133-137.

Elliot, D.H. (1965) Structure and Function of Mammalian Tendon. *Biol. Rev.* **40**, pp. 392-421.

Flint, M. (1982) Interrelationships of Mucopolysaccharides and Collagen in Connective Tissue Remodelling. *J. Embryol. Exp. Morph.* **27**, pp. 481-495.

Frank, C.B. and Hart, D.A. (1990) The Biology of Tendons and Ligaments. *Biomechanics of Diathrodial Joints* (eds. Mow, V. C., Ratcliffe, A., and Woo, S.L.Y.). Springer Verlag, New York. pp. 39-62.

Gelberman, R.H., Goldberg, V.M., An, K.N., and Banes, A.J. (1988) Tendon. *Injury and Repair of the Musculo-skeletal Soft Tissues* (eds. Woo, S. L.Y. and Buckwalter, J. A.). American Academy of Orthopaedic Surgeons, Park Ridge, IL. pp. 5-40.

Gillard, G.C., Reilly, H.C., Bell-Booth, P G., and Flint, M.H. (1979) The Influence of Mechanical Forces on the Glycosaminoglycan Content of the Rabbit Flexor Digitorum Profundus Tendon. *Connect. Tissue Res.* **7**, pp. 37-46.

Herrick, R. and Herrick, S. (1987) Ruptured Triceps in a Powerlifter Presenting as Cubital Tunnel Syndrome: A Case Report. *Am. J. Sports Med.* **15 (5)**, pp. 514-516.

Hill, A.V. (1938) The Heat of Shortening and the Dynamic Constants of Muscle. *Proc. Royal Soc.* London. **126 (B)**, pp. 136-195.

Hof, A.L., Goelen, B.A., and van den Berg, J. (1983) Calf Muscle Moment, Work and Efficiency in Level Walking; Roles of Series Elasticity. *J. Biomechanics.* **16 (7)**, pp. 523-537.

Hooley, C.J. and McCrum, N.G. (1980) The Viscoelastic Deformation of Tendon. *J. Biomechanics.* **13 (6)**, pp. 521-528.

Jozsa, L., Kannus, P., Balint, J.B., and Reffy, A. (1991) Three-dimensional Ultrastructure of Human Tendons. *Acta. Anat.* **142**, pp. 306-312.

Kastelic, J., Galeski, A., and Baer, E. (1978) The Multicomposite Structure of Tendon. *Connect. Tissue Res.* **6**, pp. 11-23.

Ker, R.F. (1981) Dynamic Tensile Properties of Sheep Plantaris Tendon (Ovis Aries). *J. of Exp. Biol.* **93**, pp. 283-302.

Ker, R.F., Dimery, N.J., and Alexander, R.McN. (1986) The Role of Tendon Elasticity in a Hopping Wallaby (Macropus Rufogriseus). *J. of Zool.* London. **208 (A)**, pp. 417-428.

Ker, R.F., Alexander, R.McN., and Bennett, M.B. (1988) Why are Mammalian Tendons so Thick? *J. of Zool.* London. **216**, pp. 309-324.

Kiiskinen, A. (1977) Physical Training and Connective Tissues in Young Mice - Physical Properties of Achilles Tendons and Long Bones. *Growth.* **41**, pp. 123-137.

Klein, L., Dawson, M.H., and Heiple, K.G. (1977) Turnover of Collagen in the Adult Rat after Denervation. *J. Bone Jt. Surg.* **59 (A)**.

Köelliker, R.A. (1850) *Mikroskopische Anatomie oder Gewebelehre des Menschen.* Leipzig.

Köelliker, R.A. (1855) *Handbuch der Gewebelehre des Menschen.* Leipzig.

Komi, P.V. and Bosco, C. (1978) Utilization of Stored Elastic Energy in Leg Extensor Muscles by Men and Women. *Med. Science Sports Exerc.* **10**, pp. 261-265.

Kramhoft, M. and Solgaard, S. (1986) Spontaneous Rupture of the Extensor Pollicis Longus Tendon after Anabolic Steroids. *J. Hand Surg.* **11 (B)**.

Lundborg, G., Holm, S., and Myrhage, R. (1980) The Role of the Synovial Fluid and Tendon Sheath for Flexor Tendon Nutrition: An Experimental Tracer Study on Diffusional Pathways in Dogs. *Scand. J. Plast. Reconstr. Surg.* **14**, pp. 99-107.

Lundborg, G. and Rank, F. (1978) Experimental Intrinsic Healing of Flexor Tendons Based upon Synovial Fluid Nutrition. *J. Hand Surg.* **3**, pp. 21-31.

Manske, P.R. and Lesker, P.A. (1982) Nutrient Pathways of Flexor Tendons in Primates. *J. Hand Surg.* **7**, pp. 436-457.

Merrilees, M.J. and Flint, M.H. (1980) Ultrastructural Study of Tension and Pressure Zones in a Rabbit Flexor Tendon. *Am. J. Anat.* **157**, pp. 87-106.

Michna, H. and Stang-Voss, C. (1983) The Predisposition to Tendon Rupture After Doping with Anabolic Steroids. *Int. J. Sports Med.* **4**, p. 59.

Noyes, F.R., DeLucas, J.L., and Torvik, P.J. (1974) Biomechanics of Anterior Cruciate Ligament Failure: An Analysis of Strain-rate Sensitivity and Mechanisms of Failure in Primates. *J. Bone Jt. Surg.* **56 (A-2)**, pp. 236-253.

Pollock, C. (1991) *Body Mass and Elastic Strain in Tendons.* M.Sc. Thesis, University of Calgary. Nat'l Library of Canada, Ottawa.

Rigby, B.J., Hirai, N., Spikes, J.D., and Eyring, H. (1959) The Mechanical Properties of Rat Tail Tendon. *J. of Gen. Physiol.* **43**, pp. 265-283.

Rollhäuser, H. (1950) Konstruktions- und Altersunterschiede in Festigkeit kollagener Fibrillen. *Gegenbauers Morph. Jahrb.* **90**, pp. 157-179.

Rowe, R.W.D. (1985) The Structure of Rat Tail Tendon. *Connect. Tissue Res.* **14**, pp. 9-20.

Scapinelli, R. and Little, K. (1970). Observations on the Mechanically Induced Differentiation of Cartilage from Fibrous Connective Tissue. *J. Pathol.* **101**, pp. 85-91.

Schmidt, R.F. (1978) Motor Systems. *Fundamentals of Neurophysiology* (ed. Schmidt, R.F.). Springer Verlag, New York. pp. 158-161.

Tidball, J.G. (1983) The Geometry of Actin Filament-membrane Associations can Modify Adhesive Strength of the Myotendinous Junction. *Cell Motility.* **3**, pp. 439-447.

Tidball, J.G. (1984) Myotendinous Junction: Morphological Changes and Mechanical Failure Associated with Muscle Cell Atrophy. *Exp. and Mole. Pathol.* **40**, pp. 1-12.

Trotter, J.A., Hsi, K., and Samora, A. (1985) A Morphometric Analysis of the Muscle-tendon Junction. *Anat. Rec.* **213**, pp. 26-32.

Vailas, A.C., Pedrini, V.A., Pedrini-Mille, A., and Holloszy, J.O. (1985) Patellar Tendon Matrix Changes Associated with Aging and Voluntary Exercise. *J. Appl. Physiol.* **58**, pp. 1572-1576.

van Ingen Schenau, G.J. (1984) An Alternative View of the Concept of Utilisation of Elastic Energy in Human Movement. *Human Movement Science.* **3**, pp. 301-336.

Viidik, A. (1979) Connective Tissues: Possible Implications of the Temporal Changes for the Aging Processes. *Mech. Aging Dev.* **9**, pp. 267-285.

Viidik, A. (1990) Structure and Function of Normal and Healing Tendons and Ligaments. *Biomechanics of Diathrodial Joints* (eds. Mow, V. C., Ratcliffe, A., and Woo, S.L.Y.). Springer Verlag, New York. pp. 3-38.

Wainwright, S.A., Biggs, W.D., Currey, J.D., and Gosline, J.M. (1982) *Mechanical Design in Organisms.* Princeton University Press, Princeton, NJ. pp. 88-93.

Wertheim, M.G. (1847) Mémoire sur l'élasticité et la Cohésion des Principaux Tissues du Corps Humain. *Ann. Chim. (Phys.)* **21**, pp. 385-414.

Wickoff, R.G.W. (1949) *Electron Microscopy: Techniques and Applications.* Interscience Publications, New York.

Woo, S.L.Y., Gomez, M.A., Woo, Y.K., and Akeson, W.H. (1982) Mechanical Properties of Tendons and Ligaments: II. The Relationships of Immobilization and Exercise on Tissue Remodelling. *Biorheology.* **19**, pp. 397-408.

Woo, S.L.Y., Maynard, J., Butler, D.L., Lyon, R., Torzilli, P.A., Akeson, W.H., Cooper, R. R., and Oakes, B. (1988) Ligament, Tendon, and Joint Capsule Insertions into Bone. *Injury and Repair of the Musculo-skeletal Soft Tissues* (eds. Woo, S.L.Y. and Buckwalter, J.A.). American Academy of Orthopaedic Surgeons, Park Ridge, IL. pp. 133-166.

Woo, S.L.Y., Ritter, M.A., and Amiel, D. (1980) The Biomechanical and Biochemical Properties of Swine Tendons: Long Term Effects of Exercise on the Digital Extensors. *Connect. Tissue Res.* **7,** pp. 177-183.

Wood, T.O., Cooke, P.H., and Goodship, A.E. (1988) The Effect of Exercise and Anabolic Steroids on the Mechanical Properties and Crimp Morphology of the Rat Tendon. *Am. J. Sports Med.* **16,** pp. 153-158.

Zajac, F.E. (1989) Muscle and Tendon: Properties, Models, Scaling, and Application to Biomechanics and Motor Control. *Crit. Rev. Biomed. Eng.* **17 (4),** pp. 359-411.

2.5 MUSCLE

HERZOG, W.

2.5.1 DEFINITIONS AND COMMENTS

Golgi tendon organ:

A sensory organ consisting of nerve fibres located near the muscle-tendon junction. Golgi tendon organs lie in series with the muscle fibres and are sensitive to tension generated in the muscle.

Isokinetic contraction:

A contraction of a muscle in which the rate of change in length (or the speed of contraction) is constant. The rate of change in length (or the speed) referred to may be associated with the joint angle, the muscle-tendon complex, or the sarcomere, depending on the context.

Isometric contraction:

A contraction of a muscle in which length remains constant. The length referred to may be that of the muscle-tendon complex, the fibre, or the sarcomere, depending on the context.

Isotonic contraction:

A contraction of a muscle in which the resistive force is kept constant throughout the contraction.

Motor neuron:

A neural cell (neuron) that conveys impulses that initiate muscular contraction.

Motor unit:

A set of muscle fibres that are innervated by the same motor neuron. It represents the smallest unit of control of muscular force.

Muscle:

(*L. musculus*) A type of tissue composed of contractile cells or fibres, which effects movement of an organ or part of the body. The outstanding characteristic of muscular tissue is its ability to shorten or contract (Taber's Cyclopedic Medical Dictionary, 1981).

Muscle fibres:

Threadlike cells of the muscle.

Muscle spindle:

Sensory organ consisting of adapted muscle fibres enclosed in a capsule and lying parallel to the normal muscle fibres. Muscle spindles are sensitive to length changes and the rate of change in length of muscle fibres.

Myofibril:

A tiny fibril containing myofilaments that runs parallel to the long axis of the fibre (cell).

Neuromuscular junction:

Junction where the motor neuron meets the muscle and where impulses from the motor neuron are transmitted to the muscle.

Sarcomere:	The basic contractile unit of striated muscle. It contains the region of a myofibril that lies between two adjacent dark lines called the Z-lines.
Thin filament:	Filament predominantly made up of actin, tropomyosin, and troponin proteins.
Thick filament:	Filament predominantly made up of myosin proteins.

2.5.2 SELECTED HISTORICAL HIGHLIGHTS

Milon of Kroton was probably the most famous athlete of the ancient Olympic games, winning 6 Olympic titles in wrestling in an athletic career that spanned more than 30 years. His incredible feats of strength are told in many stories, and one of these describes his exercising of the body by carrying a young calf every day until it had grown up. In our present vocabulary of strength training, we may consider Milon of Kroton the first documented athlete who used variable resistance strength training, since the calf grew heavier as he became stronger.

384-322 B.C. Aristotle	The origins of discovery of muscles as the organ of force and movement production may be found in ancient Greece. Aristotle (De Motu Animalium) describes the interrelation of breath, brain, and blood vessels in the production of movement. According to him, movement was associated with thrusting and pulling, and thus, the organ producing movement must be capable of contracting and expanding, like the pneuma (spirits) (Farquharson, 1912; Needham, 1971).
129-201 Galen	The discovery that muscles are the true organs of voluntary movements must be accredited to Galen (De Tremore). Galen was born in Pergamon, the son of a mathematician. He received a broad philosophical education and started to study medicine at the age of 16, first in his home town, then in Smyrna and Alexandria. His teachings and writings of medicine surpassed those of previous physicians and remained the foundation of medical practice for the following one and a half millennia. His detailed dissection of muscles, and his discovery that arteries contain blood (not air or spirits) were considered the first attempts of establishing a "science of muscles" (myology). It took the Renaissance to revive the sciences and the interest in how the human body functions (Daremberg, 1854-1857; Needham, 1971).
1543 Vesalius	(De Humano Corporis Fabrica) Was the first to break away systematically from the medical teachings of Galen. He discovered that the contractile power

resides in the actual muscle substance, and he identified individual structural components of muscles. About 100 years after Vesalius, there were a series of scientists concerned with the functioning of muscles.

1663	Swammerdam	Constancy of muscular volume during contractions was supported by elegant experiments on frog and human muscles (Needham, 1971).
1664	Croone	(De Ratione Motus Musculorum) Concluded from nerve section experiments that the brain must send a signal to the muscles to cause contraction.
1664	Stensen	(De Musculis et Glandulis Observationem Specimen, 1664; and Elementorum Myologiae Specimen sen Musculi Descripto Geometrica, 1667) Gave precise descriptions of muscular structures, showed that muscle fibres connect to tendons, and stated that contractions may occur without changes in muscular volume (Kardel, 1990).
1680	Borelli	Following these works concerned with the structure and function of muscles, Borelli (De Motu Animalium) incorporated the muscles systematically into the skeletal system to study the effects of external and muscular forces on the way the body works.
1682	van Leeuwenhoek	The major steps in identifying the structure of muscles and developing corresponding theories of muscular contraction were associated with the discovery of the light and electron microscopes. Van Leeuwenhoek performed (light) microscopic examinations and discovered the cross-striation of skeletal muscles. His descriptions of muscular structure dominated the following century (Needham, 1971; Pollack, 1990).
1939	Engelhardt	At about the same time that electron microscopes were first used for systematic studies of the structure of muscles, researchers in muscle biochemistry were able to associate actin and myosin proteins with the thin and thick myofilaments, respectively, and most importantly, discover the ATPase activity of myosin (Engelhardt and Lyubimova, 1939).
1954	Huxley, A.F. Huxley, H.E.	These biochemical discoveries, together with the structural findings using electron-microscopy, led to the theory of sliding filaments during muscular contraction (A.F. Huxley and Niedergerke, 1954; H.E. Huxley and Hanson, 1954) and finally the Cross-bridge Theory (A.F. Huxley, 1957; A.F. Huxley and Simmons, 1971). The Cross-bridge Theory has

become the accepted paradigm for muscular force production, and with few exceptions, has not been challenged or questioned seriously in the past three decades.

2.5.3 INTRODUCTION

Probably the most basic property of muscle is its ability to produce force. However, despite centuries of research on muscle and its contractile behaviour, some aspects of muscular force production have still not been resolved. For example, the precise mechanism of cross-bridge attachment and cross-bridge movement that are believed to cause relative movements of the myofilaments, and so produce force, are not clearly understood. Some scientists (e.g., Iwazumi, 1978; Pollack, 1990) propose mechanisms of force production that do not agree with the most currently popular paradigm of muscular force production, the Cross-bridge Theory (Huxley, 1957; Huxley and Simmons, 1971). In this chapter, muscular force production and the mechanical properties of muscles will be associated with the assumptions and predictions underlying the Cross-bridge Theory. However, the reader should at all times be aware that not all of these assumptions and predictions have been tested and accepted unanimously in the scientific community.

Since this book is aimed at the student of biomechanics, muscles will be viewed from a mechanical point of view. However, because of its contractile properties, muscle is less easily associated with strictly mechanical properties than are bones or ligaments, for which force-elongation or stress-strain relations are well defined. Dealing with the mechanics of muscle, its physiological and biochemical properties must always be kept in mind.

Muscles exert force and produce movement and so may be considered the basic elements of movement mechanics in humans and animals. At the same time, force-time histories of muscles during movements are like small windows to the brain that may produce insight into the mechanisms of movement control. It is precisely the duality of mechanics and control function that splits biomechanists into two groups: those who study the force output of muscles to determine the movement and loading effects on skeletal systems, in particular on joints; and those who determine the force output of muscles for varying movement conditions to study the possible mechanisms that may be responsible for movement control.

Most researchers in muscle mechanics, be it on the molecular level of isolated myofilaments or the entire muscle level in an intact system, are interested in muscular force production and its effects on the biological system. Methods of muscle force measurements are well established for acute, in-vitro or in-situ experiments; however, they are much less established and are rarely used for chronic, in-vivo experiments. Walmsley et al. (1978) performed pioneering work by measuring multiple muscle forces (soleus and medial gastrocnemius) in the cat hindlimb for a variety of locomotor conditions. Similar experiments were performed later by Hodgson (1983) and Whiting et al. (1984), and all these experiments taken together gave a clear picture of the force-sharing between soleus and medial gastrocnemius muscles for different movement conditions. Most recently, direct muscle force measurements were expanded to all muscles comprised in a functional group (i.e., cat gastrocnemius, soleus, and plantaris muscles) and a corresponding functional antagonist (tibialis anterior), to expand the understanding of force-sharing among agonist and antagonist muscles for movements ranging from slow walking to trotting (Herzog and Leonard, 1991; Herzog et al., 1993).

Most often, however, intact in-vivo muscle forces are estimated indirectly, using either theoretical approaches (Crowninshield and Brand, 1981b) or electromyographical (EMG) recordings. For either approach, the muscle forces obtained must be considered estimates. Since direct muscle force measurements are rarely used and indirect determination of muscle forces are inaccurate at present, the area of research dealing with muscular force predictions, and interactions of agonist/antagonist muscles in intact biological systems is wide open.

In this chapter, muscle is viewed as a biomaterial. Interactions of muscles during movements will be discussed later in this chapter (section 2.5.6) and in chapter 4.5, in particular when discussing underdetermined mechanical systems.

2.5.4 MORPHOLOGY

Morphology is the science of structure and form without regard to function. It will become apparent in this chapter that muscle is a highly structured and organized material and that every structure and each organization may be associated with functions of the tissue. Thus, this section is the building block for section 2.5.5, where selected functional properties of skeletal muscles are discussed.

Generally, muscles are grouped into striated and non-striated muscles, and striated muscles are further subdivided into skeletal and cardiac muscles. Cardiac and non-striated muscles are controlled by the autonomic nervous system, and in contrast to skeletal muscles, are not under direct voluntary control. In biomechanics, we are mostly (but by no means exclusively) concerned with skeletal muscles, and so, we will concentrate on this type of muscle in this chapter. The reader should realize that, although some of the force-generating mechanisms are similar for the different muscle types, there are also significant differences among them. Therefore, when studying cardiac or non-striated muscle, further references dealing specifically with these types of muscles should be consulted.

Skeletal muscle may be thought of in structural units of decreasing size. The entire muscle is typically surrounded by a fascia and a further connective tissue sheath known as the epimysium, which consists of irregularly distributed collagenous, reticular, and elastic fibres, connective tissue cells, and fat cells (Fig. 2.5.1). The next smaller structure is the muscle bundle (or fascicle), which consists of a number of muscle fibres surrounded by a connective tissue sheath called perimysium. Then comes the muscle fibre, an individual muscle cell surrounded by the endomysium, a thin sheath of connective tissue that consists principally of reticular fibres, which binds individual fibres together within a fascicle. Muscle fibres are cells within a delicate membrane, the sarcolemma.

Muscle fibres are made up of myofibrils lying parallel to one another. The systematic arrangement of the myofibrils gives skeletal muscle its typical striated pattern, which is visible under the light microscope. The repeat unit in this pattern is called a sarcomere, and it is the basic contractile unit of a muscle. Sarcomeres are bordered by the so-called Z-lines (Zwischenscheibe) and contain thin (actin) and thick (myosin) filaments. These filaments, which are mostly made up of the protein molecules that give them their names, lie parallel to one another (Fig. 2.5.2).

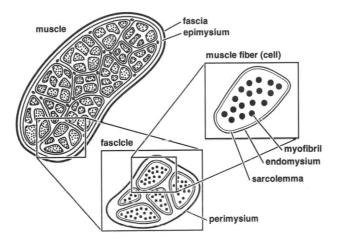

Figure 2.5.1   Schematic illustration of different structures and sub-structures of a muscle.

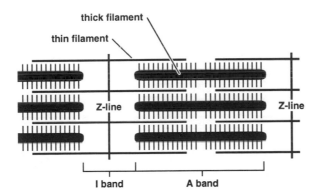

Figure 2.5.2   Schematic illustration of the basic contractile unit of the muscle, the sarcomere (from Pollack, 1990, with permission).

Thick filaments are typically located in the centre of the sarcomere. Some recent evidence, however, suggests that thick filaments may move from the centre towards the side of a sarcomere during prolonged contraction (Horowits, 1992). Thick filaments are responsible for the dark areas of the striation pattern, the so-called A-bands (A = anisotropic) (Fig. 2.5.2), and they are primarily composed of myosin molecules, a large protein. A myosin molecule contains a long tail portion, known as the light meromyosin, and a globular head attached to the tail, called heavy meromyosin (Fig. 2.5.3). The head portion extends outward from the thick filament. It contains a binding site for actin and an enzymatic site that catalyses the hydrolysis of ATP that releases the energy needed for muscular

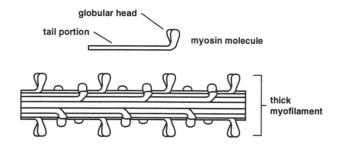

Figure 2.5.3   Schematic illustration of the thick myofilament (from Seeley et al., 1989, with permission).

contraction. The myosin molecules in each half of the thick filament are arranged in such a way that their tail ends are directed towards the centre of the filament. Therefore, the head portions are oriented in opposite directions for the two halves of the filament, and when forming cross-bridges (i.e., when the myosin heads attach to the thin filament) the myosin heads are thought to pull the actin filaments towards the centre of the sarcomere.

The cross-bridges on the thick filament are believed to be 14.3 nm apart in the longitudinal direction and 60° offset relative to one another (Fig. 2.5.4). Since cross-bridges are

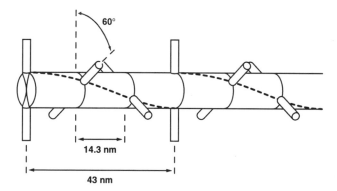

Figure 2.5.4   Schematic illustration of the arrangement of the cross-bridges on the thick filament (from Pollack, 1990, with permission).

thought to come in pairs offset by 180°, two cross-bridges with identical orientation are believed to be approximately 43 nm apart (3 x 14.3 nm).

The thin filaments are located on either side of the Z-lines within the sarcomeres (Fig. 2.5.2). They make up the light pattern of skeletal muscle striation, the so-called I-band (I = isotropic). The backbone of thin filaments is composed of two chains of serially-linked actin globules (Fig. 2.5.5). The diameter of each actin globule is about 5-6 nm. According to Pollack (1990), the strands of serially-linked globules cross over one another

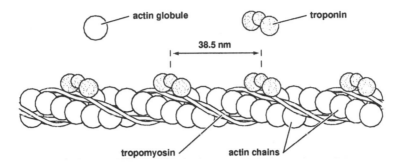

Figure 2.5.5   Schematic illustration of the thin myofilament, composed of two chains of serially-linked actin globules (from Seeley et al., 1989, with permission).

every 5 to 8 units in a somewhat random pattern. Thin filaments further contain tropomyosin, troponin, and "actin-binding proteins". Tropomyosin is a long, fibrous protein that is believed to lie in the groove formed by the actin chains (Fig. 2.5.5). Actin-binding proteins and troponin are located at intervals of 38.5 nm along the thin filament.

Thick filament lengths appear to be remarkably constant among animal species (≈ 1.6 μm); however, thin filament lengths have been reported to vary from 0.925 μm in frogs to 1.27 μm in humans (Walker and Schrodt, 1973). These lengths correspond to 24 and 33 periodicities of actin-binding proteins (38.5 nm intervals) on the thin filament, respectively. The influence of these differences in thin filament length on functional properties of the muscular tissue are discussed in the next chapter.

We have seen that skeletal muscle has an intriguing and systematic structural organization along the longitudinal axis of the muscle. This systematic organization is also observed in the cross-striation of the fibres and myofibrils, the regular arrangement of the cross-bridges on the thick filaments, and the periodicity of the troponin and actin-binding proteins on the thin filaments. The same regularity of structure may be observed in a cross-sectional plane (Fig. 2.5.6). Each thick filament within a myofibril is surrounded by 6 thin filaments in the area of filament overlap. The cross-sections of thick and thin filaments are approximately 12 nm and 7 nm, respectively, and the distance from thick to thick filament measures about 42 nm (Iwazumi, 1979).

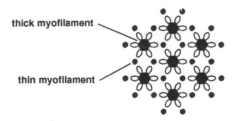

Figure 2.5.6   Schematic illustration of thick and thin filament arrangement in a cross-sectional view.

Motor units

Skeletal muscle is organized into motor units. A motor unit is defined as a set of muscle fibres that are all innervated by the same motor neuron. A small motor unit of a small muscle requiring extremely fine control may consist of a few muscle fibres only, while a motor unit of a large human skeletal muscle may contain more than 2000 individual muscle fibres. When a motor neuron is stimulated strongly enough to cause contraction, all fibres of the motor unit will contract. Since the force of contraction, among other parameters, depends on the number of fibres activated, a large motor unit (one that contains a large number of muscle fibres) can exert more force than a small motor unit.

In a detailed study of motor units in cat soleus (McPhedran et al., 1965) and cat medial gastrocnemius muscles (Wuerker et al., 1965), it was shown that there are large differences in motor unit size not only from one muscle to the other but also within the same muscle. In parallel with the differences in the size of motor units of a given muscle, there were differences in the contractile characteristics. Small motor units tended to be composed of slow twitch fibres and large units of fast twitch fibres, in particular in the mixed fibred medial gastrocnemius.

The observation that there are differences in motor unit size and that these differences are related to fibre type contents of a motor unit is interesting from a morphological point of view. It was in a series of two studies described by Henneman et al. (1965) and Henneman and Olson (1965) that the functional significance of this arrangement was revealed completely. These researchers showed that motor neurons with thin axons were the most excitable (i.e., were excited first in a graded muscular contraction) and belonged to small motor units (i.e., motor units with a small number of fibres, that were typically of the slow type), and motor neurons with thick axons were the least excitable and belonged to large motor units. From this observation they concluded that graded contractions of muscles are achieved by first recruiting small motor units and then progressively larger motor units as more force is required.

Most skeletal muscles are composed of slow and fast twitch motor units. The proportion of slow and fast motor units determines the force-velocity properties of a muscle (e.g., Hill, 1970). In terms of energy requirements, fast contractions are more costly than slow contractions (Hill, 1938). Therefore, the design and use of muscle may be seen to have taken into account the advantages and limitations of speed versus economy. In the cat triceps surae group this accounting has been accomplished by dividing the functional tasks among muscles. The soleus muscle is almost exclusively comprised of slow twitch fibres (Ariano et al., 1973) that are dominant in situations of low force requirements, whereas, the gastrocnemius muscle is composed of about 25% slow twitch and 75% fast twitch fibres, making it suitable to react to large and sudden force demands.

Fibre arrangement within muscle

Depending on the arrangement of fibres within a muscle, muscles are referred to as parallel-fibred, fusiform, or pennate. Among the pennate muscles, a further subdivision may be made to distinguish among unipennate, bipennate, and multipennate muscles (Fig. 2.5.7). In a parallel-fibred and fusiform muscle, fibres run parallel to the line of action of the entire muscle, which is typically defined as the line connecting the distal and proximal tendons. In a unipennate muscle, fibres run at a distinct angle to the line of action

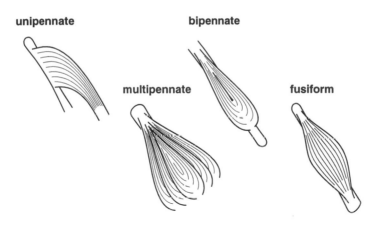

Figure 2.5.7   Classification of muscles into fusiform, unipennate, bipennate, and multipennate, depending on the arrangement of fibres within a muscle.

of the entire muscle, and all fibres are approximately parallel to one another. In a bi-(multi-) pennate muscle, fibres also run at a distinct angle to the line of action of the entire muscle, and there are two (or more) bundles of fibres that run in different directions. The different arrangements of fibres within a muscle influence some of the functional characteristics significantly, so an index of architecture was proposed to quantify the structure of a muscle. This index of architecture is defined as the ratio of muscle fibre to muscle belly length at (an assumed) optimal length of all fibres (Woittiez et al., 1984).

The contractile properties of a muscle as a function of the index of architecture are described here briefly in the form of a small example. Imagine that you have a certain volume of muscle available that you may fill with contractile material. Assuming that all sarcomeres take up the same volume, there is a precisely defined number of sarcomeres that may be used to fill the available volume. However, these sarcomeres may be assorted in series so that they form long or short fibres. Long fibres, in contrast to short fibres, take up a large volume, and then, the number of long fibres that may be arranged in parallel in a given muscle volume is smaller than the corresponding number of small fibres. Assuming further that all sarcomeres can shorten and elongate the same amount, the muscle with the long fibres can exert forces over a larger range of absolute muscle length than the short-fibred muscle. Assuming further that all sarcomeres produce the same amount of force, and that each fibre produces the same amount of force independent of its resting length, the short-fibred muscle will be stronger than the long-fibred muscle because of the larger number of fibres that it may accommodate in parallel (Fig. 2.5.8). Thus, the muscle depicted in Fig. 2.5.8a (long fibres) has a larger range of active force production but a smaller peak force potential than the muscle shown in Fig. 2.5.8b (short fibres). The maximal possible work the short- and long-fibred muscles can produce is the same, because they contain the same number of sarcomeres (equal volume), and each sarcomere can produce a precisely defined amount of work (Gordon et al., 1966).

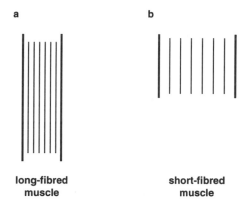

a   b

long-fibred
muscle

short-fibred
muscle

Figure 2.5.8   Comparison of long (a) and short (b) fibred muscles with respect to range of active force production and force potential.

Muscular contraction

Skeletal muscles contract in response to electrochemical stimuli. Specialized nerve cells, called motor neurons, propagate action potentials to skeletal muscle fibres. When reaching the muscle, the axons of motor neurons divide into small branches, each going to one muscle fibre. Typically, the motor neuron reaches a muscle fibre near its centre, where it forms the so-called neuromuscular junction or synapse (Fig. 2.5.9).

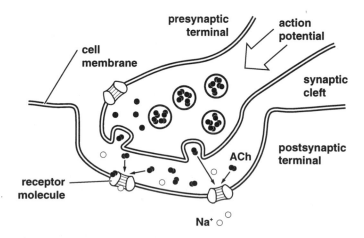

Figure 2.5.9   Neuromuscular junction with motor neuron and muscle cell membrane.

The neuromuscular junction is formed by an enlarged nerve terminal known as the presynaptic terminal that is embedded in small invaginations of the cell membrane, the motor endplate, or postsynaptic terminal. The space between presynaptic and postsynaptic terminal is the synaptic cleft.

When an action potential of a motor neuron reaches the presynaptic terminal, a series of chemical reactions take place that culminate in the release of acetylcholine (ACh) from synaptic vesicles located in the presynaptic terminal. Acetylcholine diffuses across the synaptic cleft, binds to receptor molecules of the membrane on the postsynaptic terminal, and causes an increase in permeability of the membrane to sodium (Na^+) ions. If the depolarization of the membrane due to sodium ion diffusion exceeds a critical threshold, then an action potential will travel along the stimulated muscle fibre. In order to prevent continuous stimulation of muscle fibres, acetylcholine is rapidly broken down into acetic acid and choline by acetylcholinesterase.

Excitation-contraction coupling

As an action potential of the motor neuron reaches the muscle fibres, depolarization of the fibre membrane occurs and reaches the interior of the muscle fibre at invaginations of the cell membrane called T-tubules. Depolarization causes the release of calcium (Ca^{2+}) ions from the sarcoplasmic reticulum into the sarcoplasm surrounding the myofibrils. Ca^{2+} ions are thought to bind to specialized binding sites on the troponin molecules of the thin myofilaments, and so, remove an inhibitory mechanism that prevents cross-bridge formations in the relaxed state. Cross-bridges then attach to the active sites of the thin filaments, and through the breakdown of ATP into ADP plus a phosphate ion, the necessary energy is provided to cause the cross-bridge head to move and so pull the thin filaments past the thick filaments. At the end of the cross-bridge movement, an ATP molecule is thought to attach to the myosin portion of the cross-bridge so that the cross-bridge can release from its attachment site, go back into its original configuration, and be ready for a new cycle of attachment. This cycle repeats itself as long as the muscle fibre is stimulated. When stimulation stops, Ca^{2+} ions are actively transported back into the sarcoplasmic reticulum, resulting in a decrease of Ca^{2+} ions in the sarcoplasm. As a consequence, Ca^{2+} ions diffuse away from the binding sites on the troponin molecule, and cross-bridge attachments are not possible in this state.

During the contractile process, thick and thin filaments are believed to remain at a constant length. Sarcomere, and thus fibre shortening, is associated with relative movements of the myofilaments. The relative movement of the thick and thin filaments during contraction causes shortening of the I-band of the sarcomere, whereas the A-band remains at a constant length.

Force production in a muscle is regulated in two basic ways. First, a muscle fibre or motor unit may be activated at different frequencies. When a single action potential activates a muscle fibre, force will rise and fall in a so-called twitch. Since muscle fibres, in contrast, for example, to nerve fibres, do not have a refractory period, a second stimulus may produce fibre contraction before the first twitch has subsided (Fig. 2.5.10). The higher the frequency of stimulation, the more force is produced, up to a level where a fibre contracts tetanically and further increases in the stimulation frequency do not cause a corresponding increase in muscular force. Tetanic contractions of slow-twitch fibres in cat soleus muscle are achieved at frequencies of about 25-30 Hz, whereas frequencies of 60-80 Hz may be required to produce a tetanic contraction in a fast-twitch fibre in cat gastrocnemius muscle.

The second way of increasing muscular force is by recruiting an increasingly larger number of motor units. During a graded muscular contraction only a few, small motor

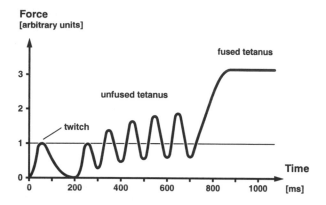

Figure 2.5.10   Schematic representation of force production regulated by changes in the frequency of stimulation of muscle fibres or motor units.

units are activated initially. As muscular force increases, more and more progressively larger motor units are recruited. In a normal muscular contraction, force regulation occurs using frequency control of motor units and changing numbers of activated motor units simultaneously.

Sensory organs

In order to understand the local control of motor neurons, it is necessary to describe the muscle sensory organs that influence motor neurons. First, however, it is necessary to understand that local control is not only regulated by the higher levels sending messages from the brain to the local level, but also by afferent fibres in muscles, tendons, skin, and joints, which may activate or inhibit motor neurons. The synaptic input to motor neurons, be it from afferent neurons or the descending pathways, is typically not direct but is channelled through interneurons. These interneurons may be taken as switches that can turn signals on or off. Afferent fibres from muscle sensory organs connect with local interneurons, with the exception of the stretch reflex. Information carried by these fibres then influences the action of (a) the muscle from which the afferent information is received, (b) corresponding agonistic, and (c) corresponding antagonistic muscles.

The first sensory organ to be discussed is the muscle spindle. Muscles spindles consist of endings of afferent nerve fibres that are wrapped around modified muscle fibres, so-called spindle or intrafusal muscle fibres (Fig. 2.5.11). Typically, several fibres are enclosed in a connective tissue capsule that makes up the muscle spindle. Muscle spindles are arranged in parallel with the muscle fibres and respond to stretch and the speed of stretch of the muscle.

When a muscle is stretched, the afferent fibres of the muscle spindle send a signal to the central nervous system. This signal is divided into different branches that can take several pathways. One pathway directly stimulates the motor neurons going back to the muscle that was stretched. This particular pathway is known as the stretch reflex arc. A second pathway for the branches of the afferent fibres from muscle spindles connects with inter-

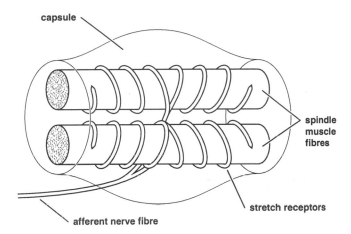

Figure 2.5.11 **Schematic structure of a muscle spindle.**

neurons that activate motor neurons of agonistic muscles and inhibit motor neurons of antagonistic muscles of the stretched muscle. A final pathway synapses with interneurons that convey information to centres of motor control in the brain.

Muscle spindles are able to contract in the end regions and can shorten at the same time when the muscle is shortening. This arrangement allows for transmission of information of muscle length and speed of contraction at any time to higher centres of motor control.

The fibres of the muscle spindle have separate motor neurons from the actual muscle fibres, i.e., the extrafusal fibres. The motor neurons controlling the muscle fibres are larger and are called alpha motor neurons, whereas, the motor neurons that innervate the muscle spindles are known as gamma motor neurons. The interaction of the alpha and gamma systems during muscle shortening is important, because the central or sensory part of the intrafusal fibres must not become slack at any time. One way of preventing intrafusal fibres from becoming slack during normal movement is to link the activation of the alpha- and gamma-motor neurons of the same muscle (e.g., Vallbo et al., 1979). However, experimental evidence obtained in the past decade suggests that there is a distinct difference between the control of gamma fusimotor neurons and alpha-motor neurons (e.g., Loeb, 1984; Murphy et al., 1984; Prochazka et al., 1985).

The second sensory organ associated tightly with muscular control is the golgi tendon organ. Golgi tendon organs are located near the muscle-tendon junctions. They consist of afferent nerve fibres wrapped around collagen bundles in the tendon, and they monitor tension. When the muscle contracts, forces are transmitted through the tendon, and afferent signals from the golgi tendon organs are transmitted to interneurons on the spinal level. Golgi tendon signals supply the motor control centres in the brain with continuous information about muscular tension. Furthermore, branches of the afferent neuron inhibit motor neurons of the contracting muscle and activate motor neurons of antagonistic muscles, thus, releasing force on tendons. This mechanism protects the muscle-tendon complex from injuries caused by excessive contractile forces.

2.5.5 PHYSICAL PROPERTIES

Since muscles are active force-producing structures, one may argue that they do not have unique material or mechanical properties, or if basic material properties such as force-elongation relations are measured passively, they are typically not very meaningful for the understanding of muscular function. In contrast, the force-elongation properties of passive musculo-skeletal structures such as ligaments, bones, and cartilage are essential to an understanding of their functions in the intact biological system. Nevertheless, two properties of muscles are repeatedly used in biomechanical experiments involving muscles or the musculo-skeletal system. These properties are the force-length and the force-velocity relation of muscles, and they are discussed here.

Force-length and force-velocity relations of skeletal muscular tissues have been determined on the sarcomere, the isolated fibre, the isolated muscle, and the intact muscle level; and depending on the level of interest, these relations may have to be interpreted differently. Furthermore, the terms "force-length" and "force-velocity" relations suggest an experimental procedure or a theoretical thinking governed by defined conditions. For example, it is implied here that force-length relations of a muscle are obtained under isometric conditions with the muscle maximally activated.

Maximal muscular activation is a term that may have to be treated quite liberally in this context. In an in-vitro preparation of a single muscle fibre, activation can be adjusted until maximal activation (i.e., the highest possible force) is achieved. In experiments involving intact human skeletal muscles, maximal activation often is associated with maximal voluntary efforts, which in the strictest sense are most likely not maximal in terms of absolute muscular activation.

Force-length and force-velocity properties differ dramatically among muscles. It is speculated that these differences are reflections of the functional demands imposed on muscles during everyday activities. On occasion, therefore, we discuss muscular properties as they relate to the functional demands, and go somewhat beyond just describing the material properties of muscular tissue.

FORCE-LENGTH RELATION

Force-length relations describe the relation between the maximal force a muscle (or fibre, or sarcomere) can exert and its length. Force-length relations are obtained under isometric conditions and for maximal activation of the muscle. Isometric may refer to the length of the entire muscle, the length of a fibre, or the length of a sarcomere, depending on the system that is studied.

Blix (1894) described almost a century ago that the force a muscle can exert depends on its length. In 1966, Gordon, A.F. Huxley, and Julian published the results of a classic study in which they showed the dependence of force production in isolated fibres of frog skeletal muscle on sarcomere length. Their results were in close agreement with theoretical predictions that were based on the Cross-bridge Theory, and they helped to establish the Cross-bridge Theory as the primary paradigm to describe muscular force production.

According to the Cross-bridge Theory, cross-bridges extend from thick to thin filaments and cause sliding of the myofilaments past one another. Each cross-bridge is assumed to generate the same amount of force and work independently of the remaining cross-bridges. Since the cross-bridges are believed to be arranged at equal distances along the thick filament, overlap between thick and thin filaments determines the number of possible cross-bridge formations, and thus, the total force that may be exerted.

For frog skeletal muscle, thick and thin filament lengths are reported to be about 1.6 µm and 0.95 µm, respectively (Page and H.E. Huxley, 1963; Walker and Schrodt, 1973). If the width of the Z-disc is about 0.1 µm and the H-zone (cross-bridge free zone in the middle of the thick filament) is 0.17 µm, a theoretical force-length relation for frog sarcomeres may be calculated (Fig. 2.5.12). At long sarcomere lengths, thick and thin filaments cease to overlap and no cross-bridges can be formed. The corresponding force must be zero. For frog striated muscle, zero force is reached at a sarcomere length of about 3.6 µm (= thick filament length (1.6 µm), plus twice the thin filament length (1.9 µm), plus the width of the Z-disc (0.1 µm)). This sarcomere length is indicated in Fig. 2.5.12 with label 5.

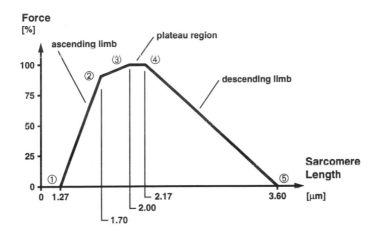

Figure 2.5.12   **Theoretical force-length relation of isolated fibres of frog skeletal muscle.**

Shortening of the sarcomeres increases the number of potential cross-bridge formations in a linear fashion with sarcomere length, or correspondingly, thick and thin filament overlap, until a maximal number of cross-bridge formations are possible (Fig. 2.5.12, label 4). This corresponds to a sarcomere length of 2.17 µm in frog muscle (i.e., twice the thin filament length (1.9 µm), plus the width of the Z-disc (0.1 µm), plus the width of the H-zone (0.17 µm)). Further sarcomere shortening to 2.0 µm (Fig. 2.5.12, label 3; twice the thin filament length (1.9 µm), plus the width of the Z-disc (0.1 µm)) increases the area of overlap between thick and thin myofilaments but does not increase the number of possible cross-bridge formations, since the middle part of the thick filament does not contain cross-bridges. Therefore, the force remains constant between 2.0 and 2.17 µm.

Sarcomere shortening below 2.0 µm has been associated with a decrease in force caused by interference of thin myofilaments as they start to overlap. Below 1.7 µm (Fig. 2.5.12, label 2; thick filament length (1.6 µm), plus width of Z-disc (0.1 µm)) the rate of decrease in force becomes higher than between 1.7 and 2.0 µm. This steeper decline in force for a given amount of sarcomere shortening has traditionally been associated with the force required to deform the thick filament. At a sarcomere length of 1.27 μm, forces determined experimentally in frog striated muscle became zero (Gordon et al., 1966).

Probably the most important support for the Cross-bridge Theory was the linear relation between force and length for sarcomere lengths between 2.17 and 3.6 μm (Fig. 2.5.12). However, the strict linearity of this relation was questioned by several investigators, who showed non-linear force-length behaviour on this so-called descending limb of the force-length relation (e.g., ter Keurs et al., 1978). The difference between studies showing linear and non-linear force-length behaviour was in the length control of the sarcomeres. Those studies showing a linear relation kept sarcomeres at a constant (controlled) length, whereas, those showing a non-linear relation kept fibre length at a constant length but allowed for non-uniform changes in sarcomere length. This latter situation appears to approach actual physiologic conditions more appropriately, and thus, may be more relevant in studying intact skeletal muscles. The mechanisms that allow for the enhancement of fibre force when sarcomere lengths are changing in an isometric fibre preparation have not been elucidated in detail.

Traditionally, the decrease in external force of frog skeletal muscle below 2.0 μm has been associated with the double overlap of thin myofilaments and the corresponding interference caused by this arrangement, and the decrease in force below 1.7 μm has been related to forces that are required to deform the thick myofilament. This view of force decrease on the so-called ascending limb of the force-length relation has been challenged by experiments that showed that Ca^{2+} release from the sarcoplasmic reticulum is length dependent. Ruedel and Taylor (1971) in experiments on skeletal muscle and Fabiato and Fabiato (1975) in experiments on cardiac muscle showed that Ca^{2+} release from the sarcoplasmic reticulum was decreased at muscle fibre lengths shorter than optimal length. Addition of caffeine to the activating solution enhanced Ca^{2+} release at short muscle fibre lengths and increased maximal force substantially. These results suggest that incomplete activation (i.e., decreased Ca^{2+} release from the sarcoplasmic reticulum) plays as much a role in decreased force production at fibre lengths below optimal length as the factors typically associated with this decrease that are based on the Cross-bridge Theory.

According to the Cross-bridge Theory, force-length relations may be determined mathematically if the lengths of thick and thin myofilaments are known. There is general agreement that thick myofilament lengths are nearly constant among many animal species (approximately 1.6 μm). However, thin myofilament lengths vary significantly among animals (Table 2.5.1) and sometimes even within the same animal. The influence of these differences in thin myofilament lengths on theoretically derived force-length properties of frog, cat, and human skeletal muscles is shown in Fig. 2.5.13. The plateau regions and descending limbs of these curves were obtained strictly according to predictions of the Cross-bridge Theory; the ascending limbs were determined assuming that interference of thin myofilaments at sarcomere lengths below the plateau region (Fig. 2.5.12) had the same effect on the rate of force decrease, and that zero force was reached at sarcomere lengths of 1.27 μm for all fibre preparations.

The plateau regions and the descending limbs (Fig. 2.5.12) are identical for frog, cat, and human skeletal muscles except for a shift along the sarcomere length axis. The width of the plateau corresponds to the width of the cross-bridge free zone in the middle of the thick myofilaments (i.e., the H-zone, 0.17 μm). The length of the descending limb corresponds to the length of the thick myofilament minus the H-zone (i.e., 1.60 μm - 0.17 μm = 1.43 μm) and is identical for all three muscles shown. The shift of the plateau and

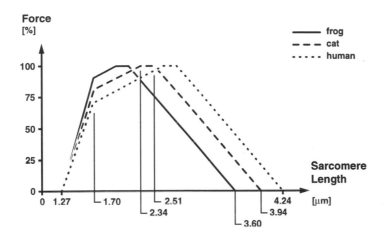

Figure 2.5.13   Influence of differences in thin myofilament lengths on theoretically derived force-length properties of frog, cat, and human skeletal muscles.

descending limb regions along the sarcomere length axis between muscles from different animals is caused by the differences in thin myofilament lengths and corresponds to twice the difference in thin myofilament length. For example, thin myofilament lengths in human and frog skeletal muscle differ by 0.32 μm (i.e., 1.27 μm - 0.95 μm, Table 2.5.1), causing a corresponding shift of 0.64 μm.

Table 2.5.1   Differences in thin myofilament length, L_{thin}, among animals from three different sources: Page and Huxley, 1963; Walker and Schrodt, 1973; and Herzog et al., 1992; with permission.

ANIMAL	L_{thin} PAGE & HUXLEY [μm]	L_{thin} WALKER & SCHRODT [μm]	L_{thin} HERZOG ET AL. [μm]
CAT	-	-	1.12
RAT	-	1.04	1.09
RABBIT	1.07	-	1.09
FROG	0.975	0.925	-
MONKEY	-	1.16	-
HUMAN	-	1.27	-

Differences in thin myofilament lengths among animals cause corresponding differences in force-length properties. If we assume that during evolution muscular systems have optimized their properties to satisfy everyday functional requirements, one may speculate on the differences of functional requirements between, for example, human and frog skeletal muscle. The range of active force production appears to be larger for human compared to frog skeletal muscle. From this observation, it may be speculated that fibre lengths in human skeletal muscles are shorter relative to their normal, everyday operating range than frog muscles, and that this "disadvantage" is partially offset by a sarcomere design that can produce active force over a large range of lengths. Support for this speculation was given in a study in which muscle fibre lengths of many animals were related to the animal's size. It was found that fibre lengths did not scale proportionally with animal size, but were relatively shorter for large animals compared to small animals (Pollock, 1991).

Another difference between human and frog sarcomere force-length relations is that the ascending limb (Fig. 2.5.13) of the frog curve is small compared to that for humans. Furthermore, there is a faster decrease in force per unit of sarcomere shortening for frog compared to human muscle. Therefore, it appears that human skeletal muscles may be better suited than frog muscles to operate on the ascending part of the force-length relation. Studies aimed at determining what part of the force-length relation is actually being used by a muscle during normal, everyday movement tasks are rare and cannot be used as conclusive support for the speculation made above. However, the little data that is available tends to fit the speculation. For example, Mai and Lieber (1990) reported that frog semitendinosus muscle works almost exclusively on the descending limb for a jumping movement (an everyday locomotor activity for a frog), whereas, we found force-length relations of intact human rectus femoris muscles that appear to be on the ascending and descending limbs of the force-length curves for anatomical ranges of knee and hip joint configurations (Herzog and ter Keurs, 1988). Intact human gastrocnemius muscle were found to operate virtually exclusively on the ascending limb of the force-length relation within the anatomical range of ankle and knee joint angles (Herzog et al., 1991a).

Force-length relations have typically been treated as a constant, invariant property of muscles. However, if we speculate that force-length properties have evolved according to the functional requirements imposed on muscles, it may be possible that such evolutions occur rapidly (i.e., within the lifetime of the animal of interest). Support for the notion that force-length properties are associated with functional demands may be found in studies in which force-length relations have been determined from muscles comprising a functional unit. Such muscles must satisfy similar functional demands, and thus, may be speculated to have similar force-length properties. Studies on the triceps surae and plantaris group of the striped skunk (Goslow and van de Graaf, 1982) and the cat (Herzog et al., 1992) show that the force-length properties of these muscles are similar when normalized to peak force and when expressed in terms of joint angles, thus, supporting the above hypothesis.

In a recent study, force-length properties of intact human rectus femoris muscles of high performance runners and cyclists were determined (Herzog et al., 1991b). Runners use the rectus femoris in an elongated position in training, compared to cyclists, because running is performed with the hip joint extended, whereas, cycling is performed, typically, with a greatly flexed hip joint. The corresponding knee angles in the two activities go through about the same range. Thus, the demands on the rectus femoris of a runner are substantially different than those on the same muscle for a cyclist. Corresponding differences were also found in the respective force-length properties. Rectus femoris muscles of runners tended to be strong at relatively long muscle lengths and weak at short muscle lengths, whereas, this muscle was weak at relatively long muscle length and strong at short muscle lengths in the cyclists. The differences were statistically significant despite the limited number of subjects (4 runners, 3 cyclists). It may, therefore, be speculated that the demands imposed by high performance training are sufficient to alter force-length properties of intact human skeletal muscles significantly.

Since the force-length properties of intact human skeletal muscles in the above study were obtained using maximal, voluntary effort contractions, it cannot be concluded from the data at hand whether the differences in force-length properties between the two groups of athletes were caused by neurophysiological or mechanical mechanisms. A possible mechanical mechanism could be the addition or deletion of sarcomeres in series in muscle fibres.

FORCE-VELOCITY RELATION

Force-velocity relations are defined here as the relation that exists between the maximal force of a muscle (or fibre) and its instantaneous rate of change in length. Force-velocity properties are determined for maximal activation conditions of the muscle, and are typically obtained at optimal length of the sarcomeres.

At the beginning of this century it was recognized that the efficiency of human movement varied as a function of movement speed (Laulanie, 1905, see Hill, 1970). At that time, it was found that for a given amount of work, the energy used (efficiency measure) increased with increasing speeds of muscular contraction (i.e., efficiency decreased). Fenn and Marsh (1935) were the first to perform experiments and report results on force-velocity properties of muscles, and their work was followed by the classic study of Hill (1938), who said himself that he "stumbled" over the force-velocity relation while working on the heat production of isolated frog skeletal muscle. Hill, and probably most muscle physiologists after him, tended to think of the force-velocity property of a muscle as applying a force to a muscle and measuring the corresponding velocity of shortening (i.e., equation (2.5.1)):

$$v = b(F_0 - F) / (F + a) \qquad (2.5.1)$$

where:

v = velocity of shortening
F_0 = maximal force at zero velocity and optimal sarcomere length
F = instantaneous force
a, b = constants with units of force and velocity, respectively

In biomechanics, we often think of the force-velocity relation as imposing the speed (independent variable) of movement and measuring the corresponding force (dependent variable). Many experiments on intact human skeletal muscles have been performed based on this idea using so-called "isokinetic" strength testing machines (e.g., Thorstensson et al., 1976; Perrine and Edgerton, 1978). In such cases, one may want to solve equation (2.5.1) for, F, to illustrate the fact that, v, is the independent and, F, the dependent variable (equation (2.5.2)):

$$F = (F_0 b - av) / (b + v) \qquad (2.5.2)$$

Setting, v, equal to zero in equation (2.5.2) corresponds to measuring the force under isometric conditions. For this situation F becomes equal to, F_0.

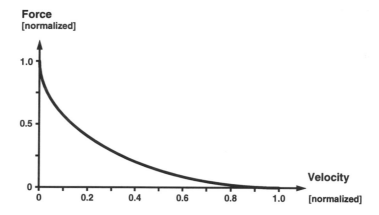

Figure 2.5.14   Normalized force-velocity relationship of concentrically contracting skeletal muscle.

Setting the external force that is acting on the muscle (F) equal to zero, we can solve equation (2.5.2) for, v, which under these circumstances corresponds to the maximal velocity of shortening (v_o):

$$v_o \quad = \quad b\,(F_o/a) \tag{2.5.3}$$

Equation (2.5.3) may be rearranged to yield:

$$a/F_o \quad = \quad b/v_o \tag{2.5.4}$$

Typical values for a/F_o have been reported to be about 0.25 for skeletal muscles from a variety of animals including frog (Hill, 1938), rat (Close, 1964), and kittens (Close and Hoh, 1967).

Equations (2.5.1) or (2.5.2) can typically be obtained easily for in-vitro fibre or muscle preparations by determining F_o, and then F and v, for a variety of different velocities of contraction. The constants "a" and "b" may then be determined in such a way that the corresponding equation gives a best fit to the experimental data.

In biomechanics, it is often of interest to describe force-velocity properties of intact human skeletal muscles. Since the experimental approach is limited for this situation, force-velocity relations may be obtained by first estimating F_o and v_o, and then solving equation (2.5.4) for the constants "a" and "b". Once "a" and "b" have been determined, equations (2.5.1) and (2.5.2) may be used with input of forces to calculate corresponding velocities or with input of velocities to calculate corresponding forces.

In order to estimate force-velocity properties of intact human skeletal muscle, it is necessary to know the physiological cross-sectional area (PCSA) and the average, optimal fibre length (l_o) of the muscle of interest, since these two values appear to be directly related to F_o and v_o, respectively.

Let us assume, for example, that we would like to estimate the force-velocity property of human vastus lateralis muscle and that average values from a variety of literature sources are 50 cm² for PCSA and 12 cm for l_0. Furthermore, research on mammalian skeletal muscles indicates that:

$$F_0 \approx 25 N/cm^2 \cdot PCSA \qquad (2.5.5)$$

and further, that:

$$v_0 \approx 6 \cdot l_0/s \qquad (2.5.6)$$

for muscles predominantly comprised of slow twitch fibres, and:

$$v_0 \approx 16 \cdot l_0/s \qquad (2.5.7)$$

for muscles predominantly comprised of fast twitch fibres (e.g., Spector et al., 1980). Therefore, $F_0 = 1250$ N and $v_0 = 72$ cm/s or 192 cm/s, depending on the fibre type content of the vastus lateralis. Since human skeletal muscles are typically of mixed fibre type composition, one may choose a statistical approach to derive force-velocity relations. Force-velocity properties of a muscle of mixed fibre composition may be calculated by separating the whole muscle into units of slow and fast fibres, or better still, into a continuum of fibres ranging in properties from slow to fast, and weighing their contribution to the total force-velocity behaviour according to information available on the distribution of fibre types within the muscle (Hill, 1970).

From the results of equations (2.5.5), (2.5.6), and (2.5.7), constants "a" and "b" may be determined using:

$$a/F_0 = b/v_0 = 0.25 \qquad (2.5.8)$$

for slow twitch fibres:

$$
\begin{aligned}
a &= 0.25 \cdot 1250 \text{ N} \\
b &= 0.25 \cdot 72 \text{ cm/s}
\end{aligned}
$$

for fast twitch fibres:

$$
\begin{aligned}
a &= 0.25 \cdot 1250 \text{ N} \\
b &= 0.25 \cdot 192 \text{ cm/s}
\end{aligned}
$$

Since Hill's equation (equation (2.5.1)) was originally derived at muscle temperatures of 0°C, the question arises whether equation (2.5.4) also holds for physiological muscle temperatures (i.e., 37°C in humans). Values for a/F_0 appear to be largely temperature independent (Hill, 1938), whereas "b" and "v_0" change as a function of temperature. As a first approximation, one may assume, however, that the ratio of b/v_0 remains about constant for a wide range of muscle temperatures.

Muscular power (P) is defined as the product of force and velocity ($F \cdot v$). Therefore, for a given force-velocity relation of a muscle, its instantaneous power as a function of velocity of contraction (P(v)) may be determined throughout the range of shortening speeds (Fig. 2.5.15). For many practical applications it is of interest to calculate at what speed of shortening, absolute maximal power, P_0, is reached.

By definition:

$$P(v) \quad = \quad F(v) \cdot v \tag{2.5.9}$$

where:

$$\frac{dP(v)}{dv} \quad = \quad \frac{dF}{dv}v \quad + \quad F(v) \tag{2.5.10}$$

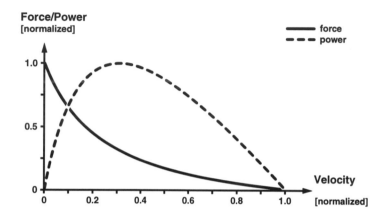

Figure 2.5.15   Schematic representation of the normalized force-velocity and the corresponding power-velocity relationships of skeletal muscle.

and using equation (2.5.2) yields:

$$\frac{dP(v)}{dv} \quad = \quad \frac{(F_o + a)\, b^2 - a\, (v + b)^2}{(v + b)^2} \tag{2.5.11}$$

Realizing that dP(v)/dv needs to be zero for P(v) to become maximal (i.e., P_o), we have:

$$0 \quad = \quad \frac{(F_o + a)\, b^2 - a\, (v + b)^2}{(v + b)^2} \tag{2.5.12}$$

Solving equation (2.5.12) for the velocity (v_m) at which P_o occurs yields:

$$v_m \quad = \quad b\,(\sqrt{(F_o/a) + 1} - 1) \tag{2.5.13}$$

Solving equation (2.5.4) for "a" and "b" and substituting into equation (2.5.13) gives:

$$v_m \quad = \quad \frac{v_o}{4}\,(\sqrt{4+1} - 1) \tag{2.5.14}$$

or:

$$v_m \quad\approx\quad 0.31\, v_o \tag{2.5.15}$$

which means that the speed of shortening at which maximal muscular power may be produced is about 31% of the maximal speed of shortening.

Substituting equation (2.5.14) into equation (2.5.2) and rearranging the terms, it is possible to calculate the force produced at the speed of shortening of v_m:

$$F(v_m) \quad=\quad \frac{F_o}{4} (\sqrt{4+1} - 1) \tag{2.5.16}$$

or:

$$F(v_m) \quad\approx\quad 0.31\, F_o \tag{2.5.17}$$

The maximum power (P_o) may then be determined using equations (2.5.9), (2.5.14), and (2.5.16):

$$P_o \quad=\quad \frac{(\sqrt{4+1} - 1)^2}{16} F_o \cdot v_o \tag{2.5.18}$$

or:

$$P_o \quad=\quad 0.095 \cdot F_o \cdot v_o \tag{2.5.19}$$

Hill's (1938) force-velocity relation has been used for the past 50 years. Recent evidence from experiments on single fibres of frog semitendinosus and tibialis anterior muscles suggests, however, that only the range of about 5-80% of the isometric force (F_o) may be approximated well with Hill's (1938) equation (Edman, 1979). Using Hill's equation, the maximal isometric force typically overestimates the actual value of F_o by about 25%, and the maximal velocity of shortening under zero load condition, using Hill's equation, underestimates the actual maximal shortening velocity considerably (Fig. 2.5.16). Since it is difficult to obtain v_o experimentally, mathematical approximations of the force-velocity properties of muscles have often been used to predict v_o theoretically. Based on the findings of Edman (1979), this approach may significantly underestimate the actual shortening velocity at zero load. The discrepancy might be accounted for by arguing that v_o obtained experimentally is strictly a measurement of the fastest contracting fibres in a entire muscle, whereas, the theoretically determined value for v_o represents an estimate of the average maximal velocity of shortening of many fibres in a muscle (Hill, 1970). However, this argument does not explain the discrepancy found by Edman (1979) on single fibre preparations.

The discrepancy between predicted and experimentally determined values for F_o in single fibre preparations has been explained by Edman (1979), as reflecting an inability of all possible cross-bridges to attach simultaneously during isometric contractions, because of the different repeat patterns of cross-bridge binding sites on the thin myofilaments and cross-bridges on the thick filaments. Edman further argues that this factor becomes less relevant at high velocities of shortening because cross-bridges, in this situation, attach and

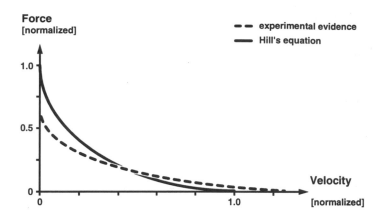

Figure 2.5.16   Schematic illustration of the difference between theoretical estimation and experimental determination (e.g., Edman, 1979) of the force-velocity relationships.

detach continuously. More attachment sites are therefore available for cross-bridges at high, compared to low (or zero), velocities of shortening.

The velocity of unloaded shortening (v_o) has been reported to remain virtually constant for a large range of sarcomere lengths (i.e., between about 1.65 and 2.7 μm for frog skeletal muscle, Edman, 1979). Since isometric tension varies considerably between 1.65 and 2.70 μm in frog skeletal muscles, this finding illustrates that v_o appears to be independent of the number of attached cross-bridges.

In frog skeletal muscle, maximal isometric force starts to decline at sarcomere lengths of about 2.0 μm (Fig. 2.5.12). Traditionally, it has been assumed that the decrease in force below sarcomere lengths of 2.0 μm was associated with some internal resistive force (e.g., Gordon et al., 1966). However, if this were the case, one would expect v_o to decrease below sarcomere lengths of 2.0 μm. It appears, therefore, that the loss of isometric force in frog skeletal muscle below 2.0 μm is not a consequence of internal resistive forces, but is caused by an inability of the contractile elements to produce forces as high as F_o. This view is supported by experiments on force-length properties of muscles in which Ca^{2+} release from the sarcoplasmic reticulum was enhanced artificially, and forces remained close to F_o for sarcomere lengths below 2.0 μm. This phenomenon is described in more detail earlier in this text.

Maximal velocity of shortening decreases sharply below a sarcomere length of 1.65 μm, and increases dramatically above a sarcomere length of about 2.70 μm. These two phenomena may safely be associated with increases in internal resistance because of thick filament deformation, and with increases in muscular force caused by parallel elastic elements, at the short and long sarcomere length, respectively.

Force-velocity properties are typically obtained at optimal sarcomere lengths. However, it has been suggested that Hill's (1938) equation may still be applied at sarcomere lengths other than those corresponding to optimal length, by replacing F_o with the maximal isometric force that corresponds to the sarcomere length at which the force-velocity properties are measured (e.g., Abbott and Wilkie, 1953). If F_o is replaced in this way, however, the velocity of unloaded shortening, according to equations (2.5.2) and (2.5.3), becomes:

$$v_o \quad = \quad bF_o l\,(x)\,/a \qquad\qquad (2.5.20)$$

where:

$l(x) \quad = \quad$ a value between 0 and 1.0, representing the normalized force as a function of sarcomere length

Since the value for "$l(x)$" in equation (2.5.20) decreases as soon as sarcomere lengths deviate from optimal length, and since "F_o", "b", and "a" are all constants, v_o will be smaller for non-optimal compared to optimal sarcomere lengths according to equation (2.5.20). However, according to Edman's (1979) experiments on single frog skeletal fibres, v_o is not influenced by sarcomere length for a range of 1.65-2.70 μm. Thus, rather than replacing F_o by the maximal isometric force at the sarcomere length of interest ($= l(x) \cdot F_o$), the entire right hand side of equation (2.5.2) may be multiplied by "$l(x)$":

$$F \quad = \quad [\,(F_o b - av)\,/\,(b+v)\,]\,l\,(x) \qquad\qquad (2.5.21)$$

This equation appears to approximate experimental observations better than the one suggested above. In particular, maximal velocity of unloaded shortening would remain as shown in equation (2.5.3), and the maximal isometric force at the sarcomere length of interest would become $F = F_o l(x)$.

When referring to "velocity of shortening" of a muscle or a fibre, it is typically implied that this is an average velocity. However, it has been argued that fibres in muscles may not shorten uniformly, and further, that sarcomeres within a fibre have a distinct maximal velocity of shortening (Edman and Reggiani, 1983). Therefore, the concept of uniform fibre or sarcomere shortening may not be adequate and may influence the force-velocity properties of a muscle. Further research in this area is needed.

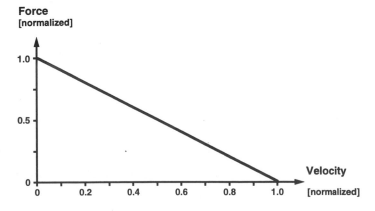

Figure 2.5.17   Force-velocity relation, expressing force as the force exerted per cross-bridge and velocity as the shortening velocity of a sarcomere.

In many instances, force-velocity properties of skeletal muscle are studied in order to explore mechanisms underlying muscular force production. However, forces produced by muscles are internal, molecular events, whereas, muscular forces are typically recorded (or applied) external to the muscle or fibre preparation. Furthermore, sarcomere shortening velocities along a fibre have been found to be non-uniform (Edman and Reggiani, 1983), and thus, measurements of shortening velocities of muscles and fibres represent an average of the velocities encountered on the sarcomere level. Most recently, ter Keurs and associates have investigated the force-velocity relation, expressing force as the force exerted per cross-bridge and the velocity as the shortening velocity of a sarcomere. They found that force and velocity, when expressed in this way, have a strong, inverse, linear relation (Fig. 2.5.17; ter Keurs, personal communication). It will be interesting to integrate the results of ter Keurs with the more traditional force-velocity relations.

2.5.6 APPLICATIONS

Muscle mechanics, or muscle physiology (depending on the preferred point of view) has countless practical applications. Experiments on muscles, single fibres, or even single myofilaments are often aimed at studying the mechanisms underlying muscular force production. Most researchers may consider this research muscle physiology, rather than muscle (or bio-) mechanics.

The field of muscle mechanics could be extremely useful in sports biomechanics for predicting optimal performances of athletes or improving the outcome of an athletic activity. It appears, however, that basic muscle mechanical approaches in sports biomechanics are virtually non-existent. There are, for example, a large number of papers related to the study of track and field, yet one would be hard-pressed to identify articles that attempt to maximize performances in track and field based on principles of muscle mechanics.

One of the few sports that has received systematic attention from a muscle mechanical point of view is bicycling. Bicycling is cyclic, virtually planar, and kinematically easy to describe; it offers itself better than most other sports to the study of muscle mechanics. Electromyographic work has given insight into muscular coordination for submaximal efforts of bicycling (e.g., Houtz and Fischer, 1959; Gregor et al., 1985; Hull and Jorge, 1985). Furthermore, muscular action, and in particular inter-action and co-contraction of two-joint antagonist pairs of muscles, has been studied using bicycling as a model (Andrews, 1987); and recently, the performance of maximal effort sprint bicycling was optimized based on principles of muscle mechanics (Yoshihuku and Herzog, 1990).

However, most of the applied studies in muscle mechanics appear to be aimed at two major problems: (1) the determination of loads acting on the musculo-skeletal system, in particular in joints of the human body; and (2) the study of muscular force interactions during movement.

DETERMINATION OF LOADS ACTING ON THE MUSCULO-SKELETAL SYSTEM

The precise mathematics underlying the determination of loads acting on the musculo-skeletal system will be discussed in a later chapter of this book (section 3.5). Here, it suffices to realize that one of the major forces influencing so-called internal forces (forces acting on the musculo-skeletal system, e.g., joint articular cartilage, ligaments, bones, etc.) are muscular forces. Paul (1965) and Morrison (1968) analysed joint articular forces in the knee and hip joint, respectively. They solved the so-called inverse dynamics approach (Andrews, 1974), calculated forces of entire functional groups of muscles, and from this information calculated joint articular forces.

These initial works on joint articular forces were followed by many others, which often used mathematical optimization theory to solve for forces in individual muscles rather than in entire functional groups of muscles (e.g., Seireg and Arvikar, 1973; Penrod et al., 1974; Crowninshield, 1978; Crowninshield and Brand, 1981a). Although individual muscle forces need not be known in order to calculate joint articular forces, they are required for determining local force and stress distributions in and around joints. Such information is typically relevant when approaching problems of artificial joint design, and tendon or ligament reconstructive surgery.

Since muscular forces cannot be measured directly in humans at present, studies predicting such forces are typically theoretical or based on indirect measures of muscular activation (i.e., electromyography). Most recently, we compared the predictions of a series of theoretical models aimed at estimating individual muscle forces to experimental muscle force measurements in an animal model, and concluded that none of the theoretical models agreed well with the experimental results (Herzog and Leonard, 1991). It will be a challenge for the future to derive models that accurately predict individual muscle forces in an intact biological system.

STUDIES OF MOVEMENT CONTROL

Knowing the force time history of a muscle (or even better, of all muscles in the system of interest) during movement, is like having a tiny window to the brain and its complex organization of voluntary movement. Thus, force-time histories of muscles have been predicted theoretically in the same way as described in the previous paragraphs, except with a different purpose in mind: the study of movement control (e.g., Pedotti et al., 1978; Dul et al., 1984; Herzog, 1987). Also, direct experimental force measurements have been performed in order to study interactions of muscles during movement. This type of work was pioneered by Walmsley et al. (1978), who studied the interaction between cat soleus and medial gastrocnemius muscles for a variety of locomotor activities. This study was followed by a series of similar experiments (e.g., Hodgson, 1983; Whiting et al., 1984), probably because these two muscles are easily accessible for chronic force measurements, and because the force-time behaviour of the one-joint, predominately slow twitch, parallel fibred soleus muscle turned out to be quite different (and thus interesting) from the force-time histories of the two-joint, predominately fast twitch, pennate fibred medial gastrocnemius muscle. Abraham and Loeb (1985) measured forces from cat hindlimb muscles other than just soleus and medial gastrocnemius, but never for more than two muscles simultaneously.

In order to gain further insight into the force-sharing behaviour of muscles during locomotion, we decided to study all the muscles in a functional group, plus one antagonist muscle, simultaneously. In our experimental protocol, forces from gastrocnemius, soleus, plantaris, and tibialis anterior muscles of the cat hindlimb were measured using standard tendon force transducers (Herzog and Leonard, 1991; Herzog et al., 1992; Herzog et al., 1993). The first three of these muscles make up the Achilles tendon, and so are the major plantar flexors of the ankle joint in the cat; the tibialis anterior is one of the primary dorsi-flexors of the cat ankle joint and may be considered a direct antagonist of the one-joint soleus muscle.

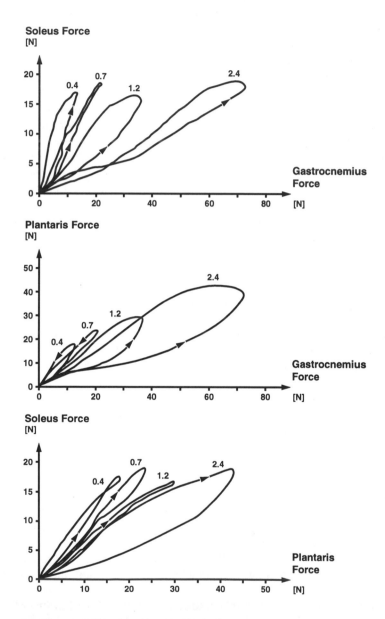

Figure 2.5.18   Representative plots of force-sharing between muscles of the ankle plantar flexor group in the cat when walking at nominal speeds of 0.4, 0.7, 1.2 m/s and trotting at 2.4 m/s. All curves shown are averages obtained over ten consecutive and complete step cycles (from Herzog et al., 1991, with permission of Pergamon Press Ltd., Headington Hill Hall, Oxford 0X3 0BW, UK).

Fig. 2.5.18 shows representative plots of force-sharing between two muscles of the plantar flexor group for walking at nominal speeds of 0.4, 0.7, 1.2 m/s and trotting at 2.4 m/s. All force-sharing loops shown at each speed are averages that were obtained over ten consecutive step cycles. The arrows indicate the direction of force build-up and decay. Average peak gastrocnemius and plantaris forces increase from about 15 N and 18 N, respectively, at a walking speed of 0.4 m/s, to corresponding values of approximately 75 N and 45 N for trotting at a speed of 2.4 m/s. This change in force corresponds to a 5-fold increase for gastrocnemius and a 2.5-fold increase for average peak plantaris forces from the slowest to the fastest experimental speeds shown. Corresponding average peak soleus forces remain between 17-20 N for all speeds. Therefore, it appears that increased force demands on the Achilles tendon caused by increased speeds of locomotion are accompanied by dramatic changes in force-sharing among the muscles involved.

In order to test if these changes in force-sharing among the cat ankle plantar flexor muscles are speed- or effort-dependent, muscle force measurements were also performed at a constant walking speed (0.7 m/s) but different external resistances. Changes in resistance were introduced by changing the slope of the walking surface from 10° downhill (D), to level (L), to 10° uphill (U). Conceptually, the results obtained using this protocol were similar to those obtained for level walking at different speeds (Fig. 2.5.19). Therefore, changes in force-sharing among cat gastrocnemius, soleus, and plantaris muscles appear to be effort- rather than just speed-dependent.

From the previous two figures, it may appear that gastrocnemius and plantaris forces change significantly in response to changes in speed or resistance of locomotion, whereas, forces of the soleus muscle remain relatively constant. This observation is not quite correct. When analyzing a series of consecutive steps at the same nominal speed, changes in soleus forces from step to step may vary considerably. For example, Fig. 2.5.20 shows the mean force-time curves (M) for one animal walking at a nominal speed of 1.2 m/s on the treadmill. Means were obtained from 43 step cycles obtained during one experimental test (i.e., about 30 seconds). Also shown are the means of the three stride cycles yielding the highest (H) and lowest (L) peak forces in those same 43 steps. Gastrocnemius and plantaris forces range from approximately 30-45 N and from 27-33 N, much less than the range of peak forces measured for different walking speeds (Fig. 2.5.18). However, peak soleus forces range from about 15-22 N, a range larger than corresponding changes of the average peak values for speeds of locomotion ranging from 0.4 m/s to 2.4 m/s. It appears, therefore, that soleus forces are sensitive to small variations in kinematics from one stride cycle to the next, whereas, corresponding average values are not sensitive to large changes in the speed of locomotion.

Changes in muscular forces for walking at a constant nominal speed of 1.2 m/s were related to stride cycle time. Peak forces of gastrocnemius and plantaris muscles tended to be small for long stride cycle times, whereas, corresponding peak forces for soleus tended to be large, compared to short stride cycle times (Fig. 2.5.21). This depression of the peak forces of soleus muscle for short stride cycle times must be associated with central control phenomena, since it is obvious from the force records at higher speeds of the same animal (i.e., 2.4 m/s trotting) that peak forces are not constrained by peripheral events such as the force-velocity relation or the time available to build up the active state.

For slow walking (0.7 m/s), force-sharing between the antagonist pair of soleus and tibialis anterior muscles follows a negative correlation (Fig. 2.5.22); i.e., an increase in soleus force is typically associated with a decrease in tibialis anterior force and vice versa,

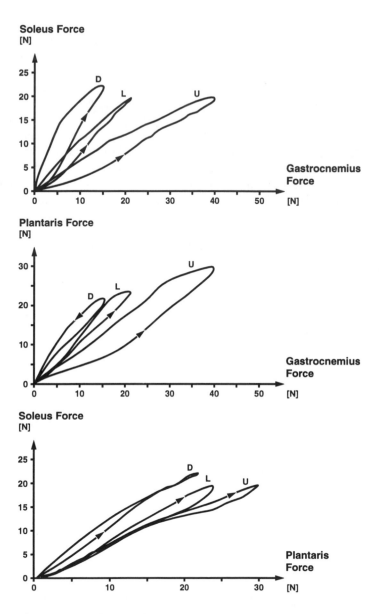

Figure 2.5.19   Representative plots of force-sharing between muscles of the ankle-plantar flexor group in the cat when walking at a speed of 0.7 m/s at different external resistances. Changes in external resistance were introduced by changing the slope of the walking surface from 10° downhill (D), to level (L), to 10° uphill (U). All curves shown are averages obtained over ten consecutive and complete step cycles.

especially during the stance phase of the stride cycle (label S to label M_S, Fig. 2.5.22). This negative correlation is exemplified by the very low tibialis anterior force at the instant of peak soleus force (P_S), and the very low soleus force at the instant of peak tibialis anterior force (P_{TA}, Fig. 2.5.22).

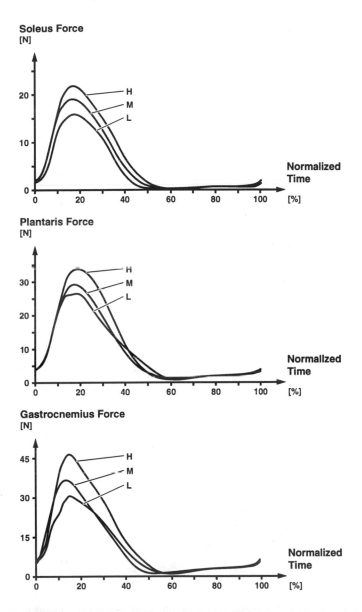

Soleus Force
[N]

Normalized
Time
[%]

Plantaris Force
[N]

Normalized
Time
[%]

Gastrocnemius Force
[N]

Normalized
Time
[%]

Figure 2.5.20   Mean force-time curves (M) for one animal walking at a nominal speed of 1.2 m/s on a motor-driven treadmill. Means were obtained over 43 step cycles. Also shown are the mean force-time histories of the three steps yielding the highest (H) and lowest (L) peak forces in these 43 steps (from Herzog et al., submitted, with permission).

For trotting (2.4 m/s), there is a simultaneous increase in soleus and tibialis anterior forces just prior to paw contact with the ground (Fig. 2.5.23, arrow), which is not seen as dramatically at the slower speed. It is hypothesized that this co-contraction of the antagonist pair causes an increase in ankle joint stiffness, which may be necessary for accomplishing the increased force requirements for the trotting compared to the walking modes

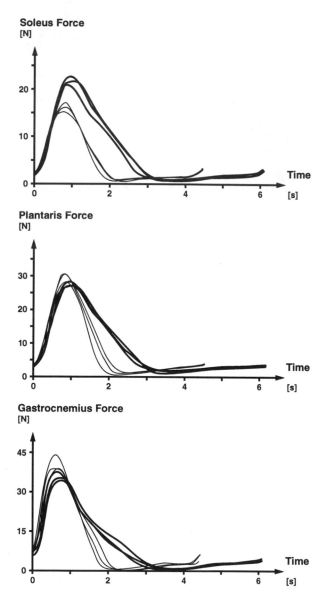

Figure 2.5.21   Force-time curves for soleus, plantaris, and gastrocnemius muscles of the cat when walking at a nominal speed of 1.2 m/s. Variations in stride cycle time (horizontal axis) were associated with corresponding peak forces for the three muscles. Soleus forces tended to increase with increasing stride cycle times, whereas, forces in gastrocnemius and plantaris tended to decrease with increasing stride cycle times (from Herzog et al., submitted, with permission).

of locomotion. In the future, force-sharing between antagonist pairs of muscles should include even faster speeds of locomotion than shown here. It is speculated that co-contraction of soleus and tibialis anterior just prior to paw contact would be even more pro-

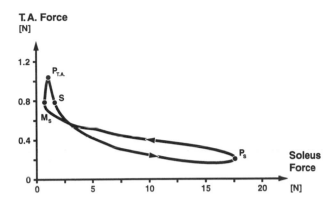

Figure 2.5.22   Force-sharing between the antagonist pair of soleus and tibialis anterior muscles in the cat when walking at a nominal speed of 0.7 m/s.

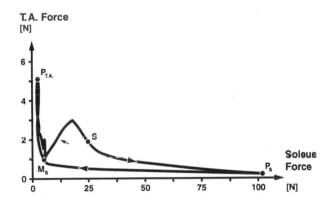

Figure 2.5.23   Force-sharing between the antagonist pair of soleus and tibialis anterior muscles in the cat when trotting at a speed of 2.4 m/s.

nounced than the co-contraction shown for trotting (Fig. 2.5.23) under such locomotor conditions.

2.5.7 REFERENCES

Abbott, B.C. and Wilkie, D.R. (1953) The Relationship Between Velocity of Shortening and the Tension-length Curve of Skeletal Muscle. *J. Physiol.* **120**, pp. 214-223.

Abraham, L.D. and Loeb, G.E. (1985) The Distal Hindlimb Musculature of the Cat (Patterns of Normal Use). *Exp. Brain Res.* **58**, pp. 580-593.

Andrews, J.G. (1974) Biomechanical Analysis of Human Motion. *Kinesiology.* Amer. Assoc. for Health, Phys. Ed., and Rec., Washington, D.C. **IV**, pp. 32-42.

Andrews, J.G. (1987) The Functional Roles of the Hamstrings and Quadriceps during Cycling. Lombard's Paradox Revisited. *J. Biomechanics.* **20,** pp. 565-575.

Ariano, M.A., Armstrong, R.B., and Edgerton, V.R. (1973) Hindlimb Muscle Fibre Populations of Five Mammals. *J. Histochem. Cytochem.* **21 (1),** pp. 51-55.

Blix, M. (1894) Die Laenge und die Spannung des Muskels. *Skand. Arch. Physiol.* **5,** pp. 149-206.

Borelli, A. (1680-1681) *De Motu Animalium.* Rome, Italy.

Close R. (1964) Dynamic Properties of Fast and Slow Skeletal Muscles of the Rat During Development. *J. Physiol.* **173,** pp. 4-95.

Close, R. and Hoh, J.F.Y. (1967) Force: Velocity Properties of Kitten Muscles. *J. Physiol.* **192,** pp. 815-822.

Croone, W. (1664) *De Ratione Motus Musculorum.* London, England.

Crowninshield, R.D. (1978) Use of Optimization Techniques to Predict Muscle Forces. *J. Biomech. Eng.* **100,** pp. 88-92.

Crowninshield, R.D. and Brand, R.A. (1981a) A Physiologically Based Criterion of Muscle Force Prediction in Locomotion. *J. Biomechanics.* **14 (11),** pp. 793-801.

Crowninshield, R.D. and Brand, R.A. (1981b) The Prediction of Forces in Joint Structures: Distribution of Intersegmental Resultants. *Exercise and Sport Sciences Reviews* (ed. Miller, D.I.). The Franklin Institute Press, Philadelphia. **9,** pp. 159-181.

Daremberg, C.V. (1854-1857) *Oeuvres Anatomiques, Physiologiques et Medicales de Galen.* Paris.

Dul, J., Johnson, G.E., Shiavi, R., and Townsend, M.A. (1984) Muscular Synergism - II: A Minimum Fatigue Criterion for Load Sharing Between Synergistic Muscles. *J. Biomechanics.* **17 (9),** pp. 675-684.

Edman, K.A.P. (1979) The Velocity of Unloaded Shortening and its Relation to Sarcomere Length and Isometric Force in Vertebrate Muscle Fibres. *J. Physiol.* **291,** pp. 143-159.

Edman, K.A.P. and Reggiani (1983) Length-tension-velocity Relationships Studied in Short Consecutive Segments of Intact Muscle Fibres of the Frog. *Contractile Mechanisms of Muscle. Mechanics, Energetics, and Molecular Models Volume II* (eds. Pollack, G.H. and Sugi, H.). Plenum Press, New York. pp. 495-510.

Engelhardt, V.A. and Lyubimova, M.N. (1939) Myosin and Adenosinetriphosphatase. *Nature.* **144,** p. 668.

Fabiato, A. and Fabiato, F. (1975) Dependence of the Contractile Activation of Skinned Cardiac Cells on the Sarcomere Length. *Nature.* **256,** pp. 54-56.

Farquharson, A.S.L. (1912) De Motu Animalium (Aristotle, translated). *The Works of Aristotle Volume 5* (eds. Smith, J.A. and Ross, W.D.). Clarendon Press, Oxford.

Fenn, W.O. and Marsh, B.S. (1935) Muscular Force at Different Speeds of Shortening. *J. Physiol.* **85,** pp. 277-296.

Gordon, A.M., Huxley, A.F., and Julian, F.J. (1966) The Variation in Isometric Tension with Sarcomere Length in Vertebrate Muscle Fibres. *J. Physiol.* **184,** pp. 170-192.

Goslow Jr., G.E. and van de Graaf, K.M. (1982) Hindlimb Joint Angle Changes and Action of the Primary Ankle Extensor Muscles During Posture and Locomotion in the Striped Skunk (Mephitis). *J. Zool.* **197,** pp. 405-419.

Gregor, R.J., Cavanagh, P.R., and LaFortune, M. (1985) Knee Flexor Moments During Propulsion in Cycling: A Creative Solution to Lombard's Paradox. *J. Biomechanics.* **18 (5),** pp. 307-316.

Henneman, E. and Olson, C.B. (1965) Relations Between Structure and Function in the Design of Skeletal Muscles. *J. Neurophysiol.* **28,** pp. 581-598.

Henneman, E., Somjen, G., and Carpenter, D.O. (1965) Functional Significance of Cell Size in Spinal Motoneurons. *J. Neurophysiol.* **28,** pp. 560-580.

Herzog, W. (1987) Individual Muscle Force Estimations Using a Non-linear Optimal Design. *J. Neurosci. Methods.* **21,** pp. 167-179.

Herzog, W. and ter Keurs, H.E.D.J. (1988) Force-length Relation of In Vivo Human Rectus Femoris Muscles. *Eur. J. Physiol.* **411,** pp. 642-647.

Herzog, W. and Leonard, T.R. (1991) Validation of Optimization Models that Estimate the Forces Exerted by Synergistic Muscles. *J. Biomechanics.* **24 (S1),** pp. 31-39.

Herzog, W., Read, L., and ter Keurs, H.E.D.J. (1991a) Experimental Determination of Force-length Relations of Intact Human Gastrocnemius Muscles. *Clin. Biomech.* **6,** pp. 230-238.

Herzog, W., Guimaraes, A.C., Anton, M.G., and Carter-Erdman, K.A. (1991b) Moment-length Relations of Rectus Femoris Muscles of Speed Skaters, Cyclists, and Runners. *Med. Sci. Sports Exerc.* **23 (11)**, pp. 1289-1296.

Herzog, W., Leonard, T., Renaud, J.M., Wallace, J., Chaki, G., and Bornemisza, S. (1992) Force-length Properties and Functional Demands of Cat Gastrocnemius, Soleus, and Plantaris Muscles. *J. Biomechanics.* **25 (11)**, pp. 1329-1335.

Herzog, W., Kamal, S., and Clarke, H.D. (1992) Myofilament Lengths of Cat Skeletal Muscle: Theoretical Considerations and Functional Implications. *J. Biomechanics.* **8 (25)**, pp. 945-948.

Herzog, W., Leonard, T.R., and Guimaraes, A.C.S. (1993) Forces in Gastrocnemius, Soleus, and Plantaris Muscles of the Freely Moving Cat. *Journal of Biomechanics.* **26**, pp. 945-953.

Herzog, W., Zatsiorsky, V., Prilutsky, B.I., and Leonard, T.R. (Submitted) Variations in Force-time Histories of Cat Gastrocnemius, Soleus, and Plantaris Muscles for Consecutive Walking Steps. *Journal of Experimental Biology.*

Hill, A.V. (1938) The Heat of Shortening and the Dynamic Constants of Muscle. *Proc. Royal Soc.* **126 (B)**, pp. 136-195.

Hill, A.V. (1970) *First and Last Experiments in Muscle Mechanics.* Cambridge University Press, Cambridge.

Hodgson, J.A. (1983) The Relationship Between Soleus and Gastrocnemius Muscle Activity in Conscious Cats: A Model for Motor Unit Recruitment. *J. Physiol.* **337**, pp. 553-562.

Horowits, R. (1992) Passive Force Generation and Titin Isoforms in Mammalian Skeletal Muscle. *Biophys. J.* **61 (2)**, pp. 392-398.

Houtz, S.A. and Fischer, F.J. (1959) An Analysis of Muscle Action and Joint Excursion During Exercise on a Stationary Bicycle. *J. Bone Jt. Surg.* **41 (A)**, pp. 123-131.

Hull, M.L. and Jorge, M. (1985) A Method for Biomechanical Analysis of Bicycle Pedalling. *J. Biomechanics.* **18 (19)**, pp. 631-644.

Huxley, A.F. (1957) Muscle Structure and Theories of Contraction. *Prog. Biophys. Chem.* **7**, pp. 255-318.

Huxley, A.F. and Niedergerke, R. (1954) Structural Changes in Muscle During Contraction. Interference Microscopy of Living Muscle Fibres. *Nature.* **173**, p. 971.

Huxley, A.F. and Simmons, R.M. (1971) Proposed Mechanism of Force Generation in Striated Muscle. *Nature.* **233**, pp. 533-538.

Huxley, H.E. and Hanson, J. (1954) Changes in the Cross-striations of Muscle During Contraction and Stretch and Their Structural Interpretation. *Nature.* **173**, p. 973.

Iwazumi, T. (1978) Molecular Mechanism of Muscle Contraction: Another View. *Cardiovascular System Dynamics* (eds. Baan, J., Noordergraaf, A., and Raines, J.). MIT Press, Cambridge. pp. 11-21.

Iwazumi, T. (1979) A New Field Theory of Muscle Contraction. *Cross-bridge Mechanism in Muscle Contraction* (eds. Sugi H. and Pollack G.H.). University of Tokyo Press, Tokyo. pp. 611-632.

Kardel, T. (1990) Niels Stensen's Geometrical Theory of Muscle Contraction (1667): A Reappraisal. *J. Biomechanics.* **23**, pp. 953-965.

Loeb, G.E. (1984) The Control and Responses of Mammalian Muscle Spindles During Normally Executed Motor Tasks. *Exerc. Sports Sci. Rev.* (ed. Terjung, R.L.). D.C. Health & Co., Lexington. **12**, pp. 157-204.

Mai, M.T. and Lieber, R.L. (1990) A Model of Semitendinosus Muscle Sarcomere Length, Knee, and Hip Joint Interaction in the Frog Hindlimb. *J. Biomechanics.* **23 (3)**, pp. 271-279.

McPhedran, A.M., Wuerker, R.B., and Henneman E. (1965) Properties of Motor Units in a Homogeneous Red Muscle (Soleus) of the Cat. *J. Neurophysiol.* **28**, pp. 71-84.

Morrison, J.B. (1968) Bioengineering Analysis of Force Actions Transmitted by the Knee Joint. *Biomed. Eng.* **3**, pp. 164-170.

Murphy, P.R., Stein, R.B., and Taylor, J. (1984) Phasic and Tonic Modulation of Impulse Rate in γ Motoneurons During Locomotion in Premammilliary Cats. *J. Neurophysiol.* **52**, pp. 228-243.

Needham, D.M. (1971) *Machina Carnis,* Cambridge University Press, Cambridge.

Page, S.G. and Huxley, H.E. (1963) Filament Lengths in Striated Muscle. *J. Cell Biol.* **19**, pp. 369-390.

Paul, J.P. (1965) Bioengineering Studies of the Forces Transmitted by Joints - II. *Biomechanics and Related Bioengineering Topics* (ed. Kenedi, R.M.). Pergamon Press, Oxford.

Pedotti, A., Krishnan, V.V., and Stark, L. (1978) Optimization of Muscle Force Sequencing in Human Locomotion. *Math. Biosci.* **38**, pp. 57-76.

Penrod, D.D., Davy, D.T., and Singh, D.P. (1974) An Optimization Approach to Tendon Force Analysis. *J. Biomechanics.* **7 (3)**, pp. 123-130.

Perrine, J.J. and Edgerton, V.R. (1978) Muscle Force-velocity and Power-velocity Relationships Under Isokinetic Loading. *Med. Sci. Sports Exerc.* **10 (3)**, pp. 159-166.

Pollack, G.H. (1990) *Muscles and Molecules: Uncovering the Principles of Biological Motion.* Ebner & Sons, Seattle, WA.

Pollock, C.M. (1991) *The Relationship Between Body Mass and the Capacity for Storage of Elastic Strain Energy in Mammalian Limb Tendons.* Ph.D. Thesis. University of Calgary. Nat'l Library of Canada, Ottawa.

Prochazka, A., Hulliger, M., Zangger, P., and Appenteng, K. (1985) Fusimotor Set: New Evidence for α Independent Control of γ Motoneurons During Movement in the Awake Cat. *Brain Res.* **339,** pp. 136-140.

Ruedel, R. and Taylor, S.R. (1971) Striated Muscle Fibres: Facilitation of Contraction at Short Lengths by Caffeine. *Science.* **172,** pp. 387-388.

Seeley, R.R., Stephens, T.D., and Tate, P. (1989) Anatomy and Physiology. Times Mirror/Mosby College Publishing, Toronto. Mosby-Year Book, Inc., St. Louis.

Seireg, A. and Arvikar, R.J. (1973) A Mathematical Model for Evaluation of Force in Lower Extremities of the Musculo-skeletal System. *J. Biomechanics.* **6 (3),** pp. 313-326.

Spector, S.A., Gardiner, P.F., Zernicke, R.F., Roy, R.R., and Edgerton, V.R. (1980) Muscle Architecture and Force-velocity Characteristics of Cat Soleus and Medial Gastrocnemius: Implications for Motor Control. *J. Neurophysiol.* **44,** pp. 951-960.

Swammerdam, J. (1758) *The Book of Nature. The History of Insects* (trans. Flloyd, T.). Seyffert, London.

Taber, C.W. (ed.) (1981) Taber's Cyclopedic Medical Dictionary. Davis, Philadelphia.

ter Keurs, H.E.D.J., Iwazumi, T., and Pollack, G.H. (1978) The Sarcomere Length-tension Relation in Skeletal Muscle. *J. Gen. Physiol.* **72,** pp. 565-592.

Thorstensson, A., Grimby, G., and Karlsson, J. (1976) Force-velocity Relations and Fibre Composition in Human Knee Extensor Muscles. *J. Appl. Physiol.* **40 (1)**, pp. 12-16.

Vallbo, A.B., Hagbarth, K.E., Torebjork, H.E., and Wallin, B.G. (1979) Somatosensory, Proprioceptive, and Sympathetic Activity in Human Peripheral Nerves. *Physiol. Rev.* **59,** pp. 919-957.

Vesalius, A. (1543) *De Humani Corporis Fabrica.* Hoffman La-Roche, Basel.

Walker, S.M. and Schrodt, G.R. (1973) I-segment Lengths and Thin Filament Periods in Skeletal Muscle Fibres of the Rhesus Monkey and the Human. *Anat. Rec.* **178,** pp. 63-82.

Walmsley, B., Hodgson, J.A., and Burke, R.E. (1978) Forces Produced by Medial Gastrocnemius and Soleus Muscles During Locomotion in Freely Moving Cats. *J. Neurophysiol.* **41 (5),** pp. 1203-1216.

Whiting, W.C., Gregor, R.J., and Edgerton, V.R. (1984) A Technique for Estimating Mechanical Work of Individual Muscles in the Cat During Treadmill Locomotion. *J. Biomechanics.* **17 (9),** pp. 685-694.

Woittiez, R.D., Huijing, P.A., Boom, H.B.K., and Rozendal, R.H. (1984) A Three-dimensional Muscle Model: A Quantified Relation Between Form and Function of Skeletal Muscles. *J. Morphol.* **182,** pp. 95-113.

Wuerker, R.B., McPhedran, A.M., and Henneman, E. (1965) Properties of Motor Units in a Heterogeneous Pale Muscle (M. Gastrocnemius) of the Cat. *J. Neurophysiol.* **28,** pp. 85-99.

Yoshihuku, Y. and Herzog, W. (1990) Optimal Design Parameters of the Bicycle-rider System for Maximal Muscle Power Output. *J. Biomechanics.* **23 (10),** pp. 1069-1079.

2.6 JOINTS

SHRIVE, N.G.
FRANK, C.B.

2.6.1 DEFINITIONS AND COMMENTS

Anatomy

Joint:	Junction of two or more bones of the human skeleton. Comment: joints are also called articulations.
• Cartilaginous joint::	Joints where the bones are connected by cartilage.
• Fibrous joint::	Joints where the bones are connected by connective fibrous tissue.
• Synovial joint:	Joints where the bones articulate relative to one another. They are corrected by ligaments.
Synovial fluid:	Transparent alkaline viscous fluid. Comment: synovial fluid is also called synovia.

Mechanics

Degrees of freedom (DOF):	Number of variables (co-ordinates) needed to describe a body's motion.
Properties:	
• Material:	Properties of a material which describe its general behaviour without including any information about its size and shape. Comment: material properties include stress and strain, etc.
• Physical:	Properties of a material which relate to its physics. Comment: physical properties include density, specific weight, etc.
• Structural:	Properties of a specific sample of a material which describe the behaviour of that sample including effects of its size and shape. Comment: structural properties include force to failure, deformation, etc.
Rotation:	Movement of a body about an axis.
Translation:	Movement of a body along a rectilinear path.

2.6.2 FUNCTION

Over time various musculo-skeletal structures have developed such as:

- Bone,
- Cartilage,
- Ligament,
- Tendon, and
- Muscle.

The functional, morphological, histological, and mechanical properties of the above biomaterials have been discussed in chapters 2.1 to 2.5. However, these biomaterials all function integrally together in complex fashions of *joints*. Of course, joints are not biomaterials, but biomaterials are structured so that joints can function the way that they do. For this reason, selected aspects of joints are discussed in this section.

A joint is the junction between two or more bones of the human or animal skeleton. The major functions of joints include:

- To permit or inhibit *movement*, and
- To *transfer forces* from one bone to another.

The skeletal system permits movement through the existence of joints which serve as the mechanical junctions between bones. Joints are made up of multiple tissues and are well designed for their mechanical purposes. Human and animal joints, unlike their man-made equivalents, function under very demanding conditions successfully for an average of 70 or 80 years, frequently without any major malfunction.

2.6.3 CLASSIFICATION OF JOINTS

Joints are classified into three subtypes: fibrous, cartilaginous, and synovial (Table 2.6.1).

In *fibrous joints,* the bones are connected by connective fibrous tissue. Fibrous joints have the appearance of hair-line cracks and are designed for stability. The skull contains fibrous joints.

In *cartilaginous joints*, cartilage layers on the end surfaces of bones are in contact and the bones are connected by ligaments. These joints allow limited movement, are relatively stiff, and are designed to provide stability and to transfer forces. A single cartilaginous joint permits only limited movement, but a combination of such joints may facilitate a large range of motion. A cat arching its back, for example, manifests a tremendous curvature of the spine. The spinal vertebrae are typical examples of cartilaginous joints.

In *synovial joints*, the bones are in contact but are not connected. These joints place few constraints on movement within a certain range, produce little friction (see chapter 2.4), and are designed to facilitate considerable motion and to transfer forces. The hip, knee, elbow, and shoulder joints are typical synovial joints, permitting a large range of motion and transfer of forces. The two most significant aspects of synovial joints, low friction and high force transfer, are illustrated in two examples. First, a marathon runner does not develop *hot* joints during a race because the friction in a synovial joint is about one-tenth of the friction between an ice skater and ice. Second, for the knee joint, the maximal forces that develop during walking at 1.5 m/s, have been estimated to be nearly three times body weight, and for the hip joint, about seven times body weight (Paul, 1976). From a biomechanical point of view the synovial joints are of considerable interest because they permit locomotion.

Table 2.6.1   **Description, function, movement, and examples for the three major joint types in the human and animal body, i.e., fibrous, cartilaginous, and synovial joints.**

TYPE	DESCRIPTION	FUNCTION	MOVEMENT	EXAMPLES
FIBROUS	Bones connected by fibrous (connective) tissue • syndesmosis • suture	stability	small non	tibia/fibula skull
CARTILAGINOUS	Bones connected by cartilage • synchondrosis • symphysis	bending	small small	sterno costalis connection sym. pubica spinal vertebrae
SYNOVIAL	Bones connected by ligaments	movement	small transl. large rotation	knee hip

2.6.4 DEGREES OF FREEDOM OF JOINTS

The different possibilities for movement of a joint can be described using independent variables. At most, six variables are needed to describe a possible movement (degrees of freedom = DOF) of the joint, three for translation, and three for rotation.

> **Degrees of freedom (DOF) are the number of variables (co-ordinates) needed to describe a body's motion.**

The DOF of a system of rigid segments can be determined using the general relationship:

DOF　　　　= number of generalized coordinates − number of constraints

For the three-dimensional and two-dimensional cases, the relationships read as:

3-D: DOF　=　6N − C

2-D: DOF　=　3N − C

where:

DOF　　　　= degrees of freedom

$$N \qquad = \quad \text{number of segments}$$
$$C \qquad = \quad \text{constraints}$$

Three examples are discussed which are illustrated in Fig. 2.6.1.

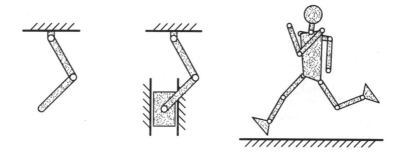

Figure 2.6.1   Schematic illustration of three examples for the determination of the degree of freedom, a two-dimensional double pendulum fixed with a hinge joint on a rigid structure (left), a two-dimensional piston fixed with a hinge joint on a rigid structure (middle), and a schematic illustration of a human body with 12 segments in a three-dimensional space (right).

EXAMPLE 1

Question:

Determine the degree of freedom of a planar double pendulum fixed at a rigid structure:

Solution 1:

For a two-dimensional case the DOF is determined by:

$$DOF_2 = 3N - C$$

$$N = 2$$

constraints:

$$C = C_1 + C_2$$

where:

C_1 = constraints for segment 1

C_2 = constraints for segment 2

A segment in two-dimensional space has two translational and one rotational degree of freedom.

A (non-sliding) hinge joint has no translational DOF and one rotational DOF. There-fore:

$$C_1 = 2$$

$$C_2 = 2$$

$$C = 4$$

therefore:

$$DOF_2 = 6 - 4$$

$$DOF_2 = 2$$

Solution 2:

For a two-dimensional case the DOF is determined by:

$$DOF_2 = 3N - C$$

$$N = 2$$

constraints:

In general, a (non-sliding) hinge joint has two constraints (no translation). Since the example has two hinge joints, the total number of constraints is four:

$$C = 4$$

therefore:

$$DOF_2 = 2$$

EXAMPLE 2

Question:

Determine the DOF for a piston in a three-dimensional space. Assume that the upper segment of the piston is fixed with a (non-sliding) hinge joint to a rigid structure, and that the joint between the two segments is a (non-sliding) hinge joint:

Solution 1:

$$DOF_2 = 3N - C$$

$$N = 2$$

constraints:

$$C = C_1 + C_2$$

$$C_1 = 2$$

$$C_2 = 3$$

The second segment cannot move at all once segment 1 is positioned because of the piston and the hinge joint.

$$C = 5$$

therefore:

$$DOF_2 = 1$$

Solution 2:

$$DOF_2 = 3N - C$$

$$N = 2$$

constraints:

The example includes two-dimensional (non-sliding) hinge joints which corresponds to four constraints, and one piston which restricts one translational DOF.

$$C = 5$$

therefore:

$$DOF_2 = 1$$

EXAMPLE 3

Question:

Determine the DOF for a schematic representation of a human as illustrated in Fig. 2.6.1. Assume that all the segments included (twelve) are rigid, and assume that the knee and elbow joints are (non-sliding) hinge joints, and all the other joints are spherical joints.

Solution:

$$DOF_3 = 6N - C$$

$$N = 12$$

constraints:

The system of interest includes four (non-sliding) hinge and seven spherical joints. A (non- sliding) hinge joint in three-dimensional space has five constraints, a spherical joint in three-dimensional space has three constraints.

$$C = 4 \cdot 5 + 7 \cdot 3 = 41$$

therefore:

$$DOF_3 = 72 - 41$$
$$DOF_3 = 31$$

2.6.5 TYPES OF JOINTS

Joints can be classified based on their translational and/or rotational motion. In the theoretical classification, joints are described in an idealistic manner assuming geometrically ideal surfaces. Such a classification includes:

- Spherical joints.
- Elliptical joints.
- (Non-sliding) hinge joints.
- (Sliding) hinge joints.
- Saddle joints.

and many other forms of joints. The rotational and translational DOF for these joints are summarized in Table 2.6.2.

The DOF provides information about the possible mobility of a joint, and the area of contact provides information about possible excessive stresses in a joint, aspects that are summarized in Table 2.6.2.

Table 2.6.2   **Summary of rotational and translational DOF, and area of contact of selected joints. Note: the areas of contact are indicated for ideal joint surfaces which are rigid. For deformable cartilage the results are different.**

JOINT	ROTATIONAL DOF	TRANSLATIONAL DOF	AREA OF CONTACT
SPHERICAL	3	0	constant
ELLIPTICAL	2	0	not constant
NON-SLIDING HINGE	1	0	constant
SLIDING HINGE	1	1	constant
SADDLE	2	2	not constant

Ideal spherical joints have three rotational and no translational DOF. The area of contact is constant.

The determination of the DOF for an ideal elliptical joint is more complex. In the following example, let us assume that the elliptical joint of interest has an elliptic cross-section for two principal axes, and a circular cross-section for one principal axis. For this assumption, the elliptical joint can rotate about an axis which is perpendicular to the circular cross-section providing:

- One rotational, and
- No translational DOF.

with the contact remaining constant. However, the two segments can also rotate about an axis which is perpendicular to an elliptical cross-section. In this case, the joint has:

- Two rotational, and

- No translational DOF.

with the contact area not remaining constant. As a matter of fact, the contact area changes to one contact point!

Hinge joints (non-sliding) have one rotational and no translational DOF with the contact area remaining constant. Sliding hinge joints have one rotational and one translational DOF with the contact area remaining constant only if the sliding rod is long enough.

Saddle joints have two rotational and two translational DOF. The contact area is not constant and may, under certain conditions, consist of just a point.

One would expect joints with a constant area of contact to have fairly constant stress distribution. A joint without a constant area of contact would be expected to show excessive stresses in certain positions, a tendency that could become critical for cartilage loading and possible damage.

Real human and animal joints are not the ideal theoretical joints as listed in Table 2.6.2. Real joints combine structures to allow combinations of movements that are not discussed with respect to ideal theoretical joints. The knee joint, for example, permits a great deal of rotation in flexion/extension, but permits only a limited amount of ad-abduction and internal/external rotation. In addition to a rolling motion, the knee has a sliding movement, especially in the a-p direction, which allows for translation of up to 2 or 3 cm. As well, the gross geometry of opposing bones and the overlying cartilage at joints is such that the two surfaces are not congruent. In the knee joint, for example, the femoral condyles are shaped like rollers, allowing rolling, gliding, and twisting movement, on the tibial surfaces. The hip joint, as another example, is a ball and socket joint, but its articular surfaces are not completely spherical, therefore, they are incongruent during motion (Bullough et al., 1968).

2.6.6 REFERENCES

Bullough, P.G., Goodfellow, J., Greenwald, A.S., and O'Connor, J.J. (1968) Incongruent Surfaces in the Human Hip Joint. *Nature.* **217,** p. 1290.

Paul, J.P. (1976) Approaches to Design. Force Actions Transmitted by the Joints in the Human Body. *Proc. R. Soc. Lond.* **192 (B),** pp. 163-172.

3   MEASURING TECHNIQUES

Many biomechanical research activities concentrate on experimental work that needs measuring techniques. Several measuring methods were established and further developed at the end of the 19th and during the 20th century. The methodologies developed and described by Marey at the end of the 19th century are still a milestone in biomechanical measuring techniques. Their principles still apply for many of todays methodologies.

Today, many measuring techniques are highly developed and allow the quantification of kinetic, kinematic, and other aspects of human or animal movement. This chapter describes some of the most important methods for movement analysis used in biomechanical research.

3.1 FORCE

NIGG, B. M.

3.1.1 DEFINITIONS AND COMMENTS

Crosstalk:	A signal at the output of a transducer caused by a variable not allocated to this particular output. Comment: the common crosstalk in force sensors occurs in all channels. It is, for instance, possible for the vertical force channel to show a signal when only horizontal forces are acting on a sensor.
Drift:	Change in output of the transducer over time that is not a function of the measurand.
Force:	Force cannot be defined. However, effects produced by forces can be described.
• Active:	Active forces in human locomotion are forces generated by movement that is entirely controlled by muscular activity.
• Dynamic:	If "something" is able to accelerate a mass, this "something" is called a force in a dynamic sense.
• Impact:	Impact forces in human locomotion are forces that result from a collision of two objects and that reach their maximum earlier than 50 milliseconds after the first contact of the two objects.
• Normal:	A force perpendicular to the surface of interest.
• Shear:	A force parallel to the surface of interest.
• Static:	If "something" is able to keep a spring deformed, this "something" is called a force in a static sense.
Frequency:	Number of oscillations per time unit.
• Natural frequency:	A frequency of vibration that corresponds to the elasticity and mass of a system under the influence of its internal forces and damping. Comment: the lowest natural frequency of a system is known as the fundamental frequency. Higher natural frequencies are harmonics of the fundamental value. Some systems exhibit several coupled or uncoupled modes of vibration.
• Resonance frequency:	The frequency of a forced vibration input to a system that corresponds to a natural frequency of the system.
Hysteresis:	The biggest difference in the output signal for any value of the measured variable within the specified

range when this value is reached with increasing or diminishing measurand.

Linearity: Amplitude of output signal in direct proportion to amplitude of input signal.
                      Comment: linearity is not necessary to obtain accurate measurements.

Measurand: Quantity to be measured.

Moment: Turning effect of a force, F, applied at point P about a base point O with:

$$M_O = r_{PO} \times F$$

where:

M_O =   moment (or moment of force) about point O
r_{PO} =   position vector from O to P
F =   force

Noise: Any unwanted disturbance of the original signal in a frequency band of interest.
                      Comment: one distinguishes between *background noise* and *random noise*.

Overload: The highest value of a measured variable that the transducer can sustain without its measuring properties being changed beyond the specified tolerances.

Range: The upper and lower limits of a variable for which a transducer is intended.

Rigidity: Measured deflection as a result of a defined acting force.

Sensitivity: The ratio between the change in the transducer output signal and the corresponding change in the measured value.
                      Comment: sensitivity can change over the possible range of measurements.

Threshold: The smallest change in the measured variable that causes a measurable alteration of the transducer output signal.

Torque: Result of a force couple.

3.1.2 SELECTED HISTORICAL HIGHLIGHTS

The techniques used to quantify forces were developed in previous centuries. However, the first attempts to quantify forces related to biomechanical questions were only made at the end of the 19th century, and force measuring devices for biomechanical purposes have only been commercially available since the late 1960's.

1686	Newton	Published the three laws in "Philosophiae Naturalis Principia Mathematica" (The Mathematical Principles of Natural Science) that are the cornerstone of all subsequent force measurements.
1743	Borelli	Published the first part of "De Motu Animalium" (The Movement of Animals), which included suggestions on how to determine forces on and in a biological system.
1856	Lord Kelvin	Demonstrated that the resistance of copper and iron wire change when subject to strain.
1872	Carlet	Developed a pneumatic measuring device to quantify the forces between foot and ground.
1873	Marey	Developed a portable pneumatic device to quantify forces between foot and ground of a human subject.
1880	Curie	The Curie brothers (Pierre and Jacques) discovered the piezoelectric effect, which was later used to quantify forces. The piezoelectric effect was put to practical use in the 1920's and 1930's.
1931	Carlson	Constructed the first unbonded resistance strain gauge.
1935	Bloach	Constructed the first bonded carbon film strain gauge.
1938	Elftman	Published measurements of the ground reaction forces and the centre of force for human locomotion. The measurements were obtained using a force plate.
1969	KISTLER	The KISTLER Company constructed the first commercially available piezoelectric force plate for gait analysis for the biomechanics laboratory of the ETH Zürich.
1976	AMTI	The AMTI Company constructed the first commercially available strain gauge force plate for gait analysis for the biomechanics laboratory of the Boston Children's Hospital.

Today, commercial force plates are used as a basic tool for biomechanical tests and measurements. However, it has only been a few years since the first commercial force plate was built. It goes back to an initiative of J. Wartenweiler, a professor at the ETH Zürich, Switzerland. He was serving as an officer in the Swiss cavalry when he discussed with another officer of his unit the comparative advantages of various horse shoes. As a biomechanist he discussed the external and internal forces resulting from the different horse shoes. It so happened that the other officer (H.C. Sonderegger) owned a company that specialized in precision force and acceleration measurements for industry using piezoelectric elements for their measurements. Consequently, H.C. Sonderegger built a force plate and charge amplifiers that were sold to the biomechanics laboratory of the ETH Zürich for gait analysis measurements with humans and animals (1969). As a matter of fact, the force plate is still operational in that laboratory.

Currently, force plates and force sensors for biomechanical application are commercially available. The two most used force measuring sensors in biomechanical and clinical-biomechanical applications are piezoelectric and strain gauge sensors.

3.1.3 MEASURING POSSIBILITIES

Force cannot be defined. However, effects produced by forces can be described:

> **If "something" is able to keep a spring deformed, this "something" is called a force in a static sense.**
>
> **If "something" is able to accelerate a mass, this "something" is called a force in a dynamic sense.**

Various mechanical elements are able to quantify the effects of a force. They include:

- Air balloon,
- Capacitor,
- Conductor,
- Inductive sensor,
- Piezoelectric sensor,
- Pyramid system,
- Spring, and
- Strain gauge.

These possibilities are discussed in the following paragraphs in two groups. The first group, including capacitor, conductor, piezoelectric sensor, and strain gauge, is discussed in more detail, since these sensors are often used in biomechanical force measurements. The second group, including the remaining methods, is discussed in less detail since these sensors are currently less frequently used in biomechanical force measurements.

CAPACITOR SENSORS

An electric capacitor consists of two electrically conducting plates that lie parallel to each other, separated by a distance that is small compared to the linear dimensions of the plates. The space between the plates is, for the purpose of biomechanical force measurements, filled with a non-conducting elastic material which is called *dielectric*. If a normal force is applied to a capacitor, the distance between the plate changes. This process can be described mathematically for a parallel capacitor with a normal force, F, acting as:

$$C \quad = \quad \frac{Q}{U}$$

and:

$$C \quad = \quad \varepsilon_0 \cdot \varepsilon_1 \cdot \frac{A}{d}$$

where:

C	=	capacitance of the capacitor
Q	=	magnitude of the total charge on each plate
U	=	potential difference between the two plates
A	=	area of one of the two identical capacitor plates
ε_0	=	permittivity of the free space
ε_1	=	permittivity of the dielectric
d	=	distance between the plates

consequently:

$$Q \cdot d \quad = \quad \varepsilon_0 \cdot \varepsilon_1 \cdot A \cdot U$$

Assuming that the relationship between an acting force, F, and the resulting distance, d, can be experimentally determined, for instance:

$$F \quad = \quad c_0 \cdot d_0 + c_1 \cdot d$$

the relation between an acting force and the resulting change in charge, Q, can be quantified:

$$F \quad = \quad c_0 \cdot d_0 + \frac{c_1}{Q} \cdot \varepsilon_0 \cdot \varepsilon_1 \cdot A \cdot U$$

where:

d_0	=	constant
c_0, c_1	=	constants

A change in force produces a change in the thickness of the dielectric material which is indirectly proportional to a current which can be measured.

CONDUCTOR SENSORS

Sensors based on the conductivity principle consist of two layers of conductive material (top and bottom) and a conductive material in between, a construction which is similar to the construction of a capacitor. However, the material between the two "plates" is electrically conductive. Resistive sensors rely on conductive elastomers as material between the plates. As a force is applied to the two plates, the elastomer between the plates is deformed, thereby, reducing the electrical resistance between the two plates. By examining the current output, the applied force can be determined using Ohm's law:

$$V \quad = \quad I \cdot R$$

where:

V	=	voltage
I	=	current
R	=	resistance

PIEZOELECTRIC SENSORS

The piezoelectric effect was discovered by the Curie brothers in 1880. A piezoelectric material is a non-conducting crystal that exhibits the property of generating an electrical charge when subjected to mechanical strain. Several materials have piezoelectric properties, quartz being one of them. Quartz is the most stable of the piezoelectric materials. Its' output is low compared to the output of ceramic materials. However, it is adequate for biomechanical measurements. The electric charge that appears on the surfaces of a quartz when strained is a result of relative movement of electric charges. This movement is illustrated in a simplified form in Fig. 3.1.1.

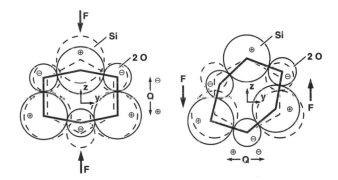

Figure 3.1.1	Schematic illustration of the piezoelectric effect (from Martini, 1983, with permission of the Instrument Society of America). Compressive forces produce a change in the electric charges on the surfaces where the force has been applied (left). Skew forces produce a change in charges on the surfaces perpendicular to the applied skew force direction (right).

For biomechanical force sensors, quartz crystals are commonly cut into disks perpendicular to their crystallographic x or y axis. If such disks are subjected to force they yield an electrical charge of 2.26 pC/N (pico Coulombs per Newton) and 4.52 pC/N respectively, which is temporarily stored in the element's inherent capacitance, C_T. As with all capacitors, however, the charge dissipates with time due to leakage. Electronic circuits have been developed to minimize this shortcoming. The current charge amplifiers overcome this problem and dynamic as well as quasi-static measurements can be performed. Further information about piezoelectric force measuring sensors can be found in Gohlke (1959), Cady (1964), Tichy and Gautschi (1980), and Martini (1983).

One major advantage of piezoelectric force plates is their wide range. The same force plate can be used to measure the forces between the ground and the hoof of a horse trotting or galloping and to measure human microvibrations (Nigg, 1977) such as forces on the ground due to the heart beat.

STRAIN GAUGE SENSORS

Strain measurements are discussed later in section 3.5. Here, only selected aspects of measuring strain are discussed. All structures deform to some extent when subjected to external forces. The deformation results in a change in length which is, when normalized, called "strain".

$$\varepsilon_a = \frac{(L_2 - L_o)}{L_o} = \frac{\Delta L}{L_o}$$

where:

$$
\begin{aligned}
\varepsilon_a &= \text{average axial strain} \\
L_o &= \text{original length} \\
L_2 &= \text{final (strained) length}
\end{aligned}
$$

Lateral strain, as described by the Poisson ratio, v, results if a structure is subjected to simple uniaxial stress:

$$v = \frac{-\varepsilon_L}{\varepsilon_a}$$

where:

$$
\begin{aligned}
v &= \text{Poisson's ratio} \\
\varepsilon_L &= \text{lateral strain}
\end{aligned}
$$

The three-dimensional relations for simultaneous stress acting in all three axis direction are:

$$\varepsilon_x = \left[\frac{1}{E}\right][\sigma_x - v(\sigma_y + \sigma_z)]$$

$$\varepsilon_y = \left[\frac{1}{E}\right][\sigma_y - v(\sigma_z + \sigma_x)]$$

$$\varepsilon_z = \left[\frac{1}{E}\right][\sigma_z - v(\sigma_x + \sigma_y)]$$

The stress strain relationship for uniaxial loading condition is:

$$\sigma_a = E \cdot \varepsilon_a$$

where:

$$
\begin{aligned}
\sigma_a &= \text{average uniaxial stress} \\
\varepsilon_a &= \text{average strain in direction of the stress}
\end{aligned}
$$

E   =   Young's modulus

Most strain measurement methods used in biomechanics are based on electrical methods. Electrical type strain gauges use resistive, piezoresistive, piezoelectric, capacitive, inductive, and/or photoelectric principles. The following discussions concentrate on the first two, the resistive and the piezoresistive methods.

The *electrical resistance transducers* use the principle that the application of stress to a structure changes its geometry. If, for instance, a wire is stretched the cross sectional area of the wire will diminish which changes the electric conductivity. The change in resistance of the stretched structure can be calibrated.

The *piezoresistive transducers* function in principle the same way as the electric resistance transducers. The difference lies in the material used. The piezoresistive transducers use semiconductor materials, such as silicon, that have a higher sensitivity than the normal electrical resistance transducers (gauge factor > 100). However, piezoresistive transducers are non-linear with strain, the sensitivity is temperature dependent (also low temperatures must be avoided to prevent operational failure), and their strain range is significantly lower than the strain rate for conventional strain gauge sensors.

The transducers are mounted on a structure that deforms if subjected to stress. These structures are usually used independently for the different axes directions, since crosstalk would be too big if the transducers were applied to one structure.

The transducers are arranged to maximize the electric output. Strategies to achieve this maximization include the application of one or several sensors on the top and bottom of a deformable structure. The sensors are electrically connected by a bridge arrangement as explained and illustrated in section 3.5, Fig. 3.5.2.

SELECTED CHARACTERISTICS OF FORCE SENSORS

Specific characteristics of force sensors that are sold commercially can be found in brochures and specific information sheets which are available from companies. This section (Table 3.1.1) attempts to summarize and to compare selected characteristics of force measuring sensors as they apply to biomechanical force measuring instruments. The assessment and the comparison dwell on information from major companies and on practical experiences in our research activities. It was attempted to concentrate on those aspects which do not depend on short technical developments but on aspects which are associated with the principles of each methodology.

One may want to divide the sensors into two groups with generally similar advantages and disadvantages. The first group would include the capacitance and conductive sensors, and the second group the piezoelectric and strain gauge sensors. Force measuring devices using capacitance and conductive sensors seem to be advantageous for measuring forces on soft or uneven surfaces and for pressure distribution measurements. They are usually less accurate than sensors from the second group. Force measuring devices using piezoelectric or strain gauge sensors seem to be advantageous for measuring forces on rigid structures (e.g., force plates) and for measurements where the sensor can be rather stiff. They are usually of relatively high accuracy.

Force measuring devices using capacitance or conductive sensors are typically used in pressure measuring devices (for instance in shoes or on sport equipment). Force measuring devices using piezoelectric or strain gauge sensors are typically used for force plates.

Table 3.1.1   **Summary table for selected characteristics of force sensors typically used in biomechanical experiments. Specific data have been extracted from sales brochures of major suppliers and/ or are based on our own experience. FSO = full scale output. * if corrected with software; if not, errors may be substantially bigger.**

CHARACTERISTIC	SENSOR TYPE			
	CAPACITOR	CONDUCTOR	PIEZOELECTRIC	STRAIN GAUGE
RANGE	limited	limited	nearly unlimited for biomech. applications	nearly unlimited for biomech. applications
LINEARITY	depends on dielectricum mostly not linear		> 99.5 % of FSO highly linear	> 96 % of FSO good linearity
CROSSTALK	depends on construction can be high		rather low	rather low
HYSTERESIS	small < 3 %	small < 3 %	very small < 0.5 %	small < 4 %
THRESHOLD	a few N	a few N	< 5 mN	50 to 100 mN
TEMPERATURE SENSITIVITY	small	small	small	small
ACCURACY	errors up to 20 % depending on application	errors up to 20 % depending on application*	errors up to 5 %	errors up to 5 %
DRIFT	none	none	≈ 0.01 N/s	none
COST	high	low	high	low

OTHER METHODS

If a *balloon* filled with fluid or air is subjected to a force, some particular pressure in the air balloon is necessary to balance the force. The pressure in the balloon can be calibrated for the applied external force and the system can be used for force assessment. This principle was used initially by Carlet (1872) in a circular walkway with the air balloon under the foot connected by long rubber hoses to a central recording instrument; and by Marey (1873), who modified Carlet's apparatus by making it portable to quantify the forces under the hoof of a horse. Currently, this principle of force measurement is used in internal force/pressure measurements in animals (e.g., between heart and surrounding tissue). In biomechanical and clinical research and monitoring, hydro-fluid balloons are used to quantify the forces between soft tissue and the environment - for example, between foot and shoe. These sensors show high linearity, small hysteresis, little drift, and high frequency response.

Forces, especially local forces, may be quantified by using the "pyramid system". A mat, which is flat at the top and has small pyramid type elements at the bottom is positioned on a glass plate (Fig. 3.1.2). If a force is applied to the top of the mat the pyramids

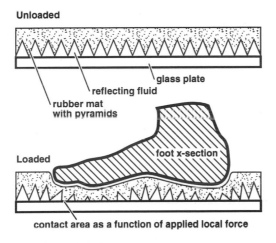

Figure 3.1.2   Illustration of local force assessment with the pyramid system as used by Miura et al. (1974), with permission.

will be deformed to a degree that depends on the magnitude of the force applied. The deformation results in an increased contact area between pyramid and glass plate (Fig. 3.1.2), which can be quantified optically (film or video). This force measuring principle was used by Miura et al. (1974) to assess pressure distribution under the foot during walking.

Inductive type force transducers measure the relative movement of one or several coils with respect to a core. This type of force transducer has only seen limited use in biomechanics.

Force measuring devices may have some mechanical elastic parts - for instance, *springs*. A force applied to the elastic part results in a deflection that is usually linear. The deflection can be used as a measure of the applied force using, in the linear case:

$$F = -kx$$

where:

F	=	applied force
k	=	deflection (spring) constant
x	=	deflection

Spring force measuring devices are not widely used in biomechanical research.

3.1.4 APPLICATIONS

CAPACITANCE TRANSDUCERS

The most frequent application of capacitance transducers in biomechanics of locomotion is in the assessment of the pressure distribution (see chapter 3.2) using insoles and/or plates. Additionally, capacitance transducers are used in the assessment of forces between two soft surfaces (Nicol and Hennig, 1976).

CONDUCTIVE TRANSDUCERS

Transducers based on conductor technology are currently used in two main applications. The first application includes the measurement of pressure distribution close to the human body (see chapter 3.2). The sensor is in the form of an insole for measurements between foot and shoe or in a special shape for specific applications. The second application includes the combination of this sensor in an array of rows and columns with force plates. The fine grid of this pressure distribution device can be used to get an instantaneous determination of the position of the foot on the force plate, a procedure which is time consuming and often not accurate with optical methods.

PIEZOELECTRIC TRANSDUCERS

Piezoelectric transducers have been used extensively for dynamic force measurements. The sensors were initially developed for measurements in industrial applications including force or pressure measurements in combustion engines, on conventional shoe brakes of trains, on turbine blades, and on machine tools during operation. In biomechanics, piezoelectric force sensors were used the first time in the late 1960's for force plates to quantify the ground reaction forces during human locomotion. Currently, force plates based on the piezoelectric principle are available in different sizes, and are made from different materials. Additionally, piezoelectric force transducers are used in many biomechanical applications, including transducers for force-deformation assessment and/or for drop tests.

STRAIN GAUGE TRANSDUCERS

Strain gauge transducers have played an important role in the development of biomechanical knowledge in the last few decades (Canderale, 1983). Assessment of the ground reaction forces during human locomotion has been done using force plates wielding strain gauge transducers. Additionally, *in vitro* stress and strain measurements on bone, specifically the femur and the stem, have been performed (Weightman, 1976; Breyer et al., 1979; Canderale et al., 1979; Huiskes and Slooff, 1979; Jakob and Huggler, 1979; Huiskes, 1980). Similar stress and strain measurement experiments have also been performed for the tibia (van Campen et al., 1979). Asang (1974) took mechanical properties of the human tibia, as they related to alpine skiing, and assessed these properties by using the transducers. Furthermore, *in vivo* forces have been quantified on various bones and tendons of animals and humans. By implanting a femoral prosthesis with strain gauge transducers, forces in the human hip joint have been quantified (English and Kilvington, 1979; Kotzar et al., 1991; Bergmann et al., 1993). Semiconductor transducers have been used to quantify the forces in the thoracic vertebrae of sheep (Lanyon, 1972).

Because strain gauge transducers can easily be placed on human or animal bone, assessment of forces for bones is simpler than for tendons or ligaments. Tendon or ligament forces are more difficult to measure because these tissues are usually not accessible.

With the help of buckle transducers, measuring multiple muscle forces has been quantified on the tendons of animals (Walmsley et al., 1978). Buckle transducers have also been used to quantify the forces in the human Achilles tendon during walking, running, and other sporting activities (Komi et al., 1987).

Finally, strain gauge transducers have been used in evaluating sporting activities, such as rowing, for taking force measurements from the oar (Nolte, 1979; Schneider, 1980).

FORCE MEASUREMENTS IN TENDONS AND LIGAMENTS

Several sensors have been developed to quantify forces in tendons and ligaments in vivo or in vitro. They include:

- Foil strain gauge transducer,
- Liquid metal strain gauge transducer,
- Buckle transducer,
- Hall effect strain transducer, and
- Implantable force transducer.

Force measuring transducers are commonly classified as either direct, meaning that force is measured directly, or indirect, meaning that force is measured indirectly (for instance, via strain). Buckle transducers and implantable force transducers are generally considered direct force transducers. Strain gauge transducers, video dimension analysis, and Hall effect transducers are generally considered indirect force transducers. This categorization, however, seems artificial, since every possible force measuring device must quantify force indirectly. A buckle transducer, for instance, uses deformation and, therefore, strain to determine force. For further information see section 2.5.

Foil strain gauge transducers typically consist of (White and Raphael, 1972):

- A foil gauge.
- A stainless steel shim that is sutured to the tendon or ligament.

Theoretically these transducers are highly accurate, but problems sometimes arise because the sutures are unable to act as rigid displacement transducers between the tissue and the shim (Shrive et al., 1992).

Liquid metal gauge transducers consist of (Brown et al., 1986):

- A gauge body of silastic or silicone tubing.
- A liquid metal (purified mercury, indium, or gallium).
- Two insulated wires plugging the ends of the gauge tubing.
- Two end seals of silicone heat shrink tubing attaching and sealing the gauge tube to the wires.

Axial strain in the tendon or ligament causes the gauge to deform by increasing the column length of the mercury and decreasing the cross-sectional area, thus, effecting a net increase in electrical resistance. The liquid metal strain transducer is small enough to fit into individual fibre bundles. Because it is not very stiff, the liquid metal strain transducer interferes minimally with the mechanics of the bone.

Buckle transducers consist of (Salmons, 1969):

- A rectangular, stainless steel frame element for supporting a portion of the tendon or ligament.
- A stainless steel beam mounted transversely between the tendon and the rectangular frame element.
- Two conventional foil strain gauges mounted on the buckle for measuring deformation (Fig. 3.1.3).

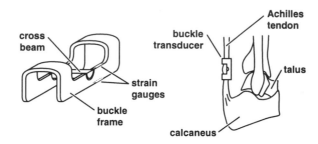

Figure 3.1.3   **Schematic illustration of a buckle transducer (left) and a possible arrangement on an Achilles tendon (right) (from Salmons, 1969, with permission).**

Under axial loading, the tendon or ligament lengthens, producing a transverse force and a resulting deformation of the beam element, both of which can be quantified. The buckle transducer is a relatively stable measuring device for chronic, in vivo experiments. Typical insertion times are 3 to 7 days (Komi et al., 1987; Komi, 1990; Herzog et al., 1993). Buckle transducers require a finite tendon or ligament length and distance between bony structures and the sensor. These requirements render it applicable only to a few structures in human and animal bodies.

Hall effect strain transducers consist of (Arms et al., 1983):

- A magnetized cylindrical rod.
- A tube mounted with a Hall generator.

The device is fixed to the tendon or ligament by barbs attached to the magnet and the tube. As the tendon or the ligament is strained, the rod displaces with respect to the tube, and the Hall generator produces a proportional voltage. The Hall effect strain transducer may be firmly anchored to the tissue and does not load the tissue during strain monitoring. It is small, and minimal surgical intervention is required to affix it to tissue. Hall effect strain transducers measure local forces. Typical insertion times are about 7 days (Platt et al., 1992).

Implantable force transducers have been described in literature, one based on the deflection of a metallic element (Xu et al., 1992) and one based on a modification of a miniature, circular pressure transducer (Holden et al., 1991). The former, the implantable force transducer based on the deflection of a metallic element, is described here. This force transducer consists of:

- A curved steel spring with foil strain gauges mounted on the upper and lower side (Fig. 3.1.4).

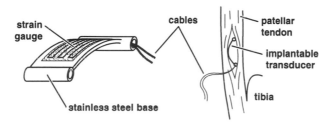

Figure 3.1.4   **Schematic illustration of an implantable force transducer (left) and a possible arrangement in a patellar tendon (right) (from Xu et al., 1992, with permission).**

This transducer is inserted directly into a longitudinal slit (pocket) in the tendon or ligament mid-substance and is typically held in place with sutures through the adjacent tissue fibres and at the slit edges. Implantable force transducers allow an absolute force and an absolute zero line to be determined.

FORCE PLATES

Force plates have been used in biomechanics for many years for quantifying external forces during human and animal locomotion. Publications over the last 50 to 100 years illustrate that force plates have been one of the most important measuring devices in biomechanics. Marey (1895) built the first force plate, which consisted of spirals of India rubber tubes installed on a wooden frame. Elftman (1938) used a plate moving vertically against four springs upon which it was supported, with the vertical displacement recorded optically to quantify the vertical component of the resultant ground reaction force and the path of the centre of force. Cavagna (1964) used force plates to quantify external vertical impact force peaks, a quantity which was frequently used in the late 1970's and thereafter in sport shoe and sport surface research. With the development of commercially available force plates, the number of researchers working in the field of locomotion analysis increased significantly, and resultant ground reaction forces were assessed for many different movements. A comprehensive discussion of ground reaction forces in distance running was published in 1980 (Cavanagh and Lafortune).

Force plates currently on the market use a construction in which the plate is rectangular and is supported at four points (Fig. 3.1.5). Such a measuring system is mechanically overdetermined, which in some cases may lead to accuracy problems (Bobbert and

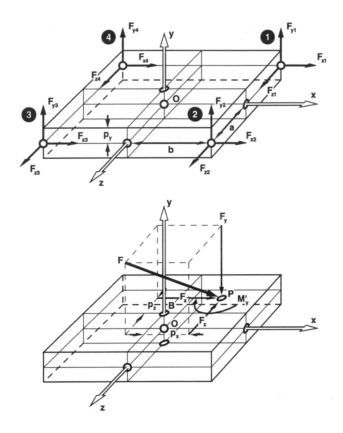

Figure 3.1.5   Illustration of the forces and moments applied to the force plate (top) and the forces registered by the force transducers in the four corners of the force plate. The force transducers are assumed to be located in the corners of the indicated frame.

Schamhardt, 1990). Force transducers for each axis direction are mounted in each corner. The signals from these transducers are:

F_{x1}	corner 1
F_{x2}	corner 2
F_{x3}	corner 3
F_{x4}	corner 4
F_{y1}	corner 1
F_{y2}	corner 2
F_{y3}	corner 3
F_{y4}	corner 4
F_{z1}	corner 1
F_{z2}	corner 2
F_{z3}	corner 3
F_{z4}	corner 4

where:

$$F_x \quad = \quad F_{x1} + F_{x2} + F_{x3} + F_{x4}$$

$$F_y \quad = \quad -(F_{y1} + F_{y2} + F_{y3} + F_{y4})$$

$$F_z \quad = \quad -(F_{z1} + F_{z2} + F_{z3} + F_{z4})$$

$$F \quad = \quad \begin{bmatrix} F_x \\ F_y \\ F_z \end{bmatrix}$$

The force components, F_x, F_y, and F_z, are normally used for biomechanical analysis of locomotion. For this text the following conventions are used when force plates are used for locomotion analysis:

(a)   x   =   direction of movement (for gait anterior-posterior)
            y   =   vertical direction
            z   =   direction perpendicular to movement (for gait medio-lateral)

(b)   In order to make possible a comparison of the force results for the medio-lateral direction, the medial direction may be defined as positive and the lateral direction as negative. Using this convention, the force time curves for the left and right foot can be averaged if needed. However, this convention is only used for the graphical illustration of medio-lateral force components and for the numerical analysis of these curves. If the force results are used for further force analysis, e.g., inverse dynamics, the conventional right turning coordinate system is used.

(c)   These conventions are chosen to comply with the conventions proposed by the ISB. However, it is obvious that the results are not different if another convention is used, as long as right turning coordinate systems are used. The convention should, therefore, not be overemphasized.

Force plates provide a resultant force that is an integral quantity. The vertical component describes the change in momentum of the centre of mass of the test subject in the vertical direction, the a-p and the m-l components correspond in the two horizontal directions. This can be used for speed control of test movements. If a person must run or walk with constant speed the integral of the a-p force component over time must be zero.

$$\int_{t_1}^{t_2} F_{ap}(t)\, dt \quad = \quad mv_2 - mv_1 \quad = \quad \int_{t_1}^{t_2} F_x(t)\, dt$$

where:

t_1 =   first contact with the force plate

t_2 = last contact with the force plate
m   = mass of the test subject
v_1 = horizontal speed of the centre of mass of the test subject before landing
v_2 = horizontal speed of centre of mass of the test subject after take-off

The measurements of the force sensors in the four corners of a force plate may be used to determine the centre of pressure, and the free moment of rotation about a vertical axis through the centre of pressure. In determining these variables, the following conventions are used:

O   = origin of the force plate, arbitrarily assumed in the centre of the force plate
P   = point where the force, **F**, applies on the top surface of the force plate
B   = origin of the surface of the force plate
a   = distance of the force transducers from the x axis
b   = distance of the force transducers from the z axis
p_y = distance between O and B
M'_y = external moment component applied to the force plate with respect to a vertical axis through P

For simplicity, the horizontal force components can be combined this way:

$$F_{x14} \quad = \quad F_{x1} + F_{x4}$$

$$F_{x23} \quad = \quad F_{x2} + F_{x3}$$

$$F_{z12} \quad = \quad F_{z1} + F_{z2}$$

$$F_{z34} \quad = \quad F_{z3} + F_{z4}$$

In determining the mathematical function for the x and z coordinates of the centre of force, p_x and p_z, two sets of moment equations can be formulated. The first set describes the moments that the forces measured with the transducers produce with respect to the laboratory coordinate system with origin in O (Fig. 3.1.5):

$$M_x \quad = \quad a\,(F_{y1} - F_{y2} - F_{y3} + F_{y4})$$

$$M_y \quad = \quad b\,(-F_{z12} + F_{z34}) + a\,(-F_{x14} + F_{x23})$$

$$M_z \quad = \quad b\,(F_{y1} + F_{y2} - F_{y3} - F_{y4})$$

The second set of equations includes the moments of the force components with respect to the laboratory coordinate system with origin in O (Fig. 3.1.5):

$$M_x \quad = \quad -p_z F_y - p_y F_z$$

$$M_y \quad = \quad -p_z F_x + p_x F_z + M_y'$$

$$M_z \quad = \quad -p_x F_y - p_y F_x$$

Note that the free moments M_x' and M_z' are zero as long as the shoe or foot is not sticking to the ground, e.g., for locomotion. The equations can be solved for the coordinates of the centre of force, p_x and p_z, and the free moment of rotation, M_y':

$$p_x \quad = \quad \left[\frac{-1}{F_y}\right] [b\,(F_{y1} + F_{y2} - F_{y3} - F_{y4}) + p_y\,(F_{x14} + F_{x23})]$$

$$p_z \quad = \quad \left[\frac{-1}{F_y}\right] [a\,(F_{y1} - F_{y2} - F_{y3} + F_{y4}) - p_y\,(F_{z12} + F_{z34})]$$

and:

$$M_y' \quad = \quad -b\,(- F_{z12} + F_{z34}) + a\,(-F_{x14} + F_{x23})$$

$$+ p_z\,(F_{x14} + F_{x23}) + p_x\,(F_{z12} + F_{z34})$$

The output of the force transducers on the four corners can be used to calculate the centre of force and the free moment of rotation applied to the force plate. However, the determination of the coordinates of the centre of force is connected with technical problems. The coordinates are calculated by dividing forces by forces. The calculation is sensitive to noise when these forces are small (division of a small quantity by a small quantity). As a result, the coordinates of the centre of force, $p_x(t)$ and $p_z(t)$, are typically inaccurate for small forces at the beginning and the end of the stance phase.

An example on how shoe construction can influence the path of the centre of force is illustrated in Fig. 3.1.6. It shows results of the centre of force for the initial phase of ground contact during heel-toe running (v = 4.5 m/s) for one test subject in three different running shoes. The shoes differed only in one aspect of construction, the shape of the heel on the lateral side of the rearfoot. One shoe had a flared, one a neutral, and one had a rounded heel construction. Other than that the shoes were identical. For illustration purposes the outline of the shoe sole and an arbitrarily defined line which may represent a projection of the "subtalar joint axis" are drawn. The example illustrates how even small alterations in shoe construction can be used to influence/change the centre of force path.

Locomotion studies often quantify the force between the subject or animal and the ground (ground reaction forces). For humans, the most frequently analysed locomotion is walking and running. In the following pages the ground reaction forces for running and walking are depicted and the typical characteristics are discussed. The ground reaction forces for heel-toe running (Fig. 3.1.7) are used to define impact and active forces.

The vertical force component for heel-toe running usually has two peaks. The first vertical force peak occurs about 5 to 30 milliseconds after first ground contact and is called *vertical impact force peak* and is referred to in this text as F_{yi}. Similar impact force peaks

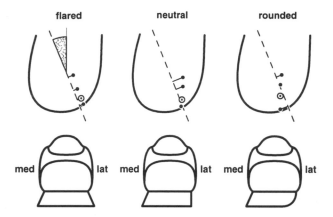

Figure 3.1.6   Illustration of changes in the centre of force path for one subject, running heel-toe (v = 4.5 m/s) in three different running shoes; a shoe with a flared (left), a neutral (centre), and a rounded lateral heel (right). The outline of the shoe during stance and an arbitrary selected projection of a possible "subtalar joint axis" are drawn to allow the reader an easy comparison between the three situations.

may occur in the two horizontal force time curves. In general, impact forces can be defined as follows:

> **Impact forces in human locomotion are forces that result from a collision of two objects reaching their maximum earlier than 50 milliseconds after the first contact of the two objects.**

In running, impact forces occur when the foot lands on the ground. The time occurrence and the magnitude of the impact force peaks depend on various factors, including running speed, material properties of heel and shoe sole, geometrical construction of the shoe sole (Nigg et al., 1987), and running style (Cavanagh and Lafortune, 1980). The vertical impact force peaks are smaller for toe landing than for heel landing, a result which is discussed further in chapter 4 (effective mass). The vertical impact force peaks are earlier for barefoot running (5 to 10 milliseconds after first contact) than for running with shoes, and earlier for running with harder shoe soles than running with softer shoe soles. Some individuals show more than one impact peak, one resulting when the heel hits the ground and one when the forefoot does. Impact forces in walking are not always evident. Some subjects show impact peaks, some do not.

The second peak in the vertical force time curve is called the *active vertical force peak* and is referred to in this text as F_{ya}. Similar peaks may occur in the horizontal force time curves. Active forces can be defined as follows:

> **Active forces in human locomotion are forces generated by movement that is entirely controlled by muscular activity.**

Vertical active force peaks from running at a speed of 4 m/s are about two to three times body weight. In running, F_{ya} occurs in the middle of the stance phase, which is

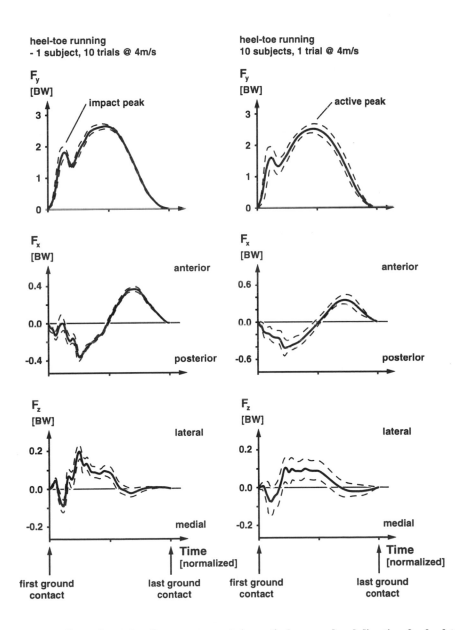

Figure 3.1.7   Ground reaction force components in vertical, a-p and m-l direction for heel toe running in units of body weight, BW. Mean and standard error for one subject and 10 trials (left) and 10 subjects with one trial each (right).

about 100 to 300 milliseconds after first ground contact (Fig. 3.1.7 and Fig. 3.1.8).

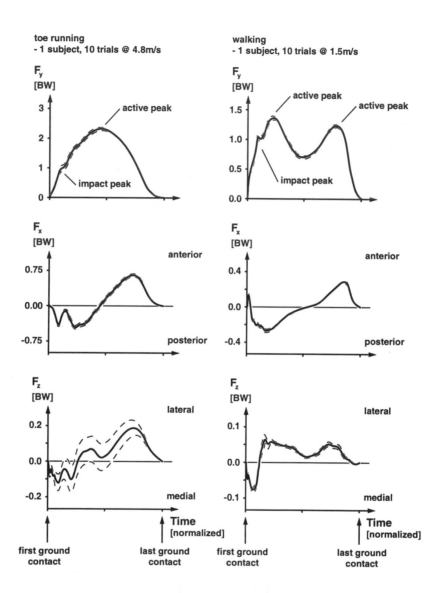

Figure 3.1.8   Ground reaction force components in vertical, a-p and m-l direction for toe running (left) and for walking (right) in units of body weight, BW. Mean and standard error for one subject and 10 trials.

In walking (Fig. 3.1.8), two active force peaks are present, the first being associated with deceleration and the second with the acceleration phase. The force component in the a-p direction has two active parts, which are similar in walking and running. In the first half of ground contact, the foot pushes in the anterior direction. Consequently, the reaction force from the force plate is directed in the posterior direction (backwards). In the second half of ground contact the foot pushes in the posterior direction. Consequently, the reaction force from the force plate is in the anterior direction. The force component in the m-l (medio-lateral) direction is less consistent intra- and inter-individually. The medio-lateral component often shows an initial reaction force in lateral direction that results from a medial (inward) movement of the foot during landing. This initial lateral force is usually shorter than 20% of the total contact time. It is usually followed by a reaction force in the medial direction that is often present during the rest of the ground contact time and that is usually smaller than the initial lateral force.

The intra- and inter-individual variability is much bigger for the medio-lateral than for the vertical and the anterior-posterior force-time curves. Furthermore, substantial differences may exist in the ground reaction force components between the left and the right foot fall for one subject.

Values of impact and active ground reaction force peaks during different movements have been reported by various authors. The summary of these results (Fig. 3.1.9) illustrates that the maximal external impact forces can exceed 10 BW (body weight) while the maximal external active forces do not exceed 5 BW.

Table 3.1.2   Selected experimental results for external vertical impact force peaks in walking and running (from Nigg, 1985, with permission).

MOVEMENT	FOOTWEAR	v [m/s]	F(max) [N]	F(max)/ BW	AUTHOR	YEAR
WALKING	barefoot	1.3	*385	0.55	Cavanagh	1981
	army boots	1.3	*259	0.37	"	"
	street shoes	1.3	*189	0.27	"	"
RUNNING (HEEL)	run. shoe	4.5	*1540	2.2	Cavanagh	1980
	run. shoe "hard"	4.0	2000	*2.9	Nigg	1980
	run. shoe "soft"	4.0	1100	*1.6	"	"
	run. shoe	3.4	1365	2.0	Frederick	1981
	run. shoe	3.8	1590	2.3	"	"
	run. shoe	4.5	1963	2.9	"	"
	run. shoe	3.0	1345	*2.0	Nigg	1987
	run. shoe	4.0	1521	*2.2	"	"
	run. shoe	5.0	1799	*2.6	"	"
	run. shoe	6.0	2070	*3.0	"	"
RUNNING (TOE)	run. shoe	4.0	300	*0.4	Denoth	1980
TAKE-OFF FOR JUMP	spikes	2.0	1000	*1.4	Nigg	1981
	spikes	4.0	2300	*3.3	"	"
	spikes	6.0	3700	*5.4	"	"
	spikes	8.0	5700	*8.3	"	"
	run. shoe	2.0	1400	*2.0	"	"
	run. shoe	4.0	2000	*2.9	"	"
	run. shoe	7.0	2900	*4.2	"	"

The quantification and analysis of impact forces in connection with running shoe research developed in the late 1970's. Several authors suggested that impact forces were of special importance in the etiology of injuries, even though there was no conclusive evidence for this speculation. Selected external impact forces as reported in the literature are summarized in Table 3.1.2.

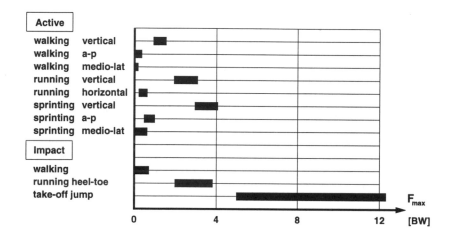

Figure 3.1.9   **Magnitude of actual impact and active ground reaction force peaks measured with a force plate during various activities (from Nigg, 1988, with permission).**

The results indicate that several factors influence the magnitude of the external impact force peaks including footwear, speed of movement, and type of movement. Externally measured forces are also used to estimate internal forces (see chapter 4, Modelling).

3.1.5 REFERENCES

Arms, S.W., Boyle, J., Johnson, R.J., and Pope, M.H. (1983) Strain Measurements in the Medial Collateral Ligament of the Human Knee: An Autopsy Study. *J. Biomechanics.* **16** (7), pp. 491-496.

Asang, E. (1974) Biomechanics of the Human Leg in Alpine Skiing. *Biomechanics IV* (eds. Nelson, R.C. and Morehouse, C.A.). University Park Press, Baltimore. pp. 236-242.

Beckwith, T.G., Buck, N.L., and Marangoni, R.D. (1982) *Mechanical Measurements* (3rd Ed.). Addison-Wesley, Reading, MS.

Bergmann, G., Graichen, F., and Rohlmann, A. (1993) Hip Joint Loading During Walking and Running, Measured in Two Patients. *J. Biomechanics.* **26** (8), pp. 969-990.

Bloach, A. (1935) New Methods for Measuring Mechanical Stress at Higher Frequencies. *Nature.* **136,** pp. 223-224.

Bobbert, M.F. and Schamhardt, H.C. (1990) Accuracy of Determining the Point of Force Application with Piezoelectric Force Plates. *J. Biomechanics.* **23** (7), pp. 705-710.

Breyer, H.G., Enes-Gaiao, F., and Kuhn, H.J. (1979) Biegebeanspruchung des Femurkopfprosthesenschaftes bei hypertrochanter Osteotomie. *Proceedings: Pauwels Symposium.* Free University Berlin, Berlin. pp. 136-143.

Brown, T.D., Sigal, L., Njus, G.O., Njus, N.M., Singerman, R.J., and Brand, R.A. (1986) Dynamic Performance Characteristics of the Liquid Metal Strain Gage. *J. Biomechanics.* **19** (2), pp. 165-173.

Cady, W.G. (1964) *Piezoelectricity: An Introduction to the Theory and Applications of Electromechanical Phenomena in Crystals* (2nd Ed.). Dover Publications, New York.

Canderale, P.M., Gola, M.M., and Gugliotta, A. (1979) New Theoretical and Experimental Developments in the Mechanical Design of Implant Stems of Stem-femur Coupling: New Improved Procedures. *Proceedings: Pauwels Symposium.* Free University Berlin, Berlin. pp. 199-208.

Canderale, P.M. (1983) The Use of Strain Gauges for Biometrics in Europe. *Engineering in Med.* **12 (3)**, pp. 117-133.

Carlet, G. (1872) Sur la Locomotion Humaine. *Ann. Sci. Naturelles.* Serie 5, pp. 1-92.

Cavagna, C.A. (1964) Mechanical Work in Running. *J. Applied Physiology.* **19 (2)**, pp. 249-256.

Cavanagh, P.R. and Lafortune, M.A. (1980) Ground Reaction Forces in Distance Running. *J. Biomechanics.* **13**, pp. 397-406.

Cavanagh, P.R., Williams, K.R., and Clarke, T.E. (1981) A Comparison of Ground Reaction Forces During Walking Barefoot and in Shoes. *Biomechanics VII-B* (eds. Morecki, A., Fidelus, K., Kedzior, K., and Wit, A.). University Park Press, Baltimore. pp. 151-156.

Denoth, J. (1980) Materialeigenschaften. *Sportplatzbeläge* (eds. Nigg, B.M. and Denoth, J.). Juris Verlag, Zürich. pp. 54-66.

Eaton, E.C. (1931) Resistance Strain Gauge Measures Stresses in Concrete. *Eng. News Records.* **107**, pp. 615-616.

Elftman, H. (1938) The Force Exerted by the Ground in Walking. *Arbeitsphysiologie.* **10**, pp. 485-491.

English, T.A. and Kilvington, M. (1979) In Vivo Records of Hip Load Using a Femoral Implant with Telemetric Output (A Preliminary Report). *J. Biomedical Eng.* **1**, pp. 111-115.

Frederick, E.C., Hagy, J.L., and Mann, R.A. (1981) Prediction of Vertical Impact Forces During Running. *J. Biomechanics.* **14 (7)**, p. 498.

Gohlke, W. (1959) *Einfüehrung in die piezoelektrische Messtechnik.* 2. Auflage, Akad. Verlagsges, Leipzig.

Herzog, W., Stano, A., and Leonard, T.R. (1993) Telemetry System to Record Force and EMG Recording from Cat Ankle Extensor and Tibialis Anterior Muscles. *J. Biomechanics.* (Submitted).

Holden, J.P., Grood, E.S., and Cummings, J.F. (1991) The Effect of Flexion Angle on Measurement of Anteromedial Band Force in the Goat ACL. *Proc. 37th Meeting of the Orthopedic Res. Soc.* p. 588.

Huiskes, R. (1980) Some Fundamental Aspects of Joint Replacement. *Acta. Orthop. Scand.* Suppl. p. 185.

Huiskes, R. and Sloot, T.J. (1979) Experimentelle und technerische Spannungsanalyse der Hüftgelenkverankerung. *Proceedings: Pauwels Symposium.* Free University Berlin, Berlin. pp. 173-198.

Jakob, H.A.C. and Huggler, A.H. (1979) Spannungsanalysen an Kunststoffmodellen des menschlichen Beckens sowie des proximalen Femurendes mit und ohne Prothese. *Proceedings: Pauwels Symposium.* Free University Berlin, Berlin. pp. 118-135.

Komi, P.V., Salonen, M., Jarvinen, N., and Kokko, O. (1987) In Vivo Registration of Achilles Tendon Forces in Man. *Int. J. Sports Med.* **8**, pp. 3-8.

Komi, P.V. (1990) Relevance of In Vivo Force Measurements to Human Biomechanics. *J. Biomechanics.* **23 (S-1)**, pp. 23-24.

Kotzar, G.M., Davy, D.T., Goldberg, V.M., Heiple, K.G., Berilla, J., Heiple Jr., K.G., Brown, R.H., and Burstein, A.H. (1991) Telemeterized in Vivo Hip Joint Force Data: A Report on Two Patients After Total Hip Surgery. *J. Orthopaedic Res.* **9 (15)**, pp. 621-633.

Lanyon, L.E. (1972) In Vivo Bone Strain Recorded from Thoracic Vertebrae of Sheep. *J. Biomechanics.* **12**, pp. 593-600.

Marey, E.J. (1972) *Movement.* Arno, New York. (Original 1895).

Marey, M. (1873) De la Locomotion Terrestre chez les Bipedes et les Quadrupedes. *J. de l'Anat. et de la Physiol.* **9**, p. 42.

Martini, K.H. (1983) Multicomponent Dynamometers Using Quartz Crystals as Sensing Elements. *ISA Transactions.* **22 (1)**, pp. 35-46.

Miura, M., Miashita, M., Matsui, H., and Sodeyama, H. (1974) Photographic Method of Analyzing the Pressure Distribution of the Foot Against the Ground. *Biomechanics IV* (eds. Nelson, R.C. and Morehouse, C.A.). University Park Press, Baltimore. pp. 482-487.

Nicol, K. and Hennig, E.M. (1976) Time Dependent Method for Measuring Force Distribution Using a Flexible Mat as a Capacitor. *Biomechanics V-B* (ed. Komi, P.V.). University Park Press, Baltimore. pp. 433-440.

Nigg, B.M. (1977) *Menschliche Microvibrationen.* Birkhäuser Verlag, Basel.

Nigg, B.M and Denoth, J. (1980) *Sportplatzbeläge*. Juris Verlag, Zürich.

Nigg, B.M., Denoth, J., and Neukomm, P.A. (1981) Quantifying the Load on the Human Body: Problems and Some Possible Solutions. *Biomechanics VII-B* (eds. Morecki, A., Fidelus, K., Kedzior, K., and Wit, A.). University Park Press, Baltimore. pp. 88-99.

Nigg, B.M. (1985) Loads in Selected Sports Activities: An Overview. *Biomechanics IX-B* (eds. Winter, D.A., Norman, R.W., Wells, R.P., Hayes, K.C., and Patla, A.E.). Human Kinetics Pub., Champaign, IL. pp. 91-96.

Nigg, B.M., Bahlsen, H.A., Lüthi, S.M., and Stokes, S. (1987) The Influence of Running Velocity and Midsole Hardness on External Impact Forces in Heel-toe Running. *J. Biomechanics.* **20 (10)**, pp. 951-959.

Nigg, B.M. (1988) The Assessment of Loads Acting on the Locomotor System in Running and Other Sport Activities. *Seminars in Orthopaedics Vol. 3.* **4 (Dec.)**, pp. 197-206.

Nolte, V. (1979) *Die Handschrift des Ruderers*. Messtechnische Briefe. **15**, pp. 49-53.

Platt, D.M., Wilson, A.M., and Goodship, A.E. (1992) Techniques for Tendon Force Measurement. *Experimental Mechanics: Technology Transfer Between High Tech Engineering and Biomechanics* (ed. Little, E.G.). Elsevier, New York. pp. 389-398.

Schneider, E. (1980) *Leistungsanalyse bei Rudermannschaften*. Limpert Verlag, Bad Homburg.

Salmons, S. (1969) The 8th Annual International Conference on Medical and Biological Engineering: Meeting Report. *Biomed. Eng.* **4**, pp. 467-474.

Shrive, N.G., Damson, E., and Frank, C.D. (1992) Technology Transfer Regarding the Measurement of Strain on Flexible Materials with Special Reference to Soft Tissues. *Experimental Mechanics: Technology Transfer Between High Tech Engineering and Biomechanics* (ed. Little, E.G.). Elsevier, New York. pp. 121-130.

Tichy, J. and Gautschi, G. (1980) *Piezoelektrische Messtechnik*. Springer Verlag, Berlin.

van Campen, D.H., Croon, H.W., and Lindwer, J. (1979) Mechanical Loosening of Knee-endoprostheses Intermedullary Stems: Influence of Dynamic Loading. *Proceedings 25th Annual ORS.* **98**, pp. 20-22.

Walmsley, B., Hodgson, J.A., and Burke, R.E. (1978) Forces Produced by Medial Gastrocnemius and Soleus Muscles During Locomotion in Freely Moving Cats. *J. Neurophysiology.* **41 (5)**, pp. 1203-1216.

Weightman, B. (1976) The Stress in Total Hip Prosthesis Femoral Stems: A Comparative Experimental Study. *Engineering in Medicine.* **2**, pp. 138-147.

White, A.A. and Raphael, I.G (1972) The Effect of Quadriceps Loads and Knee Position on Strain Measurements of the Tibial Collateral Ligament. *Acta. Orthop. Scand.* **43**, pp. 176-187.

Xu, W.S., Butler, D.L., Stouffer, D.C., Grood, E.S., and Glos, D.L. (1992) Theoretical Analysis of an Implantable Force Transducer for Tendon and Ligament Structures. *J. Biomechanical Eng.* **114 (2)**, pp. 171-177.

3.2 PRESSURE DISTRIBUTION

NIGG, B. M.

3.2.1 DEFINITIONS AND COMMENTS

Centre of force:	Point in a plane of interest through which the line of action of the resultant force vector passes. Comment: the plane of interest is, for instance, the force plate.
Crosstalk:	A signal at the output of a transducer caused by a variable not allocated to this particular output. Comment: the common crosstalk in force sensors occurs in all channels. It is, for instance, possible for the vertical force channel to show a signal when only horizontal forces are acting on the sensor.
Drift:	Change in output of the transducer over time that is not a function of the measurand.
Frequency:	Number of ostillations per time unit.
• Natural frequency:	A frequency of vibration that corresponds to the elasticity and mass of a system under the influence of its internal forces and damping. Comment: the lowest natural frequency of a system is known as the fundamental frequency. Higher natural frequencies are harmonics of the fundamental value. Some systems exhibit several coupled or uncoupled modes of vibration.
• Resonance frequency:	The frequency of a forced vibration input to a system that corresponds to a natural frequency of the system.
Hysteresis:	The biggest difference in the output signal for any value of the measured variable within the specified range when this value is reached with increasing or diminishing measurand.
Linearity:	Output signal strength in direct proportion to input signal strength. Comment: the property of linearity is not necessarily essential for obtaining accurate measurements.
Moment:	Turning effect of a force, $\mathbf{F}$, applied at point P about a base point O with:

$$\mathbf{M_O} = \mathbf{r_{PO}} \times \mathbf{F}$$

where:

	M_O = moment (or moment of force) about point O
	r_{PO} = position vector from O to P
	F = force

Noise:
Any unwanted disturbance of the original signal in a frequency band of interest.
Comment: one distinguishes between *background noise* and *random noise*.

Normal force:
Force perpendicular to the surface.

Overload:
The highest value of a measured variable that the transducer can sustain without its measuring properties being changed beyond the specified tolerances.

Pressure:
Force perpendicular to the surface of a sensor per unit area of this sensor where:

p = pressure
F = force perpendicular to the surface of a sensor
A = area of this sensor

with the unit:

$$[p] = Pa = N \cdot m^{-2}$$

Pressure is a scalar quantity.

Range:
The upper and lower limits of a variable for which a transducer is intended.

Rigidity:
Measured deflection as a result of a defined acting force.

Sensitivity:
The ratio between the change in the transducer output signal and the corresponding change in the measured value.
Comment: sensitivity can change over the possible range of measurements.

Threshold:
The smallest change in the measured variable that causes a measurable alteration of the transducer output signal.

Torque:
Result of a force couple.

The force plates described in chapter 4 measure a resultant ground reaction force which is a summation of all the forces acting from the ground on the test subject's feet.

Assume a test subject standing (bipedal stance) on a force plate with the inside of the feet 20 cm apart. For this example, the force plate will indicate a resultant force vector with the magnitude of body weight, acting in a vertical direction, with the line of action of the force vector passing through a point on the surface of the force plate between the two feet (Fig. 3.2.1). This point on the force plate is called "point of application", "centre of

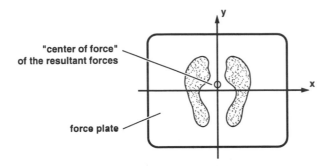

Figure 3.2.1   **Illustration of the possible location of the centre of force in bipedal standing.**

pressure", or "centre of force". The "resultant force" is obviously not a real force that can be felt by a specific part of the test subject's body. It is the vector sum of all the local forces acting from the force plate onto the feet. Similarly, the point where the resultant force vector intersects with the force plate is not a point where an actual force is acting but the point through which a theoretical resultant force would pass. Fig. 3.2.1 shows that this point may not even be a contact point between test subject and force plate. Consequently, it is suggested that the use of the term *centre of force* would be more appropriate than the use of the term point of application.

The use of a resultant force with a centre of force is often appropriate in the analysis of human motion and/or behaviour. However, in some applications information about the local forces acting between the force measuring device and parts of the contacting body may be of importance, thus, information about the local pressure distribution is needed. This chapter discusses selected aspects relevant to pressure distribution measurements.

3.2.2 SELECTED HISTORICAL HIGHLIGHTS

Attempts to quantify pressure distribution began to be made at the end of the 19th century. However, significant progress in the development of commercially available measuring devices was only made between 1980 and 1990.

1882	Beely	Used subjects standing on a sack of plaster of Paris. Impressions were assumed to be related to maximal local forces.
1927	Abramson	Used steel shot underneath a soft lead sheet. Penetration of steel shot into lead sheet indicated amount of maximal local force.
1927	Basler	Used force plate beams suspended on metal wires. Frequency of wires was assumed to be proportional to the force acting on the beam.

1930	Morton	Used a rubber mat with longitudinal ridges on top of an inked ribbon. The colouring of inked ribbon was proportional to the maximal local forces.
1934	Elftman	Used a rubber mat with pyramidal projections on its undersurface mounted on a rigid glass plate with reflecting fluid in the spaces between the mat and the glass plate. Filming the deformation of the rubber pyramids provided information on the dynamic pressure distribution.
1954	Barnett	Developed a "plastic pedobarograph", an instrument with 640 vertically mounted beams on a rubber support. The movement of the beams was filmed to determine the dynamic pressure distribution.
1963	Baumann	Used capacitive force transducers for selected local forces between foot and shoe.
1972	Hutton	Developed a strain gauge force plate divided into beams.
1976	Arcan	Used a sandwich of reflective and polarizing material with interference techniques to assess the plantar pressure distribution in standing individuals.
1976	Nicol	Developed and used capacitive force transducers and multiplexing techniques for the assessment of pressure distribution with flexible mats.
1976	Scranton	Developed a transducer to quantify local plantar forces using shear sensitive liquid crystals.
1976	Theyson	Used an array of pins on high precision springs and video analysis for clinical gait analysis.
1982	Hennig	Used piezoceramic transducers in a rubber matrix and applied them for insoles.
1983	Aritomi	Developed force detecting sheets (Fuji foils) for maximal local pressure measurements.
1985	Diebschlag	Used inductive sensors embedded in a resilient material for pressure distribution measurement in an insole.
1987	Maness	Developed ultra thin conductive type transducers to quantify pressure distribution in various applications.
1990	Nicol	Developed strain gauge sensors on a silicon membrane for pressure distribution measurements on soft surfaces.

3.2.3 MEASURING POSSIBILITIES

The possibilities for quantifying forces have been discussed in chapter 4 and will not be repeated here. However, those methods which have been used in pressure distribution measurements are listed in this chapter, and their characteristics as they relate to pressure distribution measurements are discussed. The main emphasis of this section is on measuring devices that have been used in pressure distribution plates and in devices that allow the quantification of pressure distribution between two surfaces (e.g., foot/shoe, teeth, leg/shoe shaft). Single force sensors used for the assessment of local pressures are discussed in less detail.

The measuring elements used in pressure distribution measurements (in alphabetical order) are:

- Capacitor.
- Conductor.
- Critical light reflection.
- Force sheet (Fuji foils).
- Inductive sensor.
- Piezoceramic element.
- Reflecting/polarizing sheet.
- Rod/spring elements.
- Strain gauge element.

Their main characteristics are discussed in the following paragraphs. The discussion is grouped into two parts. The first part covers those measuring techniques which have been used frequently in biomechanical research (capacitor, conductor, force sheet, and piezoceramic element). For those methods, the discussion is detailed, and specific characteristics are listed. The second part lists the remaining methods, which, in principle, work but are less frequently used in biomechanical research.

CAPACITOR SENSORS

Pressure distribution plates and/or insoles on the basis of capacitor elements consist of n stripes of conducting material arranged in rows on the top, m stripes made of conducting material arranged in columns on the bottom, and a resilient dielectric material between the two layers of stripes. The stripes of row i and column k overlap and form a condenser, C_{ik}. This construction consists of m times n condenser elements (Nicol and Hennig, 1976).

The dielectric between the "condenser plates" is an elastic material, and it's elastic material properties are of critical importance for the quality of the sensors. Ideally, dielectric materials are required that allow compression to about 50% of the unloaded thickness under maximal force. Dielectric materials with visco-elastic properties produce hysteresis in the force signals and should be avoided.

In most applications, a multiplexing technique is used to assess the forces acting on each element C_{ik}. Alternating current is sequentially connected to each row by a de-multiplexer switch, whereas, the columns are connected to a resistor via a multiplexer switch (Nicol and Hennig, 1976). Thus, each measuring capacitor, C_{ik}, is an element of a voltage divider. The advantage of such a multiplexer circuit is that only (m + n) channels and the same number of cables are required for m times n sensors. A pressure distribution insole, for instance, with 10 rows and 30 columns would have 40 cables and 300 sensors, which may provide adequate information on pressure distribution underneath the foot and provides a technical solution with acceptable interference to the test subject. If, however, each sensor would require a cable connection, 300 cables would be needed for the same arrangement which could encumber the test subject substantially.

CONDUCTOR SENSORS

Force transducers based on the principle of conductivity have been described in chapter 4. Pressure distribution measuring devices, based on conductivity, use the same construction and multiplexing ideas as were discussed for the pressure distribution measuring devices based on capacitance. Rows and columns of conductive material are used in the top and bottom layer of the system. However, the material between the rows and columns functions as a resistance (not a dielectric, as before) and consists of force sensitive ink (Maness et al., 1987) or of carbon-impregnated conductive polyurethane foam (Shereff et al., 1985). As force is applied to a sensor, the electric resistance of the material between the rows and columns is changed. The change in resistance corresponds to the applied force.

PRESSURE SHEET (FUJI FOIL)

Pressure sheets consist of two sheets A and C (Fig. 3.2.2). The upper side of sheet A is

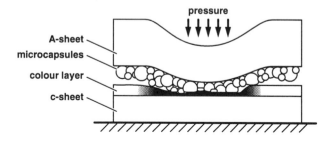

Figure 3.2.2   **Illustration for the force sheet method for maximal pressure distribution measurement as presented by Aritomi et al. (1983), with permission.**

in contact with the force applying object (e.g., the human foot). Microcapsules which contain a colour producing agent are coated onto the lower side of sheet A. Sheet A lies on top of sheet C. When force is applied locally, capsules are ruptured and the colour-producing agents colour sheet C. The colour intensity corresponds to the maximal force applied locally, and the intensity of the coloured sheet C provides a picture for the maximal local forces. With the optical density:

D   =   log (1/transmission)

the colour intensity can be quantified. The correlation between optical density, D, and local force, F_{loc}, is linear in the middle part but non-linear in the end parts of the applicable range of the foil (Hehne, 1983). In the case of a dynamic measurement, the maximal forces at different locations may not occur at the same time (Aritomi et al., 1983).

PIEZOCERAMIC SENSORS

Hennig et al. (1981) used 499 piezoceramic transducers in their first attempt to develop a measuring device to quantify the pressure distribution between foot and shoe. The elements had a surface area of 4.78 mm^2 and were embedded in a 3 to 4 mm thick layer of highly resilient silicon rubber. The embedment was electrically isolating and impermeable to moisture. Each sensor was connected through a cable to a data collection unit. In the first attempt, 499 cables connected the measuring insole with the data collection unit.

Table 3.2.1   Selected characteristics for pressure distribution sensors based on data from Nicol and Hennig, 1976; Hennig et al., 1981; Aritomi et al., 1983; Cavanagh et al., 1983; Hehne, 1983; Maness et al., 1987; with permission, and on own experiences from experimental projects. For general characteristics of force sensors see chapter 4.

CHARACTERISTIC	SENSOR TYPE			
	CAPACITOR	CONDUCTOR	PRESSURE SHEET	PIEZO-CERAMIC
FLEXIBILITY (INSOLES)	limited due to dimensions	high but may wrinkle	high but may wrinkle	good
TEMPERATURE SENSITIVITY	not sensitive	not sensitive	sensitive	must be taken into account
DYNAMIC MEASUREMENTS	standard	standard	not possible yet	standard
MAJOR LIMITATIONS	• only perpendicular forces • limited flexibility	• only perpendicular forces • changing sensitivity • wrinkle	• force direction not known • only static • wrinkle • only max. pressure	• only perpendicular forces • temperature sensitivity
COSTS	high	low	low	high

The use of piezoceramic transducers is not without problems. The fact that each transducer is connected directly with the outside, requires special shoe adaptations for the measurements to provide room for cables or holes to take the cables out of the shoe at different locations. Consequently, piezoceramic measuring devices are limited in the number of measuring cells because of the number of cables.

The piezoceramic transducers were used for the first comprehensive pressure distribution measurements in shoes (Hennig et al., 1981; Cavanagh et al., 1983), providing initial insight into the foot-shoe interaction. However, this technology was not further developed and is not widely used in biomechanical research at the present time.

A comparison of selected characteristics of the most important pressure distribution sensors is summarized in Table 3.2.1.

OTHER METHODS

In addition to the above-mentioned methods to quantify pressure distribution, various other methods were developed in the past that seem to be less often applied in current biomechanical research and testing.

The *critical light reflecting technique* was used by Chodera and Lord (1979) to assess the pressure distribution between the foot and the ground. The hardware consisted of a plastic sheet, a glass plate, a video camera, and a microcomputer. The plastic sheet is flat on the upper side, which is in contact with the foot, and has an array of small knobs on the bottom side, which is in contact with the glass plate. The single knobs have a point contact with the glass when no force is applied to the plastic. However, the knobs flatten against the glass surface when a local force is applied. The glass plate is illuminated from the side. The light rays passing through the glass are totally reflected if the neighboring medium on top of the glass is air. The light is partially reflected if the neighboring medium is plastic. If the forces applied to the plastic increase, the surface area that is in contact with plastic increases and the intensity of the reflected light can be used as an indicator of the locally applied force. The system used by Chodera and Lord (1979) has the limitation that only pressure distribution components perpendicular to the measuring surface can be quantified. However, a system based on the critical light reflecting technique that is able to analyse the pressure distribution triaxially is reported to be under construction by Chao et al. (Alexander et al., 1990).

Inductive sensors have been used by Diebschlag et al. (1985). In their construction, a spring suspended metal pin is inserted into a coil with pin and coil embedded into a flexible rubber material. If force is applied, the pin penetrates the coil and inductive current can be measured. A second order multiplexer technique was used to scan the sensors.

Arcan and Bull (1976) used a sandwich of a *reflective and polarizing material* placed onto a rigid transparent plate. Depending on local forces interference patterns were generated and used for the qualitative analysis of dynamic pressure distribution in locomotion (Lord, 1981).

Theyson et al. (1979) suspended about 1000 pins on high precision springs (*rod/spring elements*), recorded the shadows of the springs on videotape, and performed a computer analysis. Cavanagh and Ae (1978) further developed this methodology to provide qualitative (visual) and quantitative information on pressure distribution during locomotion. The rod/spring system is used currently in routine gait analysis in Germany and is built in small numbers commercially.

Nicol (personal communication) used *strain gauge sensors* on a silicon membrane to quantify dynamic pressure distribution on soft surfaces, which provided improved flexibility when compared to the earlier methods. These sensors seem more accurate than comparable single foil capacitor sensors used for soft surfaces.

New developments include *piezosheets* or *piezofoils*, extremely thin measuring sheets that allow static and dynamic quantification. One major application seems to be pressure distribution measurements in joints.

3.2.4 APPLICATIONS

Pressure distribution measuring devices are currently used in several major biomechanical and clinical fields of application.

Plates measuring pressure distribution are currently on the market with a sensor size in the order of less than 0.5 cm². Plates of different construction were used in gait analysis (Elftman, 1934; Grundy et al., 1975; Cavanagh and Ae, 1978; Clarke, 1980; Hutton and Dhanendran, 1981) and in the assessment of the diabetic foot (Stokes et al., 1975; Rhodes et al., 1988). General gait analysis concentrated on the analysis of temporal plantar pressure distribution patterns for different foot regions, the "movement" of the centre of force, and the plantar pressure distribution patterns for specific foot types (Clarke, 1980). Additionally, pressure distribution plates may have been used to describe foot types dynamically.

Research activities initiated by sport shoe manufacturers were partially responsible for the development of *shoe insole* devices that assess the pressure distribution between the plantar aspect of the foot and the shoe insole (Hennig et al., 1981; Cavanagh et al., 1983). The devices were used to study the effect of specific shoe constructions on the pressure distribution at the shoe-sole interface (Soames, 1985), the pressure distribution at the ski boot-shaft interface (Hauser and Schaff, 1987), or the estimation of internal forces in the human foot (Morlock and Nigg, 1991). Additionally, insoles were used to study the effect of foot type and/or orthotics on pressure distribution (Fig. 3.2.3).

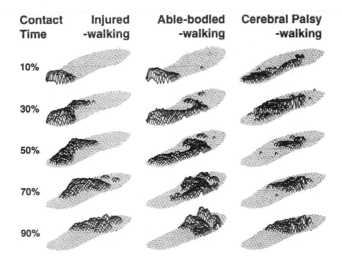

| Contact Time | Injured -walking | Able-bodied -walking | Cerebral Palsy -walking |

Figure 3.2.3   **Illustration of pressure distribution measurements between foot and shoe sole for selected applications measured with a pressure distribution insole. The pictures for injured-walking and CP-walking are for the left foot and the picture for the able-bodied walking is for the right foot.**

Additionally, pressure distribution measuring devices have been used for *local pressure distribution* measurements in specific applications. Hehne (1983) used Fuji foils to determine experimentally maximal pressure distribution between patella and femur in cadaver knees. Capacitive or conducting devices have been used to measure pressure distribution between residual stump and prosthesis of amputees, and to monitor changes in the points of tooth contact during jaw closure (Podoloff and Benjamin, 1989).

A further (indirect) application of pressure distribution measuring devices is the combination of thin conductive type transducer sheets with force plates where the sheets are mounted on top of the force plate. In this application the pressure distribution sensor is primarily used to determine the location of the foot on the force plate. This combination may prove powerful in clinical applications in which the combination of force and foot is of interest.

In addition to the extensively discussed pressure distribution plates and insoles, the literature provides many studies in which local forces have been quantified with single sensors where the sensors often were held in place with adhesive tape (Schwartz and Heath, 1937; Bauman and Brand, 1963; Shereff et al., 1985; Soames, 1985).

The main advantage of these techniques is the limited electronic and computer power required. The main criticism of this technology is the potential variation of the positioning of the transducers during movement, a concern which may be of importance for clinical and sport applications.

The methods that use plates or insoles for pressure distribution measurements can also be used to determine the path of the centre of force during one ground contact. The calculation of the centre of force using this method is less sensitive to noise than the method described for the force plates in chapter 4. An example for the path of the centre of force during a walking stride is illustrated in Fig. 3.2.4. It is measured with a pressure distribu-

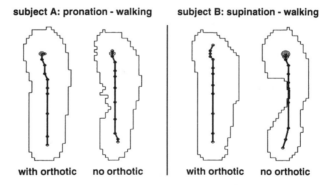

subject A: pronation - walking   **subject B: supination - walking**

with orthotic   no orthotic   with orthotic   no orthotic

Figure 3.2.4   **Path of centre of force for one test subject during one ground contact for walking with a sport shoe (left) and walking with the same shoe and an orthotic (right).**

tion insole and shows the influence of an orthotic in a running shoe on the path of the centre of force.

The methods for quantifying pressure distribution between foot and shoe are relatively new, and reliable instrumentation has only been on the market for a few years. However, such devices have good potential to be applied in various specific tasks since they provide on-line information which seems to be easily understandable. Groups that may use such devices include gait analysis laboratories in hospitals, sport medicine centres, podiatry centres, (sport) shoe stores, sport training centres (e.g., golf), and others. However, specific research is needed to provide the necessary background for these applications.

The various techniques for quantifying pressure distribution discussed in this section have been used primarily for applications related to the human foot, which is where biomechanical research applications have concentrated in the last few decades. However, the methods are applicable to most other biomechanical studies in which pressure distribution is the appropriate technique to be applied.

3.2.5 REFERENCES

Alexander, I.J., Chao, E.Y.S., and Johnson, K.A. (1990) The Assessment of Dynamic Foot-to-ground Contact Forces and Plantar Pressure Distribution: A Review of the Evolution of Current Techniques and Clinical Applications. *Foot & Ankle.* **11** (3), pp. 152-167.

Arcan, M. and Brull, M.A. (1976) A Fundamental Characteristic of the Human Body and Foot, the Foot-Ground Pressure Pattern. *J. Biomechanics.* **9** (7), pp. 435-457.

Aritomi, H., Morita, M., and Yonemoto, K. (1983) A Simple Method of Measuring the Footsole Pressure of Normal Subjects Using Prescale Pressure-detecting Sheets. *J. Biomechanics.* **16** (2), pp. 157-165.

Bauman, J.H. and Barnd, P.W. (1963) Measurement of Pressure Between Foot and Shoe. *The Lancet.* **1**, pp. 629-633.

Cavanagh, P.R. and Ae, M. (1978) A Technique for the Display of Pressure Paths from a Force Platform. *J. Biomechanics.* **11**, pp. 487-491.

Cavanagh, P.R., Hennig, E.M., Bunch, R.P., and Macmillan, N.H. (1983) A New Device for the Measurement of Pressure Distribution Inside the Shoe (eds. Matsui, H. and Kobayashi, K.). *Biomechanics VIII-B.* Human Kinetics, Champaign, IL. pp. 1089-1096.

Chodera, J.D. and Lord, M. (1979) Pedobarographic Foot Pressure Measurement and the Applications (eds. Kenedi, R.M., Paul, J.P. and Hughes, J.). *Disability.* MacMillan, London. pp. 173-181.

Clarke, T.E. (1980) The Pressure Distribution Under the Foot During Barefoot Walking. (Unpublished doctoral dissertation) Pennsylvania State University, University Park, PA.

Elftman, H. (1934) A Cinematic Study of the Distribution of Pressure in the Human Foot. *Anatomical Record.* **59** (4), pp. 481-487.

Grundy, M., Blackburn, P.A., Tosh, P.A., McLeish, R.D., and Smith, L. (1975) An Investigation of the Centres of Pressure Under the Foot While Walking. *J. Bone Jt. Surg.* **57** (B-1), pp. 98-103.

Hauser, W. and Schaff, P. (1987) Ski Boots: Biomechanical Issues Regarding Skiing Safety and Performance. *Int. J. of Sport Biomechanics.* **3**, pp. 326-344.

Hehne, H.J. (1983) *Das Patellofemoralgelenk.* Enke Verlag, Stuttgart.

Hennig, E.M., Cavanagh, P.R., Albert, H.T., and Macmillan, N.H. (1981) A Piezoelectric Method of Measuring the Vertical Contact Stress Beneath the Human Foot. *J. Biomed. Engng.* **4**, pp. 213-222.

Hennig, E.M., Cavanagh, P.R., and Macmillan, N.H. (1983) Pressure Distribution by High Precision Piezoelectric Ceramic Force Transducers. *Biomechanics VIII-B* (eds. Matsui, H. and Kobayashi, K.). Human Kinetics, Champaign, IL. pp. 1081-1088.

Hutton, W.C. and Dhanandran, M. (1981) The Mechanics of Normal and Halux Valgus Feet: A Quantitative Study. *Clin. Orthop.* **157**, pp. 7-13.

Lord, M. (1981) Foot Pressure Measurement: A Review of Methodology. *J. Biomed. Eng.* **3**, pp. 91-99.

Maness, L.W., Benjamin, M., Podoloff, R., Bobick, A., and Golden, R.F. (1987) Computerized Occlusal Analysis: A New Technology. *Quintessence International.* **18** (4).

Morlock, M. and Nigg, B.M. (1991) Theoretical Considerations and Practical Results on the Influence of the Representation of the Foot for the Estimation of Internal Forces with Models. *Clin. Biomechanics.* **6**, pp. 3-13.

Morton, D.J. (1930) Structural Factors in Static Disorders of the Foot. *American J. of Surgery.* **9**, pp. 315-326.

Nicol, K. and Hennig, E.M. (1976) Time Dependent Method for Measuring Force Distribution Using a Flexible Mat as a Capacitor. *Biomechanics V-B* (ed. Komi, P.V.). University Park Press, Baltimore. pp. 433-440.

Nicol, K. and Rusteberg, D. (1991) Measurement of Pressure Distribution on Curved and Soft Surfaces: Applications in Medicine. Institut für Sport und Sportwissenschatt der Johann Wolfgang Goethe-Universität. (Submitted).

Podoloff, R.M. and Benjamin, M. (1989) A Tactile Sensor for Analyzing Dental Occlusion. *Sensors.* **6,** pp. 41-47.

Rhodes, A., Sherk, H.H., Black, J., and Margulies, C. (1988) High Resolution Analysis of Ground Foot Reaction Forces. *Foot & Ankle.* **9 (3),** pp. 135-138.

Schwartz, R.P. and Heath, A.L. (1937) Some Factors Which Influence the Balance of the Foot in Walking: The Stance Phase of Gait. *J. Bone Jt. Surg.* **19 (2),** pp. 431-442.

Shereff, M.J., Bregman, A.M., and Kummer, F.J. (1985) The Effect of Immobilization Devices on the Load Distribution Under the Foot. *Clin. Orthop.* **192,** pp. 260-267.

Soames, R.W. (1985) Foot Pressure Patterns During Gait. *J. Biomech. Eng.* **7,** pp. 120-126.

Stokes, I.A.F., Faris, I.B., and Hutton, W.C. (1975) The Neuropathic Ulcer and Loads on the Foot in Diabetic Patients. *Acta. Orthop. Scand.* **46,** pp. 839-847.

Theyson, H., Güth, V., and Abbink, F. (1979) Fotogrammetrische Methoden zur 3-dimensionalen Vermessung der Fußsohle sowie eine Methode zur exakten Bestimmung des Druckablaufs unter den verschiedenen Bereichen der Fußsohle beim Gehen. *Funktionelle Diagnostik in der Orthopädie.* Stuttgart. pp. 87-90.

3.3 ACCELERATION

NIGG, B.M.

3.3.1 DEFINITIONS AND COMMENTS

Acceleration:	The time derivative of the velocity (vector) or the second time derivative of the position (vector).

$$a \quad = \quad \frac{d\mathbf{v}}{dt} \quad = \quad \frac{d^2\mathbf{x}}{dt^2}$$

Accelerometer:	A device used to measure acceleration.
Critical damping:	The amount of damping required to return a displaced elastic system to its initial position in minimal time without oscillation or overshoot.
Damping:	The dissipation of energy with distance or time.
Damping factor:	The ratio of actual damping to critical damping.
Natural frequency:	A frequency of vibration that correspond to the elasticity and mass of a system under the influence of its internal forces and damping.
	Comment: the lowest natural frequency of a system is known as the fundamental frequency. Higher natural frequencies of a system are harmonics of the fundamental value. Some systems exhibit several coupled or uncoupled modes of vibration.
Resonance frequency:	The frequency of a forced vibration input to a system corresponding to a natural frequency of the system.
Shock rating:	Highest input for which the transducer is not damaged.

3.3.2 SELECTED HISTORICAL HIGHLIGHTS

Accelerometry, the measurement of acceleration with electronic equipment, is relatively new in biomechanics. Acceleration measurement methods were developed parallel to force measuring techniques. These days, accelerometers can be bought commercially. However, the use of accelerometers is not widespread in biomechanical applications, possibly due to the problems associated with these measurements, as discussed in the following sections.

1967 Gage	Used accelerometers to determine the vertical and horizontal accelerations of the trunk as well as the angular acceleration of the shank analyzing human gait.

1972	Prokop	Used accelerometers mounted in the shoe sole of spike shoes on various track surfaces.
1973	Morris	Used five accelerometers to quantify the three-dimensional movement of the shank assuming that the transverse rotations of the shank are small and can be neglected.
1973	Nigg	Used accelerometers mounted at the head, hip, and shank during alpine skiing.
1974	Unold	Used accelerometers mounted at the head, hip and shank during walking and running with different footwear on various surfaces.
1977	Saha	Studied the effect of soft tissue in vibration tests using skin mounted accelerometers.
1978	Chao	Proposed an experimental protocol for the quantification of joint kinematics using systems of accelerometers.
1979	Light	Measured skeletal accelerations at the tibia using bone mounted accelerometers.
1979	Ziegert	Studied the effect of soft tissue on skin mounted accelerometer measurements.
1980	Denoth	Used acceleration measurements in an effective mass model to determine bone-to-bone impact forces in the ankle and knee joint.
1980	Denoth	Used accelerometers to determine *in vivo* force deformation diagrams of human heels.
1980	Light	Compared results from bone and skin mounted accelerometers to find a loss of high frequency content and a phase shift for skin mounted accelerometers.
1983	Voloshin	Studied the shock (impact force) absorbing capacity of the leg using skin mounted accelerometers in conjunction with force plate measurements.
1987	Valiant	Estimated a magnification factor for skin mounted accelerometers using a linear spring damper model.
1991	Lafortune	Described the contribution of angular motion and gravity to tibial acceleration during walking and running.

3.3.3 MEASURING POSSIBILITIES

Acceleration can be determined by using the measuring techniques described in chapter 3 or by using positional information (from film or video measurements) and calculating the second time derivative. Sensors described as *accelerometers* belong to the first category. They include strain gauge, piezoresistive, piezoelectric, and inductive transducers and are discussed in the following paragraphs.

STRAIN GAUGE AND PIEZORESISTIVE ACCELEROMETERS

A strain gauge accelerometer consists of a number of strain sensitive wires (often four) attached to a cantilevered mass element that is mounted on a fixed base. The wires are connected to an electric Wheatstone bridge circuit. If the base is accelerated the mass element causes a deformation due to its inertia. Deformation of the mass element causes a change of the strain in the wires, thus, changing their resistance and, consequently, changing the balance of the bridge circuit. The result is an electric output proportional to the acceleration of the base.

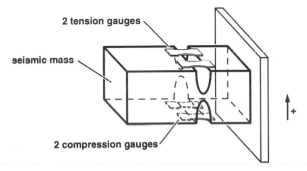

Figure 3.3.1   **Illustration of a piezoresistive accelerometer. From Instruction Manual for Endevco Piezoresistive Accelerometers (1978), with permission.**

A piezoresistive accelerometer is based on the same principle as a strain gauge accelerometer, except that piezoresistive instead of wire strain gauge elements are used. The transducers are typically solid state, single silicon crystals which change their electric resistance in proportion to the applied mechanical stress. The design of a piezoresistive accelerometer is illustrated in Fig. 3.3.1. For an acceleration in the illustrated direction, which corresponds in this example to an axis of measurement, and for the corresponding displacement of the mass element, the two top gauge elements operate in tension and the two bottom gauge elements in compression (Fig. 3.3.2).

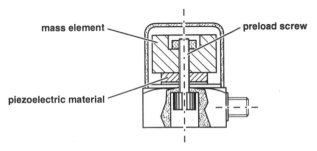

Figure 3.3.2   **Schematic illustration of the construction of a piezoelectric accelerometer (from Bouche, 1974, with permission).**

PIEZOELECTRIC ACCELEROMETERS

Piezoelectric accelerometers work on the same principles as outlined earlier (chapter 3.1). They often use ceramic materials with piezoelectric properties, producing an electric output in response to a stress. In a typical set-up, the piezoelectric material is positioned and pre-loaded in compression between the base of the accelerometer and a mass element (see section 3.3.2). Vibration of the base and the inertia of the mass element create dynamic stress and deform the piezoelectric material, resulting in an electric output.

INDUCTIVE ACCELEROMETERS

An inductive accelerometer consists of a mass element positioned and magnetically coupled between a pair of coils attached to the accelerometer base (Fig. 3.3.3). When the mass is accelerated the magnetic coupling is altered. This changes the inductive current of the coils and is measured as a change in the electric output of the coils.

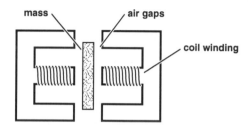

Figure 3.3.3   **Schematic illustration of the construction of an inductive accelerometer.**

The characteristics of accelerometers, of course, correspond to the characteristics of the force sensors constructed on the same principles. However, due to the specific use of accelerometers in biomechanics, accelerometer-specific characteristics can be discussed. Factors of specific interest for accelerometers include mass, natural frequency, frequency response, shock rating, and range (Table 3.3.1).

Most accelerometers available for the measurements in biomechanics are extremely light and weigh on a few grams. The natural frequency is typically high and the frequency response is sufficiently high for the frequencies relevant for biomechanical measurements (an exception are the inductive accelerometers). The shock rating for strain gauges and piezoresistive accelerometers is typically about one magnitude higher than the upper limit of the range, and may be a safety problem for accelerometers with low measuring ranges (dropping of such accelerometers usually results in the destruction of the sensor). Piezo-electric accelerometers have the advantage that their range is excellent while the range for strain gauges and piezoresistive accelerometers is typical about one magnitude. Acceler-ometers used for biomechanical measurements apply typically strain gauge techniques with conventional or piezoresistive strain wires or piezoelectric accelerometers. Strain gauges using the piezoelectric principle are advantageous for highly dynamic acceleration measurements. However, they are not an ideal solution for static or quasi-static measurements. Inductive accelerometers are rarely used in biomechanical applications.

Table 3.3.1   Summary table for selected characteristics of accelerometers typically used in biomechanical experiments. Specific data was extracted from sales brochures of major suppliers and/or from our own experience.

CHARACTERISTIC	SENSOR TYPE			
	STRAIN GAUGE	PIEZO-RESISTIVE	PIEZO-ELECTRIC	INDUCTIVE
MASS	low 1-2 g	low 1-2 g	low 1-2 g	higher several grams
NATURAL FREQUENCY	2000 Hz to 5000 Hz	2000 Hz to 5000 Hz	20,000 to 30,000 Hz	200 to 400 Hz
FREQUENCY RESPONSE	0 to 1000 Hz	0 to 1000 Hz	0 to 5000 Hz	low no shock measurements
SHOCK RATING	one magnitude higher than upper limit of range	one magnitude higher than upper limit of range	several magnitudes higher than upper limit of range	several magnitudes higher than upper limit of range
RANGE	limited to about one magnitude	limited to about one magnitude	0.01 to 10,000 g	about 30 g
ADVANTAGES	measure static and dynamic accelerations	measure static and dynamic accelerations	excellent range	accuracy in low frequency accelerations
DISADVANTAGES	limited range	limited range	limited use for static measurements	limited range

OTHER METHODS

The methods described above quantify the linear acceleration of an idealized mass point. However, in some specific applications, angular acceleration may be of interest. A magneto-hydrodynamic acceleration sensor has been proposed for assessing angular acceleration. The principle of operation was described by Laughlin. This description is a summary of his publication (Laughlin, 1989) and the general description that appeared in the journal *Sensor* (September 1989, pp. 32-37): A permanent magnet near a constrained annulus of conductive fluid produces a constant magnetic field with flux lines perpendicular to the fluid annulus (Fig. 3.3.4). As the sensor moves, the fixed permanent magnet moves with the case. The fluid tends to remain inertially stable about its rotational sensi-

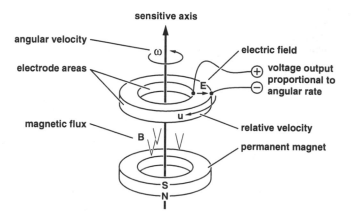

Figure 3.3.4   **Schematic illustration of a magneto-hydrodynamic angular acceleration sensor (from Laughlin, 1989, with permission).**

tive axis but tends to move with the sensor in translation and cross-axial rotation. A rotation of the sensor about its sensitive axis results in a velocity difference, **u**, between the fluid and the applied magnetic flux density, **B**. This induces an electric field, **E**, that is radial in direction and proportional in amplitude to, **u**, and, **B**:

$$\mathbf{E} \quad = \quad \mathbf{u} \cdot \mathbf{B}$$

The circumferential velocity, **u**, is determined by:

$$\mathbf{u} \quad = \quad r \cdot \omega \qquad\qquad (3.3.1)$$

where:

    r   =   root mean square radius of the channel
    ω =   input angular velocity of the sensor

The voltage, **V**, generated radially across the annulus of fluid is the integration of the induced electric field, **E**, over the width of the fluid channel:

$$\mathbf{V} \quad = \quad \int_{r_i}^{r_o} \mathbf{E}(r)\, dr$$

where:

    V   =   voltage across channel
    r_i =   inner channel wall radius
    r_o =   outer channel wall radius

The voltage is electronically amplified and is used to quantify angular acceleration. A special arrangement of three such instruments can be used to quantify three-dimensional angular acceleration and has been used in car crash experiments.

MEASURING SEGMENTAL ACCELERATION WITH ACCELEROMETERS

In experiments with human test subjects where accelerations of selected body segments ought to be measured, accelerometers may be mounted in various ways and at different locations to the segment of interest. However, this segment consists of rigid and soft tissue. For such measurements several questions are of interest such as:

- Which acceleration should be determined: the acceleration of a specific rigid part of the segment, the acceleration of a specific soft tissue part, or an average acceleration of rigid and soft tissue?
- How well does the measured acceleration correspond to the actual acceleration of interest?

Relevant acceleration

The acceleration at different locations of a bone is usually different as demonstrated with axial *in vitro* acceleration measurements on a tibia and a femur mounted on a shaker (Chu et al., 1986). Additionally, rotational aspects may become important. Consider, for instance, a steel beam landing in an inclined position on a steel plate. The accelerations of the contacting end, the centre of mass, or the other end of that beam are different. Consequently, it is important to decide which acceleration is relevant to answer the question of interest for a specific research project.

This question may become important when acceleration measurements are used to determine impact ground reaction forces. In principle, the (impact) ground reaction force can be determined by adding all the mass times acceleration terms for each segment of the body and subtracting the body weight. However, the rigid and the non-rigid masses of a segment do not have the same acceleration. One may solve the problem by introducing an *effective acceleration*, a_{eff}, which is the acceleration with which the segmental mass would have to be multiplied to describe the actual inertial contribution of the specific segment. The equation of motion would then read:

$$\sum m_i \cdot (a_{eff})_i = F_{ground}(t) - \sum m_i \cdot g$$

However, this would not solve the question of how this effective acceleration should or could be measured.

Error in the measured acceleration

The magnitude of the acceleration measured with an accelerometer depends on:

- Bone acceleration.
- Mounting interaction.

- Angular motion.
- Gravity.

The influence of these factors is discussed in the following paragraphs.

Bone acceleration:

Accelerometers may be mounted by screwing them onto the bone of interest (Light et al., 1980; Lafortune and Cavanagh, 1987) or by strapping them to the segment of interest at a location with minimal soft tissue between the accelerometer and the bone of interest. In any mounting case, the acceleration measured represents not "the bone acceleration" but rather the acceleration of a specific mass element at the surface of the bone or even of a point outside the bone (if the accelerometer is screwed onto the bone). Acceleration of a specific bone location can then be determined mathematically from several acceleration measurements or from additional measurements (with film for instance).

Mounting interaction:

Accelerometers screwed onto the bone measure the acceleration outside the bone at the location of the accelerometer. Since the connection of the screw to the bone is usually rigid, resonance problems are minimal. The only correction that must be included is the reduction of the signal to the location of interest (e.g., the proximal end of the tibia).

However, accelerations measured with accelerometers strapped to the segment of interest with soft tissue between the mounted accelerometer and the underlying bone may not reflect the acceleration of the underlying bone. A simple experiment should illustrate this statement. Two accelerometers were mounted on a wooden rod. The first accelerometer was screwed to the rod, providing the true acceleration of the rod, a_{true}. The second accelerometer was strapped with rubber bands to the rod with a water bag between the accelerometer and the rod (Fig. 3.3.5), providing the acceleration measured with the "skin

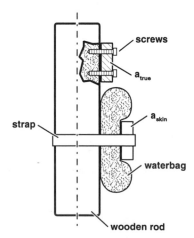

Figure 3.3.5   Schematic illustration of an experiment using two accelerometers mounted to wooden rod, the first screwed to the rod and the second strapped to the rod with a water bag between the accelerometer and the rod.

mounted" accelerometer, a_{skin}. The strapping of this second accelerometer was arbitrarily defined as light, medium, and strong, based on the subjective feeling of the experimenter. The rod was dropped from several different heights onto three different surfaces: linoleum with the thickness of 0.3 cm, artificial sport surface (tartan) with the thickness of 1.2 cm, and foam rubber with the thickness of 6 cm, all of them mounted on concrete. During ground contact the maximal accelerations, a_{true} and a_{skin}, were determined for each drop and the two measured values were represented as one data point in a "a_{skin}-a_{true} diagram" (Fig. 3.3.6). Additionally, the line "$a_{skin} = a_{true}$" is indicated with a dotted line. For the

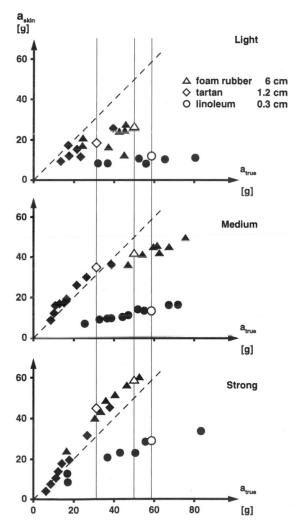

Figure 3.3.6   Experimental acceleration results for a_{true} and a_{skin} for three different strapping procedures, light, medium, and strong, for drops onto three different surfaces.

ideal case where the two accelerations were to provide identical results, all experimental data points would be on this dotted line.

The results of this experiment can be summarized as follows:

(1)   The two actually measured acceleration amplitudes, a_{true} and a_{skin}, were only identical in a few cases. The biggest difference between a_{true} and a_{skin} was about 700%.

(2)   A specific result for a_{skin} can be bigger, equal, or smaller than the corresponding result for a_{true}. The acceleration amplitudes for a_{skin} were smaller than or equal to a_{true} for the light strapping, and bigger as well as smaller for the strong strapping.

(3)   The measured skin accelerations, a_{skin}, were influenced by the surface on which the rod was dropped. For the hardest surface (linoleum) all amplitudes of a_{skin} were smaller than the amplitudes of a_{true}. For the softest surface (foam rubber) about 50% of all amplitudes of a_{skin} were larger than the amplitudes of a_{true}.

The results may, simplistically, be explained as follows: the accelerometer strapped onto the rod with a water bag between the rod and the accelerometer is, mechanically speaking, a vibrating system with the characteristics of such a system such as stiffness, damping, and natural and resonance frequency. The impact force at the moment when the rod contacts the surface may be considered an excitation signal. Due to this impact force the strapped on accelerometer may perform translational and rotational movements with respect to the rod. Several assumptions are used for the following considerations (Fig. 3.3.6):

• The relative movement between skin mounted accelerometer and the rod is translational.
• The excitation signal has one frequency (instead of a frequency spectrum).
• An increase in the tightness of the fixation from light to medium to strong corresponds to an increase in stiffness and a decrease in damping.

The idealized "relative amplitude - frequency" diagram for the three fixations used, with these assumptions in effect, is shown in Fig. 3.3.7. The forces during one specific contact of the rod with the surface for one selected dropping height correspond to excitation signals with different frequencies. The signal for linoleum has the highest and for foam rubber the lowest frequency. For illustration purposes results with similar "input signals" a_{true} from Fig. 3.3.6 are illustrated in Fig. 3.3.7. The selected experimental result for linoleum for the light fixation is indicated with a number 1, for the medium fixation with a number 2, and for the strong fixation with a number 3. The results for the surfaces tartan and foam rubber are numbered in a similar way. The schematic illustration in Fig. 3.3.7 contains many messages which seem to be relevant for acceleration measurements:

• The tightness with which an accelerometer is strapped to a human or animal segment influences stiffness and damping which in turn affect the amplitude of the measured acceleration.
• Accelerations measured with strapped on accelerometers, a_{skin}, for high excitation frequencies are often lower than the accelerations measured with screwed on accelerometers, a_{true}.
• Accelerations measured with strapped on accelerometers depend on the excitation frequency of the impact force responsible for the acceleration.

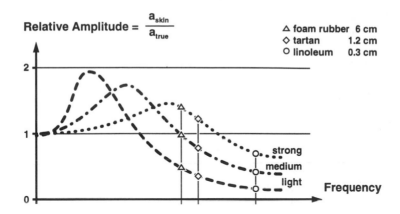

Relative Amplitude = $\dfrac{a_{skin}}{a_{true}}$

△ foam rubber 6 cm
◇ tartan   1.2 cm
○ linoleum   0.3 cm

Figure 3.3.7   Schematic illustration of a "relative amplitude (a_{skin}: a_{true}) - frequency" diagram for different fixation modes of a set-up as illustrated in Fig. 3.3.5. The idealized excitation frequencies for landing on the different surfaces have been arbitrarily selected.

- An increase in the excitation frequency relates to a decrease in the acceleration measured with a strapped on accelerometer, a_{skin}, if the excitation frequencies are above the natural frequency of the mounted accelerometer system.

As mentioned earlier, this is a simplistic way to describe the actual situation. In reality, the situation is more complex. Two important differences are that the movement of the strapped on accelerometer is translational and rotational and that the excitation signal has a frequency spectrum. However, the basic message does not change for the more complex, actual situation.

The discussed problems may suggest that acceleration measurements with strapped on accelerometers should be treated with caution, and one may wonder what strategies could be applied to allow a restricted use of acceleration results in biomechanical experiments. Some suggestions are discussed in the following.

Understanding the direction of the error

The knowledge of the tendency in which the error of the measured acceleration develops may help to support acceleration results in applications where accelerations are compared for different situations.

EXAMPLE 1

Let us assume that acceleration measurements at the crest of the tibia have been made for running with shoes with a soft, a medium, and a hard midsole. Let us assume that the mean results were 10 g for the soft, 13 g for the medium, and 16 g for the hard midsole and that the differences were significant. Let us further assume that the natural frequency of the mounting was 5 Hz and that the excitation frequency spectrum had, in all cases, a maximum higher than 10 Hz. These assumptions correspond approximately to the medium strapping in Fig. 3.3.7 and the three surfaces. The three measurements consequently correspond in principle to the three points 8 (for soft), 5 (for medium), and 2 (for hard). The conclusions of this consideration are, therefore, that the measured acceleration for the medium shoe sole is underestimated compared to the soft shoe sole, and that the measured acceleration for the hard shoe sole is underestimated compared to the soft and the medium shoe sole. In other words, the actual accelerations of the crest of the tibia are such that the actual differences between the analysed cases are even bigger than the ones shown in the measurements: note that nothing has been said about the biological relevance of this statement.

Appropriate stiffness/damping combinations

Use stiffness/damping combinations with a relatively flat "relative amplitude - frequency" diagram for the dominant frequency content of the excitation signal (Neukomm and Denoth, 1978).

EXAMPLE 2

Assume that the dominant frequency content of the acceleration signal of interest is between 20 and 40 Hz. A mounting technique with a damping coefficient of $c = 200$ s^{-1}, and a natural frequency of 50 Hz provides a "relative amplitude - frequency" diagram that is relatively flat in the frequency region between 20 and 40 Hz (Fig. 3.3.8). Consequently, the error is constant (about + 10 to 15%) and comparisons may be less critical.

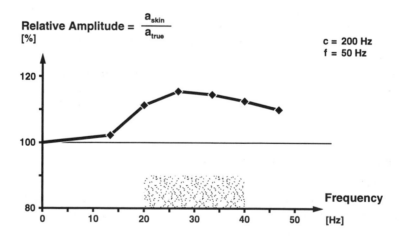

Figure 3.3.8   Schematic "relative amplitude - frequency" diagram for a damping coefficient of c = 200 s^{-1}, and a natural frequency of 50 Hz (from Neukomm and Denoth, 1978, with permission).

Similar experimental and theoretical considerations were presented by Valiant et al. (1987). Using a simplistic linear model, they estimated a magnification factor for the measured acceleration amplitudes and came to the conclusion that in their specific case (damping ratio of 0.4 for an accelerometer mass of 4.4 g) the acceleration amplitudes had been

overestimated by 20 to 30% and that the overestimation was higher for the barefoot measures than for the measures with running shoes. This result is in agreement with the results
of the general considerations of this book.

Translational, rotational, and gravitational components of acceleration

Acceleration measurements on a segment of the human body provide a signal which is
composed of a translational, a rotational, and a gravitational component (Winter, 1979;
Lafortune and Hennig, 1989). A point at the tibia during landing (e.g., the tibial tuberos-
ity), for instance, experiences a rotational acceleration due to the rotation around (about)
the ankle joint as well as a translational acceleration due to the deceleration of the foot and
lower leg during initial contact with the ground.

$$a_{tot} \quad = \quad a_{tr} + a_{rot} + a_{gra}$$

where:

a_{tot} = total acceleration measured with an accelerometer mounted to the rigid
structure

a_{tr} = contribution to the total acceleration due to the translational acceleration of a point with no rotation (e.g., contact point)

a_{rot} = contribution to the total acceleration due to the rotation of the rigid
structure

a_{gra} = contribution to the total acceleration due to gravity

Accelerations measured during human or animal locomotion have different combinations of the three acceleration components, depending on the actual movement. Actual
contributions of translational, rotational, and gravitational acceleration components have
been presented for human walking and running (Lafortune and Hennig, 1989). Accelerations at the human tibia were measured with triaxial strain gauge accelerometers secured
to a traction pin which was inserted in the tibia 3 cm below the proximal articular surface.
During impact in walking, the dominant contribution originated from the translational
component. A second but smaller contribution originated from the gravitational component (about 15 to 20%). During impact in running, the dominant contributions originated
from the translational and the rotational components. The rotational and the gravitational
contributions in this part of the movement were about 45%. Lafortune and Hennig (1989)
suggest that the relatively low correlation between the maximal impact acceleration and
the maximal vertical impact force (0.76) is related to the rotational and gravitational acceleration components during impact in running.

3.3.4 APPLICATIONS

ORDER OF MAGNITUDE OF MEASURED ACCELERATIONS

The peak accelerations measured during selected activities are summarized in
Table 3.3.2.

The results illustrate several factors that influence the acceleration or deceleration of a body segment. They can be summarized as follows:

- Accelerations seem to be smaller for increasing body masses.
- Accelerations seem to be smaller with increasing number of joints between the input force and the measurement location.
- Accelerations seem to be smaller for soft than for hard surfaces.

Table 3.3.2   Order of magnitude of peak acceleration values measured in selected activities based on data from Nigg et al., 1974; Unold, 1974; Voloshin and Wosk, 1982; Lafortune and Hennig, 1991; with permission. The acceleration values are not corrected for mounting artifacts and/or for gravitational or rotational components.

MOVEMENT	SPECIFIC COMMENTS	ORDER OF PEAK MAGNITUDE ACCELERATIONS OF		
		HEAD [g]	HIP [g]	TIBIA [g]
SKIING:				
POWDER SNOW	10 m/s	1	2	4-6
PACKED RUN	10 m/s	2	3	30-60
PACKED RUN	15 m/s	-	-	60-120
PACKED RUN	25 m/s	-	-	100-200
WALKING		1	1	2-5
RUNNING:				
HEEL-TOE	on asphalt	1-3	2-4	5-17
HEEL-TOE	on grass	1-3	2-4	5-10
TOE	on asphalt	1-3	2-4	5-12
GYMNASTICS:				
LANDING FROM 1.5M	on 7 cm mat	3-7	8-14	25-35
LANDING FROM 1.5M	on 40 cm mat	2	5	8
ROUND OFF SOMERSAULT	on 7 cm mat	3	14	24
STRADDLE DISMOUNT	on 40 cm mat	3	8	10
TAKE OFF MINITRAMP		3	7	9

Acceleration values of 200 g during alpine skiing may be impressive and one may consider such values as indicators of high loading. However, the magnitude of the acceleration is not the only important factor in the analysis of a load to which the human body is exposed. Knocking with the knuckle of the index finger at a door may correspond to accelerations of the finger joint of 100 to 200 g. However, this action is not considered dangerous. Consequently, one may dispute the relevance of simple acceleration measurements in the context of load analysis.

SHOCK ABSORPTION IN THE HUMAN BODY

Comparison of acceleration measurements during locomotion at selected locations of the human body have been used to discuss possible etiology of pain and/or injuries. Acceleration amplitudes at the medial femoral condyle and at the forehead during walking have been compared for patients with low back pain and with no pain (Voloshin and Wosk, 1982). The quotient of the femoral and the head acceleration has been determined as a measure for "shock attenuation", and it has been demonstrated that the "shock attenuation" was better for healthy subjects than for subjects with low back pain. Similar studies have been performed on landing from a jump (Gross and Nelson, 1988) and of artificial shock absorbers in shoes (Voloshin and Wosk, 1981).

IMPACT ACCELERATIONS IN THE ASSESSMENT OF PROTECTIVE EQUIP-MENT

Accelerometers have been used extensively in the assessment of appropriate head protection with helmets for various human activities. Aircrew helmets (Norman et al., 1979), American football helmets (Bishop et al., 1984), and bicycling helmets (Bishop and Briard, 1984) have all been studied. Accelerations of standardized head dummies were used to assess the protective qualities of such helmets. Similar procedures are applied in car crash experiments. Additionally, acceleration measurements have been used to compare the vibrational characteristics of tennis racquets (Elliott et al., 1980).

VIBROMYOGRAPHY

Vibromyography (VMG) represents a simple, non-invasive technique to assess a mechanical signal in the low frequency range which is produced by the contracting muscle. Grimaldi was the first to report muscle sound (Grimaldi, 1665). Wollastone recognized that muscle sound increases with increasing muscle activity (Wollaston, 1810). Recently, vibrations due to muscle contraction have been studied by many investigators (Barry, 1987; Oster and Jaffe, 1980; Stokes and Dalton, 1991; Zhang et al., 1992; Orizio, 1993). Vibromyography measurements use miniature accelerometers because of their small size, small mass (less than 1.5 g), good frequency response, and appropriate sensitivity. VMG frequencies lie in the range of the infrasonic (< 20 Hz) to the low end of the audible sound.

Comparisons show many similarities between EMG and VMG measurements. However, differences in the frequency content of these two signals suggest differences in the mechanical and the electrical responses of muscles to activation.

VMG techniques have been proposed for quantifying muscle fatigue, for automatic control of powered prostheses (Barry, 1987; Barry et al., 1992), and for diagnostic purposes for subjects with neuromuscular disorders (Rhatigan et al., 1986).

The VMG signals are affected by different variables such as the length of the muscle, the temperature, muscle properties, and possibly fibre type distribution (Barry, 1987; Frangioni, 1987; Zhang et al., 1990; Zhang et al., 1992).

CRITICAL COMMENTS

The appropriate use of accelerometers is a major concern in all applications of this measuring instrument. The author suggests that of all the measuring techniques discussed in this book, the appropriate use of accelerometers is one of the most difficult ones. The fact that an accelerometer provides on-line information may sometimes cloud the need to study carefully the following aspects of the situation:

- How well does the measured acceleration actually represent the acceleration in which one is interested?
- Is the measured acceleration the variable that answers the question of interest?
- What does the measured acceleration mean mechanically and biologically?

Certainly, these questions are relevant for all other applications of measuring techniques. However, it seems that they are more easily answered for other measuring techniques than for accelerometry.

3.3.5 REFERENCES

Barry, D.T. (1987) Acoustic Signals from Frog Skeletal Muscle. *Biophys. J.* **51**, pp. 769-773.

Barry, D.T., Hill, T., and Im, D. (1992) Muscle Fatigue Measured with Evoked Muscle Vibrations. *Muscle & Nerve.* **15**, p. 303.

Bishop, P.J., Norman, R.W., and Kozey, J.W. (1984) An Evaluation of Football Helmets Under Impact Conditions. *Am. J. of Sports Medicine.* **12** (3), pp. 233-236.

Bishop, P.J. and Briard, B.D. (1984) Impact Performance of Bicycle Helmets. *Can. J. of Applied Sports Science.* **9** (2), pp. 94-101.

Bouche, R.R. (1974) Vibration Measurements. *Process Instruments and Controls Handbook* (2nd Ed.) (ed. Considine, D.M.). McGraw-Hill, New York. **9**, pp. 68-74.

Chao, E.Y. (1978) Experimental Methods for Biomechanical Measurements of Joint Kinematics. *CRC Handbook of Engineering in Medicine and Biology.* CTC Press, West Palm Beach, FL. pp. 385-409.

Chu, M.L., Yazadani-Ardakani, S., Gradisar, I.A., and Askew, M.J. (1986) An In Vitro Simulation Study of Impulsive Force Transmission Along the Lower Skeletal Extremity. *J. Biomechanics.* **19**, pp. 979-987.

Denoth, J. (1980) Ein mechanisches Modell zur Beschreibung von passiven Belastungen. *Sportplatzbeläge* (eds. Nigg, B.M. and Denoth J.). Juris Verlag, Zürich. pp. 45-67.

Elliott, B.C., Blanksby, B.A., and Ellis, R. (1980) Vibration and Rebound Characteristics of Conventional Oversized Tennis Rackets. *Research Quarterly for Exercise and Sport.* **51** (4), pp. 608-615.

Frangioni, J.V., Kwan-Gett, T.S., Dobrunz, L.E., and McMahon, T.A. (1987) The Mechanism of Low-frequency Sound Production in Muscle. *Biophys. J.* **51**, pp. 775-783

Gage, H. (1967) *Accelerographic Analysis of Human Gait.* Biomechanics Monograph (eds. Byars, E.E., Contini, R., and Roberts, V.L.). ASME, New York.

Grimaldi, F.M. (1665) *Physicomathesis de Lumine.* Bologna, Italy.

Gross, T.S. and Nelson, R.C. (1988) The Shock Attenuation Role of the Ankle During Landing from a Vertical Jump. *Med. and Science in Sports and Exercise.* **20** (5), pp. 506-514.

Lafortune, M.A. and Cavanagh, P.R. (1987) Three-dimensional Kinematics of the Patella During Walking. *Biomechanics X-A* (ed. Jonsson, B.). Human Kinetics, Champaign, IL. pp. 337-341.

Lafortune, M.A. and Hennig, E.M. (1989) Contribution of Angular Motion and Gravity to Tibial Acceleration. *Med. and Science in Sports and Exercise.* **23** (3), pp. 360-363.

Laughlin, D.R. (1989) A Magneto-hydrodynamic Angular Motion Sensor for Anthropomorphic Test Device Instrumentation. *Proc. 33rd Stapp Car Crash Conf.* Appl. Tech. Assoc., Albequerque, New Mexico. Paper 12SC31.

Light, L.H., McLellan, G.E., and Klenerman, L. (1980) Skeletal Transients on Heel Strike in Normal Walking with Different Footwear. *J. Biomechanics.* **13**, pp. 477-480.

Morris, J.R. (1973) Accelerometry: A Technique for the Measurement of Human Body Movements. *J. Biomechanics.* **6** (6), pp. 729-736.

Neukomm, P.A. and Denoth, J. (1978) Messmethoden. *Biomechanische Aspekte zu Sportplatzbelägen* (ed. Nigg, B.M.). Juris Verlag, Zürich. pp. 28-45.

Nigg, B.M. and Neukomm P.A. (1973) Erschütterungsmessungen beim Skifahren. *Med. Welt.* **24 (48)**, pp. 1883-1885.

Nigg, B.M., Neukomm, P.A., and Unold, E. (1974) Biomechanik und Sport: Über Beschleunigungen die am menschlichen Körper bei verschiedenen Bewegungen auf verschiedenen Unterlagen auftreten. *Orthopäde.* **3**, pp. 140-147.

Norman, R.W., Bishop, P.J., Pierrynowski, M.R., and Pezzack, J.C. (1979) Aircrew Helmet Protection Against Potential Cerebral Concussion in Low Magnitude Impacts. *Aviat., Space, and Environ. Med.* **50 (6)**, pp. 553-561.

Orizio, C. (1993) Muscle Sound: Bases for the Introduction of a Mechanomyographic Signal in Muscle Studies. *CRC Reviews in Biomedical Engineering.* **21 (3)**, pp. 201-243.

Oster, G. and Jaffe, J.S. (1980) Low Frequency Sounds from Sustained Contraction of Human Skeletal Muscle. *Biophysical. J.* **30**, pp. 119-128.

Prokop, L. (1972) Die Auswirkungen von Kunststoffbahnen auf den Bewegungsapparat. *Oesterr. J. für Sportmedizin.* **2**, pp. 3-19.

Rhatigan, B.A., Mylrea, K.C., Lonsdale, E., and Stern, L.Z. (1986) Investigation of Sounds Produced by Healthy and Diseased Human Muscular Contraction. *IEEE Trans. Biomed. Eng.* **BME-33**, pp. 967-971.

Saha, S. and Lakes, R.S. (1979) The Effect of Soft Tissue on Wave Propagation and Vibration Tests for Determining the In Vivo Properties of Bone. *J. Biomechanics.* **10 (7)**, pp. 393-401.

Stokes, M.J. and Dalton, P.A. (1991) Acoustic Myographic Activity Increases Linearly up to Maximal Voluntary Isometric Force in the Human Quadriceps Muscle. *J. of the Neurological Sciences.* **101**, pp. 163-167.

Unold, E. (1974) Erschüetterungsmessungen beim Gehen und Laufen auf verschiedenen Unterlagen mit verschiedenem Schuhwerk. *Jugend und Sport.* **8**, 289-292.

Valiant, G.A., McMahon, T.A., and Frederick, E.C. (1987) A New Test to Evaluate the Cushioning Properties of Athletic Shoes. *Biomechanics X-B* (ed. Jonsson, B.). Human Kinetics, Champaign, IL. pp. 937-941.

Voloshin, A.S. and Wosk, J. (1981) Influence of Artificial Shock Absorbers on Human Gait. *Clinical Orthopaedics.* **160**, pp. 52-56.

Voloshin, A.S. and Wosk, J. (1982) An In Vivo Study of Low Back Pain and Shock Absorption in the Human Locomotor System. *J. Biomechanics.* **15 (1)**, pp. 21-27.

Voloshin, A.S., Burger, C.P., Wosk, J., and Arcan, M. (1985) An In Vivo Evaluation of the Leg's Shock Absorbing Capacity. *Biomechanics IX-B* (eds. Winter, D.A., Norman, R.W., Wells, R.P., Hayes, K.C., and Patla, A.E.). Human Kinetics, Champaign, IL. pp. 112-116.

Winter, D.A. (1979) *Biomechanics of Human Movement.* John Wiley & Sons, New York.

Wollaston, W.H. (1810) On the Duration of Muscle Action. *Philos. Trans. R. Soc.* pp. 1-5.

Zhang, Y.T., Ladly, K.O., Liu, Z.Q., Tavathia, S., Rangayyan, R.M., Frank, C.B., and Bell, G.D. (1990) Interference in Displacement Vibroarthrography and its Adaptive Cancellation. *Proc. 16th Can. Med. and Biol. Engineering Conf.* pp. 107-108.

Zhang, Y.T., Ladly, K.O., Rangayyan, R.M., Frank, C.B., Bell, G.D., and Liu, Z.Q. (1990) Muscle Contraction Interference in Acceleration Vibroarthrography. *Proc. the IEEE/EMBS 12th Annual International Conference.* pp. 2150-2151.

Zhang, Y.T., Frank, C.B., Rangayyan, R.M., and Bell, G.D. (1992) A Comparative Study of Simultaneous Vibromyography and Electromyography with Active Human Quadriceps. *IEEE Trans. on Biomedical Engineering.* **39 (10)**, pp. 1045-1052.

Ziegert, J.C. and Lewis, J.L. (1979) The Effect of Soft Tissue on Measurements of Vibrational Bone Motion by Skin-mounted Accelerometers. *Transactions of the ASME.* **101**, pp. 218-220.

3.4   OPTICAL METHODS

NIGG, B.M.
COLE, G.K.

3.4.1 DEFINITIONS AND COMMENTS

f/stop setting:	Quantification of lens opening with:

$$f/\text{stop} \quad = \quad \frac{f_L}{\text{diameter of lens opening}}$$

Field of focus:	(depth of field) The range of distance within which objects in a picture appear in focus (sharp).
Focal length:	Focal length is defined by the lens equation:

$$\frac{1}{f_L} \quad = \quad \frac{1}{a} + \frac{1}{b}$$

where:

f_L = focal length
a = distance from object to lens
b = distance from lens to film (image)

Kinematics:	The study of the geometry of motion.
Sampling frequency:	Number of measurements of the same variable made during a one second time interval, or the ratio of the number of measurements of a variable to the time interval in which the measurements were taken.
Shutter:	Mechanical device which allows light to pass through the lens onto the film for a specified period of time.
Shutter factor:	Fraction of the total circle of a circular shutter which is available to let the light pass through the lens to the film.

3.4.2 SELECTED HISTORICAL HIGHLIGHTS

1876 Janssen	Developed a sequential photographic instrument for taking pictures of the movement of the planet Venus in front of the sun.
1878 Muybridge	Published photos of movement in the journal *Nature*.
1882 Marey	Photographic documentation of movement with the photographic rifle (la fusil photographique), an instrument based on the ideas of Janssen.

1882	Marey	Published the first chronophotographic pictures of a marching human and a horse jumping an obstacle.
1887	Muybridge	Published the 11 Volume photographic work, *Animal Locomotion*, considered one of the treasures of early photography of human and animal movement.
1974	SELSPOT	A SELective light SPOT recognition system which became commercially available using LED's as active markers (Lindholm and Ocberg, 1974).
1982	VICON	A television-based system for movement analysis (VICON) which became commercially available. Since 1982, many other systems have been developed and are currently on the market.

3.4.3 MEASURING POSSIBILITIES

The possibilities of measuring human or animal kinematics with optical techniques are discussed in the following three sections. The first section summarizes selected aspects of the theory of optics. The second section discusses selected camera and marker systems. The third section addresses three- and two-dimensional motion analysis as it is currently used in biomechanical research.

OPTICAL THEORY

Kinematic information is used in biomechanical applications to describe movement and/or estimate, in conjunction with other information, forces in internal structures (Winter, 1979). This section concentrates on the most important aspects of how optical systems can be used to determine kinematic quantities. Further information about optics may be gathered from textbooks on that topic.

Lenses

A lens is an optical system made up of two refracting surfaces. Thin lenses have an optical axis which correlates to the axis of symmetry of the two spherical surfaces. Parallel rays passing through such a lens converge at a point "F" on the optical axis. This point is called the focal point (Fig. 3.4.1). A converging lens creates an inverted picture of an object at a distance "a" from the lens, and this picture is in focus at a distance "b" from the lens (Fig. 3.4.1).

The relation among the distance of the object of interest from the lens, a, the distance between the lens and the location where the image is in focus, b, and the focal length, f_L, is described by the lens equation:

$$\frac{1}{f_L} = \frac{1}{a} + \frac{1}{b}$$

where:

$$\frac{b}{a} \quad = \quad \text{magnification}$$

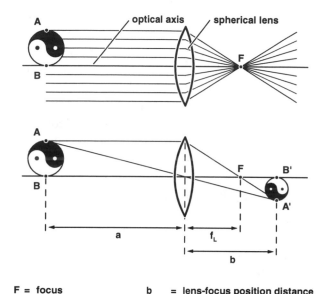

F = focus   b   = lens-focus position distance
f_L = focal length   A,B = object points
a = lens-object distance   A',B' = object points on focused image

Figure 3.4.1   Schematic illustration of the focal point of a lens (top), and schematic illustration of an object-lens-image relationship to focal length (bottom).

If an object is far away (a $\Rightarrow \infty$) one can approximate the focal length to:

$$f_L \quad \approx \quad b$$

Lenses are generally classified as wide angle, normal, or telephoto lenses. A rough classification by which to define these lenses is:

wide angle f_L = 10 to 25 mm
normal f_L = 25 to 60 mm
telephoto f_L = 60 to 200 mm

Lens opening - lens aperture

A mechanical device controls the amount of light that enters the lens. It provides a circular opening through which the light can pass. The size of the lens opening is expressed in terms of the *f/stop* setting where:

$$f/\text{stop} = \frac{f_L}{\text{diameter of lens opening}}$$

EXAMPLE 1

The f/stop value for a focal length of 20 mm and a diameter of lens opening of 5 mm is:

> f/stop = 4 or f/4

Typical f/stop settings for cameras are 22, 16, 11, 8, 5.6, 4, 2.8, and 2. The f/stop settings are constructed so that each successive setting doubles the amount of light entering the camera. For instance, while f/11 has double the diameter of f/22, it actually admits 4 times the light.

Field of focus or depth of field

The field of focus (or depth of field) is defined as the range of distance within which objects in a picture appear to be in focus. Field of focus is determined by the angle between the lines AA' and FA' (Fig. 3.4.1). Field of focus is large if this angle is small, and small if this angle is large.

The field of focus depends on f/stop and the distance of the object to the lens. The field of focus will increase if:

- f/stop increases (by either decreasing the lens opening or increasing the focal length, f_L), and/or
- The distance between object and lens, a, increases.

The distance between object and lens, a, and the focal length, f_L, are often predetermined and do not allow much room for adjustment for a given camera and environment. However, changing the lens opening produces other effects which will be discussed later.

Shutter opening

When an object moves during the time of exposure, a given point on the object may appear in different locations in the developed picture. The result is a blurred image. Image blur depends on:

v	=	speed of the moving object,
Δt	=	duration of exposure, and
b/a	=	image magnification.

For good image quality, the blur is confined to a value of less than 0.05 mm.

Of the three previous variables, the exposure time is most easily controlled by the photographer. Many optical systems have a shutter, a circular structure that rotates in front of the registration medium. A section of the shutter (defined by the shutter ratio) is open. The exposure time, Δt, is determined by:

$$\Delta t = \frac{1}{s \cdot f}$$

where:

$$f \quad = \quad \text{frame rate}$$

$$s \quad = \quad \text{shutter ratio}$$

with:

$$s \quad = \quad \frac{360°}{\text{shutter opening}} \quad = \quad \frac{360°}{\text{shutter angle}}$$

EXAMPLE 2

A shutter opening (angle) of 90° with a frame rate of 100 Hz gives the exposure time of 1/400 s or 0.0025 s.

Exposure time can also be determined from the blur restriction. For an image blur of less than 0.05 mm, the exposure time is determined by:

$$\Delta t \quad = \quad \frac{w}{200 \cdot v \cdot \cos(90 - \varphi)}$$

where:

$$\Delta t \quad = \quad \text{exposure time}$$
$$w \quad = \quad \text{width of field}$$
$$v \quad = \quad \text{speed of the moving object}$$
$$\varphi \quad = \quad \text{angle between camera axis and direction of motion}$$

EXAMPLE 3

For:

$$w_1 \quad = \quad 5 \text{ m}$$
$$w_2 \quad = \quad 10 \text{ m}$$
$$v \quad = \quad 10 \text{ m/s}$$
$$\varphi \quad = \quad 90°$$

The exposure times for unblurred pictures are:

$$\Delta t_1 \quad = \quad 0.0025 \text{ s} \quad = \quad 1/400 \text{ s}$$
$$\Delta t_2 \quad = \quad 0.005 \text{ s} \quad = \quad 1/200 \text{ s}$$

These exposure times can be realized by changing the sampling frequency or shutter ratio.

Sampling frequency

The sampling frequency is determined by various factors, including:

- The movement to be analysed: walking requires a different sampling frequency than running.
- The variable of interest: the determination of stride length, for instance, requires a different frequency than the determination of the angular velocity of the ankle joint during running.

- The amount of light available: in practice, the sampling frequency may be reduced to assure that a sufficient amount of light passes through the lens.

The sampling frequency necessary to provide an unblurred image is determined by:

$$f \quad = \quad \frac{200 \cdot v \cdot \cos(90 - \varphi)}{s \cdot w}$$

where:

f	=	frame rate
v	=	speed of the moving object
φ	=	angle between camera axis and direction of motion
s	=	shutter ratio = 360° / shutter opening (angle)
w	=	width of field

Furthermore, the minimal sampling frequency for all sampling processes is determined by the sampling theorem. The sampling theorem states that:

A signal must be sampled at a frequency that is at least twice the highest frequency present in the signal.

Note that the sampling theorem compares the sampling frequency to the highest frequency content of the signal, and not to the highest frequency of interest. In addition, the sampling theorem insures that the frequency content of the measured signal is preserved, however, the amplitude of the signal may be substantially attenuated. Therefore, in practical applications, the appropriate sampling frequency may be considerably higher than the minimal frequency determined by the sampling theorem.

Lens errors

In theory, lenses are constructed such that the two confining surfaces are portions of ideal spheres. In reality, lens surfaces have errors that may cause errors in the final position coordinates determined from a measurement. Lens errors can be corrected by applying an internal calibration of the lens-camera system. This correction covers errors of the lens system, the detector and, where applicable, the electronics. It is assumed that the lens errors are the dominant error sources for internal calibration. One possible procedure for the internal calibration uses a defined grid that is quantified with the camera system. Since the actual position of each intersection point of the grid is known, a two-dimensional image correction vector can be determined (Ladin et al., 1990). An example of a resultant vector field for the correction of a camera system in image coordinates is illustrated in Fig. 3.4.2. The illustration indicates that substantial errors frequently occur on the outside of the field of interest (lens problems). Using such internal calibration procedures, an absolute accuracy over the measured volume of 0.2% has been documented (Ladin et al., 1990). Software packages are currently available that allow correction of lens and camera system errors. Lens errors are usually bigger for zoom lenses than for fixed lenses.

0.500E-02
MAXIMUM VECTOR

Figure 3.4.2   **Illustration of a camera system correction as a vector field of correction values in image coordinates (from Ladin et al., 1990, with permission).**

CAMERAS AND MARKERS

This section discusses the different combinations of active and passive cameras as well as the markers most frequently used in biomechanical assessment. Cameras used in biomechanical analysis include conventional high-speed film cameras, high speed video cameras, and cameras that sense electronic wave signals. Markers represent points on the system of interest such as prominent bony landmarks. They may be passive (e.g., reflective tape) or active (e.g., electronic transmitters).

Film cameras and markers

One conventional method for quantifying movement is to use a film camera (or cine or motion picture camera). For research purposes, 16 mm or 35 mm film cameras are used. Two basic types of film cameras are used for high-speed filming, the intermittent pin registered camera (Fig. 3.4.3), and the camera that uses a rotating prism (Fig. 3.4.4).

The *intermittent pin registered camera* uses mechanical devices to hold the film stationary with pins, for a fraction of a second behind the lens and shutter, in the gate. During this time the film is exposed. This process is repeated for every frame. A pin registered film camera can achieve sampling frequencies of up to 500 Hz (frames per second), adequate for most biomechanical applications.

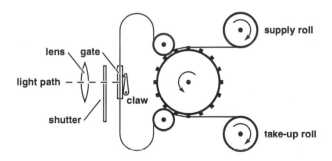

Figure 3.4.3 **Schematic illustration of an intermittent pin registered camera.**

In the *rotating prism film camera*, light passes through the lens onto a rotating prism from which the light is directed onto the film. Film and prism move constantly. Rotating prism-based cameras provide sampling frequencies of up to 10 kHz.

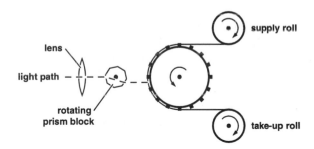

Figure 3.4.4 **Schematic representation of a rotating prism camera.**

Markers used for film analysis of motion can be stickers applied to points of interest or points painted onto the skin at the location of interest. In special cases, when soft tissue movement may interfere with the movement of interest, it may be advantageous to use a stencil of the segment of interest with defined *marker points*. This stencil can be used to determine the position of points of interest.

Time calibration of high-speed film cameras can be done internally or externally. Most cameras have internal time calibration systems that allow the determination of the film frequency. The application of time markers onto the film is usually done a few frames before or after the gate, a fact that has to be taken into consideration for the determination of the film frequency. The film frequency may also be determined using external clocks or light signals (Miller and Nelson, 1973) that appear in each frame. The determination of the actual film frequency is important since the film needs a certain time interval to reach the designated frame rate, and frequency may not remain constant after this initial time interval.

High-speed film cameras are mechanical masterpieces and are, consequently, expensive. Film cameras also use expensive film material that must be developed. In the end, economical considerations concerning the costs of film and film development may limit the progress of a planned project. However, high speed film is still one of the best methods of quantifying movement since the human eye may be able to draw important conclusions during film analysis that may be difficult to achieve with other systems. Many biomechanics laboratories use high speed film techniques in the initial phase of new projects.

Video cameras and markers

Practical problems such as film costs, the need to change film during measurements, and time-consuming film analysis have helped motivate the development of video techniques for quantifying movement in biomechanical applications. The film camera is replaced by a video camera. The basic principles of optics and camera function in video cameras are identical to those of film cameras. The major difference is that the film is replaced with an array of light sensitive cells (sensors). The process of data collection is discussed below for a black and white camera.

Like a film camera, a video camera has a shutter mechanism that opens and lets light enter for a fraction of a second. The light sensitive cells detect the incoming light, get excited, and an electrical signal is emitted. The amplitude of this signal is proportional to the intensity of the incoming light. When the shutter closes, the light sensitive cells remain excited for a limited time, and a computer-controlled multiplexer system goes through each cell and reads the electrical signal. This information is transmitted to a computer where it is stored as a set of $n \times m$ pixels. After this process is completed, the shutter opens again and the cyclic process continues.

Video cameras are predominantly used in combination with automatic or semiautomatic data analysis systems. Thus, markers are used which allow this (semi-) automatic process. The main features of markers used for this purpose are:

- Contrast,
- Shape, and
- Colour.

Markers covered with reflective material are most commonly used for black and white video cameras. Additionally, the shape of a marker may be used for identification purposes during data analysis. For colour cameras, the colour of a marker may be used as an additional feature to recognize and identify a marker during the analysis. A marker size corresponding to a field of several pixels (optimally 10 to 20) is required for optimal data analysis.

Video cameras are often less expensive than film cameras, and video tape costs much less than film.

Passive cameras using active markers

The camera-marker systems discussed above use passive markers with spatial positions of the markers registered by the passive camera system. However, one type of optical system in biomechanical applications use active markers. The most frequently used technology is based on markers that emit a signal in the infra-red (IR) part of the spectrum.

Cameras that quantify IR signals are similar, in principle, to the cameras described for film and video and are of two types; signal sensing and source sensing. In addition to the features described above, IR signal-sensing cameras may use filters that select a defined part, the IR part, of the frequency spectrum of the optical waves reaching the lens, allowing only this part to pass to the photo diode sensors (Woltring and Marsolais, 1980).

IR source sensor systems use active markers. Light emitting diodes (LED) are attached to points of interest of a segment. They emit signals with a distinct frequency in the IR part of the spectrum. Each LED attached to the segment of interest has its own specific frequency and can easily be identified at every point in time during a measurement. IR source sensors (e.g., SELSPOT or WATSMART) typically sample with a frequency of about 1 kHz. It may be of interest to note that active LED-based markers in the IR spectral wavelength are also used with video cameras for high contrast and large distance measurements.

Active cameras and passive markers

The camera systems described above are passive. Light beams from the object pass through the lens and shutter opening and produce a reaction in the picture plane (e.g., film). There exists yet another type of camera, called an "active camera", which sends a beam of light to an object. The camera has three scanners (rotating mirrors) that each sweep a narrow (~1 cm wide) planar beam of light through the space of interest. The three scanners emit the light sequentially (Fig. 3.4.5). Photo diodes are used as markers mounted

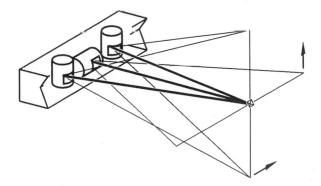

Figure 3.4.5   **Schematic illustration of an active camera system using three rotating mirrors (from Mitchelson, 1988, with permission).**

to the segment of interest. Light reaching a diode results in an electrical signal which, together with the positional information from the active mirror, is used to provide information for the reconstruction of positional data. The angular information from the three sequential mirror signals is used to reconstruct the three-dimensional coordinates of the diodes of interest (Mitchelson, 1988).

Other possibilities

There are other possibilities for quantifying movement besides film, video, IR source sensors, and active cameras. These are less frequently used and will be discussed only briefly. They include chronophotography (sometimes called chronozyklophotography), stroboscopy, and flashing light sources and are typically used in applications where a special effect should be demonstrated.

Chronophotography:

Chronophotography (or chronozyklophotography) is a method in which a conventional camera is used. A second shutter (disc with an opening) is mounted in front of the camera and rotates with a given frequency. The camera shutter is open for the period of the entire movement of interest. The resulting picture has the moving object as well as the stationary background n times exposed. Ideal set-ups for good chronocyclographic pictures require a background that is dark compared to the moving object. Chronophotography is often used to illustrate special aspects of a movement and is less used for scientific analysis in biomechanics.

Stroboscopy:

Stroboscopy is based on the same idea of multiple exposure as chronophotography. The movement is executed in a dark environment (with respect to the object of interest). The camera shutter is open for the entire duration of the movement of interest. Sequences of flashes of light are then used, and the result is a picture with the moving object in multiple exposure. Like chronophotography, stroboscopy is primarily used for illustration purposes and rarely for scientific analysis in biomechanics.

Flashing light sources:

Flashing light sources mounted to the different segments of a moving object may be used to quantify or illustrate movement. This technique is, in principle, identical to the IR source sensor method, except that the frequency spectrum used is in the visible range. This specific approach is not often used in biomechanical analysis.

THREE- AND TWO-DIMENSIONAL MOTION ANALYSIS

In this section, general comments are made about three-dimensional kinematic motion analysis. Subsequently, special, frequently occurring problems of two-dimensional analysis are added. However, these comments are implicitly already included in the information about three-dimensional motion analysis.

Calculation of three-dimensional coordinates from several cameras

Essentially, a camera provides a two-dimensional image of a three-dimensional situation on a "film" medium. The determination of three-dimensional spatial coordinates from several two-dimensional sets of information is commonly performed using the DLT (direct linear transformation) method (Abdel-Aziz and Karara, 1971). For m markers, the DLT method provides a (linear) relationship between the two-dimensional coordinates of a marker, i $(i = 1, ..., m)$, on the film and its three-dimensional location in space. For n cameras, the relationship between the coordinates of the markers on the film of camera, j $(j = 1, ..., n)$, and the spatial three-dimensional coordinates of this marker are determined by:

$$x_{ij} = \frac{a_{1j}x_i + a_{2j}y_i + a_{3j}z_i + a_{4j}}{a_{9j}x_i + a_{10j}y_i + a_{11j}z_i + 1}$$

$$y_{ij} = \frac{a_{5j}x_i + a_{6j}y_i + a_{7j}z_i + a_{8j}}{a_{9j}x_i + a_{10j}y_i + a_{11j}z_i + 1}$$

where for a given marker i:

x_{ij} = x coordinate of marker i on the film measured with camera j

y_{ij} = y coordinate of marker i on the film measured with camera j

x_i = x coordinate of marker i in the three-dimensional space

y_i = y coordinate of marker i in the three-dimensional space

z_i = z coordinate of marker i in the three-dimensional space

a_{kj} = coefficient k in the transformation formulas for marker i

A calibration of N points with known coordinates x_r, y_r, z_r, $(r = 1, ..., N)$, is used to determine the coefficients, a_{kj}, for each camera j. At least 6 calibration points (corresponding to 12 equations) are required to determine the 11 coefficients. The overdetermined system of linear equations is solved using a least squares fit technique.

Each camera is described by two equations. Consequently, the DLT method provides more equations than unknowns for the position of a marker i as long as at least two cameras are involved (can see the marker). The determination of the coordinates for each marker does not have a unique result, but is an approximation with errors Δ_j. The "norm of residuals" (NR) for one coordinate is defined as:

$$NR = \sqrt{\Delta_1^2 + \Delta_2^2 + ... + \Delta_n^2}$$

The DLT method, and any similar method, provides one distinct solution for x_i, y_i, and z_i, as long as the number of cameras involved in the process remains constant. However, in practical applications, especially when 4 to 6 cameras are used, markers are often lost by specific cameras for a certain time. Camera/lens systems have a limited accuracy. Every set of coordinates, x_{ij} and y_{ij}, is affected with some error. This error is implicit in the calculated set of coordinates, x_i, y_i, and z_i. As a result, there will be a difference in coordinate values if the set is calculated using 2, 3, 4, or more cameras. The difference is largest for the jump from 2 to 3 cameras and becomes smaller with each increase in number of cameras. This effect is illustrated in Fig. 3.4.6.

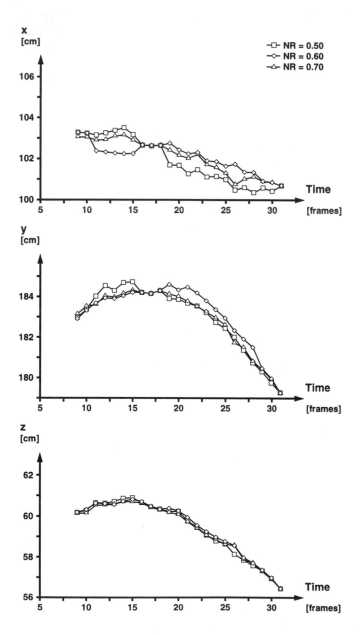

Figure 3.4.6   Difference in coordinate calculation using a DLT method. The coordinates of a marker in the laboratory coordinate system were determined using the results for 2, 3, and 4 cameras. In this case, the inclusion/exclusion of cameras has been produced by changing the "norm of residuals" (NR). Note for a given NR the number of cameras does not remain constant.

Calculation of distances in three-dimensions

The calculation of the distance between two points in a three-dimensional space uses the Pythagorean theorem:

$$d_{ik} = \sqrt{(x_i - x_k)^2 + (y_i - y_k)^2 + (z_i - z_k)^2}$$

where:

d_{ik} = distance between markers i and k
x_i = x coordinate of marker i
y_i = y coordinate of marker i
z_i = z coordinate of marker i
x_k = x coordinate of marker k
y_k = y coordinate of marker k
z_k = z coordinate of marker k

The calculation of distances in two-dimensional space uses the same theorem with only two coordinates:

$$d_{ik} = \sqrt{(x_i - x_k)^2 + (y_i - y_k)^2}$$

Calculation of rigid body orientation in three-dimensions

Several methods for representing the orientation between two body segments in three-dimensional space have been proposed (Goldstein, 1950; Chao, 1980; Grood and Suntay, 1983; Yeadon, 1990; Woltring, 1991; Cole et al., 1993). Some of them are discussed in this section using a previous publication (Cole et al., 1993).

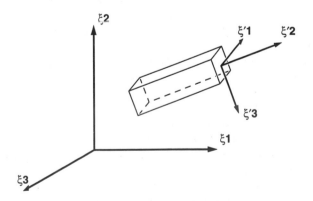

Figure 3.4.7   Orientation of a rigid body in space with an embedded Cartesian coordinate system, ξ', relative to a reference Cartesian coordinate system, ξ.

The position of a particle in three-dimensional space may be represented as a vector with three components, each representing one of the three translational degrees of freedom (DOF). The description of the position of a rigid body in three-dimensional space additionally requires information that describes its orientation in space. If a Cartesian coordinate system with axes, ξ'_i (i = 1, 2, 3), is embedded in a rigid body, then the orientation of the rigid body (Fig. 3.4.7) relative to a reference Cartesian coordinate system with axes, ξ_i (i = 1, 2, 3), is defined by a 3×3 rotation matrix, [R], with components, $R_{ij} = \cos(\xi'_i, \xi_j)$ (i, j = 1, 2, 3). This matrix is also referred to as the direction cosine matrix or attitude matrix, and it can be used as an operator to transform the components of a vector from one coordinate system to the other:

$$\{\xi'\} = [R]\{\xi\} \quad \text{and} \quad \{\xi\} = [R]^{-1}\{\xi'\}$$

where $\{\xi\}$ and $\{\xi'\}$ are a common vector represented in the two systems. The rotation matrix has the properties that its inverse is equal to its transpose, $[R]^{-1} = [R]^T$, and its determinant is equal to one, $|R| = 1$. A detailed description of a least squares method for the determination of the rotation matrix from the coordinates of landmarks fixed to a body is given in Veldpaus et al. (1988).

For the interpretation of body segment orientation in biomechanical analyses, a parametric representation of the rotation matrix is usually desired. Current parametric representations include:

- Cardan/Euler angles.
- Joint Coordinate Systems (JCS).
- Finite helical axes and rotations.
- Helical angles.

These parametric representations are discussed in this section, with emphasis given to Cardan/Euler angles and the JCS. The discussion is based on the following assumptions:

- Cartesian coordinate systems have been embedded in each body segment of interest, with the x-axis pointing anteriorly, the y axis vertically, and the z-axis to the right when the subject is in the anatomical position.
- The orientation of a body segment relative to an inertial reference system is specified by a rotation matrix.
- The orientation of one body segment with respect to another (i.e., joint orientation) is specified by a rotation matrix.

Cardan/Euler angles:

In the Cardan/Euler angle representation of three-dimensional orientation, the rotation matrix is parameterized in terms of *three independent angles,* resulting from an ordered sequence of rotations about the axes (i, j, k), of a selected Cartesian coordinate system (x_1, y_1, z_1), to obtain the attitude of a second coordinate system (x_2, y_2, z_2), relative to the first (Fig. 3.4.8). When the first and the last axes are the same, (i = k), the term "Euler angles" is used. When all three axes are different, the term "Cardan angles" is used, and it is this representation that is typically used in biomechanics (Woltring, 1991). The i and k axes correspond to axes in each of the two segment embedded coordinate systems, while the j axis is referred to as the "nodal" axis.

The parametric representation is mathematically defined by:

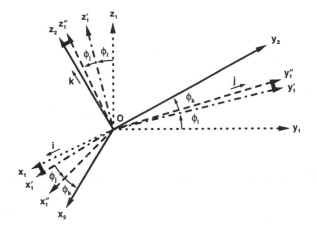

Figure 3.4.8   Sequential rotations ϕ_i, ϕ_j, and ϕ_k about the axes i, j, and k from the starting coordinate system (x_1, y_1, z_1) to obtain the orientation of the second coordinate system (x_2, y_2, z_2). The axes i, j, and k have been arbitrarily defined as the x, y, and z axes, respectively, although they are not restricted to this definition.

$$[R_{ijk}] \quad = \quad [R_k(\phi_k)] \, [R_j(\phi_j)] \, [R_i(\phi_i)]$$

where:

$[R_{ijk}]$ $=$ parametric representation of the rotation matrix for the chosen sequence of rotations

$[R_i(\phi_i)]$ $=$ parametric representation of the rotation matrix for a rotation, ϕ_i, around the specific axis i. An analogous definition is used for the rotations about the axes j and k

The parametric rotation matrices for rotations about the x, y, and z axes of a Cartesian coordinate system are defined as:

$$[R_x(\phi_x)] \quad = \quad \begin{bmatrix} 1 & 0 & 0 \\ 0 & \cos\phi_x & \sin\phi_x \\ 0 & -\sin\phi_x & \cos\phi_x \end{bmatrix}$$

$$[R_y(\phi_y)] \quad = \quad \begin{bmatrix} \cos\phi_y & 0 & -\sin\phi_y \\ 0 & 1 & 0 \\ \sin\phi_y & 0 & \cos\phi_y \end{bmatrix}$$

$$[R_z(\phi_z)] \quad = \quad \begin{bmatrix} \cos\phi_z & \sin\phi_z & 0 \\ -\sin\phi_z & \cos\phi_z & 0 \\ 0 & 0 & 1 \end{bmatrix}$$

Since matrix multiplication is not commutative, the final parametric rotation matrix, $[R_{ijk}]$, depends on which of the axes x, y, or z is chosen for the rotations i, j, and k. The parameterization of the rotation matrix into Cardan angles, therefore, is *sequence dependent*. For a given segment or joint orientation, different component values can be obtained depending on the sequence selected. Once the rotational sequence has been determined, the parametric rotation matrix can be constructed in terms of the angles ϕ_x, ϕ_y, and ϕ_z. The resulting expressions can then be equated to the numerical values of the rotation matrix for the selected orientation to solve for each angle. For example, if the axes i, j, and k are arbitrarily defined as the axes z, y, and x, respectively, the parametric rotation matrix is constructed as:

$$[R_{zyx}] \quad = \quad [R_x(\phi_x)]\,[R_y(\phi_y)]\,[R_z(\phi_z)]$$

$$[R_{zyx}] \quad =$$

$$\begin{bmatrix} (C\phi_z C\phi_y) & (S\phi_z C\phi_y) & (-S\phi_y) \\ (-S\phi_z C\phi_x + C\phi_z S\phi_y S\phi_x) & (C\phi_z C\phi_x + S\phi_z S\phi_y S\phi_x) & (C\phi_y S\phi_x) \\ (S\phi_z S\phi_x + C\phi_z S\phi_y C\phi_x) & (-C\phi_z S\phi_x + S\phi_z S\phi_y C\phi_x) & (C\phi_y C\phi_x) \end{bmatrix}$$

where:

$$S \quad = \quad \text{sine}$$
$$C \quad = \quad \text{cosine}$$

If the numerical components of the rotation matrix, $[R]$, are labeled as, R_{mn} (m = 1, 2, 3; n = 1, 2, 3), and counter-clockwise rotations are chosen as positive, the sine and cosine of each angle can be calculated from:

$$\sin\phi_y \quad = \quad -R_{13} \qquad\qquad \cos\phi_y \quad = \quad \sqrt{1 - \sin^2\phi_y}$$

$$\sin\phi_x \quad = \quad \frac{R_{23}}{\cos\phi_y} \qquad\qquad \cos\phi_x \quad = \quad \frac{R_{33}}{\cos\phi_y}$$

$$\sin\phi_z \quad = \quad \frac{R_{12}}{\cos\phi_y} \qquad\qquad \cos\phi_z \quad = \quad \frac{R_{11}}{\cos\phi_y}$$

From these equations, each angle can be determined in the range, $-\pi < \phi < \pi$, so that the time history of the angle is continuous (Yeadon, 1990).

(i) Sequence selection in Cardan/Euler angles:

The sensitivity of calculations of joint orientation to the rotational sequence is illustrated in Fig. 3.4.9. The differences in component values, $\Delta\phi$, for two different sequences of ordered rotations were determined for the ankle joint complex, with the second and third rotations being the reverse of those shown in the previous example. The results indicate that, theoretically, there are substantial differences in orientation components for different Cardanic sequences. In practical applications to human movement, however, the differences are not always relevant. Differences in relative segmental orientations of the lower extremities in running due to different sequences of ordered rotations are minimal. However, in a "side shuffle" movement, with a larger in-eversion range-of-motion, the differences can be large. In range-of-motion (ROM) measurements, the differences are substantial, and results from different sequences of ordered rotations cannot be compared (Table 3.4.1). These sensitivity considerations illustrate the need for a well-defined convention for biomechanical analysis.

| Table 3.4.1 | Comparative numerical results for measurements of angular positions for the ankle joint complex for running, "side shuffle", and range of motion (ROM) positions. Unpublished, Cole (1992), with permission. |

ACTIVITY	COMPONENT	ANGLE SEQUENCE PL-DO/AD-AB/IN-EV	ANGLE SEQUENCE PL-DO/IN-EV/AD-AB
RUNNING	DO	25	21.4
	AB	10	10.6
	EV	20	19.7
SIDE SHUFFLE	DO	20	30.3
	AB	15	18.1
	IN	35	33.6
ROM	PL	40	26.8
	AB	40	41.8
	IN	20	15.2

In biomechanics, the three Cardan angles are typically used to represent the anatomical components of joint orientation and movement (e.g., flexion-extension, adduction-abduction, and inversion-eversion). The anatomical components of movement are defined with respect to the anatomical position as follows (Moore, 1980; Snell, 1973):

- Flexion-extension is motion of a segment in a *parasagittal plane.*
- Adduction-abduction is motion of a segment *away from or toward the sagittal plane.*
- Third component is rotation of a segment about its *longitudinal axis* (Note: the terminology of the third rotational component differs depending on the joint).

For practical use in biomechanics, the sequence of ordered rotations should be chosen so that the anatomical definitions are satisfied. Rotational sequences that do not conform to these definitions should be avoided. The following argument provides evidence that the

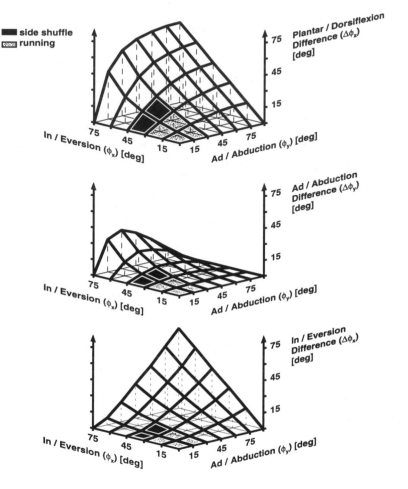

Figure 3.4.9   Differences in the calculated values for plantar-dorsiflexion (top), ad-abduction (centre), and in-eversion for two different sequences of rotation with the second and third rotation reversed. The primary sequence that is illustrated is plantar-dorsiflexion, ad-abduction, and in-eversion. ϕ_z was arbitrarily chosen as 45°. ϕ_x and ϕ_y were varied from 0° to 90° and the rotation matrices were generated based on the primary sequence. The representations of ϕ_x, ϕ_y, and ϕ_z in the secondary sequence were calculated from the generated rotation matrices and compared to the primary values (from Cole et al., 1993, with permission).

sequence of Cardan angles that satisfies the definitions of flexion-extension, adduction-abduction, and third component rotation occurs in the following order:

- Flexion-extension,
- Adduction-abduction, and
- Axial rotation.

The flexion-extension component, being first in the sequence, must occur within the parasagittal plane. The second rotation must occur in a plane perpendicular to the sagittal plane since the coordinate system is orthogonal. Therefore, the second rotation satisfies the anatomical definition of adduction-abduction. The third rotation in the sequence is axial rotation. After the first two rotations are completed, the remaining axis must correspond to the longitudinal axis of the distal segment, which again satisfies the anatomical definition for the final rotation. Consequently, the proposed sequence of (1) flexion-extension, (2) adduction-abduction, and (3) axial rotation conforms to the anatomical definitions of segment rotations.

It can easily be demonstrated that the sequence (1) flexion-extension, (2) axial rotation, and (3) adduction-abduction does not satisfy the above anatomical definitions. If axial rotation is chosen as the second rotation, the third rotation no longer conforms to the definition of adduction-abduction since it is not constrained to a plane perpendicular to the sagittal plane.

(ii) Advantages and disadvantages of Cardan angles:

(1)   Cardan angles are widely used in biomechanics because they provide a representation of joint orientation analogous to the anatomical representation that both clinicians and researchers are accustomed to using.

(2)   The fact that Cardan angles are sequence dependent has sometimes been viewed as a disadvantage. Appropriate standardization of the sequence, as proposed above, is one simple solution to this problem.

(3)   One major disadvantage of Cardan angles is gimbal lock. A mathematical singularity that occurs when the second rotation equals $\pm \pi/2$. For example, ϕ_x and ϕ_z are determined from:

$$\sin\phi_x = \frac{R_{23}}{\cos\phi_y} \quad \text{and} \quad \sin\phi_z = \frac{R_{12}}{\cos\phi_y}$$

If $\phi_y = \pm\pi/2$, then ϕ_x and ϕ_z are undefined.

(4)   Problems may result from the fact that finite rotations are not vector quantities. A simple subtraction of one orientation from another one can yield erroneous results in terms of the path of motion.

EXAMPLE 4

Consider an initial position of the arm in which it is outstretched laterally, at 90° to the vertical, with the palm facing anteriorly. This orientation is quantified as 90° of abduction of the arm from the anatomical position. The arm is then moved in a transverse plane until it is pointing anteriorly with the palm facing up. This orientation is quantified as 90° of flexion from the anatomical position. A subtraction of the two orientations suggests that the arm underwent adduction and flexion in the movement from the first to the second position. In reality, the arm axially rotated while moving through a 90° arc in a transverse plane of the body.

Joint Coordinate System (JCS):

Three-dimensional joint orientation is interpreted as a set of three rotations that occur about the axes of a "Joint Coordinate System" or JCS (Fig. 3.4.10).

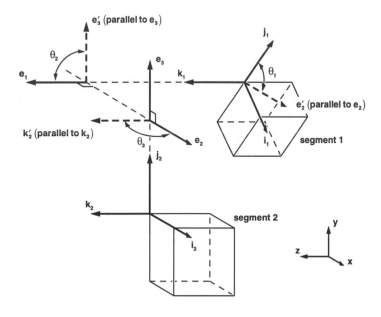

Figure 3.4.10   **Illustration for the definition of the Joint Coordinate System, JCS (from Grood and Suntay, 1983, with permission).**

(1)   One axis, with unit vector, e_1, is selected to be the medio-lateral (z) axis of the proximal segment coordinate system. This is the rotational axis for flexion-extension at the joint.

(2)   Another axis, with unit vector, e_3, is selected to be the longitudinal axis of the distal segment (e.g., y-axis in the leg, or x-axis in the foot). The axial rotation component is measured about this axis.

(3)   These two segment-fixed axes define the remaining axis, with mutually-perpendicular unit vector, e_2. This axis is the cross-product of the two segment-fixed axes, and it defines the axis of rotation for adduction-abduction at the joint.

The equations for the JCS were originally presented for the specific application to the knee joint (Grood and Suntay, 1983). These equations are presented below in a modified form to make them applicable in the general case (Cole et al., 1993). The equations cannot be expressed in general terms using the coordinate axes x, y, and z. Depending on which axes are selected for the two segment-fixed axes, it is possible to obtain both a right- and left-handed Joint Coordinate System. In order to ensure a right-handed Joint Coordinate System, an alternative labelling of segment-fixed coordinate axes is used:

F-axis

Flexion-extension axis. Chosen as the axis of the segment coordinate system that is oriented predominantly
in the medio-lateral direction. The axis is described by
the unit vector, $\mathbf{f}$.

L-axis

Longitudinal axis. Chosen as the axis of the segment
coordinate system that is oriented predominantly
lengthwise along the segment. The axis is described
by the unit vector, $\mathbf{l}$.

T-axis

Third axis. Calculated as the cross-product of the longitudinal axis and the flexion-extension axis. The axis
is defined by the cross-product:

$$\mathbf{t} = \mathbf{l} \times \mathbf{f}$$

where:

$\times$ — symbol for vector multiplication

The unit vectors, $\mathbf{l}$, $\mathbf{f}$, and $\mathbf{t}$ are obtained from the rotation matrix that specifies the orientation of the segment of interest relative to an inertial reference system. For example, if
the flexion-extension axis is chosen as the z-axis, then, $\mathbf{f} = \{R_{31}, R_{32}, R_{33}\}$. The unit
vectors that describe the orientations of the axes of the JCS between a reference segment,
i, and a target segment, j, relative to an inertial reference system are:

$$\mathbf{e}_{1_{ij}} = \mathbf{f}_i$$

$$\mathbf{e}_{2_{ij}} = \left(\frac{\mathbf{e}_{3_{ij}} \times \mathbf{e}_{1_{ij}}}{\left| \mathbf{e}_{3_{ij}} \times \mathbf{e}_{1_{ij}} \right|} \right) \cdot A$$

where :

$$A = \begin{cases} -1 \text{ if} & \left(\mathbf{e}_{3_{ij}} \times \mathbf{e}_{1_{ij}} \right) \bullet \mathbf{t}_j < 0 \\ & \text{and :} \\ +1 \text{ otherwise} & \left(\left(\mathbf{e}_{3_{ij}} \times \mathbf{e}_{1_{ij}} \right) \times \mathbf{e}_{3_{ij}} \right) \bullet \mathbf{f}_j > 0 \end{cases}$$

$\bullet$ = dot product

= scalar multiplication

$\mathbf{e}_{3_{ij}} = \mathbf{l}_j$

The three angles that represent the three-dimensional orientation of the target segment,
j, with respect to the reference segment, i, relative to a neutral position are calculated as
follows.

For the angle of rotation about the axis for flexion-extension:

$$\phi_{1_{ij}} \quad = \quad \text{arccos} \, (e_{2_{ij}} \bullet t_i) \cdot \text{sign} \, (e_{2_{ij}} \bullet l_i)$$

For the angle of rotation about the axis for adduction-abduction:

$$\phi_{2_{ij}} \quad = \quad \text{arccos} \, (r \bullet l_j) \cdot \text{sign} \, (e_{1_{ij}} \bullet e_{3_{ij}})$$

where:

$$r \quad = \quad \left(\frac{e_{1_{ij}} \times e_{2_{ij}}}{|e_{1_{ij}} \times e_{2_{ij}}|} \right)$$

For the angle of rotation about the axis for axial rotation:

$$\phi_{3_{ij}} \quad = \quad \text{arccos} \, (e_{2_{ij}} \bullet t_j) \cdot \text{sign} \, (e_{2_{ij}} \bullet f_j)$$

where:

$$\text{sign} \, (x) = \quad 1 \; \text{if} \; x \geq 0$$
$$-1 \; \text{if} \; x < 0$$

Counter-clockwise rotations relative to the neutral position have been defined as positive. In comparison to the equations originally presented by Grood and Suntay (1983), the scalar, A, has been added to ensure that each of the three angles is continuous in the range, $-\pi < \phi < \pi$, and the unit vector, r, has been added to ensure that counter-clockwise rotations about the axis, e_2, are always positive regardless of which axes are chosen for, e_1 and e_3.

A comparison of the JCS to the proposed sequence of Cardan angles shows the two parametric representations to be identical:

- The first body fixed axis, e_1, corresponds to the first rotational axis of the Cardanic sequence.
- The axis, e_2, corresponds to the "nodal" axis.
- The second body fixed axis, e_3, corresponds to the final rotational axis of the Cardanic sequence.

The three angles are calculated differently in the two methods. However, the principle remains the same and the results are identical.

(i) Advantages and disadvantages of the JCS:

(1)   The advantages and disadvantages of the JCS for representing three-dimensional joint orientation are identical to those described for Cardan angles.

(2)   The major difference between the JCS and the Cardan angle representation is conceptual. For a given joint orientation, the sequential nature of the Cardan angle approach implies that a movement from the neutral position is occurring to obtain the joint orientation of interest, which is not necessarily the case. The JCS,

on the other hand, is conceptually an actual representation of a specific joint orientation. This difference would suggest that the JCS approach is the preferable one.

Finite helical axis:

"Any finite movement from a reference position and attitude to a current one can be described in terms of a rotation, $\theta \geq 0$, and a translation, d, along a directed line in space with a unit direction vector, **n**, and with position, **s**, of some point on this line" (Woltring, 1991). This line is called the *finite helical axis*, and the motion can be compared to that of a screw. The helical axis orientation parameters, the components of the vector, **n**, and the rotation, θ, can be calculated from the rotation matrix for a selected orientation. The methods used to calculate the helical axis parameters in biomechanical applications are described in detail elsewhere (e.g., Spoor and Veldpaus, 1980; Woltring et al., 1985).

(i)   Advantages and disadvantages of finite helical axes:

(1)   When plotted relative to anatomical landmarks, finite helical axes provide a functional representation of the movement occurring at the joint, especially when the finite increments of rotation are chosen to be small.

(2)   The finite helical axis approach provides information about the actual axes of rotation in a joint, which cannot be obtained using the Cardan angle or the JCS approach.

(3)   The finite helical axis does not provide a representation of joint orientation in terms of three anatomically and clinically meaningful parameters, as do Cardan angles and the JCS.

(4)   Helical axis parameters are sensitive to noise in the spatial landmark coordinates, and to the magnitude of the finite rotation (Spoor, 1984; Woltring et al., 1985). The influence of the former, however, can be substantially reduced with appropriate smoothing of the landmark coordinates (Woltring and Huiskes, 1985; Lange et al., 1990).

Helical angles:

The concept of helical angles has recently been proposed as an alternative to Cardan or JCS angles (Woltring, 1991; Woltring, 1992). Three angles are determined from the components of the vector:

$$\theta \quad = \quad (\theta_x, \theta_y, \theta_z)^T \quad = \quad \theta\,\mathbf{n}$$

where:

n   =   unit direction vector of the finite helical axis for the orientation of a body segment relative to a reference position or to another body segment

Helical angles eliminate the problem of gimbal-lock. Additionally, for rotations about a single coordinate axis, helical angles are the same as Cardan or JCS angles. As with Cardan or JCS angles, helical angles are not additive. However, they reduce the non-linear properties of the parametric representation of joint orientation (Woltring, 1991).

Final comments:

The selection of the appropriate methodology for describing three-dimensional joint orientation must be determined by the application. Each approach has its own inherent advantages and disadvantages. The JCS approach, for example, produces a result that is easily understood in clinical and anatomical environments. The helical axis approach, on the other hand, provides results that better represent the functional movement that occurs at a joint.

Errors in three-dimensional motion analysis

Errors in three-dimensional motion analysis may occur for many different reasons. Some of them are discussed in the following sections. The discussion is divided into two parts:

- Errors in the determination of the landmark coordinates.
- Errors in representing the kinematics of body segments using markers.

Errors in landmark coordinates:

(i) Errors due to relative camera placement:

In three-dimensional motion analysis, each camera, i, is connected by an imaginary line to each marker, k. Ideally, for n cameras, these lines would intersect spatially at one point. However, in real situations, these lines do not intersect at one point because the directions of these lines are not infinitesimally accurate. Spatial coordinates of one marker point are, therefore, determined using distances between a point on a line (connecting the camera and the marker) and the assumed actual spatial position of this marker. Usually, a least squares fit method is applied to determine the location of marker, k, under the condition that:

$$\sqrt{d_{1k}^2 + d_{2k}^2 + \ldots + d_{nk}^2} \;=\; \text{minimum}$$

where:

d_{ik} = distance between a point on the line connecting camera, i, and marker, k, and the estimated spatial position of marker, k

i = symbol for the cameras

k = symbol for the markers

The quality of the determined coordinates depends, although not exclusively, on the angle of intersection between the two lines. Small errors in orientation of two perpendicular lines does not significantly influence the coordinates of the lines, whereby, the distance between these two lines is minimal. However, if the angle between two intersecting lines is small, errors in the orientation of these lines may produce substantial differences in coordinates. This suggests that cameras should be placed so that the angles between two optical axes intersect at an angle that nears 90° but is not less than 60°.

Cameras are often placed in an umbrella configuration for motion analysis of human or animal locomotion (Fig. 3.4.11). For reasons of convenience, the cameras are often

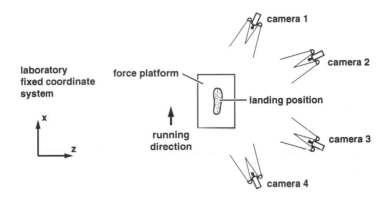

Figure 3.4.11   **Schematic illustration for a camera placement in a typical umbrella configuration, as used often in gait analysis.**

arranged close to the ground level. As a result, the coordinate of each marker within this plan should be well established. However, the coordinates in the two mutually perpendicular planes may be affected with considerable errors. An improvement in accuracy may be obtained by placing the cameras such that they are not all in the same plane.

(ii) Errors due to number of cameras:

Ideally, all cameras "see" each marker at every point in time. In real situations, this is not the case; and markers are lost from view. In film analysis, the positions of these lost markers may be determined using a best estimate based on the position of the body. In video analysis, these markers are lost until they are once again visible to the camera. The position of a marker at a given point in time is calculated using only those cameras that "see" the marker. This may be critical for certain camera placements and camera numbers.

As an example, an experiment is performed using three cameras: 1, 2, and 3. The optical axes of cameras 1 and 2 intersect at an angle of 90°, and the optical axes of cameras 1 and 3 intersect with an angle of 20°. At time, t_1, the position of marker, k, is determined using information from all three cameras; at time, t_2, information from camera 2 is missing; and at time, t_3, information from all three cameras is again available. The results for this example are illustrated for the x and y coordinates in Fig. 3.4.6.

For instance, frames 15 and 19 use (for three different norm of residuals) a different number of cameras than do frames 16, 17, and 18, (which all use the same number of cameras) to determine the position of marker, k.

Obviously, a change from 6 to 5 cameras would induce less change in the final coordinate calculation than a change from 3 to 2 cameras. In general, noise in coordinates decreases with an increasing number of cameras. However, this is not necessarily equivalent to an increase in accuracy of the determined marker positions.

(iii) Errors due to calibration:

The accuracy of the measured data depends considerably on the accuracy of the calibration procedure. It is often assumed that accuracy increases the more the calibrated volume fills the field of view. However, the situation is more complex than this. Lenses are not ideal structures. They have errors that influence the accuracy of the collected data. Consequently, data should be corrected for lens errors. Several software packages do this. If data is not corrected for lens errors, calibration of the central part of the field of view, where the lens usually has fewer errors than at its boundaries, may be better than using calibration frames that exploit the entire field of view. Additionally, errors in the determination of the position of markers depend on the accuracy of the calibration marker setup. Errors in the location of these markers affect the accuracy of marker coordinates.

(iv) Errors due to digitization:

Conventional film analysis typically uses manual digitization. Results from manual digitization may be affected by:

- Random inaccuracies of the digitizing operator in positioning the digitization pen. Such errors can be reduced by employing appropriate marker shapes (spherical), markers with dots in the middle, and multiple digitization.

Video analysis often uses automatic digitization. The accuracy of automatic digitization depends on various factors such as:

- Number of cameras (see above),
- Placement of cameras (see above),
- Marker shape (spherical markers are typically used because they have the same shape irrespective of the direction from which they are viewed),
- Size of the markers (the centroid can be determined with better accuracy for a large than for a small marker),
- Merging of markers (the centroid of a marker cannot be correctly established),
- Partial cover of a marker (the centroid of a marker cannot be correctly established), and
- Threshold level of the video system (determines how well the video screen estimates the spherical image of the marker).

Most of these problems can be solved with appropriate preparations of the setup and adequate software.

Errors in representing segment kinematics using marker information:

Markers attached to a segment of interest may not always represent true skeletal locations. These differences are referred to as relative and absolute errors. In addition, the placement of the markers can substantially influence the propagation of marker coordinate errors and relative marker movement errors when calculating segment kinematics:

- Relative marker error is defined as the relative movement of two markers with respect to each other. The relative marker error establishes how well the markers on the segment estimate the rigid body assumption. This error is caused by soft tissue movement.
- Absolute marker error is defined as the movement of one specific marker with respect to specific bony landmarks of a segment. This error establishes the actual accuracy of the measurement.

(i) Errors due to relative marker movement:

The three-dimensional motion of body segments is typically described using rigid body mechanics. The spatial coordinates of segment landmarks are used directly or indirectly to define body-fixed coordinate systems for each segment. Markers placed on the skin of a subject or animal are assumed to represent the location of bony landmarks of the segment of interest. However, skin movement and movement of underlying bony structures are not necessarily identical, and substantial errors may be introduced in the description of bone movement when using skin mounted marker arrays (Lesh et al., 1979; Ladin et al., 1990). Three major approaches have been suggested to correct skin displacement errors: invasive marker placement, data treatment, and marker attachment systems.

Invasive methods provide the most accurate results for bone movement. Bone pins have been used to assess movement of the tibia and femur (e.g., Levens et al., 1948). Knee joint kinematics have been compared for data collected from skin mounted markers and markers attached to bone pins (Ladin et al., 1990). The results for both methods showed differences of up to 50% for the knee angle, for instance. Results from invasive methods are, however, not applicable in most research and clinical settings for movement analysis, therefore, other approaches are needed.

Mathematical algorithms have been proposed for error reduction in raw data, including skin displacement effects (Plagenhoef, 1968; Miller and Nelson, 1973; Winter et al., 1974; Zernicke et al., 1976; Soudan and Dierckx, 1979; Woltring, 1985; Veldpaus et al., 1988). Smoothing algorithms are based on the assumption that the noise is additive and random with a zero mean value (Woltring, 1985). Errors due to skin displacement, however, may be correlated and may not have zero means. Mathematical algorithms as currently used do not seem to be appropriate approaches for solving the errors due to marker movement relative to bony landmarks.

Marker attachment systems (frames) have recently been developed to reduce errors due to relative marker movement (Ronsky and Nigg, 1993). The following discussion concentrates on the tibia. However, it may be used in analogy for other segments of human or animal bodies. Positional data from 4 different lightweight marker attachment systems (frames) were compared to data from skin-mounted markers. Frame 1 was constructed from thermoplastic material, frame 2 of rectangular aluminum rods, frame 3 of plastic material of medium stiffness, and frame 4 used laminated plastic strips. The 5th marker attachment system was the attachment of the markers to the skin. Skin displacement errors were analysed in terms of relative and absolute errors.

The *relative* error was determined as the change in three-dimensional length, ΔL_{ik}, between any two markers with respect to the static length, where:

$$\Delta L_{ik} \quad = \quad L_{ik}^{stat} - L_{ik}^{dyn}$$

and:

$$L_{ik} \quad = \quad \sqrt{(x_i - x_k)^2 + (y_i - y_k)^2 + (z_i - z_k)^2}$$

where:

x_i, y_i, z_i = coordinates of marker, i, at the time, t

x_k, y_k, z_k = coordinates of marker, k, at the time, t

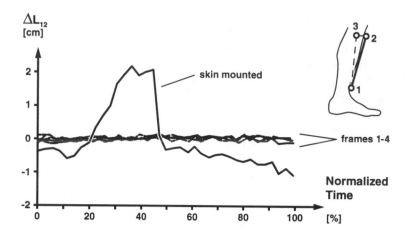

Figure 3.4.12   Actual three-dimensional length between the markers 1 and 2 during ground contact in running (time 0 corresponds to heel strike, time 1 to take-off) for 4 marker frames and for skin-mounted markers. The graph illustrates results for one subject and is representative of the general trend (from Ronsky and Nigg, 1993, with permission).

The results for the relative errors are illustrated in Fig. 3.4.12. Marker 1 was mounted close to the distal anterior part of the tibia and marker 2 at the proximal anterior part of the tibia. The 4 frames substantially reduced the relative movement of marker 1 with respect to marker 2 when compared to the results for the skin-mounted markers. The biggest length difference during stance between markers 1 and 2 was more than 3 cm in this example. It is obvious that such differences in length (as measured for the skin-mounted markers) of a structure, which is assumed to be rigid, interfere with the accuracy of the results. These results suggest that markers attached to frames may fulfil the requirements of a rigid body, and that markers attached directly to the skin may not fulfil these conditions satisfactorily.

The use of an array of more than 3 markers may be advantageous to solve for relative marker movement. In this case, the distance between all markers is monitored and the markers with the minimal changes in distance, for instance, are used for further calculation.

(ii) Errors due to absolute marker movement:

Errors due to absolute marker movement are difficult to determine without using invasive methods. Recently, an indirect method to estimate absolute marker movement was presented (Ronsky and Nigg, 1993). The results of this study were acceptable for slow movements of the tibia with no inertia effects due to impact forces. However, for movements with impact effects, the results indicated substantial inconsistencies, and it was suggested that only bone-mounted markers seem to be appropriate to quantify bone movement under all conditions.

(iii) Errors due to inadequate placement of the markers:

The placement of the markers on the segment of interest may be a further source of error in the determination of marker coordinates. Possible factors include:

- Planarity (the markers should define a plane and should not be collinear),
- Distance between markers (large distances between markers improve accuracy), and
- Marker distribution with respect to the axis of rotation of the segment (helical axes are most accurately determined when the helical axis passes directly through the centroid of the marker distribution).

Most of these factors can be controlled with a careful preparation of each segment of interest.

Errors in two-dimensional motion analysis:

Errors in the determination of length between two markers in two-dimensional analysis depend on several factors, including:

- The camera system (predominantly lens errors),
- The distance between object and camera (different plane of movement than the calibrated plane), and
- Projection errors (oblique position of the length of interest).

Errors due to the camera system have been described in this chapter.

Errors due to a difference in the distances between object and lens are schematically illustrated in Fig. 3.4.13. The error in a length measurement due to the fact that the mea-

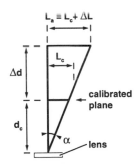
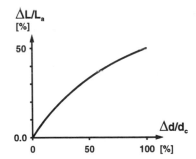

Figure 3.4.13   Illustration of selected factors influencing the accuracy of length measurements in two-dimensional motion analysis.

sured length is not in the calibrated plane is:

$$\Delta L \quad = \quad (\frac{\Delta d}{d_c + \Delta d}) \cdot L_a$$

where:

L_c = apparent length of the object in the calibrated plane

L_a = actual length of the object of interest

d_c = distance between lens and the calibrated plane

Δd = distance between the calibrated plane and the location of the object outside the calibrated plane

ΔL = difference between the actual length of the object and the length of the object measured in the calibrated plane

The graph (Fig. 3.4.13) illustrates that errors diminish if the deviation from the calibrated plane of motion, Δ_d, and/or the length of the object, L_a, are small compared to the distance, d_c, between camera lens and object.

EXAMPLE 5

A subject walks from left to right. The distance between the lens and the right arm of the subject is 5 m, and the camera axis is perpendicular to the walking direction. The width of the subject (distance between left and right arm) is 0.5 m, and the distance between the forearm and hand measures 0.5 m. On a two-dimensional photograph the left arm will appear to be 0.55 m long, or 5 cm longer than the right arm for this set of assumptions. Accordingly, the velocity and acceleration calculations will be affected by this error.

Projection errors are errors that result from an oblique position of the length to be measured with respect to the calibrated plane of motion perpendicular to the camera axis. They can be approximated for $d_c \gg L_a$ with the formula:

$$\Delta L = L_a (1 - \cos\alpha)$$

where:

ΔL = error in the length measurement

L_a = actual length of the object of interest

α = angle between the line between the two markers on the object and the calibrated plane of motion

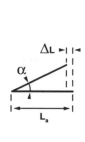
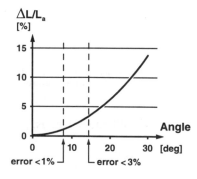

Figure 3.4.14   Error in a two-dimensional length measurement for the case where the length to be measured is not in the calibrated motion plane perpendicular to the camera axis.

The error due to an oblique position of an object with respect to the calibrated plane of motion is smaller than 1% for angles of $\alpha < 8°$ and smaller than 3% for angles of $\alpha < 14°$ (Fig. 3.4.14).

3.4.4 REFERENCES

Abdel-Aziz, Y.I. and Karara, H.M. (1971) Direct Linear Transformation from Comparator Co-ordinates Into Object Space Co-ordinates. *Proc. ASP/UI Symposium on Close-range Photogrammetry.* Am. Soc. of Photogrammetry, Falls Church, VA. pp. 1-18.

Cavanagh, P.R. (1990) The Mechanics of Distance Running: A Historical Perspective. *Biomechanics of Distance Running* (ed. Cavanagh, P.R.). Human Kinetics, Champaign, IL. pp. 1-34.

Centre Georges Pompidou, Musee National d'Art Modern (1977) *E.J. Marey, 1830/1904, La Photographie du Mouvement.*

Chao, E.Y.S. (1980) Justification of Triaxial Goniometer for the Measurement of Joint Rotation. *J. Biomechanics.* **13,** pp. 989-1006.

Cole, G.K., Nigg, B.M., Ronsky, J.L., and Yeadon, M.R. (1993) Application of the Joint Coordinate System to 3-D Joint Attitude and Movement Representation: A Standardization Proposal. *J. Biomech. Eng.* **105,** pp. 136-144.

de Lange, A., Huiskes, R., and Kauer, J.M. (1990) Effects of Data Smoothing on the Reconstruction of Helical Axis Parameters in Human Joint Kinematics. *J. Biomech. Eng.* **112,** pp. 107-113.

Goldstein, H. (1950) *Classical Mechanics.* Addison-Wesley, London.

Grood, E.S. and Suntay, W.J. (1983) A Joint Coordinate System for the Clinical Description of Three-dimensional Motions: Application to the Knee. *J. Biomech. Eng.* **105 (2),** pp. 136-144.

Ladin, Z., Mansfield, P.K., Murphy, M.C., and Mann, R.W. (1990) Segmental Analysis in Kinesiological Measurements. *Image Based Motion Measurement, SPIE* (ed. Walton, J.S.). **1356,** pp. 110-120.

Lesh, M.D., Mansour, J.M., and Simon, S.R. (1979) A Gait Analysis Subsystem for Smoothing and Differentiation of Human Motion Data. *J. Biomech. Eng.* **101 (3),** pp. 205-212.

Levens, A.S., Inman, V.T., and Blosser, J.A. (1948) Transverse Rotation of the Segments of the Lower Extremity in Locomotion. *J. Bone and Jt. Surg.* **30 (A-4),** pp. 859-872.

Lindholm, L.E. and Oeberg, K.E. (1974) An Opto-electric Instrument for Remote On-line Movement Monitoring. *Biomechanics IV* (eds. Nelson, R.C. and Morehouse, C.A.). University Park Press, Baltimore. pp. 510-512.

Miller, D.I. and Nelson, R.C. (1973) *Biomechanics of Sport.* Lea & Febiger, Philadelphia.

Mitchelson, D.L. (1988) Automated Three-dimensional Movement Analysis Using the CODA-3 System. *Biomedizinische Technik.* **33 (7-8),** pp. 179-182.

Moore, K.L. (1980) *Clinically Oriented Anatomy.* Williams & Wilkins, Baltimore.

Plagenhoef, S.C. (1968) Computer Program for Obtaining Kinetic Data on Human Movement. *J. Biomechanics.* **1,** pp. 221-234.

Ronsky, J.L. and Nigg, B.M. (1993) Error in Kinematic Data Due to Marker Attachment Methods. *J. Biomechanics.*

Snell, R.S. (1973) *Clinical Anatomy for Medical Students.* Little Brown & Company, Boston. pp. 1-5.

Soudan, K. and Dierckx, P. (1979) Calculation of Derivatives and Fourier Co-efficients of Human Motion Data, While Using Spline Functions. *J. Biomechanics.* **12 (1),** pp. 21-26.

Spoor, C.W. and Veldpaus, F.E. (1980) Rigid Body Motion Calculated from Spatial Coordinates of Markers. *J. Biomechanics.* **13,** pp. 391-393.

Spoor, C.W. (1984) Explanation, Verification, and Application of Helical-axis Error Propagation Formulas. *Human Movement Science.* **3,** pp. 95-117.

Veldpaus, F.E., Woltring, H.J., and Dortmans, L.J. (1988) A Least-squares Algorithm for the Equiform Transformation from Spatial Marker Co-ordinates. *J. Biomechanics.* **21,** pp. 45-54.

Winter, D.A. (1979) *Biomechanics of Human Movement.* John Wiley & Sons, New York.

Winter, D.A., Sidwall, H.G., and Hobson, D.A. (1974) Measurement and Reduction of Noise in Kinematics of Locomotion. *J. Biomechanics.* **7,** pp. 157-159.

Woltring, H.J. (1985) On Optimal Smoothing and Derivative Estimation from Noisy Displacement Data in Biomechanics. *Human Movement Science*. **4 (3),** pp. 229-245.

Woltring, H.J. (1991) Representation and Calculation of 3-D Joint Movement. *Human Movement Science*. **10 (5),** pp. 603-616.

Woltring, H.J. (1992) 3-D Attitude Representation: A Standardization Proposal. *J. Biomechanics*. (Submitted).

Woltring, H.J. and Marsolais, E.B. (1980) Optoelectric (Selspot) Gait Measurement in Two- and Three-dimensional Space: A Preliminary Report. *Bulletin of Prosthetics Research*. **17 (2),** pp. 46-52.

Woltring, H.J. and Huiskes, R. (1985) A Statistically Motivated Approach to Instantaneous Helical Axis Estimation from Noisy, Sampled Landmark Coordinates. *Biomechanics IX-B* (eds. Winter, D.A., Norman, R.M., Wells, R.P., Hayes, K.C., and Patla, A.E.). Human Kinetics, Champaign, IL. pp. 274-279.

Woltring, H.J., Huiskes, R., de Lange, A., and Veldpaus, F.E. (1985) Finite Centroid and Helical Axis Estimation from Noisy Landmark Measurements in the Study of Human Joint Kinematics. *J. Biomechanics*. **18 (5),** pp. 379-389.

Yeadon, M.R. (1990) The Simulation of Aerial Movement - I: The Determination of Attitude Angles from Film Data. *J. Biomechanics*. **23 (1),** pp. 59-66.

Zernicke, R., Caldwell, G., and Roberts, E.M. (1976) Fitting Biomechanical Data with Cubic Spline Functions. *Research Quarterly*. **47 (1),** pp. 9-19.

3.5   STRAIN MEASUREMENT

SHRIVE, N.

3.5.1 DEFINITIONS AND COMMENTS

Gauge factor:

Quantity with which an initial signal output is multiplied to receive the measurand of interest.

Microstrain:

Unit which is typically used for strain measurements.

$$1 \text{ microstrain} = 10^{-6} = 1\mu\varepsilon$$

Comment: $10^4\mu\varepsilon = 1\%$

Strain:

• Engineering strain:

$$\varepsilon = \frac{\Delta L}{L_o} = \frac{L - L_o}{L_o}$$

where:

ε = engineering strain

ΔL = change in length

L_o = original length

L = current length

• True strain:

$$\varepsilon_{true} = \int_{L_o}^{L} \frac{dl}{l} = \ln\left(\frac{L}{L_o}\right)$$

Comment: strain is non-dimensional. Units of strain are typically % or microstrain ($\mu\varepsilon$).

3.5.2 SELECTED HISTORICAL HIGHLIGHTS

Electrical resistance strain gauges

1870's

Unbonded wire gauges were invented.

1936-38

Bonded wire gauges developed separately and simultaneously at Caltech, MIT, and the GEC labs. The first commercial gauge was based on Prof. A.C. Ruge's gauge from MIT, with circuitry based on the Caltech technology (Stein, 1992).

1952

Foil strain gauges were patented in the UK by Peter Jackson utilizing the concept of printed circuit technology (Stein, 1992).

1960's

Semi-conductor strain gauges evolved from microelectronics industry.

1970's		Liquid mercury strain gauges introduced in biomechanics.

Mechanical strain gauges

1900's	Huggenberger	Developed a mechanical extensometer.
1953	Demec	Introduced a demountable extensometer for measurements on concrete (Morice and Base, 1953).
1960's		Development of strain gauge based cantilever extensometers.
1992	Shrive	Described soft tissue extensometers.

Optical methods

1815	Brewster	Discovered double refraction (Hendry, 1948).
1906	Coker	Used celluloid for photoelasticity.
1930	Mesnager	Photoelastic coatings were described by Mesnager, 1930.
1935	Solakian	Three-dimensional analysis by stress freezing was described.
1953	Zandman	First practical results from a photoelastic coating were reported (Redner, 1968).
1970's		Introduction of holography to biomechanics measurements.

Other techniques

1925		Brittle lacquers (Sauerwald and Wieland, 1925).
1983	Woo	Video dimension analysis was developed to determine soft tissue strain.
1984	Noyes	Used high speed film analysis to determine strain in ligaments and tendons.
1985	Arms	Developed the Hall effect displacement transducer to measure ligament strain.
1989		Introduction of SPATE to biomechanics (Friis et al., 1989; Kohles et al., 1989).

3.5.3 MEASURING POSSIBILITIES

The measurement of strain is fundamental to understanding and defining material behaviour, and to the elucidation of structural behaviour and design. Furthermore, the measurement of strain is important for the development of implants to access how the material behaves under load, and how the distribution of load in the structure is affected by the geometry and applied loading.

Strain measuring techniques for "stiff" materials such as bone, metals, and ceramics have developed over the years such that good sensitivity can be achieved fairly cheaply. The technology for measuring strain on more flexible materials - such as tendons and ligaments - has not followed the pace set by the more standard techniques. There are special problems which have proved difficult to overcome. The basic issue is frequently: has the strain being measured been changed by the very attempt to measure it?

GENERAL

Strain gauges measure engineering strain which is defined as:

$$\varepsilon = \frac{\text{change in length}}{\text{original length}} = \frac{L - L_o}{L_o} = \frac{\Delta L}{L_o}$$

where:

L = current length
L_o = original length
ΔL = change in length

True strain is defined as the integral, from the original to the current length, of the increment in length divided by the immediately preceding length.

$$\varepsilon_{true} = \int_{L_o}^{L} \frac{dl}{l}$$

$$= \ln\left(\frac{L}{L_o}\right)$$

When measurements of strain are made, the overriding principle for the measuring system is that it be highly flexible. Very little load must be required to deform the measuring device. The device will then not interfere with the strain being measured: the load that is normally carried by the material being strained, will continue to be carried by that material, rather than being transferred into the measuring device. This is in direct contrast to a load measuring device which is in series with the load being measured. Here the overriding principle is that the device be stiff. Highly stiff load cells do not interfere with the deformation characteristics of the system.

There are three basic techniques for measuring strain. The first is to measure the equivalent of strain and to convert the measurement to mechanical strain by a gauge factor. Electrical resistance strain gauges such as foil gauges or liquid mercury strain gauges in Wheatstone bridge configuration are based on this principle, as they measure electrical "strain" in the circuit.

The second technique is to measure and amplify ΔL, the change in length, and to divide by an accurate measure of L_o. Extensometers and optical interference methods are based on this principle. Methods based on these first two techniques have the potential to be both accurate and sensitive. Indeed, sensitivities in the order of one microstrain can be achieved with some semiconductor and foil gauges. While this level of accuracy and sensitivity is not normally needed in biomechanics, it is indicative of what can be achieved.

The third technique is to measure lengths and thence derive strain. Methods based on this technique will be inherently inaccurate at low strains. This is because the difference between two large numbers (the lengths) will be inaccurate when the difference is of the same order of magnitude as the error in the large numbers. Simply:

$$(L \pm E) - (L_0 \pm E_0) \;=\; (L - L_0) \pm (E + E_0) \;=\; \Delta L \pm (E + E_0)$$

where E and E_0 are the errors in the length measurements.

When ΔL is substantially larger than $(E + E_0)$ the strain reading will be meaningful. Thus, if large strains are to be measured, techniques based on length measurement may provide reasonable indications of the actual level of strain. At low strain levels, however, strain measurements from such methods will be affected with a large error.

Strain measuring devices include:

- Electrical resistance strain gauge,
- Extensometers,
- Optical methods, and
- Other methods.

They will be discussed in the following paragraphs.

ELECTRICAL RESISTANCE STRAIN GAUGES

General principles

The resistance of an electric conductor is a function of the dimensions of the conductor as well as its resistivity.

$$R \;=\; \frac{sL}{A}$$

where:

R = electrical resistance
s = resistivity
L = length
A = cross-sectional area

From the mechanics of deformable bodies, it is known that when such a conductor (like a wire) is stretched, L increases and A decreases: thus the electrical resistance R, changes. The change in electrical resistance can be related to the change in length. With appropriate circuitry:

$$\frac{\Delta R}{R_0} \;=\; G_F \frac{\Delta L}{L_0} \tag{3.5.1}$$

where:

R_0 = initial electrical resistance for the unstrained material

ΔR = change in resistance due to the strain in the material

G_F = gauge factor

L_0 = initial length of the unstrained material

ΔL = change in length of the strained material

G_F is the gauge factor which is typically about 2 for bonded wire and foil gauges in their linear range (as the strain gets higher, G_F becomes more and more non-linear with strain. Special circuitry is needed, or allowances made for the non-linearity).

The relationship of equation (3.5.1) holds so long as the conductor in the gauge stretches with the specimen. This will occur if the gauge is firmly glued to the specimen. In a strain gauge, the conductor has a very small cross-section, requiring only small forces for extension. For example, if the conductors in a gauge (Fig. 3.5.1) have a total width of 2

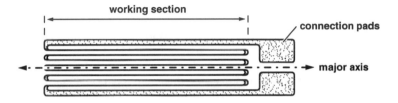

Figure 3.5.1   Schematic of a foil strain gauge. The gauge measures extensional strain in the direction of the major axis. The strain is an average measure over the length of the working section.

mm and thickness 5 micrometres (in some foil gauges, the conductor is only 2.5 micrometres thick!), less than 1 N will cause yield of a typical conductor. This level of force can easily be transmitted through a variety of adhesives, given the area of the gauge being bonded to the specimen surface.

Thus, concern about the gauge reinforcing the specimen and altering the normal strain distribution need only arise with very thin structures and/or flexible materials (for example, cartilage, ligament, or tendon in biomechanics). In such cases, consideration should be given to the level of reinforcement provided, and the alteration to the strain being measured. A related issue more pertinent to biomechanics is the potential for chemical reaction between the adhesive and the test material. Strain would be measured on this newly created material, not the material of the specimen, and the strain in the specimen would be altered by the presence of the new material.

Strain gauges measure extensional strain in the direction of the major axis of the gauge (Fig. 3.5.1). However, both the lateral strain and the shear strain in the plane of the gauge can effect the extensional strain reading. In most cases these effects are negligible. However, if they are not, one may have to compensate for them. Manufacturers normally indicate the magnitude of this "transverse sensitivity" through the K_t factor of the gauge. If this is larger than the accuracy desired, or if "rosette" gauges are being used, then the effects above may need to be considered. The extensional strain is averaged over the working area of the gauge and the division of the conductor into a series of connected "wires" in the major axis direction, amplifies the change in R over that which would be obtained by a single conductor covering the same working area as a single sheet. In gauge selection, the size of the working area of the gauge must be considered in light of the size and material of the specimen, the rate of local strain variation, and the desired information. A large gauge for example, will not reveal much value where strains vary rapidly and knowledge of the variation is desired, whereas, a small gauge will be of equally little value when an average strain over a composite is required. Foil gauges range in working length from about 0.2 mm to 100 mm, thus, providing considerable choice.

If knowledge of extensional strain in more than one direction on the surface is desired, then more than one gauge is required. Two and three gauge "rosettes" are available. For principal strains and their directions (unknown at the beginning of the test) to be determined, a three gauge rosette is required. The gauge directions are typically offset from each other by 45° or 60°.

Circuitry

Strain levels in stiff materials are typically measured in terms of microstrain. With a gauge factor of 2, the change in electrical resistance will be of the same order of magnitude (10^{-6}) compared to the original resistance (R_o) of the conductor. The Wheatstone bridge (Fig. 3.5.2) is, therefore, used to measure the small changes which occur.

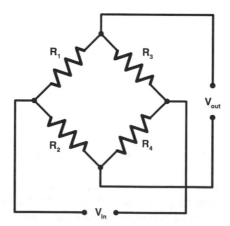

Figure 3.5.2   **Wheatstone bridge circuitry for strain gauge operation.**

The choice on a strain gauge conditioning amplifier will be for 1/4, 1/2, or full bridge operation. In 1/4 bridge operation, only one resistor (a gauge) will be active (change during the test). Let this be R_1 in Fig. 3.5.2. R_3 and R_4 in this case would be precision resistors of equal value inside the conditioning amplifier. R_2 is also in the amplifier or could be a separate (dummy) gauge. If in the amplifier, R_2 is adjusted to have the same resistance as the active gauge R_1 in the unstrained condition. Thus, the voltage drop across the centre of the bridge V_{out}, will be zero. The circuit is balanced. As soon as strain is applied, R_1 changes by ΔR and V_{out} is no longer zero. The magnitude of V_{out} depends on ΔR.

If R_2 is active as well as R_1, usually sensing strain of the opposite sign to R_1, then a half bridge circuit is being used. An example is simple bending where a gauge or one side of a beam would sense tension while one on the other side would sense compression. When all four gauges are active, with all resistance changes arranged to maximize the change in V_{out}, full bridge operation is being utilized. As an example, strain gauge based extensometers use this system.

The ability to measure strain accurately with this circuitry depends on the stability of the excitation voltage and the ability to measure V_{out}. The precision of these instruments is paramount in the strain sensitivity of the circuit. It is also important to recognize that when the excitation voltage is turned on, current flows through the arms of the bridge, heating the resistors. The circuit must be allowed to reach thermal equilibrium before measurements are taken.

In a circuit, the resistance of the lead wires to the gauges needs to be considered. If the wire is long and/or thin, then the lead wire can have sufficient resistance to desensitize the strain gauge. Basically, R_o in equation (3.5.1) is increased without an equivalent increase in ΔR.

Temperature compensation

The resistance of a strain gauge will change with temperature as the material on which the gauge is mounted expands and contracts with the change in temperature. In situations where the temperature of the specimen will change during a test, the thermal effect must be isolated from the load effect, in order for the latter to be determined. There are two basic approaches to solve this problem.

In temperature compensating circuits, the active gauge might be R_1 in Fig. 3.5.2, as before. R_2 would be a compensating dummy gauge mounted on a piece of the same material as the test specimen. This piece of material would be subject to the same thermal regime as the test specimen. Thus, the resistance of both R_1 and R_2 would change by the same amount with temperature, having a null effect on V_{out}. This technique obviously requires that the temperature of both the test specimen and the separate piece do indeed change simultaneously. Some circuits are automatically temperature compensating; for example a half-bridge where the gauges are bonded on either side of a beam being bent.

The second approach is self-temperature-compensation (STC). Certain constantan or nickel chromium alloys can be heat-treated to obtain specific thermal characteristics when rolled for incorporation in foil gauges. These gauges expand sympathetically with the material on which they are mounted and, therefore, do not measure the thermal strain. STC gauges of this type are designed to compensate for the thermal expansion of a particular material - typically steel. A steel compensating gauge will not compensate correctly for temperature changes in other materials: e.g., titanium. Temperature changes in the lead wires to the gauge must also be considered in STC circuits. A three-wire circuit is required in 1/4 bridge operation to eliminate the temperature effect in the lead wires (Fig. 3.5.3).

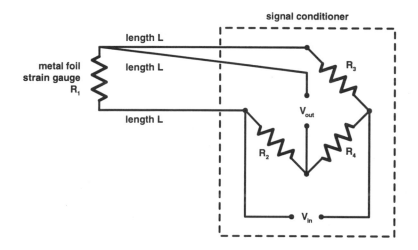

Figure 3.5.3   **Three wire arrangement for temperature compensation of the wires in quarter bridge operation with a STC gauge.**

Different types of electric resistance strain gauges

Foil gauges:

Foil strain gauges are strain measuring sensors consisting of a thin layer of a metal conductor arranged in an array of connected parallel lines (Fig. 3.5.1). These gauges have been described in the preceding section on the general principles of electric resistance strain gauges. These are the most commonly used form of strain gauge sensors. The expressions foil gauge and strain gauge are often used synonymously.

Semiconductor strain gauges:

Semiconductor strain gauges are wafers of semiconductor doped to the form of a strain gauge. They have much higher gauge factors than the foil and wire gauges above, and, therefore, have higher output for a given strain and excitation voltage. The sensitivity with currently available instrumentation is about 1 microstrain.

Liquid mercury strain gauge:

The liquid mercury strain gauge was introduced (Edwards et al., 1970) as a method for measuring strain on soft tissues. As shown in Fig. 3.5.4, the conductor which changes resistance is usually a column of mercury, or less frequently, gallium-indium electrolyte solution. The lead wires near the ends of the column are sewn to the tissue being studied. This causes a problem with the definition of the gauge length, L_0. If there were no slippage between the wire, the suture, and the tissue, the gauge length would be the distance between the two sutures. The gauge length is, therefore, not necessarily the length of the mercury column. With this method, strain measures of up to 50% have been reported.

The gauge is used as one arm of a Wheatstone bridge: sometimes it has been used in addition to a 120 Ω resistor in the arm. However, as the resistance of an LMSG is in the

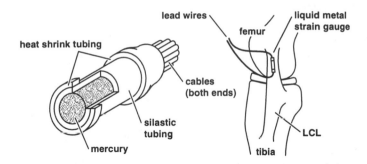

Figure 3.5.4 Schematic illustrations of a liquid metal strain gauge and its construction (left), and a possible arrangement on a lateral collateral ligament (right).

order of 1 Ω, the circuit is very insensitive. Greater sensitivity can be achieved by creating a bridge circuit with 1 Ω resistors.

LMSGs have a non-linear response to strain, and individual gauges have given reproducible results. However, there is considerable variation in response from gauge to gauge. In work on soft tissues, these gauges have been used quite widely.

EXTENSOMETERS

Extensometers were originally mechanical devices which relied on mechanical magnification of the increase in length of the gauge length by the movement of levers around pivot points. However, with the advent of strain gauges, electromechanical extensometers were introduced. The basic principle used is that the highest strain which develops in a cantilever is at the root of a cantilever (Fig. 3.5.5a). The displacement at the end of the

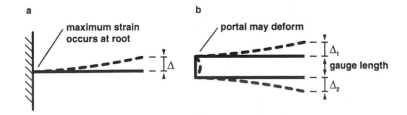

Figure 3.5.5 (a) The maximum strain in a cantilever given a simple deflection at its end occurs at the root. With two such arms (b) an extensometer is formed. The distance between them is the gauge length. Depending on the construction of the extensometer, the portal connecting the two arms may deform and be strain gauged.

cantilever is related to the strain generated. With two such arms firmly attached to a specimen, the distance between them is the gauge length, and the increase in the length of the length of the gauge length is simply the sum of the two displacements measured (Fig. 3.5.5b). This can easily be converted to specimen strain. Some extensometers have the gauges on the portal section of the portal frame made by the two arms and their connector.

The critical issues for an extensometer are the contact between the specimen and the extensometer arms, and the force required to open the arms. The latter needs to be small relative to the force being applied to the specimen, so as not to interfere with the strain being measured. An extensometer designed for large steel specimens may have too big an opening force for a thin aluminum sheet. The contact needs to be firm with no slip between arm and specimen. Thus, extensometers made for metals have very sharp edges and are held tight to the specimen through spring loading. These extensometers slice through soft tissues with considerable ease.

Thus, a variation on the extensometer theme was developed (Shrive et al., 1992 and 1993) to accommodate soft tissues (Fig. 3.5.6a). In this instance, the extensometer is not

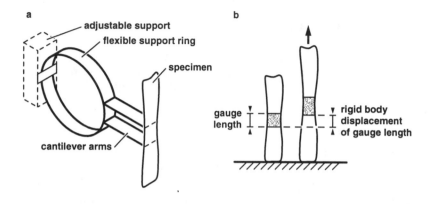

Figure 3.5.6   Schematic illustration of a soft tissue extensometer. The soft tissue extensometer uses standard principles, but is supported externally to the specimen (a). Rigid body deformation of the gauge length when the specimen extends (b), is accommodated by the support ring deforming in shear. The ring also applies the low contact forces needed to hold the tips of the arms on the specimen.

mounted on the specimen. The weight does not, therefore, distort the specimen which it would if the extensometer were suspended on the specimen as usual: here the weight is supported by a light flexible ring. Normal extensometers (held on the specimen) accommodate rigid body displacement of the gauge length, depicted in Fig. 3.5.6b, by moving with the specimen. With the new extensometer, rigid body displacement of the gauge length is accommodated by the ring rolling in shear. Low contact forces are controlled by the ring, and the tips of the cantilever are serrated to provide grip on the soft tissue.

OPTICAL TECHNIQUES

The two major optical strain measuring methods used in biomechanics are photoelasticity and direct dimensional measurement. Holography has been used in recent years and is, therefore, discussed briefly. The Moiré fringe technique, however, has been utilized only minimally in biomechanical applications and will not be discussed.

Photoelasticity

Photoelasticity is a stress induced response and, thus, is more properly called photo-elastic stress analysis. However, since directions of principal strain are the same as those of principal stress, the effect can also be considered a strain response and results related to strains. The method is based on the phenomenon that the refractive indices of certain polymers change with the level of stress. The different velocities which result in polarized light when it is shone through a piece of stressed material cause the light waves to interfere in relation to principal stresses. The interference results in patterns of dark and light fringes which can be analysed to obtain both the magnitude and direction of stress in the material.

For photoelastic analysis, equipment is needed as shown in Fig. 3.5.7. Light from a

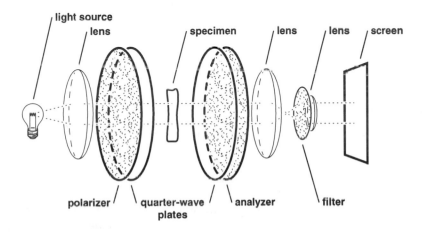

Figure 3.5.7   **Schematic of equipment for photoelastic analysis.**

light source is focused into a beam with a lens. The beam passes through a polarizing plate which transmits only those light waves in the plane of polarization. Light is an electromagnetic wave travelling along the line of intersection of the perpendicular electric and magnetic wave. In normal light, the wave pairs (electric and magnetic) are randomly oriented relative to the direction of light propagation. In polarized light, the electric wave is always in one plane and one plane only; and the magnetic wave in the perpendicular plane (Fig. 3.5.8).

The quarter wave plate splits the polarized light into two perpendicular components with a phase difference of $\pi/2$. This is called circularly polarized light. In transmission photoelasticity, the circularly polarized light passes through a stressed transparent model, where different phase shifts occur because of the different refractive indices in the differently stressed regions of the model. The second quarter wave plate converts the wave back to polarized light. The analyser combines the components of the phase shifted light into only one plane. The results are a series of dark and light fringes.

Two sets of fringe patterns can be obtained. With the quarter wave plates absent, isoclinic fringes can be obtained. Dark (isoclinic) fringes are seen where the axes of the polarizer and analyser correspond to the directions of principal stress. Rotation of the polarizer and analyser relative to the specimen will reveal the directions of principal stress through-

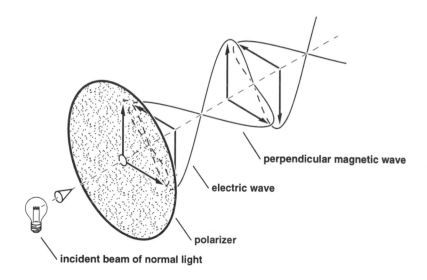

Figure 3.5.8   Schematic illustration of electric and magnetic waves for polarized light. Normal light has electric and magnetic waves in all planes, whereas, polarized light has an electric wave in one plane only, with the associated magnetic wave perpendicular to it.

out the specimen, since the dark isoclinic will move in the specimen with the rotation of the plates. With the quarter wave plates in the sequence, isoclinic are removed, and isochromatic fringes are revealed. These fringes are dependent on the difference between the principal stresses and, therefore, can be used to estimate magnitudes of stress. The zero order isochromatic fringe is black, and with white light the subsequent fringes appear as a spectra of colours which makes numerical analysis difficult. It is better to use a monochromatic source of light so that the fringes are distinct dark and light lines.

   If the model is a two-dimensional model, the fringes can be projected onto a screen to be photographed directly as the model is loaded. For three-dimensional models, a more complex procedure called stress freezing is followed. The model is loaded and the temperature raised (typically in an oven). The stiffness of the model reduces dramatically and the model deforms more. The specimen is then cooled, still under load, to room temperature. The load is removed, but the effects of the load are "frozen" into the model. Essentially what has happened is that the secondary bonds which connected the major molecules were released by thermal energy in the heating process (causing the reduction in stiffness and the realignment of the major molecules), and then new secondary bonds were formed between the major molecules on cooling. When this specimen is now sliced, and each slice viewed in the transmission polariscope, a picture of the three-dimensional stresses in the original model can be built up.

   In models where more than one material is used - for example a model of a hip implant in a femur - careful selection of the modelling materials is required. For a meaningful stress pattern to be obtained, the moduli of the materials selected to represent the bone and implant respectively, should be equally in proportion to those of the bone and the implant themselves. If stress freezing is to be employed, then the ratio of the moduli must be maintained throughout the temperature regime applied.

In photoelastic coatings, the circularly polarized light is shone onto the coating, passes through the coating, and is reflected at the surface of the object under study by an aluminium impregnated layer of cement. The light must then pass through the photoelastic layer again before being analysed. Thus, the light has been subject to a double dose of refraction in the photoelastic layer, and the equations are different to those of transmission photoelasticity by a factor of 2.

Direct dimensional measurement

Various techniques have been employed to measure lengths between markers optically.

Noyes et al. (1984) used high speed cinematography to quantify strain in ligaments and tendons. Ink lines were marked on a soft tissue and then filmed at 400 frames/s. Sequential frames of the developed film were then projected onto a digitizer plate, and points on the edges and in the middle of the ink bands were digitized. As the ink bands moved apart, the distance between the digitized points increased. The accuracy of the estimated distances between points would, therefore, depend on the accuracy of the digitizing equipment itself, and the ability of the user to select the same point on the ink mark consistently. The technique is labour intensive and was superseded by Video Dimensional Analysis which is now more automated.

Video dimensional analysis is based on a technique developed for analyzing video pictures for dimension (Yin et al., 1972). A VDA system (Fig. 3.5.9) consists of a video cam-

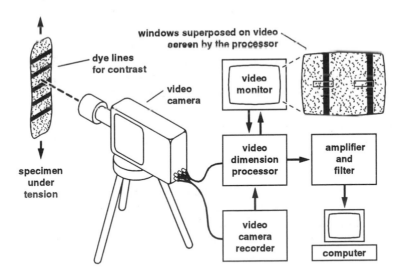

Figure 3.5.9   Schematic of a VDA System.

era which is focused on the tissue/sample being tested. The camera must be oriented such that the direction of test in the monitor view is horizontal (along a raster line). The output voltage from the camera is sent through a "video dimension analyser" micro processor as well as possibly a VCR. The signal, whether it be in real time or on replay from the VCR, can have two "windows" added to the view in the monitor (Woo et al., 1983). The operator of the system can adjust the position and length of the two windows to cover the edges of two dye lines on the specimen (what is needed is a pair of light/dark contrast lines). As the electron beam sweeps along a raster line where the windows are superimposed, the contrast in the first window (say dark to light) starts a timer. When the beam passes the sharp contrast in the second window (this time light to dark), the timer is turned off. Thus, the initial or gauge length is established as the time it takes the beam to move from one contrast to the other with the specimen in the undeformed state. As the specimen is deformed and the dye lines move apart, the time between the two contrasts increases and, therefore, can be related to the new length between the dye lines. The change in time over the original time can be related to the change in length divided by the original length; that is, strain. With long enough windows, rigid body displacement of the gauge length can easily be accommodated.

A major problem with these techniques is accuracy at low strains, where the error in the measurement of the inter-marker distance is of the same order of magnitude as the difference in length being estimated.

Variations on this video analysis theme exist, for example point markers on tissues rather than dye lines. The co-ordinates of the centres of the pixels darkened by the points are calculated, and distances between point centres determined. Again these length measuring techniques are subject to the inherent inaccuracy of the approach, for low strains.

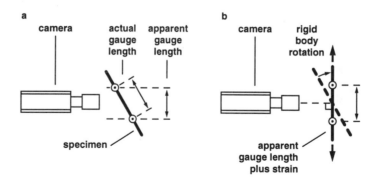

Figure 3.5.10   A source of error for VDA or film measurements is out of plane rotation (a). The specimen, not normal to the viewing plane of the camera initially, rotates into that plane without extensional deformation as the camera now sees the actual gauge length, but the system records it as the apparent gauge length plus strain (b). For example, a 5° rotation would give an apparent strain of 0.38%.

For example, the accuracy of VDA systems is 0.5% strain. This is acceptable when measuring a 10% or 40% strain at failure of a soft tissue, but may be unacceptable if subtle variations in strain are needed, or cycling at low loads/deformations is of interest. A second problem with video and film methods is that a planar view is taken. If the object moves in or out of that plane by rigid body rotation, a strain which has not actually occurred will be measured by these techniques (Fig. 3.5.10).

Holography

Holography uses the interference of two beams of monochromatic light, which began their paths simultaneously. Thus, a laser beam is split into two sub-beams; the reference beam and the object beam. The latter is reflected off the object and then interferes with the reference beam as both are projected onto photographic emulsion (Fig. 3.5.11). The emul-

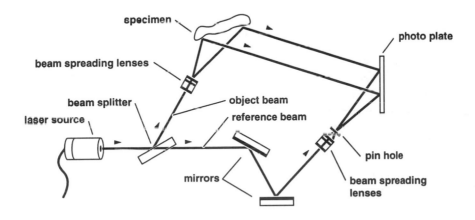

Figure 3.5.11   Schematic of equipment for holography.

sion is typically coated on a glass plate. Since the emulsion retains information not just on the intensity of light from the object (as a photograph would) but also on the phase of that light with respect to the reference beam, it is possible to reconstruct a three-dimensional image of the object.

Holographic interferometry requires the recording of two holograms. The first is recorded when the specimen is unloaded and the second after or during loading. Essentially what is being done is that the two object beams are being caused to interfere one with the other. Where the path lengths of the light beams are changed by a multiple of half a wavelength, the beams cancel each other out. Where the path lengths differ by a multiple of a whole wavelength the beams add to each other. Thus, a series of dark stripes will appear in front of the object where the beams cancel. These can be related to the deformation of the specimen.

Double exposure of the emulsion on the glass includes the object wave interference and the hologram: the fringe pattern can be viewed at will. The view, however, is for one loading condition.

If more continuous viewing during loading is required, then a hologram must first be made of the object. The hologram then replaces the photographic plate, and must do so precisely. Subsequent loading of the specimen allows the fringes to be viewed in real time, but of course there is not a permanent record of the result for any one loading position.

OTHER METHODS

Other non-standard techniques have been developed to overcome various difficulties with strain measurement in biomechanics. Three will be discussed here, brittle lacquers, the Hall Effect Displacement Transducer, and SPATE.

Brittle lacquers

Brittle lacquers are materials which can be applied to the surface of a specimen and then allowed to dry. The very thin layer of material that forms the coat provides negligible reinforcement to relativity large and stiff structures like bones. After drying, subsequent loading of the specimen in tension causes the lacquer to crack perpendicular to tensile strain. The threshold level of strain for cracking to begin can be as low as 500 $\mu\varepsilon$.

If compressive loading is to be applied, brittle lacquers will not measure compressive strain, other than indirectly. The indirect method is to apply the load, then apply the lacquer and to let it cure while the specimen remains under load. Once the lacquer is dry (some 16 hours later), the specimen can be unloaded. The removal of compression allows the specimen to expand back to its normal shape and the lacquer may crack. Problems with creep and drying of the specimen need to be considered in this type of test.

It is difficult to obtain accurate quantitative values of strain with lacquers, but they can indicate regions of high strain, if the progression of cracking is observed.

Hall effect displacement transducer

This transducer was developed for use on soft tissues (Arms et al., 1983) and consists of a piston and a cylinder (Fig. 3.5.12). On the cylinder is a Hall effect transducer, while

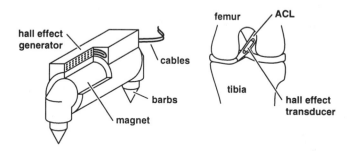

Figure 3.5.12   Schematic illustrations of a Hall effect transducer on the left, and a possible arrangement on an ACL (right).

the end of the piston is a magnet. The voltage produced by the transducer depends on the position of the magnet. In a 5 mm transducer, there is about 1 mm of travel of the piston where a linear relationship exists between the position of the piston and the Hall effect voltage.

Both the piston and the cylinder are attached to barbs which are used to fix the transducer to soft tissues. Suturing is also required to hold the two components in place. Thus, as with the LMSG, there is difficulty with definition of the initial gauge length. Subsequent straining may take the device out of the linear range. Practically, it requires considerable skill to suture the two components in place so that there is free movement of the piston in the cylinder. Misalignment can cause the piston to jam in the cylinder, especially during cyclic loading.

Axial rotation of the piston in the cylinder will also cause a change in the output voltage. Thus, situations in which there is torsional displacement as well as axial will provide inaccurate values of strain. This effect has been reduced with the more recent Differential Variable Reluctance Transducer (DVRT).

The immense advantage of these instruments is that they are small and can be implanted in vivo like the LMSG.

SPATE

Stress Pattern Analysis by Thermal Emission provides the acronym SPATE (Oliver, 1987). The method is based on the fact that materials undergo minute temperature changes when they are stressed. The thermoelastic equation for linear elasticity is:

$$\Delta T = -K_m T \Delta \sigma$$

where:

ΔT = peak to peak change in temperature in $^\circ$K due to $\Delta \sigma$

$\Delta \sigma$ = peak to peak change in the sine wave component of the sum of the principal applied stress at the reference original frequency

T = mean temperature of the specimen in $^\circ$K

K_m = thermoelastic constant

The equation is applicable to isotropic, homogeneous materials being loaded adiabatically in their linear elastic range. Cyclic loading is required and the small changes in temperature (thousandths of a degree K) are reversible. Thus, these are not the temperature changes observed in cyclic loading of viscoelastic materials, where the material heats up due to dissipation of hysteresis energy in the form of heat.

The theory has been extended to allow analysis of non-homogeneous and/or anisotropic materials. Residual stress measurement and plastic effects have also been elucidated. The equipment consists of a highly sensitive infra-red sensor. As the specimen is cyclically loaded, the sensor picks up the changes in the infra-red photon emission from the surface. With the aid of the necessary signal conditioning and analysis, a picture can be built up of the cyclic changes in temperature on the surface of the specimen. The hot spots are where the largest changes in the sum of the principal stresses occur. Methods have been developed to determine individual stress components.

3.5.4 APPLICATIONS

Strain gauges find wide application in force transducers (load cells) as well as displacement and strain transducers (extensometers). Strain gauges can be applied to metal components with relative impunity: thus, artificial joint or limb components can provide interesting data (Weightman, 1977; Little, 1985 and 1990; Bartel et al., 1986). Protection of the circuitry against fluid attack and corrosion is necessary. Gauges have also been embedded into the cement between prosthesis and bone (Lewis et al., 1982; Crownin-shield and Tolbert, 1983; Burke et al., 1984; Miles and Dall, 1985; Little and O'Keefe, 1989). Soft tissue loads have also been estimated indirectly through the use of strain gauges bonded to metal components - devices such as the buckle transducer and the IFT (see section 3.1.4).

However, the extensive difficulties in gluing strain gauges to biological materials have led to few direct applications. Strains on bones have been measured in vitro or in vivo where the short-lived adhesion (2-3 days) has not been an issue. Bonding directly to soft tissues has not been achieved, due to the high water content of these structures and the lack of knowledge of the effect of any chemical reaction of the adhesive with the tissue.

The validation of strain gauge measurements should always be a matter of concern to the user. The selection of the technique to be used should involve a careful assessment of the capabilities and limitations of the various methods available. The following points should be considered:

(1)   How well are the gauges aligned with the intended direction?

(2)   What is the effect of the strain gradient where the gauge is located - if there is one.

(3)   What happens when the circuit is turned on and heats up - this is of particular importance with non-metallic applications (ceramics, polymers).

(4)   Has the bonding or the gauge (plus bonding) affected the material being tested, chemically or through reinforcement? Thus, has the strain you wanted to measure been affected by your very attempt to measure it?

(5)   Will the temperature vary in the test from ambient and, therefore, is temperature compensation required?

(6)   Does the loading procedure affect the result?

(7)   How reproducible and repeatable are the results?

(8)   If you are strain gauging a model of a particular structure, how good is your model (appropriate relative stiffness of components, geometric errors, scaling factors, etc.)?

More detailed reviews of these factors are provided elsewhere (Little et al., 1991, Pople, 1983).

Strains in bones and bone plates have been measured by holography (Kojima et al., 1986; Shelton et al., 1990), as have strains in prostheses (Manley et al., 1987) and an external fixator (Jacquot et al., 1984), although the technique is not used widely. Photo-elastic models are more common, having been used for example, to examine stresses in models of hips (Haboush, 1952; Steen Jensen, 1978) and hip replacements (Orr et al., 1985a, b), the ankle (Kihara et al., 1987) and the knee (Chand et al., 1976). Photoelastic coatings on the other hand, appear to have been used mainly in pilot studies to determine where strain gauges should be placed in subsequent tests (Finlay et al., 1986 and 1989; Walker and Robertson, 1988). Brittle lacquers and SPATE have also been used to assess metallic and bony structures in the body (Gurdjian and Lissner, 1947; Evans and Lissner, 1948; Kalen, 1961; Harwood and Cummings, 1986; Friis et al., 1989; Kohles et al., 1989).

Soft tissue strains have proven more difficult to assess. Methods have been classified as contact or non-contact, rather than on the fundamental principle of the measurement technique. Conceptually non-contact methods do not interfere with the strain being measured since the tissue is not being "touched". However, permanent staining of a specimen through the application of dye lines or spots suggests that the dye has interacted chemically with some component of the tissue. How this affects the strain of the tissue is not known, and so leaves open the question of just how much the strain has been altered and is the method really non-contact? Contact problems and gauge length definition have not stopped researchers attempting to determine strain magnitudes in vivo with either the LMSG or the Hall effect transducer. Liquid Mercury Strain Gauges have been used in vivo on horses (Lochner et al., 1980) and ponies (Riemersma et al., 1980) as well as in vitro on human joints (Colville et al., 1990; Terry et al., 1991). The Hall effect transducer has been used in vivo on the Anterior Cruciate Ligament of humans (Howe et al., 1990). In vitro measurements have typically been made with film and video systems despite the inherent inaccuracy of most methodologies. For the large strains at soft tissue failure, the inaccuracies are not of major concern. Perhaps the soft tissue extensometer will provide the accuracy and sensitivity to determine more subtle strain effects in soft tissues.

3.5.5 REFERENCES

Arms, S., Boyle, J., Johnson, R., and Pope, M. (1983) Strain Measurement in the Medial Collateral Ligament of the Human Knee: An Autopsy Study. *Journal of Biomechanics.* **16 (7)**, pp. 491-496.

Bartel, D.L., Bicknall, V.L., and Wright, T.M. (1986) The Effect of Conformity, Thickness, and Material on Stresses in Ultra High Molecular Weight Components for Total Joint Replacement. *Journal of Bone Jt. Surgery.* **68 (A-7)**, pp. 1041-1051.

Burke, D.W., Davies, J.P., O'Connor, D.O., and Harris, W.H. (1984) Experimental Strain Analysis in the Femoral Cement Mantle of Simulated Total Hip Arthroplasties. *Trans. Orthop. Res. Soc.* **9**, p. 296.

Chand, R., Haug, E., and Rim, K. (1976) Stresses in the Human Knee Joint. *Journal of Biomechanics.* **9 (6)**, pp. 417-422.

Colville, M.R., Marder, R.A., Boyle, J.J., and Zarins, B. (1990) Strain Measurement in Lateral Ankle Ligaments. *Am. J. Sports Med.* **18 (2)**, pp. 196-200.

Crowninshield, J. and Tolbert, J. (1983) Cement Strain Measurement Surrounding Loose and Well Fixed Femoral Component Stems. *Journal of Biomed. Matls. Res.* **17**, pp. 819-828.

Edwards, R.G., Lafferty, J.F., and Lange, K.O. (1970) Ligament Strain in the Human Knee Joint. *Journal of Basic Engineering, Trans. ASME.* pp. 131-136.

Evans, F.G. and Lissner, H.R. (1948) Stresscoat Deformation Studies of the Femur Under Static Vertical Loading. *Anat. Rec.* **100**, pp. 159-190.

Finlay, J.B., Bourne, R.B., Landsberg, R.P.D., and Andreae, P. (1986) Pelvic Stresses In Vitro - I: Malsizing of Endoprostheses. *Journal of Biomechanics.* **19 (9)**, pp. 703-714.

Finlay, J.B., Rorabeck, C.H., Bourne, R.B., and Tew, W.M. (1989) In Vitro Analysis of Proximal Femoral Strains Using PCA Femoral Implants and a Hip-abductor Muscle Simulator. *Journal of Arthroplasty.* **4 (4)**, pp. 335-345.

Friis, E.A., Cooke, F.W., Henning, C.E., and Samani, D.L. (1989) Effect of Bone Block Shape on Patellar Stress in ACL Reconstruction: An Evaluation Using SPATE. *15th Annual Meeting, Society for Biomaterials.*

Gurdjian, E.S. and Lissner, H.R. (1947) Deformations of the Skull in Head Injury as Studied by the Stress-coat Technique. *Am. J. Surg.* **73**, pp. 269-281.

Haboush, E.J. (1952) Photoelastic Stress and Strain Analysis in Cervical Fractures of the Femur. *Bulletin Hospital Joint Diseases.* **13**, pp. 252-258.

Harwood, N. and Cummings, W.M. (1986) Applications of Thermoelastic Stress Analysis. *Strain.* **22 (1)**, pp. 7-12.

Hendry, A.W. (1948) *An Introduction to Photoelastic Analysis.* Blackie & Sons Ltd., Glasgow.

Howe, J.G., Wertheimer, C., Johnson, R.J., Nichols, C.E., Pope, M.H., and Beynnon, B. (1990) Arthroscopic Strain Gauge Measurement of the Normal Anterior Cruciate Ligament. *Arthroscopy: The Journal of Arthroscopic and Related Surgery.* **6 (3)**, pp. 198-204.

Jacquot, P., Rastogi, P.K., and Pflug, L. (1984) Mechanical Testing of the External Fixator by Holographic Interferometry. *Orthopaedics.* **7 (3)**, pp. 513-523.

Kalen, R. (1961) Strains and Stresses in the Upper Femur Studied by the Stresscoat Method. *Acta. Orthop. Scand.* **31 (2)**, pp. 103-113.

Kihara, T., Unno, M., Kitada, C., Kubo, H., and Nagata, R. (1987) Three-dimensional Stress Distribution Measurement in a Model of the Human Ankle Joint by Scattered-light Polarizer Photoelasticity: Part 2. *Applied Optics.* **26**, pp. 643-649.

Kohles, S.S., Vanderby Jr., R., Belloli, D.M., Thielke, R.J., Bowers, J.R., and Sandor, B.I. (1989) Differential Infrared Thermography: A Correlation with Stress and Strain in Cortical Bone. *AMD - Vol 98, Biomechanics Symposium* (eds. Torzilli, P.A. and Friedman, M.H.). ASME, pp. 81-84.

Kojima A., Ogawa, R., Izuchi, N., Matsumoto, T., Iwata, I., and Nagata, R. (1986) Holographic Investigation of the Mechanical Properties of Tibia Fixed with Internal Fixation Plates. *Biomechanics: Basic and Applied Research* (eds. Bergmann, A., Kolbel, R., and Rohlmann, A.). Martinius Nijoff. pp. 243-248.

Lewis, J., Askey, M., and Jaycox, D. (1982) A Comparative Evaluation of Tibial Component Designs of Total Knee Prostheses. *Journal of Bone Jt. Surgery.* **64 (A-1)**, pp. 129-135.

Little, E.G. (1985) A Static Experimental Stress Analysis of the Geomedic Knee Joint Using Embedded Strain Gauges. *Engineering Medicine.* **14 (2)**, pp. 69-74.

Little, E.G. (1990) Three-dimensional Strain Rosettes Applied to an Analysis of the Geomedic Knee Prosthesis and Cement Fixation: Applied Stress Analysis. *Applied Stress Analysis.* Elsevier, New York.

Little, E.G. and O'Keefe, D. (1989) An Experimental Technique for the Investigation of Three-dimensional Stress in Bone Cement Underlying a Tibial Plateau. *Proceedings Institution of Mechanical Engineers.* **203 (1)**, pp. 35-41.

Little, E.G., Tocher, D., and McTague, D. (1980) The Validation of Strain Gauge Measurements. *Strain Measurement in Biomechanics, Canadian Medical and Biological Engineering Society.* pp. 73-86.

Lochner, F.K., Milne, D.W., Mills, E.J., and Groom, J.J. (1980) In Vivo and In Vitro Measurement of Tendon Strain in the Horse. *Am. J. Vet. Res.* **41 (12)**, pp. 1929-1937.

Manley, M.T., Ovryn, B., and Stern, L.S. (1987) Evaluation of Double-exposure Holographic Interferometry for Biomechanical Measurements In Vitro. *J. Ortho. Research.* **5 (1)**, pp. 144-159.

Mesnager, M. (1930) Sur la Deter400miniation Optique des Tensions Intrieures dans les Solides Trois-dimensions. *Compt. Rendu.* **190 (22)**, pp. 1249-1250.

Miles, A.W. and Dall, D.M. (1985) An Experimental Study of Femoral Cement Stress in Total Hip Replacement: Influence of Structural Stiffness of the Femoral Stem. *Engineering Medicine.* **14 (3)**, pp. 133-135.

Morice, P.B. and Base, G.D. (1953) The Design and Use of a Demountable Mechanical Strain Gauge for Concrete Structures. *Magazine of Concrete Research.* **5 (13)**, pp. 37-42.

Noyes, R.F., Butler, D.L., Grood, E.S., Zernicke, R.F., and Hefzy, M.S. (1984) Biomechanical Analysis of Human Ligament Grafts Used in Knee Ligament Repairs and Reconstructions. *Journal of Bone Jt. Surgery.* **66 (A)**, pp. 344-352.

Oliver, D.E. (1987) Stress Pattern Analysis by Thermal Emission. *Handbook on Experimental Mechanics* (ed. Kobayashi, A.S.). pp. 610-620.

Orr, J.F., James, W.V., and Bahrani, A.S. (1985) A Preliminary Study of the Effects of Medio-lateral Rotation on Stresses in an Artificial Hip Joint. *Engineering in Medicine.* **14**, pp. 39-42.

Orr, J.F., James, W.V., and Bahrani, A.S. (1985) The Effects of Hip Prosthesis Stem Cross-sectional Profile on the Stresses Induced in Bone Cement. *Engineering in Medicine.* **15**, pp. 13-18.

Pople, J. (1983) Errors and Uncertainties in Strain Measurement. *Strain Gauge Technology.* Applied Science Publishers, New York. pp. 209-264.

Redner, S. (1968) Photoelasticity. *Encyclopedia of Polymer Science and Technology.* **9**, pp. 590-610.

Riemersma, D.J., van den Bogert, A.J., Schamhardt, H.C., and Hartman, W. (1988) Kinetics and Kinematics in the Equine Hind Limbs: In Vivo Tendon Strain and Joint Kinematics. *Am. J. Vet. Res.* **49 (8)**, pp. 1353-1359.

Sauerwald, F. and Wieland, H. (1925) Über die Kerbschlagprobe nach Schule-Moser. *Z. Metalkunde.* **17**, pp. 358-364 and 392-399.

Shelton, J.C., Gorman D., and Bonfield W. (1990) Application of Holographic Interferometry to Investigate Internal Fracture Fixations Plates. *Journal Mater. Sci. Mater. in Medicine.* **1**, pp. 146-153.

Shrive, N.G., Damson, E.L., and Frank, C.B. (1992) Technology Transfer Regarding the Measurement of Strain on Flexible Materials with Special Reference to Soft Tissues. *Experimental Mechanics: Technology Transfer Between High-tech Engineering and Biomechanics* (ed. Little, E.G.). Elsevier, New York. pp. 121-130.

Shrive, N.G., Damson, E.L., Meyer, R.A., and Iverslie, S.P. (1993) Soft Tissue Extensometer. U.S. Patent Application 08/112,841.

Solakian, A.G. (1935) A New Photoelastic Method. *Mechanical Engineering.* **57**, pp. 767-771.

Steen Jensen, J. (1978) A Photoelastic Study of a Model of the Proximal Femur. *Acta. Orthop. Scandinavia.* **49**, pp. 54-59.

Stein, P.K. (1992) The Lack of Technology Transfer Within Low-tech Engineering Disciplines: How it has Affected Experimental Stress Analysis History. *Experimental Mechanics* (ed. Little, E.G.). Elsevier, New York. pp. 201-213.

Terry, G.C., Hammon, D., France, P., and Norwood, L.A. (1991) The Stabilizing Function of Passive Shoulder Restraints. *Am. J. Sports Med.* **19 (1)**, pp. 26-34.

Walker, P.S. and Robertson, D.D. (1988) Design and Fabrication of Cementless Hip Stems. *Clin. Orthop.* **235**, pp. 25-34.

Weightman, B. (1977) Stress Analysis. *The Scientific Basis of Joint Replacement* (eds. Swanson, S.A.V. and Freeman, MAR). Pitman, Kent, England. pp. 18-45.

Woo, S.L.Y., Gomez, M.A., Seguchi, Y., Endo, C.M., and Akeson, W.H. (1983) Measurement of Mechanical Properties of Ligament Substance from a Bone-ligament Preparation. *Journal Orthopaedic Research.* **1 (1)**, pp. 22-29.

Yin, F.C.P., Tompkins, W.R., Peterson, K.L., and Intaglietta, M. (1972) A Video-dimension Analyser. *IEEE Transactions on Biomedical Engineering.* BME **(19-5)**, pp. 376-381.

3.6 EMG

HERZOG, W.
GUIMARAES, A.C.S.
ZHANG, Y.T.

3.6.1 DEFINITIONS AND COMMENTS

Adaptive filtering system:	A system whose structure (or weights of the filter) is automatically adjustable in accordance with some algorithm to produce an estimate of the desired response.
Bi-, multi-polar electrodes:	Two or more electrodes measuring EMG signals relative to a common ground electrode. The difference between the signals received by each of the electrodes, relative to the common ground electrode, is amplified and gives the EMG records.
Compound motor unit action potential:	(CMUAP) The sum of MUAP, obtained when several motor units are stimulated during a contraction.
EMG:	Spatial and temporal summation of the action potentials of motor units during muscular contraction, measured using specifically designed electrodes.
Electromechanical delay (EMD):	EMD is typically defined as the time interval between the onset of the EMG signal and the onset of the corresponding muscular force.
Indwelling electrodes:	Electrodes placed inside the muscle to measure EMG signals.
Integration:	Mathematical integration of previously rectified EMG signals over specified periods of time, or until the integrated EMG value reaches a preset limit.
LMS algorithm:	Least mean square (LMS) gradient search adaptive algorithm. The LMS algorithm is widely used for implementing the adaptive filtering process because of its simplicity and robustness.
Mono-polar electrodes:	A single electrode measuring EMG signals relative to a ground electrode.
Motor unit action potential:	(MUAP) Change in the electrical potential across the muscle fibre membranes when a motor unit is stimulated beyond a critical threshold.
Power density spectrum:	Mathematical conversion of EMG signals from the time to the frequency domain for analysis of the frequency content of the signal.

Raw EMG:	The unprocessed (typically amplified) EMG signal displayed in the time domain.
Rectification:	Elimination of all negative values from the raw EMG signal (half-wave rectification), or multiplying all negative values from the raw EMG signal by -1 (full-wave rectification).
Resting potential:	Electrical potential across the muscle fibre membrane at rest (about -90 mV inside the muscle fibre).
Smoothing:	Mathematical procedures aimed at reducing the high-frequency content of signals.
Surface electrodes:	Electrodes placed outside (on the surface) of the muscle to measure EMG signals.

3.6.2 SELECTED HISTORICAL HIGHLIGHTS

1664	Croone	(De Ratione Motus Musculorum) Concluded from nerve section experiments that the brain must send a signal to the muscles to cause contraction.
1791	Galvani	Showed that electrical stimulation of muscular tissues produces contraction and force.
1849	DuBois-Reymond	Probably was the first to discover and describe that contraction and force production of skeletal muscles were associated with electrical signals originating from the muscle.
1867	Duchenne	Probably was the first to perform systematic investigations of muscular function using an electrical stimulation approach.
1985	Basmajian De Luca	Summarized the existing knowledge and research on muscle function, as revealed by electromyographic studies.

3.6.3 INTRODUCTION

There are entire books devoted to EMG (e.g., Basmajian and De Luca, 1985; Loeb and Gans, 1986), and it would be presumptuous of us to try to rival them. However, most of these texts focus on the physiology rather than the mechanics of muscle. One of the most burning issues in biomechanics, the relation between electromyographic signal and muscular force, has not been addressed thoroughly in these texts. We will attempt to address this issue from two points of view: first, by presenting results of studies employing electroneuromuscular stimulation (ENMS) of α-motor neurons to investigate the EMG-force relation under precisely controlled conditions; and second, by discussing the findings of studies employing direct and simultaneous measurements of EMG and force in a freely moving animal model. First, however, some basic concepts of the nature of the EMG signal, its recording and processing, are described.

3.6.4 EMG SIGNAL

When recording an EMG signal from a single muscle fibre, one measures changes in electrical potential across the muscle fibre membrane. At rest, the potential of a muscle fibre is approximately -90 mV (Fig. 3.6.1). With sufficient stimulation, the potential inside

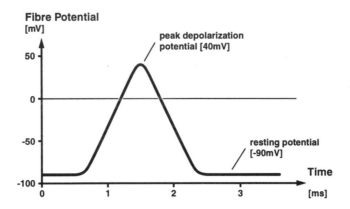

Figure 3.6.1   **Schematic representation of a muscle fibre action potential.**

the cell rises temporarily to about 30-40 mV. This change in potential, representing a fibre action potential, can be recorded. In an intact muscle, a fibre is never stimulated by itself but always together with all the other fibres that make up a motor unit. The EMG signal recorded from the depolarization of a motor unit is called a motor unit action potential (MUAP). In general, the actual EMG signal obtained from voluntarily contracting muscles is a record of many motor units firing at different mean rates. Since the placement of the recording electrode on a muscle determines its geometric relation to the motor units, recordings obtained using different electrode placements on the same muscle will, usually, be different.

The resting membrane potential is caused by the concentrations of ions inside and outside the muscle fibre. The key ions are potassium (K^+) and sodium (Na^+). There are two basic mechanisms that must be accounted for when explaining the resting membrane potential. For illustrating the first mechanism, let us concentrate on the K^+ ions exclusively. The ratio of K^+ ions inside versus outside the muscle cell is about 30:1 (Wilkie, 1968). K^+ ions diffuse out of the muscle fibre, following the concentration gradient, but when they leave the cell, the electrical capacitance of the cell increases and prevents further loss of K^+ ions. The K^+ ions reach an equilibrium state when the two opposing forces acting on them - the concentration gradient and the electrical capacitance of the cell - are equal. At equilibrium, K^+ ions neither gain nor lose energy when crossing the cell membrane.

The membrane potential can have only one value, which may be determined for the K^+ ions. However the corresponding membrane potential for the Na^+ ions cannot be equal to that of the K^+ ions, since the ratio of Na^+ inside versus outside the cell is about 1:7.7 (Wilkie, 1968). Therefore, a second mechanism must influence the resting potential of a muscle fibre. This second mechanism is associated with the fact that the muscle cell membrane is not freely permeable to all ions. For example, Na^+ ions cannot easily permeate the cell membrane, and if they do enter the cell, they are actively (i.e., at the expense of metabolic energy) pumped out again by the so-called sodium pump. The facts that the muscle cell membrane cannot be freely crossed by some ions, and that electrical and osmotic equilibrium conditions influence ion traffic through the cell membrane, must be considered when determining the resting concentration of ions.

The resting potential of a muscle fibre is about -90 mV inside the cell, and depends on the selective permeability of the cell membrane. When an action potential of a motor neuron reaches the presynaptic terminal (see chapter 2.5), a series of chemical reactions takes place, culminating in the release of acetylcholine (ACh). Acetylcholine diffuses across the synaptic cleft, binds to receptor molecules of the muscle fibre membrane, and causes a change in membrane permeability. Most importantly, permeability to Na^+ ions increases, and if the depolarization of the membrane, caused by Na^+ ion diffusion, exceeds a critical threshold, an action potential propagates along the muscle fibre. If such an action potential were measured using an electrode inside the muscle fibre, it would go from about -90 mV (resting potential) to about +40 mV (peak depolarization potential), and back again to the resting value (Fig. 3.6.1). The membrane potential and active state of the contractile machinery of the muscle fibre are tightly linked. A depolarization of the cell membrane exceeding the critical threshold will cause contraction of the muscle fibre (see chapter 2.5).

3.6.5 EMG SIGNAL RECORDING

Electromyographic signals are typically recorded by electrodes measuring differences in voltage (potential) between two points. EMG electrodes may be grouped broadly into four categories. The first two categories describe where the electrodes are placed: inside the muscle (indwelling electrodes), or on the surface of the muscle (surface electrodes). The remaining two categories describe the electrode configuration: mono-polar, or multi- (typically bi-) polar.

Indwelling electrodes come in dozens of different shapes and are often home-made. Here we describe the indwelling wire electrodes that we have used successfully for chronic recordings from cat hindlimb muscles (e.g., Herzog et al., 1993b). For further discussion of this topic see Loeb and Gans (1986).

For our indwelling, bi-polar electrodes, we use Teflon-insulated, multistranded, stainless steel biomedical wire (Bergen BW9.48, Fig. 3.6.2). For recording, two millimeters of wire are exposed at the tip by mechanically removing and/or heat-shrinking the Teflon insulation (Fig. 3.6.2, arrows). About 10 mm of the wire is bent backward to produce a permanent, hook-like deformation. A silk suture is tied to the bent area of both wires so that the distance between the two exposed tips is about 5 to 7 mm. A ground electrode is constructed from the same wire as the recording electrodes by exposing about 50 mm of the insulation on one end and forming a loop held in place by Silastic tubing.

For chronic implantation, all electrodes are autoclaved and implanted under sterile conditions. The silk sutures and electrodes are attached to a small surgical needle (MILTEX MS-140) which is pulled through the largest part of the muscle belly in the direction of the muscle fibres. Once the recording area of the electrodes in the mid-belly region of the muscle is reached, the silk sutures emerging from the entry and exit holes are tied to

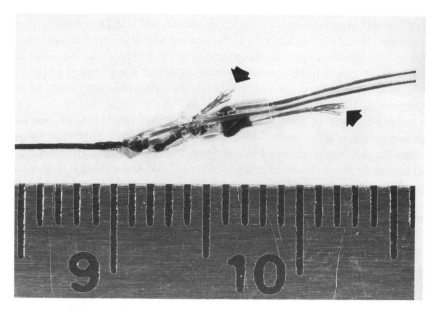

Figure 3.6.2   Indwelling, bi-polar wire electrodes used for chronic EMG recording in cat hindlimb muscles. The arrows indicate where the insulation was removed from the wires to produce the recording sites (from Herzog et al., 1993b, with permission).

the muscle fascia in order to prevent movement of the electrodes relative to the muscle. The ground electrode is typically placed on a bony surface such as the medial surface of the tibia.

When used in the cat, this type of indwelling electrode allows for recording from a specific muscle; cross-talk from other muscles may be virtually eliminated by choosing the recording site and the inter-electrode distance appropriately. Furthermore, indwelling electrodes allow measurements from small and deep-lying muscles; the problems of skin movement during locomotion of the cat, and irritation to the animal caused by fixing the electrodes on the skin, are avoided. Using the type of indwelling electrode described above allows for chronic recordings in the freely moving animal over a period of weeks, possibly months.

Surface electrodes are placed on the skin overlying the muscle of interest. Like the indwelling electrodes, surface electrodes come in a variety of types. Probably the most frequently used surface electrodes are commercially available silver-silver chloride electrodes.

Before recording with surface electrodes, the electrical impedance of the skin must be decreased by shaving the area of electrode placement and by applying rubbing alcohol or abrasive pastes to remove dead cells and oils. The recording electrodes are then attached using electrode gel and slight pressure is applied to improve contact between the electrode with the skin. Electrode gels are commercially available, and pressure may be applied by fixing the electrodes with adhesive tapes and elastic bands.

Surface electrodes are simpler to use than indwelling electrodes, and are non-invasive. Surface EMG recordings are typically used to obtain a general picture of the electrical activity of an entire muscle or muscle group. However, surface electrodes can only be used for recording from superficial, relatively large muscles.

Typically, EMG recordings are obtained using a *bi-polar electrode configuration.* A bi-polar configuration implies that there are two electrical contacts used to measure a potential, each relative to a common ground electrode. The two potentials measured by each of the electrodes are then sent to a differential amplifier, which determines (and amplifies) the voltage difference between the two electrodes. In a *mono-polar configuration,* the difference in voltage is recorded from a single recording electrode relative to the ground electrode. The mono-polar electrode configuration has the disadvantage that it detects any electrical signal in its vicinity - not only the signal originating from the muscle of interest. In a bi-polar electrode configuration, two signals are recorded relative to the ground electrode. Each of these two signals contains all the electrical activity in the vicinity of the recording electrodes, but subtracting these two signals from one another, using differential amplification, can cancel unwanted electrical activity from outside the muscle, since it appears as a similar signal at both recording electrodes. The electrical signal originating from muscular activity is received by each recording electrode as an essentially different signal and, thus, is enhanced in the differential amplification process.

3.6.6 EMG SIGNAL PROCESSING

Fig. 3.6.3a shows an unprocessed EMG signal obtained from human rectus femoris muscle during an isometric contraction at a level of 70% of the maximal voluntary knee extension force. The record was obtained using bi-polar surface electrodes. Raw EMG signals resemble noise signals with a distribution around the zero point. Any interpretation of the raw EMG signal with respect to the force production of the muscle, its relative activation, or its fatigue state, are difficult. Therefore, EMG signals are typically processed before they are used for assessments of the contractile state of the target muscles. Processing of the EMG signal may be done in the time or frequency domain.

EMG SIGNAL PROCESSING IN THE TIME DOMAIN

Rectification

Any type of averaging of the EMG signal in the time domain will yield a value of close to zero, independent of the number of motor units contributing to the signal and their mean firing rates. Therefore, before performing any type of averaging procedure, it is necessary to rectify the EMG signal. Rectification is the name of the process by which either all negative values of the EMG signal are eliminated from the analysis procedure (half-wave rectification, Fig. 3.6.3b), or only the absolute magnitudes of the signal are considered (full-wave rectification, Fig. 3.6.3c). Typically, full-wave rectification is preferred since it retains the entire signal. Often the rectified signal is used for further signal processing.

Smoothing

The rectified signal still contains the high-frequency content of the raw signal. Often it is desirable to eliminate the high-frequency content of the EMG records in order to better

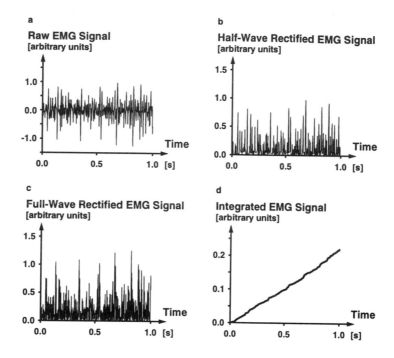

Figure 3.6.3   **Raw EMG signal from human rectus femoris muscle during an isometric knee extensor contraction at 70% of the maximal knee extensor strength (a), the corresponding half-wave rectified signal (b), the corresponding full-wave rectified signal (c), and the corresponding integrated signal (d).**

relate the EMG signal to contractile features of the muscle. Elimination of the high-frequency content of the EMG signal is accomplished using any type of low-pass filtering approach. For details of filtering procedures, see Basmajian and De Luca (1985) chapter 2.

Integration

The integrated EMG (IEMG) has been related to muscular force more often than any other form of the processed EMG (e.g., Bigland and Lippold, 1954; Bouisset and Goubel, 1971; Thorstensson et al., 1976). It is for this reason, we feel, that IEMG is in such popular use. Integration of the EMG signal is performed on the rectified form of the signal and refers to the mathematical integration of the EMG-time records with respect to time (Fig. 3.6.3d). Integration, therefore, is equivalent to calculating the area under the rectified EMG-time curve, and is defined as:

$$\text{IEMG} = \int_{t}^{t+T} |\text{EMG}(t)| \cdot dt \qquad (3.6.1)$$

Where |EMG (t)| represents the rectification of the EMG signal.

Since integration of the EMG signal gives steadily increasing values over time, integrations are typically performed either over a sufficiently small time period, T, or the integrator is reset to zero when the integrated value reaches a specified limit.

Root Mean Square

The root mean square (RMS) value of the EMG signal is an excellent indicator of the magnitude of the signal, and it is frequently used in studying muscular fatigue. RMS-values are calculated by summing the squared values of the raw EMG signal, determining the mean of the sum, and taking the square root of the mean so obtained:

$$\text{RMS} \quad = \quad \left(\frac{1}{T} \int_{t}^{t+T} EMG^2 (t) \, dt \right)^{1/2} \tag{3.6.2}$$

EMG SIGNAL ANALYSIS IN THE FREQUENCY DOMAIN

The study of the EMG signal in the frequency domain has received much attention because of the loss of the high-frequency content of the signal with increasing muscular fatigue (e.g., Lindström et al., 1977; Komi and Tesch, 1979; Bigland-Ritchie et al., 1981; Hagberg, 1981; Petrofsky et al., 1982). Power density spectra of the EMG signal can be obtained readily by using the Fast Fourier Transformation technique (Fig. 3.6.4).

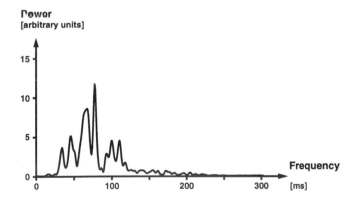

Figure 3.6.4   Power density spectrum of the EMG signal shown in Fig. 3.6.3a, obtained using a Fast Fourier Transformation (FFT).

The most important parameters for analysing the power density spectrum of the EMG signal are the mean and the median frequencies, although the bandwidth and the peak power frequency have also been used to describe the spectrum. The mean frequency is defined as the sum of the first moments of frequency divided by the area under the power-frequency curve. The median frequency is defined as the frequency that divides the power of the EMG spectrum into two equal areas. The peak power frequency is defined as the frequency at which the power of the signal is greatest; and the bandwidth of the frequency spectrum is typically defined as the range of frequencies associated with a power-value of 0.5 or higher, where the peak power is assigned the value 1.0.

3.6.7 EMG-FORCE RELATION

As stated in section 3.6.3, we are not attempting to duplicate the information that is presented in several excellent textbooks dealing exclusively with EMG. Rather, we would like to address one aspect of electromyography that has played a major role in the field of biomechanics: the relation that may exist in a muscle between EMG and force.

The idea that EMG should relate in some way to muscular force is appealing. After all, an increase in the firing rates of motor units is known to increase force and the integrated form of the EMG (Basmajian and De Luca, 1985), and recruitment of an increasing number of motor units increases muscular force and the corresponding IEMG as well (Guimaraes et al., 1992). Thus, there must be at least a qualitative relation between the EMG signal and the corresponding force of a muscle. The existence of such a relation is not questioned in the scientific community, but its precise, quantitative nature is hotly debated.

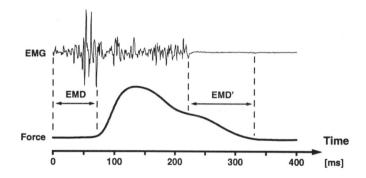

Figure 3.6.5   Force and EMG signals obtained from cat soleus muscle during walking at a nominal speed of 1.2 m/s on a motor driven treadmill. The signals represent a normal step from the onset of EMG to paw off. The electromechanical delay for this particular step was, EMD = 72 ms, and EMG finished, EMD' = 109 ms, before soleus force became zero.

The difficulty in finding a relation between the EMG signal and the corresponding force is at least partly associated with the difficulties in measuring EMG and force properly from individual muscles, and the temporal disassociation of the two signals. For example, when measuring EMG signals from a particular muscle in an intact system, there always exists the possibility that signals from surrounding muscles may also be recorded (cross-talk). Furthermore, force recordings from individual muscles may be obtained quite readily, but only in animal models, because of the legal and ethical constraints on taking such invasive measurements in humans. Finally, the temporal disassociation of the EMG and force signals is well recognized and is typically accounted for by the so-called electro-mechanical delay (EMD). EMD is normally defined as the time interval between the onset of the EMG signal and the onset of the corresponding force (e.g., Fig. 3.6.5). However, the duration of the EMG signal may differ from the duration of the corresponding force signal. For example, the EMG and force signals in the cat soleus during a single step shown in Fig. 3.6.5 have an EMD of approximately 72 ms (as defined above), but the EMG signal terminates approximately 109 ms before the force signal. The question that arises in this situation is whether the EMD is constant and may be defined by a single value, or whether it varies continuously throughout the step cycle. If the latter is assumed to be correct, then the relation between the continuously varying EMD and time must be established.

EMG-force relations of skeletal muscles have often been described for isometric contractions (e.g., Lippold, 1952; Milner-Brown and Stein, 1975; Moritani and deVries, 1978). Isometric contractions under steady state conditions (i.e., constant force) allow for the simplest comparisons between force and EMG, because (a) one value of force may be compared to one value of the processed EMG, (b) the temporal disassociation between EMG and force may be neglected in a steady state contraction lasting for a long period of time (i.e., a second or more), and (c) the instantaneous contractile conditions of the muscle may be determined easily. Despite the apparent simplicity of isometric contractions, many different relations between EMG and force have been suggested for these contractile conditions, possibly because different muscles may have different relations between EMG and force (Bigland-Ritchie et al., 1980), and possibly because forces in muscles have been estimated indirectly from the forces or moments produced by an entire group of agonistic muscles; as well, EMG signals may have been recorded, processed, and analysed in many different ways.

As a result of the difficulty in quantifying EMG-force relations in isometrically contracting muscle, descriptions of EMG-force relations for dynamically contracting muscle are rare and controversial. Most of the dynamic experiments have been performed using isokinetic contractions on strength dynamometers. These dynamometers typically enforce a relatively constant angular velocity of joint movement. Needless to say, a constant angular velocity of joint movement cannot necessarily be associated with a constant rate of change in length of the muscle-tendon complex, because of the possible variations of the muscular moment arm as a function of the joint angle, and the varying velocity (vector) of the attachment site of the muscle relative to the direction of shortening of the muscle as a function of the joint angle. Furthermore, a constant contractile speed of the muscle-tendon complex may not translate into a constant speed of shortening of the contractile elements (fibres and sarcomeres) because of the varying force in the muscle and the associated changes in length of the elastic elements arranged in series with the contractile elements. Only a few studies have attempted to relate EMG and force (moments) during normal, unrestrained movements (Hof and van den Berg, 1981a; Olney and Winter, 1985; Ruijven and Weijs, 1990).

In order to study the basic relation that may exist between the EMG signal and the force of a muscle, we felt it necessary not only to measure the force of the muscle and the uncontaminated EMG, but also to control precisely the activation to the muscle and to activate the muscle in the most physiological way possible in an attempt to produce an EMG signal indistinguishable from a signal obtained during voluntary muscle contraction. Our approach to this problem and our results are presented below.

EMG-FORCE RELATION OF CAT SOLEUS USING CONTROLLED STIMULATION

In studying the EMG-force relation of mammalian skeletal muscle, we used the cat soleus as our experimental model. Myotomy of the back muscles and laminectomy were performed, exposing the nerve roots of the seventh lumbar (L7) and the first sacral (S1) vertebrae as described by Rack and Westbury (1969). The ventral roots of L7 and S1 containing the α-motor neurons of the soleus were separated into ten filaments and placed over hook-like electrodes for stimulation. The electromyographical activity of the soleus during ventral root stimulation was measured using indwelling, bi-polar wire electrodes inserted in the mid-belly region along the axis of the soleus fibres. Soleus forces were measured by attaching the distal insertion of the soleus tendon containing a remnant piece of bone to a force transducer (Guimaraes et al., in press).

Each of the prepared ventral root filaments could be stimulated independently of the remaining filaments. Stimulations were performed using pseudo-random interstimulus intervals (with a coefficient of variation of 12.5%) based on a Gaussian distribution (Zhang et al., 1992). The independent stimulation of the ten ventral root filaments at different mean frequencies using pseudo-random interstimulus intervals allowed for the creation of EMG signals that were similar (in the time and frequency domain) to EMG signals measured during voluntary contractions of the same muscle (Fig. 3.6.6).

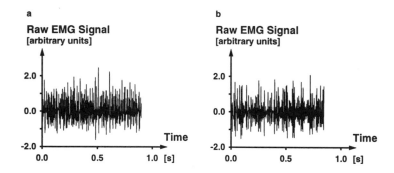

Figure 3.6.6   One of the EMG signals shown in this figure was obtained from cat soleus muscle during walking, the other signal was obtained using pseudo-random stimulation of ten ventral root filaments which contained α-motor neurons to the soleus muscle (we leave it up to the reader to decide which is which) (from Zhang et al., 1992, with permission).

A single ventral root filament prepared as described above contained an estimated 15 to 25 individual motor units. Applying a single stimulus to a single ventral root filament produced a compound motor unit action potential (CMUAP) of particular shape and duration for isometric contractions at a particular muscle length. When stimulating a single ventral root filament repeatedly and at different frequencies, the shape and duration of CMUAP was preserved, so the integration of a single full-wave rectified CMUAP remained approximately constant for rates of stimulation ranging from 5 to 50 Hz, and consequently, the corresponding IEMG was linearly related to the mean stimulation rate (Fig. 3.6.7a).

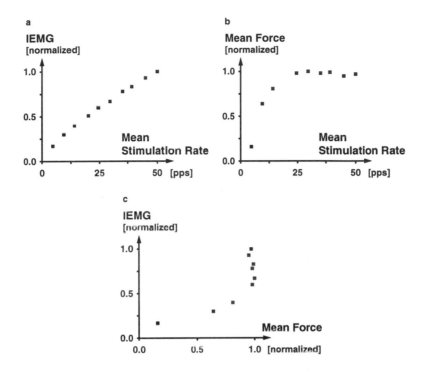

Figure 3.6.7   IEMG (a) and mean force (b) as a function of mean stimulation rate to a single ventral root filament of cat soleus muscle, and the corresponding IEMG-mean force relation (c) (from Guimaraes et al., in press, with permission).

The force response of soleus to increasing mean rates of stimulation of a single ventral root filament is shown in Fig. 3.6.7b. By contrast to the behaviour of the IEMG, mean forces were non-linearly related to the mean stimulation rate and tended to saturate beyond frequencies of approximately 25 Hz in the cat soleus. The resulting IEMG-mean force relation for stimulation of a single ventral root filament is shown in Fig. 3.6.7c.

A graded increase in the stimulation of the isometric soleus muscle was achieved by first increasing the number of stimulated ventral root filaments and simultaneously increasing the mean stimulation rate of already active ventral root units (Table 3.6.1, trials 1-10), and then by increasing the mean stimulation rates of all ventral root filaments progressively (Table 3.6.1, trials 11-27). The response of an isometrically held soleus muscle to such a stimulation protocol was a gradual increase in the IEMG (Fig. 3.6.8a) and an "S"-shaped increase in the corresponding force (Fig. 3.6.8b). For trials 1-5 (Fig. 3.6.8b), the force response remained almost zero, perhaps because the frequencies of stimulation used in these first few trials were very low, in fact, lower than the threshold frequencies observed during recruitment of motor units in skeletal muscles during voluntary movements (Burke, 1981). Therefore, trials 1-5 in Table 3.6.1 may be considered non-physiologic. In the corresponding relation between IEMG and mean force (Fig. 3.6.8c), these trials are labeled L, for Low stimulation region.

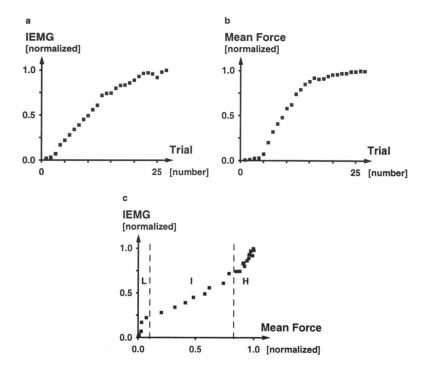

Figure 3.6.8   IEMG (a) and mean force (b) response of cat soleus muscle subjected to the stimulation protocol described in Table 3.6.1, and the corresponding IEMG-mean force relation (c). L is the low stimulation region, unphysiologic; H is the high stimulation and high force region, unphysiologic; I is the intermediate stimulation region, physiologic (from Guimaraes et al., in press, with permission).

Beyond trial 17, the increase in force with increasing stimulation starts to level off (Fig. 3.6.8b). Trial 17 is the first trial in which the majority of the ventral root filaments were stimulated at mean frequencies of 25 Hz or more. Therefore, it appears that the leveling off of the force response beyond trial 17 is associated with the saturation in muscular force observed in single ventral root filaments at stimulation frequencies of approximately 25 Hz (Fig. 3.6.7b). This particular region of force saturation is labeled H, for High stimulation region, in Fig. 3.6.7c showing the IEMG-mean force relation. The forces in soleus muscle that were reached using trials 17-27 (Table 3.6.1) are not reached typically during normal voluntary movements in the cat (Walmsley et al., 1978; Hodgson, 1983; Herzog et al., 1993a). Therefore, it has been argued that the H region shown in Fig. 3.6.8c, like the L region, is not relevant for the IEMG-force relation of cat soleus muscle during voluntary movements (Guimaraes et al., in press).

The IEMG-mean force relation for trials 6 to 16 (Table 3.6.1) could be approximated well with a straight line for all animals tested ($r^2 \geq 0.97$ in all cases). It appears, therefore, that the relation between mean force and IEMG is linear for isometrically contracting cat soleus muscle within the physiologically relevant region, i.e., the intermediate, or I region (Fig. 3.6.8c), and non-linear (i.e., S-shaped) for the whole range of stimulations used in the protocol shown in Table 3.6.1.

Table 3.6.1   Protocol used for independent stimulation of 10 ventral root filaments.

TRIAL #	VENTRAL ROOT FILAMENT #									
	1	2	3	4	5	6	7	8	9	10
1	3									
2	3	3								
3	5	5	3							
4	5	5	3	3						
5	7	7	5	5	3					
6	9	9	7	7	5	3				
7	11	11	9	9	7	5	3			
8	13	13	11	11	9	7	5	3		
9	15	15	13	13	11	9	7	5	3	
10	17	17	15	15	13	11	9	7	5	3
11	19	19	17	17	15	13	11	9	7	5
12	21	21	19	19	17	15	13	11	9	7
13	23	23	21	21	19	17	15	13	11	9
14	25	25	23	23	21	19	17	15	13	11
15	27	27	25	25	23	21	19	17	15	13
16	29	29	27	27	25	23	21	19	17	15
17	31	31	29	29	27	25	23	21	19	17
18	33	33	31	31	29	27	25	23	21	19
19	35	35	33	33	31	29	27	25	23	21
20	37	37	35	35	33	31	29	27	25	23
21	39	39	37	37	35	33	31	29	27	25
22	41	41	39	39	37	35	33	31	29	27
23	43	43	41	41	39	37	35	33	31	29
24	45	45	43	43	41	39	37	35	33	31
25	47	47	45	45	43	41	39	37	35	33
26	49	49	47	47	45	43	41	39	37	35
27	51	51	49	49	47	45	43	41	39	37

The results presented in Fig. 3.6.7 and Fig. 3.6.8 were all obtained for isometrically contracting soleus muscle at a given length. It is well accepted that the EMG-force relation of skeletal muscles under dynamic conditions needs to account for the instantaneous contractile conditions of the target muscle: the instantaneous length, rate of change in length, and activation (Hof and van den Berg, 1981a; Olney and Winter, 1985). The effect of muscle length on the EMG, force, and the corresponding IEMG versus mean force relation is shown in Fig. 3.6.9a-c for isometric contractions at three lengths of cat soleus muscle corresponding to ankle joint angles of 80°, 105°, and 130° (180° = full extension of the ankle joint, therefore, the ankle joint angle of 80° corresponds to the longest, and the angle of 130° to the shortest muscle length).

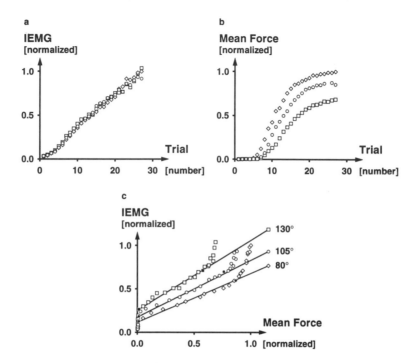

Figure 3.6.9   IEMG (a) and mean force (b) response of cat soleus subjected to the stimulation protocol described in Table 3.6.1 for three different muscle lengths, and the corresponding IEMG-mean force relation (c). The intermediate region of the IEMG-mean force relations were approximated by a straight line (by eye). The slopes of the straight lines become progressively steeper with decreasing muscle length ($180°$ = full extension of the ankle joint) (from Guimaraes et al., submitted a, with permission).

The IEMG results of the experiment shown in Fig. 3.6.9a were independent of muscle length. We feel that this independence can be obtained only if the location of the recording electrodes relative to the muscle does not change considerably during shortening and lengthening of the muscle, and if the interelectrode distance is fixed. Mean force values clearly depended on the muscle length. In this particular case, higher forces at all levels of stimulation were obtained for longer muscle lengths (Fig. 3.6.9b). This result corresponds to the observation that cat soleus muscle operates on the ascending limb of the force-length relation for the lengths tested here (Rack and Westbury, 1969; Herzog et al., 1992).

The IEMG-mean force relations of cat soleus muscle clearly depend on muscle length (Fig. 3.6.9c). They are S-shaped for each length, as described above for the single length, with an intermediate region that may be approximated adequately with a straight line. As discussed above, it is this intermediate region that appears to be most relevant for voluntary muscular contractions.

The straight-line approximations of the IEMG-mean force relations obtained in the intermediate regions of the three different muscles lengths have different slopes. The slopes become progressively steeper with decreasing muscle length, indicating that more

IEMG is required to produce a given force at short, compared to long, muscle lengths in cat soleus muscle. Similar findings have been reported during voluntary contractions of human soleus (Close et al., 1960).

In order to assess whether the differences in the forces produced at different muscle lengths depend only on muscle length or whether they depend on muscle length and the level of stimulation, the force obtained at a given level of stimulation for one muscle length was divided by the corresponding force at another muscle length. If forces were independent of the level of stimulation (and so, depended only on muscle length), one would expect the resultant force ratio to be a constant value; however, this is not the case. Fig. 3.6.10 shows the ratio that was found by dividing the forces obtained at the shortest muscle length (130°) by the corresponding forces obtained at the longest muscle length (80°) for all 27 levels of stimulation used in this protocol (Table 3.6.1). The resulting ratio was not constant, indicating that there is a non-linear scaling of the force-length property of cat soleus muscle for submaximal levels of stimulation.

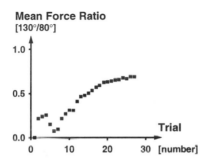

Figure 3.6.10   Ratio of the forces obtained from the shortest (ankle angle = 130°) and the longest muscle length (ankle angle = 80°) for all 27 levels of stimulations shown in Table 3.6.1 (from Guimaraes et al., submitted a, with permission).

EMG-FORCE RELATION OF CAT SOLEUS DURING UNRESTRAINED LOCO-MOTION

In order to assess the EMG-force relation of skeletal muscle during voluntary, dynamic contractions, it is necessary to measure the forces and EMGs directly from the target muscle. Simultaneous measurements of force and EMG have been performed quite frequently in cat hindlimb muscles during locomotion (e.g., Walmsley et al., 1978; Hodgson, 1983; Abraham and Loeb, 1985). However, researchers have not used such data systematically to relate the instantaneous muscular forces to the corresponding EMG records, with the exception of Sherif et al. (1983), who described qualitatively the relation between EMG and force in the cat medial gastrocnemius muscle.

Rather than using a controlled stimulation approach (as described above), when attempting to assess the EMG-force relation in muscle during locomotion in the cat, two major difficulties are added to the interpretation of the data. First, the stimulation received by the muscle at any instant in time is not known during voluntary contractions, and second, the variable contractile conditions of the muscle fibres during locomotion must be determined or estimated. It has been suggested that these contractile conditions (i.e., the length and rate of change in length of the contractile elements of the muscle) affect the EMG-force relation substantially, and, thus, must be known (e.g., Inman et al., 1952; Close et al., 1960; Heckathorne and Childress, 1981; Hof and van den Berg, 1981a, b, c, d; Basmajian and De Luca, 1985; Olney and Winter, 1985; van den Bogert et al., 1988; Ruijven and Weijs, 1990). Variable muscle fibre lengths (and corresponding rates of change in fibre length) have been estimated for cat hindlimb muscles during locomotion (Hoffer et al., 1989; Allinger and Herzog, 1992; Weytjens, 1992), however, the validity of these estimates must still be established thoroughly in further experiments.

At present, there is no theoretical paradigm that allows for the accurate prediction of muscular force from EMG records during dynamic, voluntary muscular contractions. In what follows, we present an approach that has been selected to shed some light on the EMG-force relation of skeletal muscle during voluntary movement. The main goal of this approach was to describe the EMG-force relation of cat soleus during locomotion, and to assess qualitatively the association that may exist between the EMG signals and the corresponding forces.

Force and EMG signals were obtained from soleus of three cats walking at speeds ranging from 0.4 to 1.2 m/s, and trotting at speeds varying from 1.5 to 1.8 m/s, on a motor-driven treadmill. The EMG data were full-wave rectified and integrated over time periods that varied from step to step. The time periods chosen for integration of the EMG signals were such that the integrated EMG (IEMG) would contain a frequency spectrum similar to the force signals for the corresponding steps (Guimaraes et al., submitted b). The IEMG signals calculated for each time period were then approximated using cubic interpolation splines to give a continuous IEMG-time history for each step cycle.

IEMG-time histories calculated in this way were plotted against the corresponding force-time histories. Fig. 3.6.11 shows such IEMG-force relations for three steps of three animals, all walking at a nominal speed of 0.8 m/s. Each graph starts at the onset of EMG, which corresponds to the data points aligned close and parallel to the horizontal, IEMG axis, follows the loops in the direction indicated by the arrows, and finishes at the end of soleus force production, which is indicated by the data points on the vertical axis going towards the origin of the graph.

EMG always precedes soleus force production, and the IEMG values reach peak levels before any appreciable force is developed (Fig. 3.6.11). This fast build-up of EMG activity with little or no soleus force precedes paw contact by about 60 to 90 m/s at a speed of walking of 0.8 m/s. Activation in this phase of the step cycle may be associated with preparing the contractile machinery of the muscle for force production at paw impact; that is, Ca^{2+} is released from the sarcoplasmic reticulum, attaches to troponin C, and so enables actomyosin interaction through cross-bridge formation. Cross-bridges attached in this phase, preceding paw contact, provide stiffness to the muscle, and may act as spring-like elements during and following paw impact at the beginning of the stance phase (Walmsley and Proske, 1981).

Following this initial phase, IEMGs tend to decrease, while force first increases and then decreases until the time when EMG activity becomes zero. At the end of EMG activity, soleus forces are typically still about 25 to 75% of the maximal force achieved during the step cycle (Fig. 3.6.11). In the last phase of stance, soleus forces steadily decline to zero in the absence of any EMG activity.

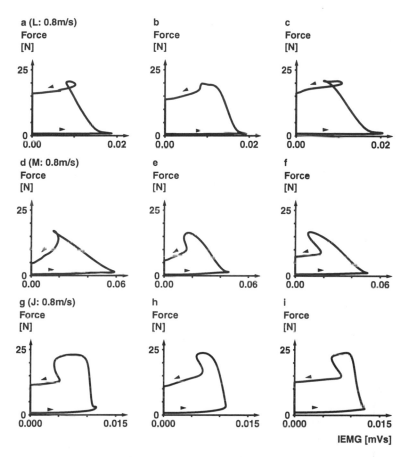

Figure 3.6.11   IEMG-force relations obtained from three animals (L, M, J) for three non-consecutive steps of walking at 0.8 m/s on a motor-driven treadmill. Data points were obtained at the same instance in absolute time, thus, these graphs do not take electromechanical delay into account. Steps of the same animal are aligned horizontally in rows (from Guimaraes et al., submitted b, with permission).

At a walking speed of 0.8 m/s, EMG activity typically terminates about 120 to 150 m/s before soleus forces become zero. Since EMG activity finishes 120 to 150 m/s before soleus force becomes zero, but only starts about 60 to 90 m/s before the onset of soleus force, it is obvious that the EMG signal has not only shifted relative to the force records, but is also shorter in duration than the corresponding force signal. Therefore, two of the difficulties of deriving an instantaneous relation between force and EMG during dynamic muscular contractions are associated with the relative shift of the EMG and the force signals, and the difference in length of the two signals.

In order to further explore the EMG-force relation of the cat soleus during locomotion, the delay between the onset of the EMG and the onset of force was accounted for by shifting the onset of the EMG signal in time to the onset of force (i.e., the electromechanical delay was offset). Furthermore, the EMG signal was linearly scaled along the time axis until the end of the EMG activity coincided with the end of soleus force production in each step cycle. In this way, a given IEMG could always be associated with a given, non-zero force of soleus during the stance phase of locomotion. Relating the IEMG obtained in this way to the corresponding soleus forces, produced the relations shown in Fig. 3.6.12.

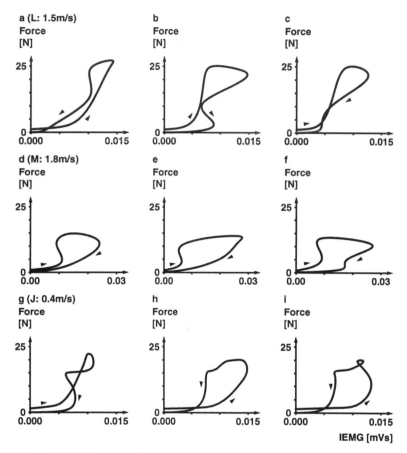

Figure 3.6.12   IEMG-force relations obtained during three consecutive steps of trotting (animals L and M) and walking (animal J) at the speeds indicated on the graphs. The electromechanical delay between the EMG and force signals was taken into account in these graphs, as described in the text (from Guimaraes et al., submitted b, with permission).

It has been suggested that any explanation of the relation between EMG and force in dynamic situations must consider the instantaneous contractile properties of the muscle (i.e., the instantaneous length and rate of change in length of the muscle or the contractile elements) (Hof and van den Berg, 1981a,b,c, d; Basmajian and De Luca, 1985; Olney and Winter, 1985). We have taken this idea a step further and have calculated a theoretical activation (A) of the muscle. This theoretical activation (Allinger and Herzog, 1992) was defined as:

$$A \quad = \quad F_i/F_{imax} \tag{3.6.3}$$

where F_i is the force required by (or measured in) the muscle, and F_{imax} is the maximal force that the muscle can produce for the given instantaneous contractile conditions. We then compared the activation calculated theoretically with the IEMG obtained experimentally for the cat soleus during locomotion. A perfect match between the theoretically calculated activation and the IEMG would indicate that the amount of IEMG signal depends precisely on the ratio of the force produced by a muscle and the maximal possible force. Fig. 3.6.13a-c shows the mean soleus forces and the corresponding theoretically determined maximal soleus forces for three animals walking at a speed of 1.2 m/s. For all animals, the measured soleus force is much smaller than the theoretically determined maximal force for the first 20 to 40% of the stance phase; however, beyond the initial 40% of the stance phase, the two forces become similar in magnitude. Sometimes the measured force even exceeds the theoretically determined maximal force.

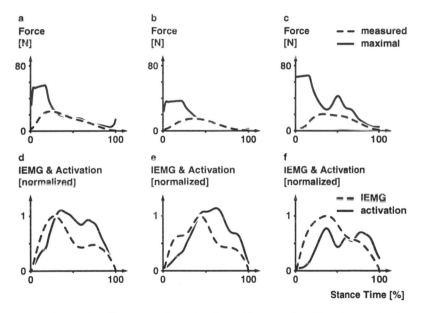

Figure 3.6.13   Force-time histories and maximal possible force-time histories of cat soleus muscle during the stance phase of walking at 1.2 m/s for three different animals (a-c), and the corresponding IEMG-time and theoretically calculated activation-time histories for the same steps as shown above (d-f) (from Guimaraes et al., submitted b, with permission).

Calculating the theoretical activation, defined in equation (3.6.3), from the results shown in Fig. 3.6.13a-c, and comparing the theoretical activation to the experimentally determined IEMG of cat soleus at a walking speed of 1.2 m/s, gave the results shown in Fig. 3.6.13d-f. The IEMG values were always larger in the initial part of the stance phase (≈ 0 to 20%) and tended to be lower in the final part of the stance phase (≈ 80 to 100%) than the corresponding activation predicted theoretically. It appears that the actual activation required in cat soleus is much higher in the initial part of the stance phase than one would expect from the maximal available force, and the forces required during this phase. This high activation is most likely associated with the initial priming of the muscle for force production; the activation of the filaments and the cross-bridges, and the corresponding stiffening of the muscle (Walmsley and Proske, 1981).

Similarly, in the final 20% of the stance phase, IEMG values are typically lower than expected theoretically. In this final phase, soleus forces are still relatively high compared to the maximal available soleus force. Nevertheless, the IEMG values are typically less than 50% of the maximum, and decrease rapidly to zero (Fig. 3.6.13d-f). As for the initial part of the stance phase, the systematic deviations of the IEMG from the theoretically calculated activation are most likely associated with the relatively slow changes in force following activation or deactivation of the muscle. Contrary to the initial part of the stance phase, in this final part of stance, the sharp decline of the IEMG is followed slowly by a decrease in muscular force, resulting in the smaller values of the IEMG compared to the corresponding values of the theoretical activation for the last 20% of the stance phase. Also, calculations of the theoretical activation (equation (3.6.3)) do not account for potentiation of force of a shortening muscle that was stretched just prior to the shortening phase (e.g., Cavagna et al., 1968). Since the soleus undergoes a stretch-shortening cycle during the stance phase of walking, force potentiation may occur during the latter stages of stance, and so, large forces may be produced with relatively little activation.

In the middle part of the stance phase (between approximately 20 to 80% of the stance time, Fig. 3.6.13d-f), the IEMG values form a bi-modal curve. The first peak of this bi-modal curve is reached between about 25 to 45% of the stance time, followed by the second, smaller peak (or sometimes just a shoulder rather than a peak) at approximately 70 to 80% of the stance time. The first peak of the IEMG appears to coincide with the end of the initial priming of the muscle for the following force production (Fig. 3.6.11). The second peak appears to be associated with a burst of EMG required near the end of the step phase to maintain soleus forces at a given level, while the contractile conditions become highly unfavorable for force production (Fig. 3.6.13a-c).

The activation-time histories predicted theoretically are typically bi-modal or multimodal, similar to the experimentally determined IEMG-time histories in the middle part of the stance phase (Fig. 3.6.13d-f). Despite this similarity between the two curves, it is apparent that the theoretical activations do not match the IEMG-time curves in detail. We speculate from these results that IEMG, and the activation calculated theoretically based on the instantaneous contractile conditions of the muscle, represent the same phenomenon, and that the theoretically predicted activation may be a possible predictor of IEMG in the middle part of the stance phase, if the instantaneous contractile conditions of the muscle can be obtained accurately.

EMG-FORCE RELATION OF CAT PLANTARIS DURING UNRESTRAINED LOCOMOTION USING AN ADAPTIVE FILTERING APPROACH

Up to this point, we have always attempted to describe or explain the EMG-force relation using basic knowledge of the physiology of force and EMG production in skeletal muscle. This approach provides insight into the behaviour of muscle, and in the end may be used to formulate a general paradigm that contains specific hypotheses which may be tested systematically. However, this approach is not the only one that may be chosen. For example, if the goal of a project is to give the best possible force predictions from EMG, without any real concern about where these two signals originated and what biological properties they may have, a purely mathematical approach linking the EMG to force may be the best and the simplest method (Zhang et al., 1993). Having explored this route, our selected results are presented below.

Before choosing a theoretical approach to predict dynamic forces from EMG exclusively, some characteristics of the force and EMG signal have to be established. Most importantly, it must be realized that the characteristics of force and EMG are time-dependent, or non-stationary, during dynamic contractions. Therefore, conventional filtering and smoothing techniques, based on a fixed (stationary) design, were rejected.

Adaptive filtering techniques can account for non-stationarities and have been used successfully in the analysis of a variety of biological signals, for example, electrocardiograms (Ferrara and Widrow, 1982; Yelderman et al., 1983), electrogastric signals (Kentie et al., 1981; Chen et al., 1990), and vibroarthrographic signals (Zhang et al., 1991). The adaptive filtering system used for plantaris force estimations from the corresponding EMG of normally walking and running cats (Herzog et al., 1993a) is shown in Fig. 3.6.14. In the figure, z^{-1} represents a one-sample delay and j stands for the time instant. The symbol d_j stands for the primary input which contains the force signal, s_{jo}, plus additive noise, n_{jo}. The reference input, x_j, which is the output of a demodulator consisting of a full-wave rectifier and a low-pass filter, contains noise, n_{j1}, and the signal s_{j1}, which is related to s_{jo}, but does not necessarily have the same waveform as the signal s_{jo}, at the primary input. The noise component, n_{jo}, is assumed to be uncorrelated with n_{j1}, s_{j1}, or s_{jo}. The output signal, y_j, is formed as the weighted sum of a set of input signal samples:

$$x_j, x_j, ..., x_j - N + 1$$

Mathematically, the output, y_j, is equal to the inner product of the input vector, X_j, and the weight vector, W_j:

$$y_j = X_j^T W_j \tag{3.6.4}$$

where the input signal vector is defined as:

$$X_j = \left[x_j, x_{j-1}, ..., x_{j-N+1}\right]^T \tag{3.6.5}$$

and the filter weight vector is:

$$W_j = \left[w_{ji}, w_{j2}, ..., w_{jN}\right]^T \tag{3.6.6}$$

During the adaptation process, the weights are adjusted according to the least mean squared (LMS) algorithm (Zhang et al., in press; Widrow, et al., 1975). Let P represent the cross-correlation vector defined by:

$$P = E\{d_j X_j\} \tag{3.6.7}$$

and let R represent the input correlation matrix defined by:

$$R = E\{X_j X^T_j\} \tag{3.6.8}$$

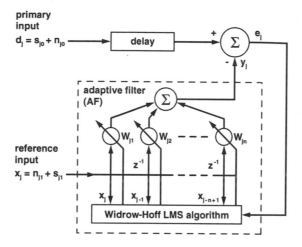

Figure 3.6.14   An adaptive filtering system with the tapped delay line. z^{-1} represents a one-sample delay and j stands for the time instant.

where E { } stands for the expectation operator. Thus, a general expression of the mean-squared error (MSE) as a function of the weight vector can be obtained as follows:

$$e_j = d_j - y_j = d_j - X^T_j W_j \tag{3.6.9}$$

$$E\{e^2_j\} = E\{d^2_j\} - 2P^T W_j + W^T_j R W_j \tag{3.6.10}$$

Using the steepest-descent method and approximating the gradient MSE by gradient squared error, we obtain:

$$W_{j+1} = W_j - \mu \Delta_j \tag{3.6.11}$$

$$\Delta_j = \frac{de^2_j}{dW_j} \tag{3.6.12}$$

and the well-known Widrow-Hoff LMS algorithm is derived as (Widrow et al., 1975):

$$W_{j+1} = W_j + 2\mu e_j X_j \tag{3.6.13}$$

where μ is the step size which controls the convergence rate of the adaptation and the stability of the system. The larger the value of μ, the larger is the gradient noise that is introduced, but the faster the algorithm converges, and vice versa. Therefore, in the non-stationary environment, step-size optimization becomes a critical issue for successful adaptive signal processing using the LMS algorithm. Generally, a compromise has to be made between a fast convergence, which is necessary to track the statistical changes in the input signal, and the slow adaptation which is needed to reduce the gradient noise (or mis-adjustment). A careful selection of performance measurements is important for step-size optimization (Zhang et al., 1991; Zhang et al., in press).

After the convergence of the system, the adaptive filter with the Widrow-Hoff LMS algorithm creates a least mean squares estimate of the force signal, n_{jo}, in the primary input by filtering the reference input, i.e., the rectified EMG signals. According to the adaptive theory (Widrow et al., 1975), the filter output is the best estimate of the signal, n_{jo}, in the sense of the mean square error. A critical requirement for the successful opera-tion of this scheme is that the primary and the reference signal components must be uncor-related with the primary noise components, but correlated with each other. This statement is explained using the following demonstration (for convenience, the subscript j is omit-ted). From Fig. 3.6.14, the system output is the error, given by:

$$e = s_0 + n_0 - y \qquad (3.6.14)$$

where s_0 represents the primary signal component. Then, the MSE can be obtained as:

$$E\{e^2\} = E\{n_0^2\} + E\{(s_0 - y)^2\} + 2E\{n_0(s_0 - y)\} \qquad (3.6.15)$$

Since n_0 and $(s_0 - y)$ are uncorrelated and have a mean of zero, $E\{n_0(s_0 - y)\} = 0$, the equation (3.6.15) becomes:

$$E\{e^2\} = E\{n_0^2\} + E\{(s_0 - y)^2\} \qquad (3.6.16)$$

Thus, minimizing the MSE for a given input can be achieved by minimizing $E\{(s_0 - y)^2\}$. That is:

$$minE\{e^2\} = minE\{(s_0 - y)^2\} + E\{n_0^2\} \qquad (3.6.17)$$

The MSE minimization yields the least mean square estimate of the signal, s_0. This result is obtained by the adaptive filtering process governed by the LMS algorithm without requiring any prior knowledge of the statistics of the signal and the noise, except the requirement that the signal components in the primary input and the reference input are correlated with each other, but are uncorrelated with the noise. According to the adaptive filtering principle, the output of the system, y_j, is the best estimate of the muscle force, s_0, in the primary input.

Fig. 3.6.15 shows the actual forces recorded in the cat plantaris muscle during eight walking steps on a $10°$ downhill slope at a speed of 0.4 m/s, and Fig. 3.6.16 shows the cor-responding EMG signals. The actual plantaris forces were used as the desired signals of the primary input to the adaptive filter, and the demodulated plantaris EMG signal was used as the reference input. The estimated force using the adaptive filter with $\mu = 0.01$ is given in Fig. 3.6.17 (dashed line), together with the corresponding actual force (solid line). As can be seen, cat plantaris forces estimated exclusively from rectified and low-pass fil-tered EMG signals were adequate after the convergence (about a quarter of the first step cycle) of the adaptive filtering process.

It has been stated that EMG-force relations of dynamically contracting muscles must contain muscle properties, such as force-length and force-velocity relations. However, instantaneous force-length-velocity properties of muscles are hard to measure in vivo. The

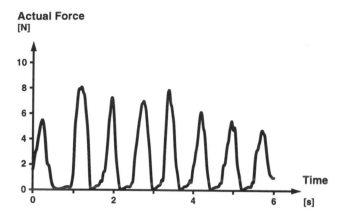

Figure 3.6.15   The actual forces recorded in the cat plantaris muscle during eight walking steps on a 10° downhill slope at a speed of 0.4 m/s.

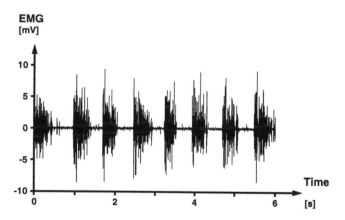

Figure 3.6.16   The EMG signals of cat plantaris corresponding to the force results of Fig. 3.6.15.

results of this study indicate that adequate estimates of dynamic muscular forces are possible without using the instantaneous contractile properties of muscles, which suggests, therefore, that an adaptive approach may provide a simple way to predict dynamic muscle forces.

As pointed out in the Introduction, the purpose of this chapter was not to copy information on EMG that is available in textbooks devoted exclusively to EMG. Rather, we attempted to present information on a specific topic of EMG, the EMG-force relation of skeletal muscles. The results of three very different and new approaches were discussed. In the first approach, we pointed out the possibility of how a controlled stimulation technique on the ventral root level could be used to create a "physiologic"-looking EMG, and how, under these controlled conditions, EMG and force of a muscle could be measured for any contractile conditions desired. The second and third approaches dealt with the relations between EMG and force of dynamically contracting muscles during locomotion. Forces and EMGs were measured directly from the target muscles in both cases. In the

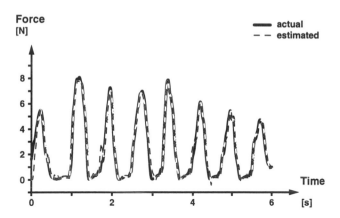

Figure 3.6.17 Estimated plantaris forces using adaptive filtering with the LMS algorithm (dashed line), and the actual plantaris forces (solid line) observed during normal cat locomotion.

second approach, the EMG-force relations were described, and then qualitatively analysed, using the known physiological properties of muscle as guidelines. The third approach was based exclusively on a theoretical input-output-type system. Physiological (or other) properties of muscle were disregarded. The origin of the input signal (EMG) and the estimated output variable (force) were of no importance. Interestingly, though this last approach often gave excellent estimates of muscular force, no insight into the system's behaviour could be gained from it.

Book chapters typically summarize the well-established facts and condense the wealth of knowledge and information in a particular area into easily digestible portions. Not so in this chapter. We have, on purpose, chosen to focus on an area of research that is hotly debated. Many of the results presented here have not been published elsewhere at the time of writing this chapter. The results shown are correct, however, many of the ideas are new and may turn out to be, in the words of A.V. Hill, "false trails". As long as they are stimulating, to the student who reads this book and the researcher who is interested in the relation between EMG and force in dynamically contracting skeletal muscles, they will have served as a significant contribution to our goal.

3.6.8 REFERENCES

Abraham, L.D. and Loeb, G.E. (1985) The Distal Hindlimb Musculature of the Cat. *Exp. Brain Res.* **58,** pp. 580-593.

Allinger, T. and Herzog, W. (1992) Calculated Fibre Lengths in Cat Gastrocnemius During Walking. *Proceedings of NACOB II.* Amer. Soc. of Biomechanics, Chicago. pp. 81-82.

Basmajian, J.V. and De Luca, C.J. (1985) *Muscles Alive. Their Functions Revealed by Electromyography* (5th Ed.). Williams & Wilkins, Baltimore.

Bigland, B. and Lippold, O.C.J. (1954) Motor Unit Activity in the Voluntary Contraction of Human Muscle. *J. Physiol.* **125,** pp. 322-335.

Bigland-Ritchie, B., Kukulka, C.G., and Woods, J.J. (1980) Surface EMG/force Relations in Human Muscle of Different Fibre Composition. *J. Physiol.* **308**, pp. 103-104.

Bigland-Ritchie, B., Donovan, E.F., and Roussos, C.S. (1981) Conduction Velocity and EMG Power Spectrum Changes in Fatigue of Sustained Maximal Efforts. *J. Appl. Physiol.* **51**, pp. 1300-1305.

Bouisset, S. and Goubel, F. (1971) Interdependence of Relations Between Integrated EMG and Diverse Biomechanical Quantities in Normal Voluntary Movements. *Activitas Nervosa Superior.* **13**, pp. 23-31.

Burke, R.E. (1981) Motor Units: Anatomy, Physiology, and Functional Organization. *Handbook of Physiology-The Nervous System II* (ed. V. Brooks). Williams & Wilkins, Baltimore. pp. 345-422.

Cavagna, G.A., Dusman, B., and Margaria, R. (1968) Positive Work Done by a Previously Stretched Muscle. *J. Appl. Physiol.* **24 (1)**, pp. 21-32.

Chen, J., Vandewalle, J., Sansen, W., Vantrappen, G., and Janssens, J. (1990) Multichannel Adaptive Enhancement of the Electrogastrogram. *IEEE Trans. on Biomedical Engineering.* BME **37 (3)**, pp. 285-294.

Close, J.R., Nickel, E.D., and Todd, F.N. (1960) Motor-unit Action Potential Counts. Their Significance in Isotonic and Isometric Contractions. *J. Bone Jt. Surg.* **42 (A-7)**, pp. 1207-1222.

Croone, W. (1664) *De Ratione Motus Musculorum.* London, England.

DuBois-Reymond, E. (1849) *Untersuchungen ueber thierische elektricitaet.* Reimer Verlag von G, Berlin. **2 (2)**.

Duchenne, G.B.A. (1959) *Physiologie des Mouvements* (trans. Kaplan, E.B.). W.B. Saunders, London. (Original 1867).

Ferrara, E.R. and Widrow, B. (1982) Fetal Electrocardiogram Enhancement by Time-sequenced Adaptive Filtering. *IEEE Trans. on Biomedical Engineering.* BME **29 (6)**, pp. 458-460.

Galvani, L. (1953) *DeViribus Electricitatis* (trans. Green, R.). University Press, Cambridge. (Original 1791).

Guimaraes, A.C.S., Herzog, W., Hulliger, M., Zhang, Y.T., and Day, S. (In press) EMG-force Relation of the Cat Soleus Muscle: Experimental Simulation of Recruitment and Rate Modulation Using Stimulation of Ventral Root Filaments. *J. Exp. Biol.*

Guimaraes, A.C.S., Herzog, W., Allinger, T.L., and Zhang, Y.T. (Submitted a) Effects of Muscle Length on the EMG-force Relation of the Cat Soleus Muscle Using Non-periodic Stimulation of Ventral Root Filaments. *J. Exp. Biol.*

Guimaraes, A.C.S., Herzog, W., Hulliger, M., Zhang, Y.T., and Day, S. (Submitted b) EMG-force Relation of the Cat Soleus Muscle During Locomotion, and its Association with Contractile Conditions. *J. Exp. Biol.*

Hagberg, M. (1981) Muscular Endurance and Surface Electromyogram in Isometric and Dynamic Exercise. *J. Appl. Physiol: Resp., Env., & Ex. Phys.* **51**, pp. 1-7.

Heckathorne, C.W. and Childress, D.S. (1981) Relationships of the Surface Electromyogram to the Force, Length, Velocity, and Contraction Rate of the Cineplastic Human Biceps. *Am. J. Phys. Med.* **60 (1)**, pp. 1-19.

Herzog, W., Leonard, T.R., Renaud, J.M., Wallace, J., Chaki, G., and Bornemisza, S. (1992) Force-length Properties and Functional Demands of Cat Gastrocnemius, Soleus, and Plantaris Muscles. *J. Biomech.* **25 (11)**, pp. 1329-1335.

Herzog, W., Leonard, T.R., and Guimaraes, A.C.S. (1993a) Forces in Gastrocnemius, Soleus, and Plantaris Tendons of the Freely Moving Cat. *J. Biomech.* **26 (8)**, pp. 945-953.

Herzog, W., Stano, A., and Leonard, T.R. (1993b) Telemetry System to Record Force and EMG from Cat Ankle Extensor and Tibialis Anterior Muscles. *J. Biomech.* **26**, pp. 1463-1471.

Hodgson, J.A. (1983) The Relationship Between Soleus and Gastrocnemius Muscle Activity in Conscious Cats - A Model for Motor Unit Recruitment? *J. Physiol.* **337**, pp. 553-562.

Hof, A.L. and van den Berg, J. (1981a) EMG to Force Processing I: An Electrical Analog of the Hill Muscle Model. *J. Biomech.* **14 (11)**, pp. 747-758.

Hof, A.L. and van den Berg, J. (1981b) EMG to Force Processing II: Estimation of Parameters of the Hill Muscle Model for the Human Triceps Surae by Means of a Calf-ergometer. *J. Biomech.* **14 (11)**, pp. 759-770.

Hof, A.L. and van den Berg, J. (1981c) EMG to Force Processing III: Estimation of Model Parameters for the Human Triceps Surae Muscle and Assessment of the Accuracy by Means of a Torque Plate. *J. Biomech.* **14**, pp. 771-785.

Hof, A.L. and van den Berg, J. (1981d) EMG to Force Processing IV: Eccentri-concentric Contractions on a Spring-flywheel Set Up. *J. Biomech.* **14 (11)**, pp. 787-792.

Hoffer, J.A., Caputi, A.A., Pose, I.E., and Griffiths, R.I. (1989) Roles of Muscle Activity and Load on the Relationship Between Muscle Spindle Length and Whole Muscle Length in the Freely Walking Cat. *Prog. Brain Res.* **80**, pp. 75-85.

Inman, V.T., Ralston, H.J., Saunders, J.B. de C.M., Feinstein, B., and Wright Jr., E.W. (1952) Relation of Human Electromyogram to Muscular Tension. *Electroenceph. Clin. Neurophysiol.* **4**, pp. 187-194.

Kentie, M.A., van der Schee, E.J., Grashuis, J.L., and Smout, A.J.P.M. (1981) Adaptive Filtering of Canine Electrogastrographic Signals. Part 2: Filter Performance. *Med. & Biol. Eng. & Comput.* **19**, pp. 765-769

Komi, P. and Tesch, P. (1979) EMG Frequency Spectrum, Muscle Structure, and Fatigue During Dynamic Contractions in Man. *Eur. J. Appl. Physiol. Occup. Physiol.* **32**, pp. 41-50.

Lindström, L., Kadefors, R., and Petersén, I. (1977) An Electromyographic Index for Localized Muscle Fatigue. *J. Appl. Physiol.* **43**, pp. 750-754.

Lippold, O.C.J. (1952) The Relation Between Integrated Action Potential in Human Muscle and its Isometric Tension. *J. Physiol.* **117 (4)**, pp. 492-499.

Loeb, G.E. and Gans, C. (1986) *Electromyography for Experimentalists.* The University of Chicago Press, Chicago, Il.,

Milner-Brown, H.S. and Stein, R.B. (1975) The Relation Between the Surface Electromyogram and Muscular Force. *J. Physiol.* **246**, pp. 549-569.

Moritani, T. and deVries, H.A. (1978) Re-examination of the Relationship Between the Surface Integrated Electromyogram and Force of Isometric Contraction. *Am. J. Phys. Med.* **57 (6)**, pp. 263-277.

Olney, S.J. and Winter, D.A. (1985) Predictions of Knee and Ankle Moments of Force in Walking from EMG and Kinematic Data. *J. Biomech.* **18 (1)**, pp. 9-20.

Petrofsky, J.S., Glaser, R.M., and Phillips, C.A. (1982) Evaluation of the Amplitude and Frequency Components of the Surface EMG as an Index of Muscle Fatigue. *Ergonomics.* **25**, pp. 213-223.

Rack, P.M.H. and Westbury, D.R. (1969) The Effects of Length and Stimulus Rate on Tension in the Isometric Cat Soleus Muscle. *J. Physiol.* **204**, pp. 443-460.

Sherif, M.H., Gregor, R.J., Liu, L.M., Roy, R.R., and Hager, C.L. (1983) Correlation of Myoelectric Activity and Muscle Force During Selected Cat Treadmill Locomotion. *J. Biomech.* **16 (9)**, pp. 691-701.

Thorstensson, A., Karlsson, J., Viitasalo, J.H.T., Luhtanen, P., and Komi, P.V. (1976) Effect of Strength Training on EMG of Human Skeletal Muscle. *Acta. Physio. Scand.* **98**, pp. 232-236.

van den Bogert, A.J., Hartman, W., Schamhardt, H.C., and Sauren, A.A.H.J. (1988) In Vivo Relationship Between Force, EMG, and Length Change in Deep Digital Flexor Muscle of the Horse. *Biomechanics XI-A* (eds. Hollander, A.P., Huijing, P.A., and van Ingen Schenau, G.J.). Free University, Amsterdam. pp. 68-74.

van Ruijven, L.J. and Weijs, W.A. (1990) A New Model for Calculating Muscle Forces from Electromyograms. *Eur. J. Appl. Physiol.* **61**, pp. 479-485.

Walmsley, B., Hodgson, J.A., and Burke, R.E. (1978) Forces Produced by Medial Gastrocnemius and Soleus Muscles During Locomotion in Freely Moving Cats. *J. Neurophysiol.* **41 (5)**, pp. 1203-1216.

Walmsley, B. and Proske, U. (1981) Comparison of Stiffness of Soleus and Medial Gastrocnemius Muscles in Cats. *J. Neurophysiol.* **46**, pp. 250-259.

Weytjens, J.L.F. (1992) Determinants of Cat Medial Gastrocnemius Muscle Force During Simulated Locomotion. *Thesis, Department of Clinical Neuroscience.* University of Calgary. Nat'l Library of Canada, Ottawa.

Widrow, B., Glover, J.R., McCool, J.M., Kaunitz, J., Williams, C.S., Hearn, R.H., Zeidler, J.R., and Goodlin, R.C. (1975) Adaptive Noise Cancelling: Principles and Applications. *Proc. IEEE.* **63**, pp. 1692-1716.

Wilkie, D.R. (1968) *Studies in Biology, No. 11, Muscle.* Edward Arnold Publishers Ltd., London.

Yelderman, M., Widrow, B., Cioffi, J.M., Hesler, E., and Leddy, J.A. (1983) ECG Enhancement by Adaptive Cancellation of Electrosurgical Interference. *IEEE Trans. on Biomed. Eng.* EMB **30 (7)**, pp. 392-398.

Zhang, Y.T., Frank, C.B., Rangayyan, R.M., Bell, G.D., and Ladly, K.O. (1991) Step Size Optimization of Non-stationary Adaptive Filtering for Knee Sound Analysis. *Med. & Biol. Eng. & Comput.* **29** (Supplement, Part 2), p. 836.

Zhang, Y.T., Herzog, W., Parker, P.A., Hulliger, M., and Guimaraes, A. (1992) Distributed Random Electrical Neuromuscular Stimulation: Dependence of EMG Median Frequency on Stimulation Statistics and Motor Unit Action Potentials. *NACOB II* (eds. Draganich, L., Wells, R., and Bechtold, J.). American Society of Biomechanics, Chicago, IL. pp. 185-186.

Zhang, Y.T., Herzog, W., Sokolosky, J., and Guimaraes, A.C.S. (1993) Adaptive Estimation of Muscular Forces from the Myoelectric Signal Obtained During Dynamically Contracting Muscle. *Proceedings, Canadian Conference on Electrical and Computer Engineering.* pp. 1305-1307.

Zhang, Y.T., Rangayyan, R.M., Frank, C.B., and Bell, G.D. (In press) Adaptive Cancellation of Muscle Contraction Interference in Vibroarthrographic Signals. *IEEE Trans. on Biomedical Engineering.*

3.7 INERTIAL PROPERTIES OF THE HUMAN OR ANIMAL BODY

NIGG, B.M.

3.7.1 DEFINITIONS AND COMMENTS

Centre of gravity:	The centre of gravity of an object is the centroid of all the gravitational forces, G, acting on its mass elements, dm, with:

$$r_{CG} = \frac{\int r \, dG}{\int dG}$$

Centre of mass:	The centre of mass of an object is the centroid of all mass elements, dm, with the position vector:

$$r_{CM} = \frac{\int r \, dm}{\int dm}$$

Centroid:	The centroid of an assemblage of n similar quantities, $\Delta_1, \Delta_2,..., \Delta_n$, situated at points $P_1, P_2,...,P_n$, and having the position vectors $r_1, r_2,..., r_n$, has a position vector **r**.

$$r = \frac{\sum_{i=1}^{n} r_i \Delta_i}{\sum_{i=1}^{n} \Delta_i}$$

Frustrum:	The solid figure formed when the top of a cone or pyramid is cut off by a plane parallel to the base.
Moment of inertia:	The moment of inertia, I, of a mass, m, is the sum of the axial moments of all its elements, dm.

$$I = \int r^2 dm$$

3.7.2 SELECTED HISTORICAL HIGHLIGHTS

1860 Harless	Used the *water immersion method* to determine the mass, the location of the centre of mass, and the moment of inertia of body segments.

1889	Braune & Fischer	Published data about the centre of gravity of the human body as a contribution to the design of military equipment.
1938	Weinbach	Used volume contour maps to determine the location of the centre of mass and moment of inertia.
1947	Bernstein	Used weight measurements in different positions to determine the mass of body segments if the location of the CM was known, or to determine the location of the CM if the relative mass of the body segments was known.
1955	Dempster	Identified mass and inertia values for all major human body segments using various methods.
1957	Hertzberg	Used a *photogrammetric method* to determine the mass, the location of the centre of mass, and the moment of inertia of body segments.
1964	Hanavan	Presented a mathematical model of the human body for determining inertial properties using symmetric mathematical volumes as a base.
1967	Bouisset & Pertuzon	Determined moments of inertia for limb segments experimentally by using the so-called "quick release method".
1969	Clauser	Contributed cadaver data for weight, volume, and centre of mass of human limb segments.
1969	Gurfinkel & Safornov	Used a *mechanical vibration method* to determine the moments of inertia of body segments.
1980	Hatze	Presented a mathematical model for the computational determination of inertial properties of anthropometric segments.
1983	Zatziorsky	Used a *gamma-scanner method* to determine the mass, the location of the centre of mass, and the moment of inertia of body segments.
1989	Martin	Used MRI to determine the mass, the location of the centre of mass, and the moment of inertia of body segments.

3.7.3 INERTIAL PROPERTIES

The inertial properties of the human body, such as mass, centre of gravity, and mass moments of inertia, are often required for quantitative analyses of human motion. This section will discuss selected techniques which have been developed for obtaining data for calculating inertial properties. The discussion concentrates on theoretical considerations, experimental methods, and theoretical methods.

THEORETICAL CONSIDERATIONS

Inertial properties are defined below for two- and three-dimensional cases. Consider first the general case (Fig. 3.7.1), in which the origin of a rectangular coordinate system (x, y, z) has been located at the centre of mass, CM, of a rigid body.

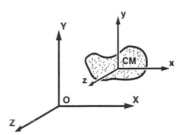

Figure 3.7.1   **A rigid body with a coordinate system having its origin in CM = (x_{cm}, y_{cm}, z_{cm}), and a general coordinate system with its origin in O – (0, 0, 0).**

In the three-dimensional case, whereby, an object moves in all three-dimensions, rotation can occur about any axis in space. For this situation, the mass moment of inertia about the centroid of the total mass is given by the matrix:

$$I^{CM} = \begin{bmatrix} I_{xx}^{CM} & -I_{xy}^{CM} & -I_{xz}^{CM} \\ -I_{yx}^{CM} & I_{yy}^{CM} & -I_{yz}^{CM} \\ -I_{zx}^{CM} & -I_{zy}^{CM} & I_{zz}^{CM} \end{bmatrix} \tag{3.7.1}$$

$I_{xx}^{CM}, I_{yy}^{CM}, I_{zz}^{CM}$ are called the moments of inertia about the x, y, and z axes, through the CM, respectively, where:

$$I_{xx}^{CM} = \int (y^2 + z^2) \, dm$$

$$I_{yy}^{CM} = \int (x^2 + z^2) \, dm$$

$$I_{zz}^{CM} = \int (x^2 + y^2) \, dm \tag{3.7.2}$$

$I_{xy}^{CM}, I_{yx}^{CM}, I_{xz}^{CM}, I_{zx}^{CM}, I_{yz}^{CM}$ and I_{zy}^{CM} are called the products of inertia, where:

$$I_{xy}^{CM} = I_{yx}^{CM} = \int xy \, dm$$

$$I_{xz}^{CM} = I_{zx}^{CM} = \int xz \, dm$$

$$I_{yz}^{CM} \quad = \quad I_{zy}^{CM} \quad = \quad \int yz \, dm$$

For a general moment of inertia for a point A with A $\neq$ CM, the moment of inertia can be determined using the parallel axis theorem.

For practical purposes one may chose a coordinate system with the axes coinciding with the principle axes. For this simplification the products of inertia become zero and the moment of inertia about the centroid (CM) becomes:

$$I^{CM} \quad = \quad \begin{bmatrix} I_{xx} & 0 & 0 \\ 0 & I_{yy} & 0 \\ 0 & 0 & I_{zz} \end{bmatrix} \tag{3.7.3}$$

To further simplify the situation, one often reduces the space of interest to two-dimensions, assuming:

- A two-dimensional problem in the x, y plane,
- A symmetrical object,
- That the moment of inertia with respect to CM is of interest, and
- That the axis of rotation is perpendicular to x, y plane.

Equation (3.7.1) can then be written as:

$$I^{CM} \quad = \quad \int r^2 \, dm \quad = \quad \int (x^2 + y^2) \, dm \tag{3.7.4}$$

The determination of the total moment of inertia about the centroid requires detailed information on the location and mass of each differential mass element, dm, under consideration. Usually, information of this detail is unavailable. Often the procedure can be further simplified by assuming that the body is made up of a homogeneous material of uniform density, p, which simplifies equation (3.7.4) to:

$$I^{CM} \quad = \quad p \int r^2 \, dV \tag{3.7.5}$$

where:

dV   =   differential volume element

The moment of inertia can be determined by direct measurement or through calculation. The various inertial parameters that may need to be determined depending upon the calculation method used are:

m   =   total mass of the object of interest
dm   =   differential mass element
r   =   perpendicular distance of a mass element from the axis under consideration
p   =   density of the material

dV   =   volume element corresponding to the mass elements

$$\left.\begin{array}{l} x_{CM} \\ y_{CM} \\ z_{CM} \end{array}\right\}$$ coordinates of the CM of the object of interest

$$\left.\begin{array}{l} x_A \\ y_A \\ z_A \end{array}\right\}$$ coordinates of an arbitrary point A

$$x_{CA} = x_{CM} - x_A$$
$$y_{CA} = y_{CM} - y_A$$
$$z_{CA} = z_{CM} - z_A$$

The next sections discuss various experimental and theoretical methods of determining inertial parameters.

EXPERIMENTAL METHODS

This section discusses experimental methods used to determine mass, centre of mass, centre of gravity, and moments of inertia. Historically, the first experimental attempts to determine these values were made on cadavers. Braune and Fischer (1889), Dempster (1955), Clauser et al. (1969), and Chandler et al. (1975) used cadaver studies to determine mass, volume, density, and centre of mass of the total human body and of selected segments.

Mass and density

The determination of the *mass* is defined as:

$$m = \frac{W}{g}$$

where:

m   =   mass of object of interest
g   =   acceleration due to gravity
W   =   weight of object of interest

In biomechanical applications, one problem lies in the definition of the object of interest. For example, the thigh of a human body is not well defined. How much of the muscles crossing the knee or the hip should be included? Should the cuts be made in an extended or a flexed position? Should the cuts be perpendicular or inclined? Many such questions can and have been answered arbitrarily in the past.

The *average density*, p, of an object is defined as:

$$p = \frac{m}{V} \qquad\qquad (3.7.6)$$

where:

$$p \quad = \quad \text{average density of the object of interest}$$
$$m \quad = \quad \text{mass of the object of interest}$$
$$V \quad = \quad \text{volume of the object of interest}$$

Two methods are commonly used to determine the average density of an object.

In the first method, the mass is determined by weighing the object in the air. By lowering the object into water and weighing the displaced water, the volume of the object can be determined (after correcting for temperature differences). The mass and volume are then used to determine the density.

In the second method the object is weighed first in air and then in water. The density can then be determined using:

$$p \quad = \quad \frac{W_{air} - W_{water}}{V \cdot g} \tag{3.7.7}$$

where:

$$p \qquad = \quad \text{density of the object of interest}$$
$$W_{air} \quad = \quad \text{weight of the object in air}$$
$$W_{water} = \quad \text{weight of the object in water}$$
$$V \qquad = \quad \text{volume of the object of interest}$$
$$g \qquad = \quad \text{acceleration due to gravity}$$

The volume displacement method was used to determine the volumes of segments of living people (Dempster, 1955; Clauser et al., 1969). The subjects lowered the limb into the measuring tank to a given depth, and the displaced water was measured. The volume displacement method proved less reliable for living people than for cadavers, with repeat measurement variations of 3-5%. These variations were attributed to difficulties in keeping the subject's limb stationary while the displaced volume was drained.

The densities determined with the methods described above provide an average density for the whole segment. The tissues that make up a human segment have different densities, though. Cadaver segments have been dissected to determine bone, muscle, and fat densities. The volumes of the individual tissues can be determined through the use of methods such as computerized axial tomography or magnetic resonance imaging. These volumes can then be used to determine overall segment masses (indicated later in Table 3.7.2).

The physical properties of muscle, skin, adipose tissue (fat), and bone are different. Experimental data from six embalmed cadavers for relative mass and density for various body segments (Clarys and Marfell-Jones, 1986) are summarized in Table 3.7.1. The density of skin is rather constant for the four locations tested and the differences are smaller than 1%. Adipose density varies by about 4% with the lowest values measured for the arm and the highest values for the foot. Muscle density varies by about 2% with the lowest values measured for the foot and the highest for the forearm. Bone density varies by 14% with the highest values measured for the forearm and the lowest for the foot. The high variation of bone density suggests that these differences should be included in precise calculations of inertia parameters.

Table 3.7.1   Relative mass and density for arm, forearm, leg, and foot based on data from six embalmed cadavers (from Clarys and Marfell-Jones, 1986, with permission).

BODY SEGMENT	TISSUE	RELATIVE MASS [%]	DENSITY [gcm^{-3}]
ARM	skin	6.9	1.050
	adipose	36.45	0.954
	muscle	41.85	1.049
	bone	14.8	1.224
FOREARM	skin	8.35	1.051
	adipose	23.55	0.961
	muscle	51.5	1.054
	bone	16.6	1.308
LEG	skin	6.0	1.055
	adipose	28.95	0.958
	muscle	42.7	1.042
	bone	22.35	1.2075
FOOT	skin	13.25	1.0586
	adipose	30.95	0.992
	muscle	24.7	1.037
	bone	31.1	1.1525

Centre of gravity and centre of mass

The centre of gravity of an object is the location of the centroid of all gravitational forces, G, of its mass elements, dm:

$$r_{CG} = \frac{\int r\, dG}{\int dG} \qquad (3.7.8)$$

The centre of mass of an object is the location of the centroid of all mass elements, dm:

$$r_{CM} = \frac{\int r\, dm}{\int dm} \qquad (3.7.9)$$

In a first approximation, the locations of the centre of gravity and the centre of mass are the same. However, a skyscraper example illustrates the difference between the centre of gravity and the centre of mass (Fig. 3.7.2).

The centre of mass of the skyscraper is at the centre of the structure. Consider the skyscraper's upper and lower half separately. Let the centre of gravity of the upper half be at CG_1 and the centre of gravity of the lower half at CG_2. Since CG_2 is closer to the earth's centre than CG_1, the force acting at the lower half is greater than the force acting at the upper half. Consequently, the CG of the whole structure is closer to the ground than the CM. In general, however, these differences are small and can be neglected.

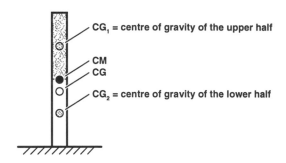

Figure 3.7.2   **Illustration of the difference between centre of gravity, CG, and centre of mass, CM, of a long object (e.g., skyscraper) exposed to gravity.**

Several methods were used to determine the centre of gravity (Table 3.7.2). Braune and Fischer (1889) used thin metal rods driven into frozen cadaver segments at right angles to the three cardinal planes. The segments were suspended from each rod and the planes of intersection of the rod with the segment marked. The intersection of the three planes defined the centre of gravity.

Dempster (1955) used a balance plate, which consisted of a plate positioned so that it could pivot around the turned down-ends of one of the diagonals, to locate the plane of the centre of gravity along the longitudinal axis relative to the ends of the segments. Similar balance plate configurations can be used on living people to determine the whole body centre of mass, but not segmental centroids.

A technique more commonly used than the balance plate is a "moment table". A table (or plate) is supported at one end by a knife edge on a metal stand, and at the other end by a knife edge placed on a scale (Fig. 3.7.3). If the weight of the object under consideration

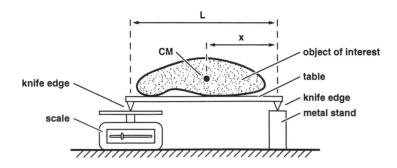

Figure 3.7.3   **Illustration of a "moment table" used to determine the centre of gravity.**

is known, the location of the centre of gravity can be found by determining moments about the metal stand end, using the force reading from the scale and the distance between the knife edges.

Table 3.7.2   Relative segment mass and location of centre of gravity from selected cadaver studies. Relative segment masses are given as a percentage of the total body mass. Location of the centre of gravity is given as a percentage of the distance along the longitudinal axis from the proximal joint.

SEGMENT	RELATIVE MASS $\left(\dfrac{m_i}{m}\right)$				RELATIVE DISTANCE $\left(\dfrac{L_i}{L}\right)$					
	HARLESS (1806)		BRAUNE & FISCHER (1889)		FISCHER (1906)		DEMPSTER (1955)		CLAUSER (1969)	
	$\dfrac{m_i}{m}$ [%]	$\dfrac{L_i}{L}$ [%]	$\dfrac{m_i}{m}$ [%]	$\dfrac{L_i}{L}$ [%]	$\dfrac{m_i}{m}$ [%]	$\dfrac{L_i}{L}$ [%]	$\dfrac{m_i}{m}$ [%]	$\dfrac{L_i}{L}$ [%]	$\dfrac{m_i}{m}$ [%]	$\dfrac{L_i}{L}$ [%]
HEAD	7.6	36.2	7.0		8.8		7.9		7.3	
TRUNK	44.2	44.8	46.1		45.2		46.9		50.7	
ARM	5.7		6.2	52.6	5.4	44.6	4.9	51.2	4.9	41.3
UPPER ARM	3.2		3.3	47.0	2.8	45.0	2.7	43.6	2.6	51.3
FOREARM & HAND	2.6		3.0	47.2	2.6	46.2	2.2	67.7	2.3	62.6
FOREARM	1.7	42.0	2.1	42.1			1.6	43.0	1.6	39.0
HAND	0.9	39.7	0.8				0.6	50.6	0.7	
LEG	18.4		17.3	40.7	17.6	41.2	15.7	43.4	16.1	38.2
UPPER LEG	11.9	48.9	10.7	43.9	11.0	43.6	9.6	43.3	10.3	37.2
LOWER LEG & FOOT	6.6		6.5	51.9	6.6	53.7	5.9	43.4	5.8	47.5
LOWER LEG	4.6	43.3	4.7	42.0	4.5	43.3	4.5	43.3	4.4	37.1
FOOT	2.0	44.4	1.7	43.4	2.1		1.4	42.9	1.5	44.9

$$x = \frac{L \cdot W_{scale}}{W_{object}}$$

where:

x =   distance between the location of the centre of gravity of the object and the joint pivot axis A

L =   distance between the knife edges

W_{scale} =   reading of the scale

W_{object} =   weight of the object of interest

Note: The scale is set to zero for the initial setup where the table is unloaded.

Van den Bogert (1989) used the photographic suspension method which was initially developed by Braune and Fischer (1889) to determine the centre of gravity. The segment is hung from a rope attached to a point on the segment. The segment and rope are then photographed and the whole process repeated using a second suspension point on the segment. The two photographs are overlaid and the centre of gravity located at the point of intersection of the extrapolations of the ropes.

Values (as found by Harless, 1860; Braune and Fischer, 1889; Fischer, 1906; Dempster, 1955; and Clauser et al., 1969) for the related masses and for the location of the centre of gravity of a segment, as a percentage of the distance along the longitudinal axis of the segment relative to the proximal end, are given in Table 3.7.2.

Moments of inertia

Several methods have been proposed to determine experimentally moments of inertia. The methods use a pendulum or a torsional pendulum approach.

The pendulum method consists of suspending the object of interest from a fixed point, setting it in motion by shifting it a few degrees from its equilibrium position, and measuring the time it takes to swing for one period of oscillation. The moment of inertia I_o about an axis through the point of suspension I_o, is given by:

$$I_o \quad = \quad \frac{W_{air} \cdot h \cdot T^2}{4\pi^2} \qquad\qquad (3.7.10)$$

where:

I_o =   moment of inertia about an axis through the point of suspension

W_{air} =   weight of the object in air

h   =   distance from the centre of gravity of the object to the point of suspension

T   =   period of one oscillation

The moment of inertia for an axis through the centre of mass, I_{CM}, can be determined using the parallel axis theorem:

$$I_{CM} \quad = \quad I_o - mh^2 \qquad\qquad (3.7.11)$$

where:

m   =   mass of the object of interest

Several authors have used pendulum methods to determine the moment of inertia of the whole body and of body segments.

Dempster (1955) screwed the proximal joint centres of a frozen limb segment onto a bar with knife edges on both ends. The segment was supported in a metal frame by the bar while the knife edges sat upon the frame. The segment was allowed to swing freely through an arc of approximately 5° while pivoting on the knife edges. The period of oscillation was measured, and the moment of inertia about the centre of gravity calculated from equations (3.7.10) and (3.7.11). The trunk, with head and neck attached, was frozen and suspended by placing the bar with the knife edges through a horizontal bore hole directed from the centre of one acetabulum to the other. The moment of inertia was determined for the segments. The shoulder portions were held in their natural configurations using plaster of paris moulds while being frozen. The frozen shoulder was then screwed to the bar at the medial end of the clavicle and the moment of inertia determined as above. A summary of the moments of inertia determined by Dempster can be found in Table 3.7.3, along with a comparison to values determined by Hatze (1980).

Table 3.7.3   Experimentally determined values for segmental transverse moment of inertia (based on data from Dempster, 1955; Hatze, 1980; with permission).

SEGMENT	DEMPSTER (1955)	HATZE (1980)
	TRANSVERSE MOMENT OF INERTIA [kgm^2]	TRANSVERSE MOMENT OF INERTIA [kgm^2]
HEAD - NECK	0.0310	0.0337
LEFT ARM	0.0222	0.0203
RIGHT ARM	0.0220	0.0229
LEFT FOREARM	0.0055	0.0086
RIGHT FOREARM	0.0072	0.0093
LEFT HAND	0.0009	0.0010
RIGHT HAND	0.0011	0.0010
LEFT LEG	0.0650	0.0798
RIGHT LEG	0.0620	0.0747
LEFT FOOT	0.0037	0.0051
RIGHT FOOT	0.0040	0.0051

Chandler et al. (1975), determined the three principle moments of inertia and the products of inertia of various body segments using a specimen holder box. The segment to be measured was frozen and strapped into the box. The moment of inertia and products of inertia were determined using the pendulum technique. By swinging the box around the coordinate system's x, y, and z axes, around an xy axis (located in the xy plane), a yz axis (located in the yz plane), and around an xz axis (located in the xz plane), inertia was determined. The distance from the centre of mass of the segment to each of the swing axes was measured using a photographic suspension method. From these measurements, and knowledge of the orientation of the box coordinate system relative to the segment coordinate system, the three principle moments of inertia and direction angles were calculated. This procedure provided an inertial tensor for the centre of gravity.

Later they evaluated the procedure using a solid aluminum bar whose inertial properties were known (Chandler et al., 1975). The six trials that were conducted produced deviations from the theoretical moment of inertia values that ranged from −10.4 to +3.8%. The deviations clustered in the range −3 to +3%, indicating a reasonable degree of accuracy for the method. Their results, however, are not included in Table 3.7.3 because they do not relate directly to the coordinate system used by other authors.

Lephart (1984) refined the method developed by Chandler et al. (1975) by correcting for a systematic error, which over-predicted the oscillation time (therefore overestimating the moment of inertia). With this correction the absolute error for the principle moments of inertia was less than 5%.

Santschi (1963) determined the moments of inertia for 66 living male subjects in eight body positions: standing, standing with arms over head, spread eagle, sitting, sitting with forearms down, sitting with thighs elevated, mercury configuration, and relaxed. The accuracy for the moment of inertia values ranged from 2 to 8%.

Van den Bogert (1989) introduced the *torsional pendulum method*, using a vertically oscillating turntable to measure the moments of inertia of horse segments. The turntable consisted of a vertical shaft, supported by bearings, upon which cross members holding the segment of interest were located. Springs attached to a fulcrum on the shaft maintained the oscillation. A segment was placed on the cross members such that the plane in which it normally moved was horizontal (sagittal plane). Photographs taken from above were used to determine the distance from the centre of gravity to the axis of rotation. The centre of gravity was determined using the photographic suspension method. The moment of inertia, I, in the sagittal plane was calculated from:

$$I_{CG} = k(T^2 - T_0^2) - md^2 \tag{3.7.12}$$

where:

I_{CG}	=	moment of inertia for the segment for an axis through the CG of the segment perpendicular to the sagittal plane
k	=	rotational spring constant divided by $4\pi r^2$
T	=	measured period of oscillation for the table and segment
T_0	=	measured period of oscillation for the unloaded table
m	=	mass of the segment
d	=	distance between the axis of the rotating table and the CG

The rotational spring constant, k, was determined by measuring the oscillation time of eight steel cylinders of known moments of inertia, and calculating the best fit to the measured T and known I values. Using the calibrated spring constant, k, moments of inertia of the calibrated objects could be determined with an accuracy of 0.8%.

Moments of inertia are often important for human movement analysis, especially during sport activities. Consequently, moments of inertia have been estimated for various body positions and body axes. Examples of such estimates (Donskoi, 1975) are illustrated in Fig. 3.7.4 for body positions and rotation axes of the human body that are typical in gymnastics.

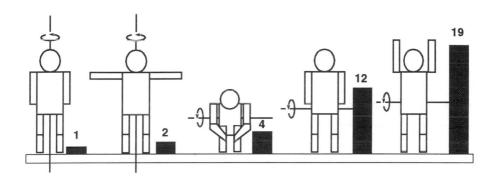

Figure 3.7.4   Moments of inertia (in kgm²) for the human body in selected positions for different axes (from Donskoi, 1975, with permission).

Comprehensive methods

Computerized axial tomography:

Computerized axial tomography (CAT) has been used for determining inertial properties of humans or animals (Rodrigue et al., 1983; Huang and Saurez, 1983). CAT takes a three-dimensional x-ray of a small section of the object of interest. Scans are taken at small intervals along the segment, generating a series of cross-sections of the segment. These cross-sections can be digitized, and the various tissue areas (generally broken down into bone, muscle, and fat) can be calculated. The cross-sectional area multiplied by section width gives section volume. Summations of section volumes yield segment volume. Densities for each tissue type can be taken from dissection measurements. For living people, densities can be taken from published values or calculated directly from the CAT number (Huang and Saurez, 1983) using the equation:

$$p \quad = \quad A \cdot N_{CAT} + B \tag{3.7.13}$$

where:

p = density of the segment of interest
N_{CAT} = CAT number of the tissue as indicated by Huang and Saurez (1983)
A, B = coefficients

A and B can be calculated during a calibration procedure (Huang and Saurez, 1983) from:

$$A = \frac{(p_2 - p_1)}{(n_2 - n_1)}$$

$$B = p_1 + n_1 \frac{(p_2 - p_1)}{(n_2 - n_1)}$$

where:

p_1, p_2 =   mass densities of the two known materials used for the calibration process

n_1, n_2 = N_{CAT} for the same materials, respectively

The densities within the cross-sections can be calculated on a point-by-point basis using this method.

Alternatively, measured or published values can be used for the densities. By combining these values with the volumes calculated from the digitizing process you can determine the inertial properties. Rodrigue and Gagnon (1983) used the CAT method to estimate the densities of 24 forearms from 6 male and 4 female human cadavers. Volume estimates for each tissue type were made by digitizing the CAT scans. These estimates were combined with density values obtained by dissection to determine the overall segment mass, segment volume, and average segment density. Estimates of segment density obtained using the CAT method were within 2% of measured values.

Quick release method:

Fenn (1938) developed the "quick release method", an angular motion method, to determine the moment of inertia of the forearm. This method requires the subject's limb to exert a constant force against a device which is then suddenly released, causing the limb to accelerate. The angular acceleration of the limb is measured and the moment of inertia determined from:

$$I = \frac{F \cdot d}{\alpha}$$

where:

F   =   applied force

d   =   distance between the point of application of the force and the pivot point

α =   measured angular acceleration of the limb

The underlying assumption is that the movement of the limb is purely inertial and that the visco-elastic properties of the limb can be neglected. The mean moment of inertia (in the sagittal plane) of the forearm, as measured by Bouisett and Pertuzon (1967) on 11 subjects, was 0.0599 kgm^2, which is comparable to other published results using different methods (Table 3.7.4). The underlying premise is, therefore, reasonably valid. The variation in results between consecutive tests was only 3%.

Table 3.7.4   **Experimentally determined moments of inertia for the forearm and hand segment about the humeral axis from selected authors using different methods.**

REFERENCES	NUMBER OF SUBJECTS	MOMENT OF INERTIA OF THE FOREARM [kgm²]	
		MEAN	RANGE
BRAUNE AND FISCHER (1889)	2	0.0505	n. a.
FENN (1938)	1	0.0590	n. a.
HILL (1940)	1	0.0277	n. a.
WILKE (1950)	1	0.0530	n. a.
DEMPSTER (1955)	8	0.0577	0.0397 - 0.0852
BOUISSET AND PERTUZON (1968)	11	0.0599	0.0430 - 0.080
KWEE (1971)	4	n.a.	0.03 - 0.051
*ALLUM AND YOUNG (1976)	4	0.0740	0.0690 - 0.0820
PEYTON (1986)	8	0.0646	0.0476 - 0.0862

*These values represent the moment of inertia of the entire limb about the humeral axis.

Relaxed oscillation method:

The basic idea behind the relaxed oscillation method is to apply a single force or an oscillating force to a resting limb, and then to determine the moment of inertia from the resulting response of the limb. The method has been applied to the forearm and is generally done on an apparatus which (for the forearm) consists of a beam upon which a handle is placed at one end, and an elbow cup at the other. The subject to be tested sits beside the apparatus, with the upper arm loosely hanging down and the elbow flexed at 90°. If a force is applied to the arm and beam, they start to oscillate (Peyton, 1986). A small aluminum strip attached to the beam acts as a spring element. A strain gauge mounted on the aluminum strip can be used to measure the movement and the oscillation time, T, which can be used to determine the moment of inertia, I, as:

$$I = \frac{k \cdot T^2}{4\pi^2}$$ (3.7.14)

where:

k   =   relaxation constant determined by statically loading the device and measuring the deflection

T   =   oscillation time (period)

The moment of inertia of the arm is determined by repeating the procedure with the apparatus empty to determine its moment of inertia, and then subtracting this value from the one determined in equation (3.7.14). The assumptions are that the device and forearm are rigidly coupled, that almost all of the elastic deformation is occurring at the spring element, and that the damping of the system is small. Errors are introduced into the method because:

- The axis of rotation of the forearm at the elbow may not coincide with the axis of rotation of the device (up to 5% error).
- Forearm oscillations will be transmitted through the upper arm to the shoulder joint, resulting in a slight underestimation of the moment of inertia.
- The forearm is not a rigid structure, but this potential source of error can be compensated for by having the subject tense the forearm muscles.

The form of the output signal indicates whether or not the subject is preparing his muscles properly (relaxing the joint and tensing the forearm). The mean and the range of forearm values determined with this method compare well to others found in literature on this subject (Table 3.7.4).

The method was also applied with the limb oscillating continuously, driven by a torque motor (Allum and Young, 1976). The lateral force at the control stick, linear (tangential) acceleration, and angular position of the beam were measured with a strain gauge, linear accelerometer, and potentiometer, respectively. The frequency of oscillation varied from 1.4 to 9.0 Hz, and the amplitudes of acceleration were chosen to increase linearly with the frequency. During testing, the subjects were to relax their shoulder musculature and maintain a firm wrist. Five cycles of oscillation in this position were recorded, and the average peak-to-peak accelerations for the five cycles were plotted on the graph. The slope of the graph was used as a measure of the subject's forelimb moment of inertia about the humeral axis. Results are compared for forearm plus hand moments of inertia in Table 3.7.4.

Gamma-scanner method:

The intensity of a gamma ray beam decreases as it passes through a substance. This intensity reduction can be measured and used to determine the density of a substance. Zatziorsky and Seluyanov (1983) measured 100 subjects using the gamma ray scanner technique. The subject was placed in a reclined position on a couch, and the entire body was scanned with a gamma ray emitter moving above the body. A collimated detector, moving underneath the body directly under the emitter, measured the reduced gamma ray radiation. A surface density profile for the entire body was produced in this way. The resulting segment mass is comparable to data presented by Dempster (1955), except for the thigh, which is considerably higher for results from Zatziorsky and Seluyanov's (1983) method (Table 3.7.5).

Table 3.7.5   Relative segment mass comparisons for different selected experimental methods. Relative segment masses are given as a percentage of the total body mass. The values for Hatze were calculated based on volumes from Hatze (1980) and average segment densities from Dempster (1955). The difference in the results between Dempster and Zatziorsky & Seluyanov may be explained by the different subjects studied. Zatziorsky & Seluyanov studied athletes.

SEGMENT	RELATIVE SEGMENT MASS [%]					
	DEMPSTER (1955)	CLAUSER ET AL. (1969)	HATZE (1980)	ZATZIORSKY & SELUYANOV (1983)	JENSEN (1986)	JENSEN (1986)
	ADULTS	ADULTS	ADULTS	ADULTS	12 YEARS OLD	15 YEARS OLD
HEAD	7.9	7.3	6.5	6.9	10.1	6.7
TRUNK	49.6	50.7	45.1	43.5	41.7	41.6
THIGH	9.7	10.3	11.9	14.2	11.0	12.1
SHANK	4.5	4.0	5.5	4.3	5.3	5.6
FOOT	1.4	1.5	1.5	1.4	2.1	2.1
UPPER ARM	2.7	2.6	2.9	2.7	3.2	3.5
LOWER ARM	1.6	1.6	1.8	1.6	1.7	1.7
HAND	0.6	0.7	0.6	0.6	0.9	0.8

Magnetic resonance imaging (MRI):

Magnetic resonance imaging (MRI) is a non-invasive, non-radioactive technique for obtaining cross-sectional images of structures. MRI uses the change in orientation of the magnetic moment of hydrogen nuclei for a particular tissue, that is generated when the tissue is placed in a magnetic field and stimulated with a radio frequency wave. The decay signal is measured by a receiver coil. The received signal is processed and an image is generated. The image shows the cross-section of the object that has been scanned. Martin et al. (1989) applied MRI to determine inertial properties for baboon segments. Eight embalmed baboon segments were procured (four forearms, two upper arms, and two lower legs), mounted in a jig parallel to the scanning direction, and longitudinally scanned in 15 mm increments. The segments were weighed in water and in air, and volumes and average densities were calculated. The centres of gravity were determined using the reaction board method, and the moment of inertia about the traverse axis was measured using the pendulum method. Coordinate data describing the perimeter of the cross-sectional areas of fat, muscle, and bone were determined by digitizing their respective boundaries from the MRI image. The cross-sectional area was converted to real area using a scaling factor obtained from an image of an object with a known size.

The muscle's cross-sectional area was determined by subtracting the fat and bone areas from the segmental area of the particular image. Image sections were represented as frustra of right circular cones in order to determine slice volume. Tissue densities were determined by dissection for muscle and bone, while fat densities were taken from Clauser et al. (1969). MRI tissue volumes multiplied by tissue densities yielded image section masses, which were summed to determine segment masses. The location of the centre of gravity with respect to the proximal end of the segment, was determined by multiplying each individual section mass by the distance from the proximal end to the section's centre of gravity. The moment of inertia of the segment about the traverse axis through the segment's centre of gravity, was determined by the sum of all the moments of inertia of the image sections about the segment's centre of gravity, using the parallel axis theorem.

MRI predicted the location of the centre of gravity relative to the proximal end well (43.3% versus 44.6% measured), and the mean segment density exactly (1.124 kg/m^3) relative to measured values. MRI overestimated segment volumes (633 cm^3 versus 595 cm^3), segment masses (770 g versus 720 g), and moments of inertia (0.00333 kgm^2 versus 0.00321 kgm^2). A thin layer of embalming fluid between the segment and the plastic bag containing the segment is believed to have introduced volumetric errors of +7%, since the fluid looked exactly like skin on the MRI image. When this overestimation was taken into account, MRI tended to underestimate slightly the moment of inertia. Additional errors could occur as a result of the difficulty in distinguishing between cancellous and cortical bone (with largely differing densities) near segment end points, and as a result of tissue movement during moment of inertia measurements.

Overall, the accuracy of the MRI method is better than, or similar to other methods discussed so far. It appears to be an accurate, non-invasive technique for determining inertial properties and will undoubtedly be used often in the future.

Volume contour mapping:

Weinbach (1938) developed volume contour maps from front and right side view photographs of an individual. Using the basic assumption that a horizontal section through the body is elliptical in shape, a contour map is prepared by plotting the cross-sectional area of a given section through the body, on the ordinate of a graph, and plotting the distance from the soles of the feet to the given section, on the abscissa of the graph. The volume of the body is given by the area under the curve. Further graphical manipulation allows the calculation of the centre of gravity and moment of inertia. The reader is referred to Weinbach (1938) for a detailed description of this method.

THEORETICAL METHODS

The experimental methods discussed in the last few pages seem appropriate to determine relative masses, location of segmental centres of mass, and volumes. However, it has been suggested (Hatze, 1986) that the experimental determination of principal moments of inertia, and the inclination of the principal axes, is better estimated using theoretical approaches and anthropomorphic models.

Instead of determining the inertial properties experimentally, the human body can be represented as a mathematical model, allowing the inertial properties to be determined mathematically. Five theoretical approaches will be discussed as examples of alternative modelling methods: the Hanavan model (Hanavan, 1964), the photogrammetric method (Jensen, 1978), the Hatze model (Hatze, 1980), the Yeadon model (Yeadon, 1989a, b), and regression equations.

Mathematical models allow for the calculation of inertial properties of body segments and of the whole body not only in one position but in many positions and, therefore, can be used to simulate human movement. In general, anthropometric measurements are taken of a subject or from photographs which are subsequently digitized, to determine the dimensions used as input into the model.

The Hanavan model

The Hanavan model (1964) is made up of fifteen simple geometric solids (Fig. 3.7.5).

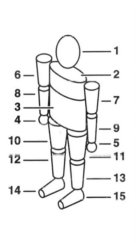

SEGMENT	NUMBER(S)	CONSTRUCTION
HEAD	1	right circular ellipsoid
UPPER TORSO	2	right elliptical cylinder
LOWER TORSO	3	right elliptical cylinder
HAND	4, 5	sphere
UPPER ARM	6, 7	frustum of right circular cone
LOWER ARM	8, 9	frustum of right circular cone
UPPER LEG	10, 11	frustum of right circular cone
LOWER LEG	12, 13	frustum of right circular cone
FOOT	14, 15	frustum of right circular cone

Figure 3.7.5   **Schematic illustration of the Hanavan model with description of specific segments. (from Hanavan, 1964, with permission).**

Twenty-five anthropometric measurements were taken from each individual subject, and used to tailor the geometric solids. Several assumptions underlie the development of the model:

- The human body can be represented by a set of rigid bodies of simple geometric shapes and uniform density.
- The regression equations used for the segment weights are representative of the spectrum of body weights of interest.
- The limbs move about fixed points when the body changes position.
- The limbs are connected by massless hinge joints.

Centre of gravity, mass, and moments of inertia of each segment are derived from the geometry of each of the simple solids. The Barter regression equations (Barter, 1957) were used to calculate the segment weights. If the calculated total body weight did not equal the measured body weight, the difference was distributed proportionally over the segments. Cardan angles were used to describe the position of each of the moveable segments of the body with respect to the body coordinate system.

The model was tested using data provided by Santschi et al. (1963). Anthropometric measurements from 66 subjects were used in the model. Centre of gravity location and principle moments of inertia were calculated for eight body positions and compared to measured data from Santschi et al. (1963). The mathematical model predicted the centre of gravity of the total body within 1.8 cm, and the moments of inertia within 10 % of measured values, which, of course, may be affected with errors.

Photogrammetric method

The human body may be modelled using elliptical zones (Jensen 1976, 1978, 1986, 1989). In this "photogrammetric method", the entire body is sectioned (mainly in the traverse plane) into 2 cm wide zones, and is represented as sixteen segments: head, neck, upper trunk, lower trunk, upper arm, lower arm, hand, upper leg, lower leg, and foot. The division of the segments follows the basic procedures described in Dempster (1955) but with sections modified to fall in the transverse plane. Each zone is represented by an ellipse, which is constructed from the major and minor axes of the two neighboring planes, and is, therefore, a cross-sectional average area of the zone. The centroid and volume of each zone can be calculated directly from the geometry of an elliptical plate. Segment volumes and whole body volumes are simply the sum of the zone volumes. The mass of the segment and whole body is the sum of the mass of each zone (determined by the average density of the zone multiplied by the volume of the zone). Two assumptions are made:

- The zone centroids lie on the link connecting the distal and proximal joint centroids.
- The reference orientations of the segment axes are parallel to the body axes.

Since the ellipses are symmetrical, the segment axes and the body axes become principle axes. The mass moment of inertia can be calculated from the moment of inertia of an elliptical plate, and the body mass moments of inertia can be determined using the parallel axis theorem to sum all the segment moments of inertia.

When the method is applied, a subject lies prone upon blocks in a manner that allows all segments to be viewed from the top and side (i.e., hyperextend the neck, planter flex the foot, and extend the fingers). The segments are positioned to be parallel to the body axis. Horizontal and vertical grids are marked on surfaces adjacent to the subject and are photographed along with the subject. The photographs are digitized and segmented. Whole body as well as segment inertial parameters are then calculated from the digitized records. Segment average densities are taken from published sources.

The body mass calculated using the photogrammetric method is generally within 2 % of the measured body mass, which indicates a high degree of model accuracy for this quantity. The results of a comparison of segment mass calculations and other studies is given in Table 3.7.5. According to Jensen, the application of the method is quick, requiring about ten minutes per subject to mark reference points on the body and to photograph the individual. Manual digitization requires about two hours.

The Hatze model

Hatze (1980) developed a 17 segment model offering the following improvements over past models. It:

- Includes the shoulders as separate entities,
- Differentiates between male and female subjects,
- Considers actual shape fluctuations of each individual segment,
- Accounts for varying densities across the cross-section and along the longitudinal axis of the segment,
- Adjusts the density of certain segments according to the value of a special subcutaneous fat indicator,
- Does not assume that segments are symmetrical,
- Accounts for body morphological changes such as obesity and pregnancy,
- Is valid for children, and
- Models the lungs at a lower density.

Representation of a subject with the model requires 242 separate anthropometric measurements. The greatest simplifying assumption used in the model is the assumption that the segments are rigid. The reader is directed to Hatze (1980) and other publications by Hatze referred to therein for a detailed description of how the segments are built up from various geometric elements, and for a sample of anthropometric measurements and corresponding calculations.

Hatze used four subjects to test the model. The mean error for total body mass predication was 0.26% with a maximum error of 0.52%. This indicates an accurate estimate of body mass. The values for moment of inertia about the transverse plane of one subject in the test group were compared to those for a similarly structured cadaver from the Dempster (1955) study. The results are comparable and appear in Table 3.7.3.

Direct comparisons made against various segment inertia properties measured directly from the subject, indicated that the overall agreement of the model with the experimental data was about 3% with a maximum error of about 5%.

The Yeadon model

Yeadon (1989a, b) described an 11 segment model (in fact, segmental inertial parameters can be calculated for 20 separate body segments) using 40 separate solids. Ninety-five anthropometric measurements were taken on an individual and used to define the shape of the model. In developing his model, Yeadon assumed:

- That the segments are rigid bodies.
- That no movement occurs at the neck, wrists, or ankles.
- That the solids comprising a segment have coinciding longitudinal axes.
- That density values are uniform across each solid.

Segment masses, locations of centroids, and principle moments of inertia about the centroids are calculated based on the geometry of the solid.

The model predicted the total body masses of three subjects to be within 2.3%. If a correction is made for air contained in the lungs (instead of uniform thoracic density) the total body mass error is reduced to approximately 1%. Information about the accuracy of other inertial properties was not supplied by Yeadon (1989a, b).

Regression equations

Regression equations can provide a quick and easy way to determine inertial parameters for an individual. Regression equations are developed from a group of baseline measurements on cadavers or living people, and are used to extrapolate various internal properties to individuals outside the baseline group, thereby, avoiding some of the time consuming procedures discussed so far. Regression equations developed by Barter (1957); Clauser et al. (1969); Hinrichs (1985); Yokoi et al. (1985); Zatziorsky and Seluyanov (1985); Ackland et al. (1988); and Yeadon and Morlock (1989) are discussed below. The actual equations and anthropometric measurements (where applicable) can be found in the relevant publications.

Barter (1957), in an attempt to overcome the limitations imposed by the small sample sizes of the Braune and Fischer (1889) and Dempster (1955) studies, combined the results of all available studies. He took the masses of selected segments, applied statistical regression analysis to them, and derived regression equations that can be used to determine the various segment masses as a function of total body mass. The standard error in his equations ranged from a low of 0.3 kg for both feet, to a high of 2.9 kg for the head, neck, and trunk.

Clauser et al. (1969) developed regression equations to determine the mass and location of the centre of gravity for a segment from various anthropometric measurements. The standard error for segment mass ranged from a low of 0.002 kg for the hand, to a high of 0.93 kg for the head and trunk. The standard error for the centre of gravity location ranged from a low of 0.16 cm for the forearm, to a high of 1.5 cm for the total leg.

Yokoi et al. (1985) applied the photogrammetric method of Jensen (1976) to 184 subjects, 93 boys and 91 girls ranging from 5 to 15 years of age, to determine inertial properties. Instead of developing regression equations, the Kaups index:

$$k = \frac{BW}{H} \cdot 1000$$

where:

k   =   Kaups index
BW =   body weight in kg
H   =   height in cm

along with age, gender, and body type (lean, normal, or overweight) was used to classify the calculated segment inertial properties (segment mass, centre of gravity, and radius of gyration about the centre of gravity) into 16 groups. The authors did not comment on the accuracy of this method except to say that using the above classifications should provide more accurate estimates for children's segmental inertial properties than would using adult properties.

Ackland et al. (1988) used the photogrammetric method to determine inertial properties. A study was done over the course of 5 years using 13 adolescent male subjects. Prediction equations were developed from the results of the photogrammetric method, allowing the determination of segmental inertial properties (mass, centre of gravity location, and moments of inertia) from anthropometric measurements. Direct comparison between the predicted and actual measured values were not made. However, the centre of gravity locations of the thigh (43.6% from proximal end of segment) and leg (41.8%) were compared to various published data.

Hinrichs (1985) developed a set of regression equations for inertial parameters based on data presented by Chandler et al. (1975). The equations predict moments of inertia about the transverse and longitudinal axes passing through the centre of mass, using various anthropometric measurements. These equations are based upon a very small sample size (6 cadavers) and, therefore, should be treated with caution.

Table 3.7.6   Comparison of number of segments, number of anthropometric measurements, errors, and number of subjects used in the original studies by Hanavan (1964), Jensen (1978), Hatze (1980), and Yeadon (1990).

VARIABLE	HANAVAN (1964)	JENSEN (1978)	HATZE (1980)	YEADON (1990)
NUMBER OF SEGMENTS	15	16	17	11
NUMBER OF ANTRHO-POM. MEASUREMENTS	25	--	242	95
NUMBER OF SUBJECTS	66	3	4	3
ERROR MASS	N/A	< 2%	0.26% 0.52%*	2.3% 1.0%**
CM I^{CM}	1.8 cm 10%	N/A N/A	N/A N/A	N/A N/A

*   maximal error
** if correction made for air contained in lungs

Yeadon and Morlock (1989) developed regression equations to determine segmental moments of inertia based on the Chandler data, using linear and non-linear regression equations. Equations developed using right limb cadaver data were tested on left limb data. As well, anthropometric measurements taken from a 10 year old boy were input into the Yeadon mathematical model and used to determine segmental inertial properties. The regression equations predicted the left arm properties with a standard error of 21% and 13% for the linear and non-linear approaches, respectively. The standard error for the modelled inertial properties for the 10 year old boy were 28.6% and 20% for the linear and non-linear approaches respectively. The error for the non-linear approach appears reasonable if one considers the degree to which a 10 year old boy differs from the sample group of 6 cadaver adults.

Zatziorsky and Seluyanov (1985) combined anthropometric measurements with the gamma ray scanner technique and developed a series of regression equations for mass, centre of gravity, and principle moment of inertia for various body segments. The equations are based on data from 100 subjects and yield a standard error of generally less than 10%.

To summarize, regression equations can provide quick values for inertial parameters using only a few anthropometric measurements. These values should, however, only be used as first approximations until they can be more thoroughly evaluated on large sample sizes of actual measurements.

A comparison of the theoretical models developed by Hanavan (1964), Jensen (1978), Hatze (1980), and Yeadon (1990) is summarized for selected variables in Table 3.7.6.

DISCUSSION

An assumption of uniform density within the segment and throughout the body, is often made while estimating inertial properties using experimental methods or mathematical models. It is important to understand the effect this assumption has on the final results. Ackland et al. (1988) conducted a study to investigate the effect a uniform density assumption has on the calculation of inertial properties. In this study, a right leg segment of a cadaver was measured using computerized tomography, and was subsequently dissected to measure mass and density properties. The tomography and dissection results showed that the leg density was not uniform either in cross-section or longitudinally. In order to determine the effect this has on inertial parameters, prior to dissection, the inertial properties, centre of mass, and moment of inertia about the transverse axis were measured using a Mettler balance and the pendulum method, respectively. The photogrammetric method developed by Jensen was used to model the leg with 5 elliptical zones, and inertial properties were calculated.

Using a uniform density obtained from Dempster (1955) the photogrammetric method overestimated the measured cadaver segment mass by 8.2% (due to an overestimation in volume - likely as a result of using only 5 zones), whereas, the segment mass was underestimated by 3.5% using the computerized tomography data. The use of a variable density value (determined from dissection) with computerized tomography data resulted in a mass underestimation of 2.8% relative to the measured data.

With computerized tomography the centre of gravity locations of the cadaver leg were 41.7% of the segment length, measured from the proximal end of the segment using variable measured densities, and 41.2% using constant densities (Dempster data). These compared favorably to the measured location of 41.2%. All locations were closer to the proximal end of the segment than those suggested by Dempster (43.3%) and Braune and Fischer (44.2%). Computer tomography scans conducted on a living person indicated centre of gravity location of 42.2% using variable density, versus 41.9% using a constant density taken from Dempster. The centre of gravity location determined by the photogrammetric method for the live person was at 40.0% of the total length.

The transverse moment of inertia of the cadaver leg using uniform density and computerized tomography data was within 3% of measured values, whereas, the photogrammetric method overestimated the moment of inertia by 13.2% (again possibly due to the

Hatze, H. (1980) A Mathematical Model for the Computational Determination of Parameter Values of Anthropometric Segments. *J. Biomechanics*. **13**, pp. 833-843.

Hatze, H. (1986) *Methoden biomechanischer Bewegungsanalyse*. Öesterreichischer Bundesverlag, Wien.

Hertzberg, H.T., Dupertius, C.V., and Emanuil, J. (1957) *J. Photogrammetric Engineering*. **23**, pp. 942-947.

Hill, A.V. (1940) The Dynamic Constants of Human Muscle. *Proc. R. Soc.* **128 (B)**, pp. 263-274.

Hinrichs, R.N. (1985) Regression Equations to Predict Segmental Moments of Inertia from Anthropometric Measurements: An Extension of the Data of Chandler et al. (1975). *J. Biomechanics*. **18 (18)**, pp. 621-624.

Huang, H.K. and Saurez, F.R. (1983) Evaluation of Cross-sectional Geometry and Mass Density Distribution of Humans and Laboratory Animals Using Computerized Tomography. *J. Biomechanics*. **16 (10)**, pp. 821-832.

Jensen, R.K. (1976) Model for Body Segment Parameters. *Biomechanics V-B* (ed. Komi, P.V.). University Park Press, Baltimore. pp. 380-386.

Jensen, R.K. (1978) Estimation of the Biomechanical Properties of Three Body Types Using a Photogrammetric Method. *J. Biomechanics*. **11**, pp. 349-358.

Jensen, R.K. (1986) Body Segment Mass, Radius, and Radius of Gyration Proportions of Children. *J. Biomechanics*. **19 (5)**, pp. 359-368.

Jensen, R.K. (1989) Changes in Segment Inertial Proportions Between 4 and 20 Years. *J. Biomechanics*. **22 (6-7)**, pp. 529 536.

Kwee, H.H. (1971) *Neuromuscular Control of Human Forearm Movements with Active Dynamic Loading*. Ph.D. Thesis. McGill University, Montreal. Nat'l Library of Canada, Ottawa.

Lephart, S.A. (1984) Measuring the Inertial Properties of Cadaver Segments. *J. Biomechanics*. **17 (7)**, pp. 537-543.

Martin, A.D. (1991) *Variability in the Measures of Body Fat*. Sport and Exercise Sciences Institute, University of Manitoba, Winnipeg.

Martin, P.E., Mungoile, M., Marzke M.W., and Longhill J.M. (1989) The Use of Magnetic Resonance Imaging for Measuring Segment Inertial Properties. *J. Biomechanics*. **22 (4)**, pp. 367-376.

Peyton, A.J. (1986) Determination of the Moment of Inertia of Limb Segments by a Simple Method. *J. Biomechanics*. **19 (5)**, pp. 405-410.

Rodrigue, D. and Gagnon, M. (1983) The Evaluation of Forearm Density with Axial Tomography. *J. Biomechanics*. **16 (11)**, pp. 907-913.

Santschi, W.R., Du Bois, J., and Omoto, C. (1963) *Moments of Inertia and Centres of Gravity of the Living Human Body*. Aerospace Medical Research Laboratory (Report No. TDR-63-36). Wright-Patterson Air Force Base, OH.

van den Bogert A.J. (1989) *Computer Simulation Locomotion in the Horse*. Ph.D. Thesis, University of Utrecht, Netherlands.

Weinbach, A.P. (1938) Contour Maps, Centre of Gravity, Moment of Inertia, and Surface are of Human Body. *Hum. Biol.* **10**, pp. 356-371.

Wilke, D.R. (1950) The Relation Between Force and Velocity in Human Muscle. *J. Physiol. Lond.* **110**, pp. 249-280.

Yeadon, M.R. and Morlock, M. (1989) The Appropriate Use of Regression Equations for the Estimation of Segmental Inertial Parameters. *J. Biomechanics*. **22 (6-7)**, pp. 683-689.

Yeadon, M.R. (1989a) The Simulation of Aerial Movement - II: A Mathematical Model of the Human Body. *J. Biomechanics*. **23 (1)**, pp. 67-74.

Yeadon, M.R. (1989b) The Simulation of Aerial Movement - III: The Determination of the Angular Moment of the Human Body. *J. Biomechanics*. **23 (1)**, pp. 75-83.

Yokoi, T., Shibukawa, K. Ae, M., Ishijima, S., and Hashihara, Y. (1985) Body Segment Parameters of Japanese Children. *Biomechanics IX-B* (eds. Winter, D.A. et al.). Human Kinetics, Champaign, IL. pp. 227-232.

Zatziorsky, V. and Seluyanov, V. (1983) The Mass and Inertia Characteristics of the Main Segments of the Human Body. *Biomechanics VIII-B* (eds. Matsui, H. and Kabayashi, K.). Human Kinetics, Champaign, IL. pp. 1152-1159.

Zatziorsky, V. and Seluyanov, V. (1985) Estimation of the Mass and Inertial Characteristics of the Human Body by Means of the Best Predictive Equations. *Biomechanics IX-B* (eds. Winter D.A., Norman, R., Wells, R.P., Hayes, K.C., and Patla, A.E.). Human Kinetics, Champaign, IL. pp. 233-239.

4   MODELLING

Modelling, the attempt to represent reality, is often used when the understanding of phenomena becomes difficult. A model seems to be a powerful tool to increase the understanding of mechanisms, and has been applied, therefore, quite frequently in many daily and/or research situations. The power of modelling is increasingly recognized in biomechanical research. Modelling, often combined with experimental data, becomes a powerful scientific tool and is discussed in this chapter.

4.1   A NEARLY POSSIBLE STORY

NIGG, B.M.

On his way to class, Expe meets his colleague Theo. While walking together, the two discuss a project currently studied by Expe. The project concerns the vertical velocity of a dropping mass before touching the ground. Researcher Expe says:

> "We finished our experiments and obtained very interesting results. For all masses used in the study, we found a non-linear relation between the dropping height, H, and the vertical velocity of the centre of mass, v, just before reaching the ground surface. The mathematical relation between these two variables is:
>
> $$v \quad = \quad const . H^{0.498}$$
>
> Isn't that interesting?".

Researcher Theo takes a piece of paper and says (fatherly):

> "Dear colleague Expe, we know that the total mechanical energy of a falling object remains constant (if we neglect air resistance). Furthermore, the total translational kinetic energy of the mass immediately before reaching the ground is $\frac{1}{2} \cdot m \cdot v^2$. The rotational energy of the body is assumed to be zero initially and remains so because of the absence of external moments. The loss of potential energy from the highest point to the position immediately before contacting the ground is $m \cdot g \cdot H$. Consequently, if you use the law of conservation of energy you get:
>
> $$m \cdot g \cdot H \quad = \quad \frac{1}{2} \cdot m \cdot v^2 \qquad or: \qquad v = \sqrt{2g} \cdot H^{0.5}$$
>
> This means that your constant is the square root of twice the gravitational acceleration, 2 g, which is about 4.43".

Researcher Expe is surprised and interested because, in fact, he found almost exactly this value for his constant. Patronizingly, his colleague, Theo, adds:

> "My dear friend, it may be of advantage to you, when planning your next project, to contact me, so we can discuss your problem for five minutes. I would be happy to help you".

In this "nearly possible story", which is partly translated and adapted from Hatze (1978), Theo represents a theoretical and Expe an experimental researcher. The story may be exaggerated. However, it contains several messages, the most important being that in some research projects a (theoretical) model may shed light on a problem or help in the prediction of numerical values for variables of interest. Specifically, the use of a mechanics-based model may be helpful in biomechanical studies. The following paragraphs discuss selected examples of biomechanical modelling.

4.2 GENERAL COMMENTS ABOUT MODELLING

NIGG, B.M.

4.2.1 DEFINITIONS AND COMMENTS

Deduction:	Logical reasoning from a known principle to an unknown, from the general to the specific (Webster).
Induction:	Logical reasoning from particular facts or individual cases to a general conclusion (Webster).
Model:	An attempt to represent reality.
Scientific research:	Investigation in some field of knowledge undertaken to discover or establish facts or principles (Webster).
Unique solution:	One and only one solution.
Validation:	(of a model) Providing evidence that a model is strong and powerful. Comment: it is recommended that this term be replaced by evaluation of a model.

4.2.2 SELECTED HISTORICAL HIGHLIGHTS

Selected historical highlights as they relate to biomechanical modelling are summarized in the following paragraphs:

1938	Elftman	Estimated internal forces in the lower extremities and energy changes during walking. Used a mechanical model to make these estimations.
1965	Paul	Developed a deterministic model to estimate forces in the hip joint region.
1971	Passerello & Huston	Presented a simulation model for astronaut reorientation.
1973	Seireg & Arvikar	Published a mathematical model to estimate forces in the lower extremities using the minimization of several objective functions.
1978	Pedotti	Published a non-linear model of the human body with biological input.
1981	Crowninshield & Brand	Presented a possible mathematical solution to the indeterminacy problem of load sharing in muscles crossing a joint, by using optimization procedures.
1981	Hatze	Presented a comprehensive simulation model of the take-off phase in the long jump that relied on mechanical, muscle-physiological, and neuro-physiological knowledge.

1984	Dul	Presented a possible mathematical indeterminacy problem using sequen procedures.
1987	Herzog	Included contractile properties of mu ing internal forces.
1990	Yeadon	Presented a simulation model for and applied it to twisting somersaults

4.2.3 GENERAL CONSIDERATIONS

The term "model" is used in many different ways in daily life. For th
"model" is defined as:

A model is an attempt to r

Usually a model is a simplified representation of reality. However, so
els seem more "complicated" than reality (see section 4.2.6).

Models are constantly developed in daily life and research, and it se
human nature to try to overcome our inability to cope with complex si
simplified models. Examples for models in a general sense can be ta
fields, including:

Daily life	- he is a bad boy
	- she is an angel
Religion	- theism
	- deism
Physics	- Coulomb friction
	- Bohr's model of the atom
Biomechanics	- effective mass model (Denoth, 1980)
	- 160 element model (Seireg and Arvikar, 197
	- simple models of walking and jumping (Ale

4.2.4 INFORMATION USED TO CONSTRUCT A MOD

The construction of models relies on two types of information: *kno*
tem being modelled, and *experimental data* that constitute system inpt
Using knowledge of the system being modelled, to move from genera
cifics, is a deductive process. Using experimental data, in an attempt to
conclusion that explains the data, is an inductive process. A deductive s
a given problem with clearly defined assumptions yields a logically de
is typically a unique solution. An inductive solution process for a give
probable result, which is typically not a unique solution, because an
models satisfy the observed input-output relationship (Fig. 4.2.1).

At first glance, Fig. 4.2.1 would seem to imply that deduction shou
as possible, since deduction typically provides a unique solution. Be
the "knowledge" on which a deductive approach is built may merely

4.2   GENERAL COMMENTS ABOUT MODELLING

NIGG, B.M.

4.2.1 DEFINITIONS AND COMMENTS

Deduction:

Logical reasoning from a known principle to an unknown, from the general to the specific (Webster).

Induction:

Logical reasoning from particular facts or individual cases to a general conclusion (Webster).

Model:

An attempt to represent reality.

Scientific research:

Investigation in some field of knowledge undertaken to discover or establish facts or principles (Webster).

Unique solution:

One and only one solution.

Validation:

(of a model) Providing evidence that a model is strong and powerful.
Comment: it is recommended that this term be replaced by evaluation of a model.

4.2.2 SELECTED HISTORICAL HIGHLIGHTS

Selected historical highlights as they relate to biomechanical modelling are summarized in the following paragraphs:

1938   Elftman

Estimated internal forces in the lower extremities and energy changes during walking. Used a mechanical model to make these estimations.

1965   Paul

Developed a deterministic model to estimate forces in the hip joint region.

1971   Passerello & Huston

Presented a simulation model for astronaut reorientation.

1973   Seireg & Arvikar

Published a mathematical model to estimate forces in the lower extremities using the minimization of several objective functions.

1978   Pedotti

Published a non-linear model of the human body with biological input.

1981   Crowninshield & Brand

Presented a possible mathematical solution to the indeterminacy problem of load sharing in muscles crossing a joint, by using optimization procedures.

1981   Hatze

Presented a comprehensive simulation model of the take-off phase in the long jump that relied on mechanical, muscle-physiological, and neuro-physiological knowledge.

1984	Dul	Presented a possible mathematica indeterminacy problem using seque procedures.
1987	Herzog	Included contractile properties of m ing internal forces.
1990	Yeadon	Presented a simulation model for and applied it to twisting somersault

4.2.3 GENERAL CONSIDERATIONS

The term "model" is used in many different ways in daily life. For t "model" is defined as:

A model is an attempt to

Usually a model is a simplified representation of reality. However, so els seem more "complicated" than reality (see section 4.2.6).

Models are constantly developed in daily life and research, and it se human nature to try to overcome our inability to cope with complex si simplified models. Examples for models in a general sense can be ta fields, including:

Daily life	- he is a bad boy
	- she is an angel
Religion	- theism
	- deism
Physics	- Coulomb friction
	- Bohr's model of the atom
Biomechanics	- effective mass model (Denoth, 1980)
	- 160 element model (Seireg and Arvikar, 197
	- simple models of walking and jumping (Alex

4.2.4 INFORMATION USED TO CONSTRUCT A MOD

The construction of models relies on two types of information: *know* tem being modelled, and *experimental data* that constitute system inpu Using knowledge of the system being modelled, to move from general cifics, is a deductive process. Using experimental data, in an attempt to a conclusion that explains the data, is an inductive process. A deductive so a given problem with clearly defined assumptions yields a logically deri is typically a unique solution. An inductive solution process for a given probable result, which is typically not a unique solution, because an i models satisfy the observed input-output relationship (Fig. 4.2.1).

At first glance, Fig. 4.2.1 would seem to imply that deduction should as possible, since deduction typically provides a unique solution. Be av the "knowledge" on which a deductive approach is built may merely a

4.2 GENERAL COMMENTS ABOUT MODELLING

NIGG, B.M.

4.2.1 DEFINITIONS AND COMMENTS

Deduction:

Logical reasoning from a known principle to an unknown, from the general to the specific (Webster).

Induction:

Logical reasoning from particular facts or individual cases to a general conclusion (Webster).

Model:

An attempt to represent reality.

Scientific research:

Investigation in some field of knowledge undertaken to discover or establish facts or principles (Webster).

Unique solution:

One and only one solution.

Validation:

(of a model) Providing evidence that a model is strong and powerful.
Comment: it is recommended that this term be replaced by evaluation of a model.

4.2.2 SELECTED HISTORICAL HIGHLIGHTS

Selected historical highlights as they relate to biomechanical modelling are summarized in the following paragraphs:

1938 Elftman

Estimated internal forces in the lower extremities and energy changes during walking. Used a mechanical model to make these estimations.

1965 Paul

Developed a deterministic model to estimate forces in the hip joint region.

1971 Passerello & Huston

Presented a simulation model for astronaut reorientation.

1973 Seireg & Arvikar

Published a mathematical model to estimate forces in the lower extremities using the minimization of several objective functions.

1978 Pedotti

Published a non-linear model of the human body with biological input.

1981 Crowninshield & Brand

Presented a possible mathematical solution to the indeterminacy problem of load sharing in muscles crossing a joint, by using optimization procedures.

1981 Hatze

Presented a comprehensive simulation model of the take-off phase in the long jump that relied on mechanical, muscle-physiological, and neuro-physiological knowledge.

1984	Dul	Presented a possible mathematical indeterminacy problem using sequent... procedures.
1987	Herzog	Included contractile properties of mu... ing internal forces.
1990	Yeadon	Presented a simulation model for a... and applied it to twisting somersaults.

4.2.3 GENERAL CONSIDERATIONS

The term "model" is used in many different ways in daily life. For th... "model" is defined as:

A model is an attempt to re...

Usually a model is a simplified representation of reality. However, som... els seem more "complicated" than reality (see section 4.2.6).

Models are constantly developed in daily life and research, and it se... human nature to try to overcome our inability to cope with complex sit... simplified models. Examples for models in a general sense can be tak... fields, including:

Daily life	- he is a bad boy
	- she is an angel
Religion	- theism
	- deism
Physics	- Coulomb friction
	- Bohr's model of the atom
Biomechanics	- effective mass model (Denoth, 1980)
	- 160 element model (Seireg and Arvikar, 197...
	- simple models of walking and jumping (Alex...

4.2.4 INFORMATION USED TO CONSTRUCT A MOD...

The construction of models relies on two types of information: *kno...* tem being modelled, and *experimental data* that constitute system inpu... Using knowledge of the system being modelled, to move from general... cifics, is a deductive process. Using experimental data, in an attempt to... conclusion that explains the data, is an inductive process. A deductive s... a given problem with clearly defined assumptions yields a logically der... is typically a unique solution. An inductive solution process for a give... probable result, which is typically not a unique solution, because an i... models satisfy the observed input-output relationship (Fig. 4.2.1).

At first glance, Fig. 4.2.1 would seem to imply that deduction shoul... as possible, since deduction typically provides a unique solution. Be a... the "knowledge" on which a deductive approach is built may merely a...

4.2   GENERAL COMMENTS ABOUT MODELLING

NIGG, B.M.

4.2.1 DEFINITIONS AND COMMENTS

Deduction:	Logical reasoning from a known principle to an unknown, from the general to the specific (Webster).
Induction:	Logical reasoning from particular facts or individual cases to a general conclusion (Webster).
Model:	An attempt to represent reality.
Scientific research:	Investigation in some field of knowledge undertaken to discover or establish facts or principles (Webster).
Unique solution:	One and only one solution.
Validation:	(of a model) Providing evidence that a model is strong and powerful. Comment: it is recommended that this term be replaced by evaluation of a model.

4.2.2 SELECTED HISTORICAL HIGHLIGHTS

Selected historical highlights as they relate to biomechanical modelling are summarized in the following paragraphs:

1938	Elftman	Estimated internal forces in the lower extremities and energy changes during walking. Used a mechanical model to make these estimations.
1965	Paul	Developed a deterministic model to estimate forces in the hip joint region.
1971	Passerello & Huston	Presented a simulation model for astronaut reorientation.
1973	Seireg & Arvikar	Published a mathematical model to estimate forces in the lower extremities using the minimization of several objective functions.
1978	Pedotti	Published a non-linear model of the human body with biological input.
1981	Crowninshield & Brand	Presented a possible mathematical solution to the indeterminacy problem of load sharing in muscles crossing a joint, by using optimization procedures.
1981	Hatze	Presented a comprehensive simulation model of the take-off phase in the long jump that relied on mechanical, muscle-physiological, and neuro-physiological knowledge.

1984	Dul	Presented a possible mathematical indeterminacy problem using sequent... procedures.
1987	Herzog	Included contractile properties of mus... ing internal forces.
1990	Yeadon	Presented a simulation model for a... and applied it to twisting somersaults.

4.2.3 GENERAL CONSIDERATIONS

The term "model" is used in many different ways in daily life. For this... "model" is defined as:

A model is an attempt to rep...

Usually a model is a simplified representation of reality. However, som... els seem more "complicated" than reality (see section 4.2.6).

Models are constantly developed in daily life and research, and it seen... human nature to try to overcome our inability to cope with complex situ... simplified models. Examples for models in a general sense can be take... fields, including:

Daily life	- he is a bad boy
	- she is an angel
Religion	- theism
	- deism
Physics	- Coulomb friction
	- Bohr's model of the atom
Biomechanics	- effective mass model (Denoth, 1980)
	- 160 element model (Seireg and Arvikar, 1973)
	- simple models of walking and jumping (Alexa...

4.2.4 INFORMATION USED TO CONSTRUCT A MODEL

The construction of models relies on two types of information: *knowle...* tem being modelled, and *experimental data* that constitute system inputs ... Using knowledge of the system being modelled, to move from general pr... cifics, is a deductive process. Using experimental data, in an attempt to ar... conclusion that explains the data, is an inductive process. A deductive solu... a given problem with clearly defined assumptions yields a logically deriv... is typically a unique solution. An inductive solution process for a given p... probable result, which is typically not a unique solution, because an infi... models satisfy the observed input-output relationship (Fig. 4.2.1).

At first glance, Fig. 4.2.1 would seem to imply that deduction should b... as possible, since deduction typically provides a unique solution. Be awa... the "knowledge" on which a deductive approach is built may merely am...

Figure 4.2.1   Schematic illustration of the connection between the types of basic information, the method used, and the expected results.

nary assumptions. Changing from inductive to deductive methods may simply move the uncertainty from, lying in the number of possible answers, to, lying in the number of possible assumptions.

4.2.5 SIMPLIFICATION

One important aspect of developing a model is to decide what should be neglected and what should be included. It seems obvious that all important aspects should be included, and that all unimportant aspects may be neglected. However, attempting to list rules and guidelines is not easy. One may suggest that, in general, simpler is better. However, there is always the danger that a simple model may not agree with reality since some aspects have been left out. The development of a model and the choice of including or excluding certain aspects seems to be more of an art than a science.

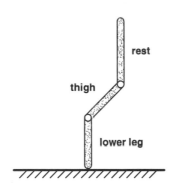

Figure 4.2.2   Illustration of a simple model for the determination of the magnitude of impact forces and moments at the tibio-femoral joint (from Denoth, 1980, with permission).

Here is an example that demonstrates simplification. Let us assume that the purpose of a project is to develop a model to estimate the magnitude of the impact forces at the tibio-femoral joint, during landing in heel-toe running. Obviously, the colour of the eyes is not important and can be neglected. It may even be assumed that the size and mass of the head and the mass of the foot, have little influence on the magnitude of the impact forces and impact moments at the tibio-femoral joint. Finally, one may assume that the muscles can be neglected when estimating the magnitude of the forces and moments at the tibio-femoral joint (Nigg, 1986). These assumptions and simplifications lead to a simple model of the "human body", a three segment model comprising a lower leg, a thigh, and a third segment, the "rest of the body" (Fig. 4.2.2) which are connected by hinge joints (Denoth, 1980).

4.2.6 THE PURPOSE OF A MODEL

Models are used for actual or theoretical (simulated) situations. Results and conclusions from models, for both applications, can be used for two specific purposes:

> **Purpose a:**   **to increase knowledge and insight about reality, and**
>
> **Purpose b:**   **to estimate or predict variables of interest.**

A model may provide insight into the "true" nature of the system of interest, and increase our understanding of the interaction of the variables important for the situation described by the model. Furthermore, a model may provide an accurate estimation of the variable(s) of interest. The two purposes of a model seem to be clear. However, there are at least three comments to be made in this context:

(1)   Purpose (a) suggests that results and conclusions from a model may be used to improve knowledge of, or insight into, a complex situation. However, knowledge and insight are essential for the development of a model. The fact that insight and knowledge are prerequisites for the development of a model, but are also the purpose of using a model, seems contradictory. To be more specific, we may state: a model provides insight into the relationships among variables, and may indicate how these relationships are governed.

(2)   A model can only provide information that has been implicitly implemented into the model. A model cannot come up with the suggestion that, for instance, a set of variables, x_i ,...., x_n, should be added to the original design of the model, if these variables have not been included in the model. However, results and conclusions from a model can be used to study how the variables that have been considered important for the question of interest, are sensitive to changes in internal or external conditions (increase the general understanding = purpose a).

(3)   Naively, one may think that a model is good if the results are accurate. In order to understand the problem associated with such a view, an example from history is discussed:

> Theories concerning the solar system have changed with time. About 300 years before Christ, Aristarchos of Samos proposed that the sun was fixed at the centre of the "universe" and the earth revolved around the sun in a circular orbit. He also suggested that the stars appeared fixed in position

because their distances from the sun were tremendous compared to the distance from the earth to the sun. Some of the early astronomers accepted this heliocentric theory of the universe. In about 150 A.D., Claudius Ptolemeus proposed a theory, called the Geocentric Theory, which suggested that the earth was at the centre of the universe and that the sun moved around the earth in a circle. This theory (model) succeeded in explaining and predicting planetary motion with the degree of measurement accuracy then possible. Ptolemeus' theory was taught and used extensively from the 2nd to the 16th century. Copernicus (1472-1543) revived and extended the heliocentric theory of Aristarchos, starting a revolution in scientific thought that was carried forward by Kepler, Galileo, and Newton. However, for a substantial period of time it was more accurate to calculate the path of planets as viewed from the earth, or one's own position in navigation, by using the "old" geocentric system than by using the "new" heliocentric system.

This example illustrates that it may be possible to calculate or estimate variables of interest sufficiently accurate by using a model whose conceptual construction does not correspond to reality, and that quality criteria for a model depend on the actual purpose of the model.

4.2.7 THE VALIDATION OF A MODEL

It is common in scientific work that a "validation" of a model is required. The validation process will be discussed in the following section.

The word *valid* derives from the Latin word validus which translates as strong or powerful.

> **To validate a model can be defined as to provide evidence that the model is strong and powerful for the task for which it has been designed.**

Validation of a model consists in the provision of a set of cases for which the results of the model correspond to reality. Such cases support the statement that the model is strong and powerful for the task for which it has been designed. The evaluation of the power of a model can be done in three ways: direct measurements, indirect measurements, and trend measurements.

DIRECT MEASUREMENTS

In some cases it is possible to measure the estimated variable in a limited set of experiments (e.g., force measurements in the Achilles tendon, Komi et al., 1987). Consequently, one can compare the estimated force with the measured force. If the experimental and the estimated results agree within an acceptable range, one may feel comfortable with the model. Agreement between estimated and measured results, with changed boundary or initial conditions, may further increase the confidence in a model. However, these agreements are no guarantee that the structure of the model is similar to the real world, even if the model is commonly used in many applications.

INDIRECT MEASUREMENTS

The comparison between estimated and measured results can often not be made because it is often impossible to make experimental measurements of internal variables (e.g., force distribution in the patella-femoral joint). Direct validation, therefore, cannot be performed. However, in some cases, measurements of another (measurable but not needed) variable may be made, and compared with the value predicted for this variable by the model. As an example, the forces in the tibio-femoral joint could be calculated by using the gravitational and the inertial forces produced by the segments above the knee. Additionally, the gravitational and inertial forces could be used to calculate the ground reaction force. This calculated ground reaction force could be compared with the measured ground reaction force. If the two ground reaction forces were close, one could feel comfortable about the potential of the model to be able to estimate the forces in the tibio-femoral joint. However, it would still be possible that the estimated forces in the tibio-femoral joint were substantially wrong. Other examples of indirect measurements include EMG measurements for the estimation of muscle forces.

The limitations of such indirect "validation" procedures is illustrated with the following example. Several models estimating forces in human skeletal muscles (Pedotti et al., 1978; Crowninshield and Brand, 1981; Dul et al., 1984; Herzog, 1987) were indirectly "validated" or evaluated when they were first presented. However, when applied to the same situation, these models predicted substantially different muscle forces (Herzog and Leonard, 1991). Most, if not all, of these models estimated muscle forces incorrectly. The example suggests that the importance of indirect "validation" should not be overestimated. It is nothing more than an often weak evaluation of the potential of a model to do the task for which it was developed.

TREND MEASUREMENTS

An agreement between facts and results from a model is not necessarily an agreement in the values of the variables. One purpose of a model is to describe the general behaviour of the system of interest. In this case, agreement may consist of similar trends and developments. For example, a model might say that if x increases linearly, y will increase quadratically. If the purpose of a model is to improve our understanding of how variables in the model interact, the quality of the model depends on how well the trends predicted, agree with the trends measured.

FINAL COMMENTS

To validate a model means to obtain evidence that the proposed model is strong and powerful for the purpose for which it has been developed. Validation may lead to increased confidence in a model, but it never confirms that the model corresponds to reality. A difference between the results of an estimation given by a model, and an experimental measurement, may indicate that the conceptual structure or the detailed composition of the model are inadequate.

The word "validation" is not well-defined and has developed different meanings over time. It may be appropriate to replace the term "validation of a model" with *"evaluation of a model"*. Each model should be evaluated by providing direct or indirect evidence that the model is strong and powerful for the purpose for which it was developed. Such an evaluation indicates whether the results of a model are strongly supported by evidence, or whether the supporting evidence is rather weak. In further chapters the expression "evaluation" is used whenever the strength and power of a model is assessed for the task it has been designed.

4.2.8 TYPES OF MODELS

A model provides information about a relationship between cause and effect. A model can be used in one of two ways, direct or inverse. In the direct use, the model proceeds from cause to effect and, typically, yields a unique solution. In the inverse use a model attempts to move from the effect to the cause(s) and, typically, yields several possible solutions (not unique).

Models can be classified in various ways. One way is to call some models intuitive, and others abstract (Fischbein, 1987). Another is to recognize inductive and deductive models (Kemeny, 1959). The taxonomy of biomechanical models used here has four groups: analytical, semi-analytical, black box, and conceptual.

ANALYTICAL (DEDUCTIVE) MODELS

Analytical models are developed on the basis of (real or speculated) knowledge and insight. Using our understanding of the human musculo-skeletal system, its physiology and physics (mechanics), a structure can be developed that can be described with a mathematically deterministic model. The advantage of such a model is that it has a unique solution that is independent of the selected mathematical procedure. The critical point in the development of such a model is the selection of the assumptions and simplifications.

Selected examples for analytical models in biomechanical applications include the model for running to determine optimal surface compliance (McMahon and Greene, 1979), the three segment model for the determination of the impact forces in the tibio-femoral joint (Denoth, 1980), the wobbling mass model used for the estimation of the actual bone-to-bone forces and joint moments of the lower extremities (Gruber et al., 1987), and the series of models for sequential joint extension in jumping (Alexander, 1989).

SEMI-ANALYTICAL MODELS

Semi-analytical models are based on knowledge and insight. However, the system of interest is too complicated to make it mathematically deterministic with the available basic information. The mathematical description of the system of interest has more unknowns than equations. Consequently, more assumptions are added.

Biomechanical examples for semi-analytical models include most models for estimating muscle forces and bone-to-bone forces from external kinematics and kinetics (Pedotti et al., 1978; Crowninshield and Brand, 1981; Winter, 1983; Dul et al., 1984; Herzog, 1987; Morlock and Nigg, 1991).

BLACK BOX MODELS

In black box models (also called "regression models"), a set of mathematical functions are used to determine the input-output relation. In the first step, an appropriate (but perhaps arbitrary) mathematical function is determined that best describes the relationship between known input and output pairs. In a second step, the determined mathematical function is used to predict output values for a given set of input values.

The black box approach has two applications. First, it can be used to estimate quantities that cannot be measured in certain situations. The aspect of interest is the accuracy of the output if compared to the correct value. The way the output is determined, the structure of the mathematical formula and the functional relationship between input and output, are not of interest. Second, the black box approach can be used to provide insight into possible functional relationships between input and output. A mathematical function is used to describe the input-output relationship. A subsequent analysis of the mathematical components that determine the input-output relationship may provide insight into the general structure of the relationship studied.

A biomechanical example for the black box model, in the sense of the first approach, could be the prediction of the length of a long jump based on variables such as approach velocity and take-off angle (Ballreich and Brüggemann, 1986). Biomechanical examples of using the black box model to explore functional relationships, between input and output, could be the determination of the dominant material properties (elasticity, visco-elasticity, etc.) of the soft tissue under the human heel performed by Denoth (1980). He used a complex non-linear mathematical function to approximate the experimentally measured force-deformation curves, and then used the dominant terms of this function to describe the material.

CONCEPTUAL MODELS

Conceptual models consist of a hypothesis (which is based on insight or speculation), and procedures capable of supporting or disproving the tested hypothesis. These procedures can be experimental or theoretical. In biological research, experimental procedures for disproving or supporting a hypothesis are frequently used. The advantage of the conceptual approach is that larger concepts can be subdivided into smaller steps that can be treated individually. The disadvantage of the conceptual approach is that a hypotheses can never be proven, only disproven. Consequently, several pieces of evidence must be accumulated to provide enough support for a concept. Although this approach is common in the biological sciences, biomechanical examples for this approach are not easy to find.

A classification of tools (a model is a kind of a tool) is associated with difficulties since many appropriate possibilities exist for an organization of these tools into groups with similar characteristics. Grouping of tools typically depends on the decision, which of the different characteristics one selects as relevant for the subdivision. In this text, the characteristics describing the mathematical solution has been arbitrarily chosen as criterion. However, other approaches may be as appropriate as the one presented here.

4.2.9 DESCRIPTIVE, EXPERIMENTAL, AND/OR ANALYTICAL RESEARCH AND MODELLING

> Science must start with facts and end with facts, no matter what theoretical structures it builds in between.

This statement by Einstein, underlines the fact that scientific activities include a sequence of steps (Kemeny, 1959). First, a scientist is an observer. Second, a scientist describes what is observed. Third, a scientist develops theories that allow predictions. Fourth, a scientist compares the predictions with the facts. This process is cyclic in its structure (Fig. 4.2.3).

The main characteristic of scientific activities is the cyclic interaction between facts and theory. In the world of biomechanics, researchers with instruments such as EMG sensors, force plates, and video cameras observe and describe as accurately as possible and/or as needed, the situation of interest. Based on these observations, theories (models) are developed using induction. Such theories may be a set of mathematical formulas that are, in turn, used deductively to predict the outcome of a theoretically possible future "experiment". The prediction will then be compared with the actual results (facts) of an experiment. This process then continues because the new facts and information are the starting point for new cycles of induction and deduction.

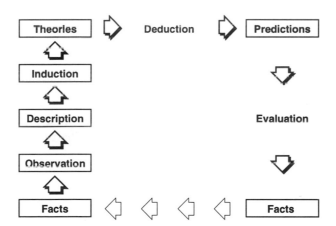

Figure 4.2.3   Schematic illustration of the cyclic interaction between facts and theory in scientific activities (from Kemeny, 1959, with permission).

Several comments may be of interest in this context:

(1)   It may be wise to start a research project by describing accurately the actual situation (in biomechanics the actual kinetics, kinematics, material properties, etc.).

Early developments in established sciences such as physics, chemistry, or biology have included long periods of observations and description of these observations. Sometimes errors have been made because observations have not been made well and were not accurately documented.

(2)   Quantum leaps in sciences were often a result of brilliant theories or models (Newton, Einstein, Bohr) some of which, of course, were based on a wealth of observation.

(3)   The actual (mathematical) development of a model in itself is not research, but corresponds to the development of a tool. Measuring devices such as force platforms, pressure distribution insoles, accelerometers, indwelling electrodes for EMG measurement, and video cameras, as well as models, are tools, and their development is technology.

4.2.10 GENERAL PROCEDURES IN MODELLING

A set of general steps are common for developing a biomechanical model. These steps are:

(1)   Definition of the question(s) to be answered.
(2)   Definition of the system of interest.
(3)   Review of existing knowledge.
(4)   Selection of procedure (model) to be applied.
(5)   Simplifications and assumptions.
(6)   Mathematical formulation.
(7)   Mathematical solution (using appropriate input data).
(8)   Evaluation of the model.
(9)   Discussion, interpretation, and application of the results.
(10)   Conclusions.

This procedure is general and can be applied for any form of models, including inverse dynamic models, black box models, simulation models, etc. The specific steps are discussed further in the next paragraphs.

STEPS (1), (2), (3), (9), AND (10)

These steps are common to every research project and are not specific for the development of a model. The definition of the question to be answered is the cornerstone of any reasonable research project. All the following steps are designed to answer this question. The definition of the "body" of interest is the next step in narrowing the field of future work.

The review of existing knowledge, which usually includes a review of the relevant literature, helps a researcher to decide whether or not further steps have to be taken to answer the question of interest, or whether the question or parts of it have already been answered. It is often suggested that a review of the literature is the most important initial step in a research project. Many researchers suggest that a thorough study of the available literature is the "conditio sine qua non", the necessary condition for a successful start of a research project. However, other researchers argue that a thorough review of the literature should only be started when your own approach to the problem (= project design) has been developed. This latter strategy, they argue, has the advantage that innovative thinking is not inhibited by the study of other solutions, and, they argue, this approach is more likely to produce new and different approaches.

Discussing and interpreting results and forming conclusions are, of course, the most exciting and interesting part of any research project, independent of the applied methodology.

STEP (4)

Based on the first three steps, a decision has to be made about what methodology to apply to solve the question of interest. The decision relevant for our purposes is to develop a model. If this is the approach taken, the type of model to be developed must be specified. There are at least two philosophies about how to proceed. One is to start with a model as simple as possible and to make it more complex when needed (i.e., if the simple model was not able to provide the answer of interest). The other is to start with a complex model and to exclude sequentially it's superfluous part(s).

Models range from simple, one-particle models to complex, multi-segment ones with physiological, anatomical, and neurological components. The selection process is crucial, since the task is not to find the most elaborate model but the most appropriate one.

STEP (5)

Making the simplifications and assumptions is a challenging and difficult step in the development of a model. They incorporate the modeller's philosophy and understanding of the situation. Two steps are needed in forming them. First, what to include and what to neglect, and what general assumptions to be made must be decided. Second, evidence and reasons why these assumptions are reasonable must be supplied. Note that evidence and support may range from experimental data to blunt statements such as "to simplify the mathematical calculations".

STEPS (6) AND (7)

The mathematical formulation and solution are usually simple, straightforward steps. A few standard methods are available that basically serve the same functions. Some are more elegant than others. However, the different mathematical methods should not usually influence the outcome of the results.

In order to make use of the model, it is necessary to have appropriate input data. This could be based upon previously collected data or may be completely hypothetical. In the latter case, some caution should be exercised to ensure that the proposed input data is reasonable. Using appropriate input data, the model is applied to produce the output data. This may be done for a large number of simulations in order to generate a general picture of how the system responds.

STEP (8)

Mathematical formulations of modelled situations can provide every possible result. It is important to provide evidence for the appropriateness of the findings derived from the model. This is not proof as discussed earlier, but rather some support for the confidence level in the output of the model. Note that output is a general term. It may be a number, a set of time histories, or a trend.

4.2.11 REFERENCES

Alexander, R.M. (1989) Sequential Joint Extension in Jumping. *Human Movement Science.* **8,** pp. 339-345.

Alexander, R.M. (1992) Simple Models of Walking and Jumping. *Human Movement Science.* **11,** pp. 3-9.

Andrews, J.G. (1974) Biomechanical Analysis of Human Motion. *Kinesiology IV.* Amer. Assoc. for Health, Phys. Ed., & Rec., Washington, D.C. pp. 32-42.

Ballreich, R. and Brüggemann, G. (1986) Biomechanik des Weitsprungs. *Biomechanik der Leichtathletik* (eds. Ballreich, R. and Kuhlow, A.). pp. 28-47.

Crowninshield, R.D. and Brand, R.A. (1981) A Physiologically Based Criterion of Muscle Force Prediction in Locomotion. *J. Biomechanics.* **14 (11),** pp. 793-801.

Dempster, W.T. (1958) Analysis of Two-handed Pulls Using Free Body Diagrams. *J. Appl. Physiology.* **13 (3),** pp. 469-480.

Dempster, W.T. (1961) Free Body Diagrams as an Approach to the Mechanics of Human Posture and Motion. *Biomechanical Studies of the Musculo-skeletal System* (ed. Evans, F.G.). Thomas, Springfield, IL. pp. 81-135.

Denoth, J. (1980) Ein Mechanisches Modell zur Beschreibung von passiven Belastungen. *Sportplatzbeläge* (eds. Nigg, B.M. and Denoth, J.). pp. 45-53.

Denoth, J. (1980) Materialeigenschaften. *Sportplatzbeläge* (eds. Nigg, B.M. and Denoth, J.). pp. 54-67.

Dul, J., Johnson, G.E., Shiavi, R., and Townsend, M.A. (1984) Muscular Synergism - II: A Minimum-fatigue Criterion for Load Sharing Between Synergistic Muscles. *J. Biomechanics.* **17 (9),** pp. 675-684.

Elftman, H. (1938) Forces and Energy Changes in the Leg During Walking. *Am. J. Physiology.* **125 (2),** pp. 339-356.

Fischbein, E. (1987) *Intuition in Sciences and Mathematics.* D. Reidel Publishing Company, Dordrecht, Netherlands.

Gruber, K., Denoth, J., Stüssi, E., and Ruder, H. (1987) The Wobbling Mass Model. *Biomechanics X-B* (ed. Jonsson, B.). pp. 1095-1099.

Gurlanik, D.B. (ed.) (1979) *Webster's New World Dictionary.* William Collins Publishers, Ohio.

Hatze, H. (1978) Sportbiomechanische Modelle und myokybernetische Bewegungsoptimierung - Gegenwartsprobleme und Zukunftsaussichten. *Sportwissenschaft.* **4,** pp. 1-13.

Hatze, H. (1981) A Comprehensive Model for Human Motion Simulation and its Application to the Take-off Phase of the Long Jump. *J. Biomechanics.* **14 (3),** pp. 135-142.

Herzog, W. (1987) Individual Muscle Force Estimations Using a Non-linear Optimal Design. *J. Neuroscience Methods.* **21,** pp. 167-179.

Herzog, W. and Leonard, T.R. (1991) Validation of Optimization Models that Estimate the Forces Exerted by Synergistic Muscles. *J. Biomechanics.* **24 (S1),** pp. 31-39.

Kemeny, J.G. (1959) *A Philosopher Looks at Science.* D. van Nostrand, Princeton, NJ.

Komi, P.V., Salonen, M., Jarvinen, N., and Kokko, O. (1987) In Vivo Registration of Achilles Tendon Forces in Man. *Int. J. Sports Medicine.* **8,** pp. 3-8.

McMahon, T.A. and Greene, P.R. (1979) Influence of Track Compliance on Running. *J. Biomechanics.* **12 (12),** pp. 893-904.

Morlock, M. and Nigg, B.M. (1991) Theoretical Considerations and Practical Results on the Influence of the Representation of the Foot for the Estimation of Internal Forces with Models. *Clinical Biomechanics.* **6,** pp. 3-13.

Nigg, B.M. (1986) Biomechanical Aspects of Running. *Biomechanics of Running Shoes* (ed. Nigg, B.M.). Human Kinetics Pub. Inc., Champaign, IL. pp. 1-26.

Passerello, C.E. and Huston, R.L. (1971) Human Attitude Control. *J. Biomechanics.* **4 (2)**, pp. 95-102.

Paul, J.P. (1965) Bioengineering Studies of the Forces Transmitted by Joints. *Engineering Analysis, Biomechanics, and Related Bioengineering Topics* (ed. Kennedy, R.M.). Pergamon Press, Oxford. pp. 369-380.

Pedotti, A., Krishnan, V.V., and Starke, L. (1978) Optimization of Muscle-force Sequencing in Human Locomotion. *Math. Biosciences.* **38,** pp. 57-76.

Roth, R. (1989) On Constructing Free Body Diagrams. *Int. J. Appl. Engineering.* **5 (5)**, pp. 565-570.

Seireg, A. and Arvikar, R.J. (1973) A Mathematical Model for the Evaluation of Forces in Lower Extremities of the Musculo-skeletal System. *J. Biomechanics.* **6 (13),** pp. 313-326.

Winter, D.A. (1983) Moments of Force and Mechanical Power in Jogging. *J. Biomechanics.* **16 (1)**, pp. 91-97.

Yeadon, M.R., Atha, J., and Hales, F.D. (1990) The Simulation of Aerial Movement - IV: A Computer Simulation Model. *J. Biomechanics.* **23 (1)**, pp. 85-89.

4.3 THE FREE BODY DIAGRAM

NIGG, B.M.

4.3.1 DEFINITIONS AND COMMENTS

Contact force:	External force resulting from physical contact between two objects. An example of a contact force is the force on the foot segment that results from contact with the floor surface.
Free body diagram (FBD):	A free body diagram for a particle consists of a sketch of the particle of interest, a representation of all the external forces acting on that particle, and a coordinate system. A free body diagram for a rigid body consists of a sketch of the rigid body of interest, a representation of all the external forces and moments acting on that rigid body, and a coordinate system. Comment: a free body diagram properly drawn allows the immediate application of Newton's second law.
Frictional joint moment:	Moment in a joint due to friction in the joint, including both surface roughness (dry) and viscous (fluid) phenomena. The frictional joint moment is often neglected or, if not, included in the resultant joint moment.
Inertial frame of reference:	Frame of reference in which Newton's laws are valid. Relative to an inertial frame of reference, a body remains at rest or in a state of rectilinear translation with constant speed when no resultant force or moment act on it.
Particle:	Matter that is assumed to occupy a single point in space. The volume of the body of interest is small compared to the space in which its behaviour is of interest.
Remote force:	External force not resulting from physical contact which one object exerts on another. An example of a remote force is the gravitational force.
Resultant joint moment:	Resultant moment with respect to a joint that is produced by all forces with lines of actions that cross the joint. Resultant joint moments may be subdivided into resultant muscle joint moments, resultant ligament joint moments, resultant bony contact force moments, etc.

Rigid body: Matter that is assumed to occupy a finite volume in space and that does not deform if subjected to external forces.

4.3.2 SELECTED HISTORICAL HIGHLIGHTS

The development of the free body diagram for the analysis of mechanical and biomechanical problems has taken decades. The most important steps in the development of the free body diagram included (Dempster, 1961):

1858	Rankine	A Scottish engineer-physicist who used a graphical technique for analyzing problems of applied mechanics.
1864	Maxwell	Showed that reciprocal figures and force polygons are related and may be used in mechanical analyses.
1875	Culmann	From the ETH Zürich. Published a book dealing with Graphical Statics and later became the leader in the application of graphical methods to engineering problems.
1875	Weisbach	A German engineer who used diagrams of bodies with forces acting on them (forces indicated by arrows). Some of these diagrams were identical to free body diagrams in the current understanding. Other diagrams only included selected forces.
1881	Gibbs	Developed the mathematics of vector analysis, which is used to solve the vector equations that result from the free body diagrams.
1910	Smith & Longley	Outlined the procedures for handling mechanical problems by using drawings and vectors.
1928	Reynolds	Used the term free body diagram and defined it as "the body isolated from all other bodies with arrows representing the forces acting upon it".
1955	Lissner	Made an attempt to interpret the mechanics of the musculo-skeletal system for physiotherapists.
1961	Dempster	Published an article on free body diagrams as an approach to the mechanics of human posture and motion.

4.3.3 INTRODUCTION

In biomechanics research, the human body is often simplified as a system of n particles, or a system of n rigid bodies connected by idealized joints. The following discussion concentrates on one selected rigid body, B_i, and may be adapted for particles. The movement of a rigid body in an inertial frame, R_i, is governed by the equations:

$$\mathbf{F}_i = m_i\mathbf{a}_i = \frac{d\mathbf{p}_i}{dt} \qquad \text{for translation}$$

$$\mathbf{M}_i = \frac{d\mathbf{H}_i}{dt} \qquad \text{for rotation}$$

where:

$\mathbf{F}_i =$ resultant external force acting on the rigid body, $\mathbf{B}_i$ (vector sum of all external forces acting on the rigid body, $\mathbf{B}_i$)

$m_i =$ mass of the rigid body, $\mathbf{B}_i$

$\mathbf{a}_i =$ acceleration of the centre of mass of the rigid body, $\mathbf{B}_i$, in the inertial frame, R

$\mathbf{p}_i =$ linear momentum $= m_i\mathbf{v}_i$

$\mathbf{M}_i =$ resultant moment about the centre of mass of all forces and couples acting on the rigid body, $\mathbf{B}_i$

$\mathbf{H}_i =$ angular momentum of the rigid body, $\mathbf{B}_i$, about the centre of mass in the inertial frame, $\mathbf{R}_i$

When performing a biomechanical analysis of a segment of the human body, which is represented as a rigid body, a free body diagram is often used. This method is so central to modelling that a special chapter is devoted to the definition and description of it.

A free body diagram (FBD) for a particle consists of a sketch of the particle of interest, a representation of all the external forces acting on that particle, and a coordinate system.

A free body diagram (FBD) for a rigid body consists of a sketch of the rigid body of interest, a representation of all the external forces and moments acting on that rigid body, and a coordinate system.

Three main steps are important for the development of a free body diagram, (a) the system of interest must be defined and sketched, (b) the external forces and moments acting on the body of interest must be represented, and (c) the coordinate system must be defined. These three steps are discussed further.

4.3.4 THE USE OF A FREE BODY DIAGRAM

THE SYSTEM OF INTEREST

In order to determine specific forces acting on a body, several different systems of interest can be defined. In a first step, the body (e.g., the human body or the leg) is subdivided into two parts. The separation is made at the location of the unknown forces. The system of interest is one of the two body parts that have been separated from each other.

Suppose one wants to determine the force in the hip joint, in the supporting leg, during running for a human being. First, the human body is separated at the hip joint. Possible systems of interest include (a) the supporting leg, (b) the supporting femur, (c) the pelvis and, (d) the rest of the body (Fig. 4.3.1). Some of these possible systems of interest may be

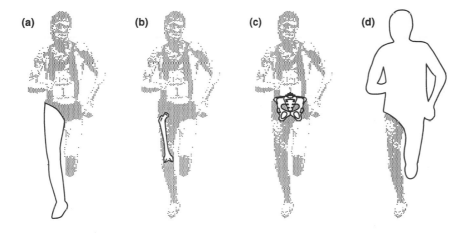

Figure 4.3.1   Examples for possible systems of interest for the estimation of the bone-to-bone forces in the human hip joint. The possible systems of interest are drawn in solid lines. The "rest of the body" is indicated in a shaded form.

more appropriate than others, depending on the assumptions and the possible measurements that can be made. For this text the following convention is applied: the system of interest is drawn with solid lines. Other parts of the total body (e.g., human body) may be drawn with dotted lines, in a shaded form, or may not be drawn at all (Fig. 4.3.1).

The selection and definition of the system of interest is an important step in the development of a free body diagram. Many errors in the solution of a mechanical problem start right at this point: the system of interest has not been defined properly.

The example above shows: different segments of the human or animal body may be selected appropriately as the system of interest for the same question (e.g., determination of the forces in the hip joint. The selection depends on the knowledge one has about the total system, the mathematical skills available to the scientist, the experimental set-up, and the variables which can be determined experimentally and can be used as input into the mathematical equations describing the problem. The selection of the system of interest often influences the elegance and ease of the mathematical solution.

EXTERNAL FORCES AND MOMENTS

There are three main possibilities for drawing a free body diagram with its representative external forces and moments. Two of them are discussed in detail in the following paragraphs while the other will be described superficially.

Approach 1

This approach uses forces and moments that could be actually measured if appropriate transducers were to be used. The "actual forces and moments approach" starts with the definition of the force-transmitting elements included in the analysis. Subsequently, all actual forces and moments are drawn into the free body diagram. They include:

(1)   Remote forces
  • Body weight.

(2)   Contact forces
  • Forces due to adjacent bones.
  • Forces due to adjacent ligaments.
  • Forces due to adjacent muscle tendon units.
  • Reaction forces due to contact with other external bodies (e.g., ground reaction force).

(3)   Frictional joint moments
  • Moments due to friction in the joints: these are usually assumed to be very small and are usually neglected.

EXAMPLE 1

Question: Draw the FBD for the foot of a person standing on the forefoot of one leg.

System of interest: Foot.

Assumptions:

(1)   The foot is a rigid structure.
(2)   The foot has one idealized joint, the "ankle joint", which is responsible for plantar- and dorsi-flexion between foot and leg.
(3)   The structures responsible for contact forces are the Achilles tendon, the tibia at the ankle joint, and the ground.
(4)   There is no friction in the ankle joint.
(5)   The weight of the foot can be neglected.
(6)   The problem can be solved two-dimensionally.
(7)   No moments act at the "toe" and the "ankle joint".

FBD:

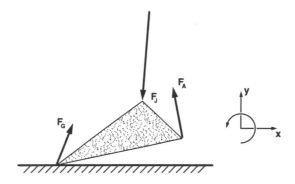

Figure 4.3.2   **Example for a free body diagram, estimating the force in the Achilles tendon when standing on the forefoot of one leg using the "actual forces and moments approach".**

The "actual forces and moments approach" is advantageous for simple applications as illustrated in the example (Fig. 4.3.2). This approach is also used for the simple estimations of internal forces or to determine the order of magnitude of internal forces. It has the advantage that one sees the "actual forces" in the drawing and that one can relate the drawing to the real situation, recognizing forces acting in selected structures. However, in applications with more complex force distributions, this approach may not be appropriate and is often replaced by another approach, the "resultant forces and moments approach".

Approach 2

This approach uses equipollent resultant forces and moments (Andrews, 1974). This approach will be named the "resultant forces and moments approach". The free body diagram includes:

(1)   Remote forces
   • Body weight: acting at the centre of mass.

(2)   Contact forces and contact moments
   • Resultant joint forces and moments: the complicated force distributions at the distal and proximal joints are replaced by the equipollent resultant joint forces and moments.
   • Resultant surface forces and moments acting on the system of interest: the sometimes complicated force distributions acting on the segment surface are replaced by an equipollent resultant surface force and moment acting at some arbitrarily, but appropriately located point.

The calculated resultant joint forces and moments do not correspond to actual forces and moments, and cannot be measured with appropriate transducers as in the previous approach. They are abstract quantities that are often not used as final results but they are used as input into a second step, the distribution of these forces and moments to the specific structures (ligaments, tendon, bone, etc.) that cross a joint.

EXAMPLE 2

Question: Draw the FBD for the foot of a person standing on the forefoot of one leg (Fig. 4.3.3).

System of interest: Foot.

Assumptions:

(1)   The foot is a rigid structure.

(2)   The foot has one idealized joint, the "ankle joint", which is responsible for plantar- and dorsi-flexion between foot and leg. The foot does not perform any in-eversion and/or ab-adduction movements.

(3)   The structures responsible for contact forces in this specific example are the Achilles tendon, the tibia at the ankle joint, the rigid foot at the ankle joint, and the ground.

(4)   There is no friction in the ankle joint.

(5)   The weight of the foot can be neglected.

(6)   The problem can be solved two-dimensionally.

FBD:

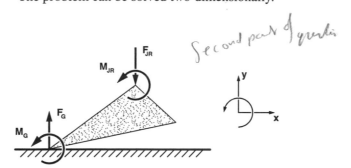

Figure 4.3.3   Example for a free body diagram, estimating the force in the Achilles tendon when standing on the forefoot of one leg using the "resultant joint forces and moments approach".

The "resultant forces and moments approach" is appropriate for situations in which the complexity of the acting forces prohibits the drawing of a simple diagram, and in which the subsequent calculations favor a two step approach (step 1: resultant joint moments and forces, step 2: distribution of them).

Approach 3

It has been suggested that free body diagrams should, in addition to the above mentioned forces, include inertia forces (Roth, 1989). Using this approach d'Alembert's principle is applied. This approach will not be discussed further. This text will concentrate on the application of Newton's laws. The reader is referred to textbooks.

The coordinate system:

Each free body diagram must include a coordinate system that defines the direction of the positive axes for translation and rotation. Often the following convention for a Cartesian coordinate system is used:

y = vertical direction
x = a-p direction
z = medio-lateral direction

In the following chapters, the "actual forces and moments approach" is applied for the simple applications, and the "resultant forces and moments approach" is applied for more complex applications. In selected cases the same problem will be solved twice, once using each approach.

4.3.5 POSSIBLE CONVENTION

The construction of a free body diagram, the formulation of the corresponding equations of motion, and the solution and interpretation of them, may be complicated for biological/biomechanical applications. Consequently, it may be advantageous to use a systematic convention for the construction of the free body diagram, and the formulation of the equations of motion. Such a convention may reduce a number of errors that occur as the free body diagram and equations of motion are produced. Here, a convention is outlined for the "resultant joint forces and moments approach" for a planar case. It can be adapted for the three-dimensional case.

Assumptions: (1)   A right handed Cartesian coordinate system is defined, with x, y, and z axes perpendicular.

(2)   Longitudinal axis of the segment remains in the x-y plane.

Proposed convention: (1)   Define the system of interest.

(2)   Use solid lines to represent the system of interest.

(3)   Use dotted lines (if you want) to represent other (adjacent) systems to illustrate a particular situation.

(4)   Number all systems. If appropriate, number the ground "0" and use a sequence from there.

(5)   Draw all force components in positive axis directions in the FBD.

(6)   Draw moments in positive x axis direction in the FBD.

(7) F_{ijx} denotes a force component in the x direction exerted by system j on system i.

(8) M_{ijx} denotes a moment component about the x axis exerted by system j on system i.

(9) $C_i = (x_i, y_i)$ is the centre of mass of segment i.

(10) $J_{ik} = (x_{ik}, y_{ik})$ is the joint centre between the segment of interest i and the neighboring segment k.

(11) m_i is the mass of segment i.

(12)   g is the acceleration due to gravity in m/s².

(13)   Use a superscript "dot" for the first time derivative, and a superscript "double dot" for the second time derivative.

(14) I_{iz} is the moment of inertia of segment i about the z-axis through its centre of mass.

(15) φ_{iz} is the angle the segment's longitudinal axis between the two joint centres, $J_{i(i-1)}$ and $J_{i(i+1)}$, makes with the positive x axis.

The corresponding free body diagram, FBD, is (Fig. 4.3.4):

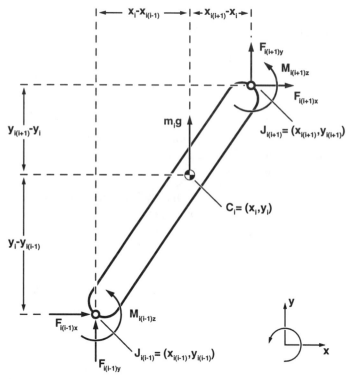

Figure 4.3.4   Standardized general free body diagram for a body i with two joint centres.

The convention proposed in this section is one of several appropriate possibilities. The proposed convention that all forces and moments should be drawn in positive axis direction, may deserve some further explanation.

There are at least three possibilities to draw the forces and moments in a free body diagram:

(1)   Draw all forces and moments the way they act.

(2)   Draw the forces and moments in an arbitrary way.

(3)   Draw all forces and moments in a positive axis direction.

For all three approaches the calculated forces and moments follow the same rule: the forces and moments act as drawn in the FBD if the calculated forces and moments are positive. The forces and moments act in opposite direction to the forces and moments drawn in the FBD if the calculated forces and moments are negative. For the three possibilities to draw forces and moments that means:

(1)   If a calculated force or moment has a negative sign the force or moment was "wrongly" drawn in the FBD, and acts in opposite direction than the one drawn.

(2)   If a calculated force or moment has a negative sign the force or moment was "wrongly" drawn in the FBD and acts in opposite direction than the one drawn. This is, of course, the same statement as the statement shown in (1) above. The difference is that for the second approach more forces and moments will (most likely) be negative.

(3)   If a calculated force or moment has a positive sign the force or moment acts in a positive axis direction. If a calculated force or moment has a negative sign the force or moment acts in a negative axis direction. The result for this third approach, therefore, relates to the selected coordinate system, and not to the drawing. It is suggested that this is a more general approach. However, it may be contra-intuitive to draw specific forces in a positive axis direction (e.g., weight).

The reader, however, should not put too much emphasis into this aspect. All approaches are appropriate if properly used and understood.

Using these conventions, the two-dimensional general equations of motion for segment i with two joints are:

Translation:

$$m_i \ddot{x}_i \quad = \quad F_{i(i+1)x} \quad + \quad F_{i(i-1)x}$$

$$m_i \ddot{y}_i \quad = \quad F_{i(i+1)y} \quad + \quad F_{i(i-1)y} \quad + m_i g$$

Rotation:

$$I_{iz} \ddot{\varphi}_{iz} \quad = \quad M_{i(i+1)z} \quad + \quad M_{i(i-1)z}$$

$$- \ (y_{i(i+1)} - y_i) F_{i(i+1)x} \ + \ (y_i - y_{i(i-1)}) F_{i(i-1)x}$$

$$+ \ (x_{i(i+1)} - x_i) F_{i(i+1)y} \ - \ (x_i - x_{i(i-1)}) F_{i(i-1)y}$$

The first two equations are merely statements of the principle of linear momentum (Euler's 1st law or principle of the motion of the mass centre, applied to bodies of finite extent) which requires that an inertial reference frame, R, be used. The third equation, the moment equation, uses the net moment, the moment of inertia, and the angular acceleration of the body relative to the inertial reference frame, R.

The proposed convention has the advantage that the equations of motion are independent of the particular (second approach) free body diagram drawn. Such a procedure is advantageous when analyzing a complex multi-link system in combination with a computer program. The next stage (after calculating the resultant joint forces and moments) is to distribute the resultant joint forces and moments to the structures that cross a particular joint.

4.3.6 REFERENCES

Alexander, R.M. (1989) Sequential Joint Extension in Jumping. *Human Movement Science.* **8,** pp. 339-345.

Alexander, R.M. (1992) Simple Models of Walking and Jumping. *Human Movement Science.* **11,** pp. 3-9.

Andrews, J.G. (1974) Biomechanical Analysis of Human Motion. *Kinesiology IV.* Amer. Assoc. for Health, Phys. Ed., & Rec., Washington, D.C. pp. 32-42.

Ballreich, R. and Brüggemann, G. (1986) Biomechanik des Weitsprungs. *Biomechanik der Leichtathletik* (eds. Ballreich, R. and Kuhlow, A.). pp. 28-47.

Crowninshield, R.D. and Brand, R.A. (1981) A Physiologically Based Criterion of Muscle Force Prediction in Locomotion. *J. Biomechanics.* **14 (11),** pp. 793-801.

Dempster, W.T. (1958) Analysis of Two-handed Pulls Using Free Body Diagrams. *J. Appl. Physiology.* **13 (3),** pp. 469-480.

Dempster, W.T. (1961) Free Body Diagrams as an Approach to the Mechanics of Human Posture and Motion. *Biomechanical Studies of the Musculo-skeletal System* (ed. Evans, F.G.). Thomas, Springfield, IL. pp. 81-135.

Denoth, J. (1980) Ein mechanisches Modell zur Beschreibung von passiven Belastungen. *Sportplatzbeläge* (eds. Nigg, B.M. and Denoth, J.). pp. 45-53.

Denoth, J. (1980) Materialeigenschaften. *Sportplatzbeläge* (eds. Nigg, B.M. and Denoth, J.). pp. 54-67.

Dul, J., Johnson, G.E., Shiavi, R., and Townsend, M.A. (1984) Muscular Synergism - II: A Minimum-fatigue Criterion for Load Sharing Between Synergistic Muscles. *J. Biomechanics.* **17 (9),** pp. 675-684.

Elftman, H. (1938) Forces and Energy Changes in the Leg During Walking. *Am. J. Physiology.* **125 (2),** pp. 339-356.

Fischbein, E. (1987) *Intuition in Sciences and Mathematics.* D. Reidel Publishing Company, Dordrecht, Netherlands.

Gruber, K., Denoth, J., Stuessi, E., and Ruder, H. (1987) The Wobbling Mass Model. *Biomechanics X-B* (ed. Jonsson, B.). pp. 1095-1099.

Guralnik, D.B. (ed.) (1979) Webster's New World Dictionary. William Collins Publishers, Ohio.

Hatze, H. (1978) Sportbiomechanische Modelle und myokybernetische Bewegungsoptimierung: Gegenwartsprobleme und Zukunftsaussichten. *Sportwissenschaft.* **4,** pp. 1-13.

Hatze, H. (1981) A Comprehensive Model for Human Motion Simulation and its Application to the Take-off Phase of the Long Jump. *J. Biomechanics.* **14 (3),** 135-142.

Herzog, W., (1987) Individual Muscle Force Estimations Using a Non-linear Optimal Design. *J. Neuroscience Methods.* **21,** pp. 167-179.

Herzog, W. and Leonard, T.R. (1991) Validation of Optimization Models that Estimate the Forces Exerted by Synergistic Muscles. *J. Biomechanics.* **24 (S-1),** pp. 31-39.

Kemeny, J.G. (1959) *A Philosopher Looks at Science.* D. van Nostrand Pub. Comp., Princeton, NJ.

Komi, P.V., Salonen, M., Jarvinen, N., and Kokko, O. (1987) In Vivo Registration of Achilles Tendon Forces in Man. *Int. J. Sports Medicine.* **8,** pp. 3-8.

McMahon, T.A. and Greene, P.R. (1979) Influence of Track Compliance on Running. *J. Biomechanics.* **12,** pp. 893-904.

Morlock, M. and Nigg, B.M. (1991) Theoretical Considerations and Practical Results on the Influence of the Representation of the Foot for the Estimation of Internal Forces with Models. *Clinical Biomechanics.* **6,** pp. 3-13.

Nigg, B.M. (1986) Biomechanical Aspects of Running. *Biomechanics of Running Shoes* (ed. Nigg, B.M.). Human Kinetics Pub. Inc., Champaign, IL. pp. 1-26.

Passerello, C.E. and Huston, R.L. (1971) Human Attitude Control. *J. Biomechanics.* **4 (2)**, pp. 95-102.

Paul, J.P. (1965) Bioengineering Studies of the Forces Transmitted by Joints. *Engineering Analysis, Biomechanics, and Related Bioengineering Topics* (ed. Kennedy, R.M.). Pergamon Press, Oxford. pp. 369-380.

Pedotti, A., Krishnan, V.V., and Starke, L. (1978) Optimization of Muscle-force Sequencing in Human Locomotion. *Math. Biosciences.* **38,** pp. 57-76.

Roth, R. (1989) On Constructing Free Body Diagrams. *Int. J. Appl. Engineering.* **5 (5)**, pp. 565-570.

Seireg, A. and Arvikar, R.J. (1973) A Mathematical Model for the Evaluation of Forces in Lower Extremities of the Musculo-skeletal System. *J. Biomechanics.* **6 (3)**, pp. 313-326.

Winter, D.A. (1983) Moments of Force and Mechanical Power in Jogging. *J. Biomechanics.* **16 (1)**, pp. 91-97.

Yeadon, M.R., Atha, J., and Hales, F.D. (1990) The Simulation of Aerial Movement - IV: A Computer Simulation Model. *J. Biomechanics.* **23 (1)**, pp. 85-89.

4.4 MATHEMATICALLY DETERMINATE SYSTEMS

NIGG, B.M.

4.4.1 DEFINITIONS AND COMMENTS

Contact force:	External force resulting from physical contact between two objects. Comment: an example of a contact force is the force on the foot segment due to contact with the floor surface.
Free body diagram (FBD):	A free body diagram for a particle consists of a sketch of the particle of interest, a representation of all the external forces acting on that particle, and a coordinate system. A free body diagram for a rigid body consists of a sketch of the rigid body of interest, a representation of all the external forces and moments acting on that rigid body, and a coordinate system. Comment: a free body diagram properly drawn allows the immediate application of Newton's second law.
Frictional joint moment:	Moment in a joint due to friction in the joint, including both surface roughness (dry) and viscous (fluid) phenomena. Comment: the frictional joint moment is often neglected or, if not, included in the resultant joint moment.
Particle:	Matter that is assumed to occupy a single point in space. Comment: the volume of the body of interest is small compared to the space in which its behaviour is of interest.
Remote force:	External force not resulting from physical contact which one object exerts on another. Comment: an example of a remote force is the gravitational force.
Resultant joint moment:	Resultant moment with respect to a joint that is produced by all forces with lines of actions that cross the joint. Comment: resultant joint moments may be subdivided into resultant muscle joint moments, resultant ligament joint moments, resultant bony contact force moments, etc.
Rigid body:	Matter that does not deform.

4.4.2 SELECTED HISTORICAL HIGHLIGHTS

For selected historical highlights see chapters 4.2 and 4.3. The models presented in this section are arbitrarily selected, and the key references discussed earlier are still the most important highlights.

4.4.3 INTRODUCTION

This section attempts to show how simple models using particles may be and have been used to discuss biomechanical situations. The discussion of what level of complexity of a biomechanical model is appropriate will probably never end. Complex models such as the one used by Yeadon (1984) to analyse airborne movements or the one by Seireg and Arvikar (1973) that contained 29 muscles per leg are certainly impressive from a mathematical point of view alone. However, one may disagree with the view that because the human body is complex, biomechanists should always reproduce as much as possible of its complexity in their models. This section attempts to illustrate McNeill Alexander's (1992) proposition that in certain cases simple models may be appropriate. The section starts with the simplest possible model, a particle with no force acting on it, and progresses stepwise to more complicated models using one or two particles. The examples are selected in an attempt to provide an insight into important aspects of modelling.

This section has many examples of models with particles. The examples start with the simplest possible case, a particle with no force acting on it. From there, the section proceeds to more complex examples. Some readers may disregard the rather trivial initial examples and jump to the more demanding ones. For others, these simple examples may provide an adequate tool to refresh aspects of mathematics: first, they attempt to illustrate different possibilities to model determinate systems of particles with simple mathematical tools, second, they attempt to discuss the physical and biological interpretation which these models may include.

Every example uses the same set-up. Each example starts with explanatory comments that provide insight into the problem. The comments may explain why the model was developed at all and/or what may be concluded from the results. The following headings are used:

Question:	Question to be answered.
Assumptions:	Assumptions used in this example.
FBD:	Free body diagram.
	Illustrations: in some cases it may be advantageous to add a drawing to illustrate the actual situation at different time points. Such drawings may or may not include forces. However, it is recommended that in addition to such drawings, a real free body diagram be added. Free body diagrams, FBDs, may also be added to the text during the solution of the problem if this sheds light on the development of additional steps in the process.
Equations of motion:	Equations of motion used in this example.

	EM for translation of particles, and EM for translation and rotation of rigid bodies.
Initial conditions:	Initial conditions for this example.
Final conditions:	Final conditions for this example.
Solution:	Solution of the problem.

At the end of each example, its results are discussed and its biomechanical importance is addressed.

4.4.4 MECHANICAL MODELS USING PARTICLES

EXAMPLE 1 (the simplest case - no force)

The simplest possible model in mechanics is a particle in space with no force acting on it. The set-up and the mathematics for this example are trivial, and the result is obvious. However, the example may serve as a simple start to a journey that leads to more complex models.

Question:

Describe the one-dimensional movement of a particle upon which no force is acting.

FBD:

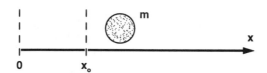

Figure 4.4.1   **Free body diagram, FBD, for one particle with no force acting on the particle.**

Equations of motion:

$$m\ddot{x}(t) = 0$$

Initial conditions:

$$x(0) = x_0$$

$$\dot{x}(0) = v_0$$

Solution:

$$\ddot{x}(t) = 0$$

Integration of this equation provides:

$$\dot{x}(t) \quad = \text{const} = c_1$$

Using the initial condition:

$$\dot{x}(0) \quad = v_0$$

The constant, c_1, can be determined as v_0:

$$\dot{x}(t) \quad = v_0$$

A second integration provides:

$$x(t) \quad = v_0 t + c_2$$

Using the initial condition:

$$x(0) \quad = x_0$$

provides the general equation for the position of the particle of interest as a function of time:

$$x(t) \quad = x_0 + v_0 t \qquad (4.4.1)$$

This is the well-known equation for uniform rectilinear motion. The position, $x(t)$, depends on the initial position, the initial velocity, v_0, and the time. Of course, it may be difficult to find a biomechanical application for this model.

EXAMPLE 2 (one particle with a constant force acting on it)

Another simple case is a particle in space with a constant force acting on it. One practical example from physics is a mass in the air with only the gravitational force acting on it. Again, although we know the result of this example in advance (i.e., the movement of a particle with constant acceleration), this example leads the way to a thorough understanding of the possibilities of modelling biomechanical problems.

Question:
Describe the one-dimensional movement of a particle upon which a constant force is acting.

Assumptions:
(1)   A constant force, F, is acting on the particle in vertical (y) direction:

$$F \quad = \quad mg$$

where:

$$g \quad = \quad 9.81 \text{ m/s}^2$$

(2)   The mass can be considered as a particle.

(3)   Air resistance can be neglected.

(4)   All forces are drawn in a positive axis direction.

FBD:

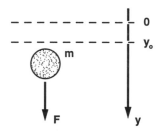

Figure 4.4.2   **FBD for a particle with a constant gravitational force, F = - mg, acting on it.**

Equations of motion:

$$m\ddot{y}\,(t) \quad = \quad mg$$

Initial conditions:

$$y\,(0) \quad = \quad y_0$$

$$\dot{y}\,(0) \quad = \quad v_0$$

Solution:

Integration of the acceleration provides the velocity:

$$\dot{y}\,(t) \quad = \quad g\,t + const$$

Using the initial condition provides the equation for the velocity:

$$\dot{y}\,(t) \quad = \quad g\,t + v_0$$

Integration of the velocity provides the position:

$$y\,(t) \quad = \quad 1/2\,g\,t^2 + v_0\,t + const$$

Using the initial condition provides the equation for the position:

$$y\,(t) \quad = \quad y_0 + v_0\,t + 1/2\,g\,t^2 \tag{4.4.2}$$

This is the equation of uniformly accelerated rectilinear motion. The equation can be used for the determination of the vertical motion of a projectile with no air resistance.

EXAMPLE 3 (force of a linear spring acting on a particle)

This example corresponds in a very simplistic way to the real-life situation of a gymnast taking off from a trampoline or a diver landing on a diving board (with gravity being neglected). The particle would correspond to the athlete and the spring to the trampoline or the diving board. The model permits the calculation of where the particle is at a certain point in time.

Question:

Describe the movement of a particle upon which the force of a linear spring is acting.

Assumptions:

(1)   The force of the linear spring is:

$$F = -k \cdot x$$

where:

$$k = \text{spring constant}$$
$$x = \text{spring deformation}$$

(2)   The mass of the spring can be neglected.

(3)   The question can be treated as a one-dimensional problem.

FBD:

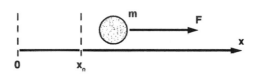

Figure 4.4.3   **Free body diagram, FBD, for a particle in contact with an ideal linear spring.**

Equations of motion:

$$m\ddot{x}(t) = F = -kx(t)$$

$$\ddot{x}(t) = -\frac{k}{m} x(t)$$

and from spring theory:

$$\frac{k}{m} = \Omega^2 \quad (\Omega = \text{circular frequency})$$

consequently:

$$\ddot{x}(t) = -\Omega^2 x(t)$$

which is a second order linear differential equation.

Initial conditions:

$$x(0) = x_0$$

$$\dot{x}(0) = v_0$$

Solution:

General solution:

$$x(t) = A\sin\omega t + B\cos\omega t$$

$$\dot{x}(t) = A\omega \cdot \cos\omega t - B\omega \cdot \sin\omega t$$

$$\ddot{x}(t) = -A\omega^2 \cdot \sin\omega t - B\omega^2 \cdot \cos\omega t$$

$$= -\omega^2(A \cdot \sin\omega t + B \cdot \cos\omega t)$$

$$\ddot{x}(t) = -\omega^2 \cdot x(t)$$

consequently:

$$\omega^2 = \Omega^2$$

and with the initial conditions:

$$x(0) = B = x_0$$

$$\dot{x}(0) = A \cdot \omega = v_0$$

$$A = \frac{v_0}{\omega} = \frac{v_0}{\Omega}$$

$$x(t) = \frac{v_0}{\Omega} \cdot \sin\Omega t + x_0 \cos\Omega t \qquad (4.4.3)$$

Equation (4.4.3) describes the position of a particle during contact with an ideal linear spring. The maximal position (deformation of the spring), velocity, and acceleration can be determined. For the sake of simplicity the initial condition for the position is simplified to:

Initial conditions:

$$x(0) = x_0 = 0$$

This provides for the position the general equation:

$$x(t) = v_0 \cdot \sqrt{\frac{m}{k}} \cdot \sin \sqrt{\frac{k}{m}} \cdot t$$

and for the minimal position (maximal deformation) the equation:

$$x_{max} = v_0 \cdot \sqrt{\frac{m}{k}}$$

The equation for the velocity, using the same initial conditions is:

$$\dot{x}(t) = v_0 \cdot \cos \sqrt{\frac{k}{m}} \cdot t$$

which provides for the maximal velocity (at the beginning and at the end of contact):

$$\dot{x}_{max} = v_0$$

The equation for the acceleration, using the same initial conditions, is:

$$\ddot{x}(t) = -v_0 \sqrt{\frac{k}{m}} \cdot \sin \sqrt{\frac{k}{m}} \cdot t$$

which provides the maximal acceleration at the time of the maximal deformation:

$$\ddot{x}_{max} = -v_0 \sqrt{\frac{k}{m}}$$

The general result is shown in Fig. 4.4.4, which illustrates the movement characteristics for position, velocity, and acceleration. Note the phase shift between position, velocity, and acceleration for the given assumptions. These considerations could be applied to the movement of a diver during the last contact with the diving board. The velocity would be zero at the lowest position of the board, and the acceleration would be maximal at this point in time. The actual situation in diving corresponds reasonably well to reality. The main difference is that the position is not a "clean" sinusoidal curve but a slightly "asymmetrical sinusoidal" curve.

EXAMPLE 4 (force of two idealized springs acting on a particle)

One could imagine a foot landing on a surface and describe the foot as a particle, with the heel pad and the surface as two springs in series. This conception of the situation is rather simplistic, but it may provide some insight into the problem of interest.

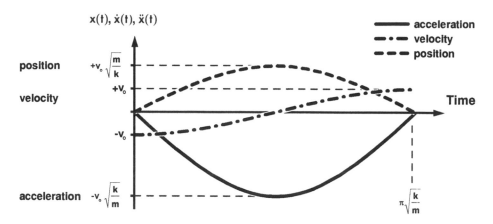

Figure 4.4.4   Position, velocity, and acceleration as functions of time for a particle dropping onto an ideal linear spring.

Question:

Describe the movement of a particle as it drops onto a system of two ideal linear springs. Specifically, answer the following questions:

(1) Determine the velocity at contact, v_{crit}, for which the first spring "bottoms out".

(2) Determine the position, $z(t)$, of the particle for the two springs.

(3) Determine the maximal acceleration (deceleration) of the particle.

(4) Draw and discuss the graphs for $z(t)$, $\dot{z}(t)$, and $\ddot{z}(t)$.

Assumptions:

(1) Spring #1 is in contact with the mass. Spring #2 is in contact with spring #1.

(2) Spring #1 is much softer than spring #2 (i.e., $k_1 \ll k_2$).

(3) Spring #1 compresses first completely to z_b before spring #2 starts to compress.

(4) Spring #1 remains compressed at position z_b while spring #2 is compressed: consequently, the movement can be subdivided into two parts, a first part in which spring #1 changes length while spring #2 remains at its original length, and a second part in which spring #1 remains compressed at z_b while spring #2 changes length. The idealized force deformation diagram for this case is illustrated in Fig. 4.4.5.

(5) The forces are drawn in positive axis direction.

(6) The two springs are ideal. The force is independent of the velocity, v_0. There is no loss of energy.

$$F_1 = -k_1 \cdot z(t)$$

$$F_2 = -k_2 \cdot z(t)$$

(7) The masses of the springs can be neglected.

(8) The numerical values are:

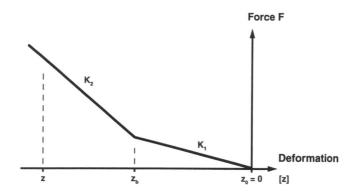

Figure 4.4.5   Assumed force deformation diagram of the two spring systems.

$$
\begin{aligned}
m &= 4 \text{ kg} &= \text{dropping mass} \\
z_b &= 0.04 \text{ m} &= \text{total compression of spring \#1} \\
k_1 &= 10^4 \text{ N/m} &= \text{spring constant \#1} \\
k_2 &= 10^6 \text{ N/m} &= \text{spring constant \#2} \\
v_0 &= 4 \text{ m/s} &= \text{initial velocity of mass (for question 4)}
\end{aligned}
$$

(9) Gravity can be neglected.

(10) The problem can be solved one-dimensionally.

Answer to question 1 - determination of critical velocity

FBD:

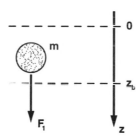

Figure 4.4.6   Free body diagram, FBD, for the first part of the problem (question 1), in which the second spring is assumed to be infinitely stiff. This FBD, of course, corresponds to Fig. 4.4.3.

Equations of motion:

Assuming that only spring #1 is compressed:

$$
m\ddot{z}(t) = F_1 = -k_1 \cdot z(t)
$$

or:

$$\ddot{z}(t) \quad = \quad -\Omega_1^2 \cdot z(t)$$

Initial conditions:

$$z(0) \quad = \quad 0$$

$$\dot{z}(0) \quad = \quad v_0$$

Solution:

$$z(t) \quad = \quad A \cdot \sin\omega_1 t + B \cdot \cos\omega_1 t$$

$$\dot{z}(t) \quad = \quad A \cdot \omega_1 \cdot \cos\omega_1 t - B\omega_1 \cdot \sin\omega_1 t$$

$$\ddot{z}(t) \quad = \quad -\omega_1^2 \cdot z(t)$$

where:

$$\omega_1^2 \quad = \quad \Omega_1^2$$

Using the initial condition:

$$z(0) \quad = \quad 0$$

provides:

$$B \quad = \quad 0$$

and the initial condition:

$$\dot{z}(0) \quad = \quad v_0$$

provides:

$$A \quad = \quad \frac{v_0}{\Omega_1} \quad = \quad \frac{v_0}{\sqrt{k_1}} \cdot \sqrt{m}$$

consequently:

$$z(t) \quad = \quad \frac{v_0}{\Omega_1} \cdot \sin\Omega_1 t$$

Calculation of the critical velocity, v_c, at which spring #1 is totally compressed ("bottoms out"):

$$v_c \quad = \quad z_b \cdot \Omega_1$$

Spring #1, the softer spring in this example, does not compress completely if the landing velocity, v_o, is smaller than the critical velocity, $(v_o < v_c)$. However, spring #1 compresses completely if $v_o \geq v_c$.

Result for the selected numerical values:

$$v_c \quad = \quad 0.04\,\mathrm{m} \sqrt{\frac{100\,\mathrm{m}}{0.04\,\mathrm{ms}^2}}$$

$$v_c \quad = \quad 2 \text{ m/s}$$

The critical velocity for this particle is 2 m/s, which corresponds to a drop from about 0.2 m (20 cm). If the mass is dropped from a height of about 0.2 m, the material of the soft spring compresses completely (bottoms out). The values used correspond to material properties of the human heel as follows: the mass of a leg is about 4 to 6 kg, which is the mass used. The spring constant of the material of the human heel was reported in the order of magnitude of about $2 \cdot 10^4$ N/m (Nigg and Denoth, 1980). Consequently, the numbers used in the numerical calculations and the subsequent results correspond in their order of magnitude to the landing of a heel on the ground. A landing velocity of the heel of about 2 m/s in barefoot running may produce complete compression of the heelpad. The human heel is, of course, not an elastic but rather a visco-elastic material. The numerical results and their translation into the practical application must, therefore, be taken with a grain of salt. More accurate and appropriate results might be determined with a more complex model that contained viscous elements (dampers) and considered the effect produced by the increasing contact area between heel and ground.

Answer to question 2 - determination of the general position of the particle

The FBD is illustrated in Fig. 4.4.7

FBD:

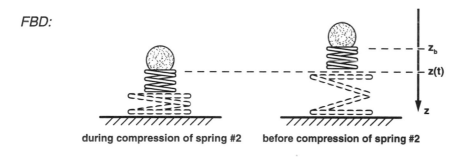

during compression of spring #2   before compression of spring #2

Figure 4.4.7   FBD for the point at which spring #1 is totally compressed. The system of interest is the mass of the particle and the totally compressed massless spring #1.

Equations of motion:

The force acting on the particle can be composed from the force needed to completely compress the first spring and the force needed to compress the second spring partially.

$$m\ddot{z}(t) \quad = \quad \text{force (spring \#2) + force (spring \#1)}$$

$$m\ddot{z}(t) \quad = \quad -k_2[z(t) - z_b] - k_1 z_b$$

Initial conditions:

$$z(t_b) \quad = \quad z_b$$

$$\dot{z}(t_b) \quad = \quad v_b$$

Solution:

$$z(t) \quad = \quad C \cdot \sin\omega_2 t + D \cdot \cos\omega_2 t + a \tag{4.4.4}$$

$$\dot{z}(t) \quad = \quad C\omega_2 \cdot \cos\omega_2 t - D\omega_2 \cdot \sin\omega_2 t \tag{4.4.5}$$

$$\ddot{z}(t) \quad = \quad -\omega_2^2[z(t) - a] \tag{4.4.6}$$

Determination of ω_2, a_1, and D:

$$-\Omega_2^2[z(t) - z_b] - \Omega_1^2 \cdot z_b \quad = \quad -\omega_2^2[z(t) - a]$$

$$-\Omega_2^2 \cdot z(t) + \Omega_2^2 \cdot z_b - \Omega_1^2 \cdot z_b \quad = \quad -\omega_2^2 \cdot z(t) + \omega_2^2 \cdot a$$

consequently:

$$\omega_2 \quad = \quad \Omega_2 \tag{4.4.7}$$

and:

$$-\Omega_2^2 \cdot z_b - \Omega_1^2 \cdot a \quad = \quad \omega_2^2 \cdot a$$

$$z_b(\Omega_2^2 - \Omega_1^2) \quad = \quad \Omega_2^2 \cdot a$$

which provides for a:

$$a \quad = \quad z_b\left[1 - \frac{k_1}{k_2}\right] \tag{4.4.8}$$

$$z_b \quad = \quad C \cdot \sin\Omega_2 t + D \cdot \Omega_2 t + z_b - z_b \cdot \frac{k_1}{k_2}$$

For the time $t_b = 0$:

$$D \quad = \quad z_b \cdot \frac{k_1}{k_2} \tag{4.4.9}$$

Determination of C:

for $\dot{z}\ (t = t_b = 0) = v_b$

$$v_b \quad = \quad C \cdot \Omega_2 \cdot \cos\Omega_2 t - D \cdot \Omega_2 \cdot \sin\Omega_2 t \tag{4.4.10}$$

$$C \quad = \quad \frac{v_b}{\Omega_2}$$

v_b can be replaced using the conservation of energy:

$$\frac{1}{2} m v_o^2 \quad = \quad \frac{1}{2} k_1 \cdot z_b^2 + \frac{1}{2} m v_b^2 \tag{4.4.11}$$

$$v_b^2 \quad = \quad v_o^2 - \Omega_1^2 \cdot z_b^2 \tag{4.4.12}$$

Substituting equation (4.4.12) in equation (4.4.11) provides for C:

$$C \quad = \quad \sqrt{\frac{v_o^2 \cdot m}{k_2^2} - \frac{k_1}{k_2} \cdot z_b^2} \tag{4.4.13}$$

Substituting equations (4.4.7), (4.4.8), (4.4.9), and (4.4.13) in equation (4.4.6) provides the position of the mass as a function of time:

$$z(t) \quad =$$

$$\sqrt{\frac{v_o^2 \cdot m}{k_2} - z_b^2 \cdot \frac{k_1}{k_2}} \cdot \sin\left[\sqrt{\frac{k_2}{m}} \cdot t\right] + z_b \cdot \frac{k_1}{k_2} \cdot \cos\left[\sqrt{\frac{k_2}{m}} \cdot t\right] + z_b\left[1 - \frac{k_1}{k_2}\right]$$

$z(t)$ is the answer to question 2. It provides the general position of the particle as a function of time, for case 2.

Answer to question 3 - determination of the maximal acceleration

Using equation (4.4.6):

$$\ddot{z}(t) \quad = \quad -\Omega_2^2 (C \cdot \sin\Omega_2 t + D \cdot \cos\Omega_2 t)$$

which can be written as:

$$\ddot{z}(t) \quad = \quad -\Omega_2^2 \cdot \sqrt{C^2 + D^2} \cdot \sin\left(\Omega_2 t + \delta\right)$$

For the maximal acceleration:

$$\ddot{z}_{max} \quad = \quad -\Omega_2^2 \cdot \sqrt{C^2 + D^2}$$

$$\ddot{z}_{max} \quad = \quad -\Omega_2^2 \sqrt{v_o^2 \cdot \frac{m}{k_2} - \frac{k_1}{k_2} \cdot z_b^2 + z_b^2 \cdot \frac{k_1^2}{k_2^2}}$$

$$\ddot{z}_{max} \quad = \quad -\sqrt{v_o^2 \cdot \frac{k_2}{m} - z_b^2 \cdot k_1 \cdot \frac{k_2}{m^2} + z_b^2 \cdot \frac{k_1^2}{m^2}} \tag{4.4.14}$$

The maximal acceleration (deceleration) depends on the landing velocity of the particle, the mass, and the two material constants of the two ideal springs. The result for the numerical values of this example (answer to question (4)) is illustrated in Fig. 4.4.8.

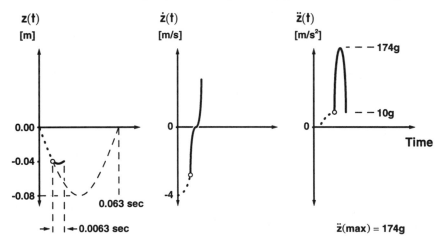

Figure 4.4.8   **Graphical illustration of position, $z(t)$, velocity, $\dot{z}(t)$, and acceleration, $\ddot{z}(t)$, for a particle dropping on two springs in series.**

The set-up and the results of this example need a few additional comments with respect to their methodology and interpretation:

(1)   In real life one rarely has a situation like the one described in this example. Usually the two materials (e.g., heel, shoe sole, sport surface, landing mats) do not

behave like linear springs, and the second material starts to deform at the same time as the first one. A real force-deformation diagram for the loading part would realistically be curved, not composed of two straight lines. As well, one should usually expect a loss of energy for the above-mentioned materials.

(2)   The mass was assumed to be a particle with no volume. In reality masses have a volume and a geometrical shape. The force-deformation diagrams of two spheres with diameters of 5 and 10 cm (if everything else is kept constant) are different. This fact is not reflected in this model.

(3)   This model should not be used to estimate accurate decelerations since the assumptions are general. The value of this model lies in explaining some general findings. It can, for instance, be used to explain the "bottoming out" phenomenon. Corners in the position-, velocity-, and acceleration-time diagram (or in the force-time diagram) indicate that a material bottoms out.

(4)   The results of this model illustrate that sport shoes or playing surfaces designed to protect the human heel should have material properties (material stiffness) in the same order of magnitude as the material constants of the human heel. If their stiffness is much higher than the stiffness of the human heel, they will only be effective after the heel pad is mostly compressed. Artificial track surfaces are generally much stiffer than the human heel. One should not expect that these surfaces would affect the impact forces significantly.

EXAMPLE 5 (force depending on deformation and velocity)

Experimental measurements from drop tests with a human heel (Nigg and Denoth, 1980; Misevich and Cavanagh, 1984) suggest that the force acting on the heel depends on the deformation $x(t)$ and on the velocity of deformation $\dot{x}(t)$. The heel pad, therefore, has visco-elastic properties.

Question:

Determine the movement of a mass (particle) dropping onto a system with visco-elastic properties.

Assumptions:

(1)   The human heel has some spring-like behaviour.

(2)   Friction is involved and, therefore, the loss of energy is not zero. The material exhibits visco-elastic behaviour.

(3)   Force $=$ $F(t)$ $=$ $F_{damp} + F_{spring}$ $=$ $-k \cdot x(t) - r \cdot \dot{x}(t)$

This mathematical description of human tissue is simplistic. A more sophisticated approach proposed by Nigg and Denoth (1980) was:

$$\sigma(x, \dot{x}) = a \cdot x^2 + b \cdot x \cdot \dot{x}$$

(4)   The problem can be treated one-dimensionally.

(5)   The mass is considered as a particle.

(6)   The spring and damping elements are arranged in parallel.

(7)   The forces are drawn in the positive axis direction.

Equations of motion:

$$m\ddot{x}(t) \quad = \quad -kx(t) - r\dot{x}(t)$$

or:

$$m\ddot{x}(t) + r \cdot \dot{x}(t) + k \cdot x(t) \quad = \quad 0$$

Initial conditions:

$$x(0) \quad = \quad 0$$

$$\dot{x}(0) \quad = \quad v_o$$

FBD:

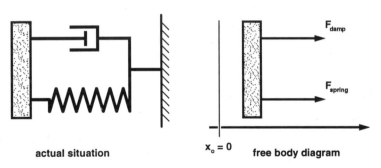

actual situation $x_o = 0$ **free body diagram**

Figure 4.4.9   **Illustration of the set-up (left) and FBD of a particle acted upon by a spring and a damper (right).**

Solution:

$$x(t) \quad = \quad A \cdot e^{-ct}$$

$$\dot{x}(t) \quad = \quad -c \cdot A \cdot e^{-ct}$$

$$\ddot{x}(t) \quad = \quad c^2 \cdot A \cdot e^{-ct}$$

$$m \cdot c^2 \cdot A \cdot e^{-ct} - r \cdot c \cdot A \cdot e^{-ct} + k \cdot A \cdot e^{-ct} \quad = \quad 0$$

$$mc^2 - rc + k \quad = \quad 0$$

$$c_{1,2} = \frac{r \pm \sqrt{r^2 - 4mk}}{2m}$$

or:

$$c_{1,2} = \frac{r}{2m} \pm \sqrt{\frac{r^2}{4m^2} - \frac{k}{m}} \qquad (4.4.15)$$

If $r^2 \neq 4mk$ the result indicates two values for c, c_1, and c_2. Therefore:

$$x(t) = A_1 \cdot e^{-c_1 t} + A_2 \cdot e^{-c_2 t}$$

$$x(0) = 0 = A_1 + A_2$$

$$A_1 = -A_2$$

$$x(t) = A_1 (e^{-c_1 t} - e^{-c_2 t})$$

$$\dot{x}(t) = -c_1 \cdot A_1 \cdot e^{-c_1 t} + c_2 \cdot A_1 \cdot e^{-c_2 t}$$

$$\dot{x}(0) = v_0 = -c_1 A_1 + c_2 A_1$$

$$A_1 = \frac{v_0}{c_2 - c_1}$$

Which provides equation (4.4.16):

$$x(t) = \frac{v_0}{c_2 - c_1} (e^{-c_1 t} - e^{-c_2 t}) \qquad (4.4.16)$$

where:

$$c_1 = \frac{r}{2m} + \sqrt{\frac{r^2}{4m^2} - \frac{k}{m}}$$

$$c_2 = \frac{r}{2m} - \sqrt{\frac{r^2}{4m^2} - \frac{k}{m}}$$

$$c_2 - c_1 = -\frac{1}{m} \sqrt{r^2 - 4mk}$$

If $r^2 = 4mk$ the general form may be shown to be:

$$x(t) = A_1 e^{-ct} + A_2 t e^{-ct}$$

where:

$$c = \frac{r}{2m}$$

$$x(0) = 0 = A_1$$

$$x(t) = A_2 t e^{-ct}$$

$$\dot{x}(t) = -cA_2 t e^{-ct} + A_2 e^{-ct}$$

$$\dot{x}(0) = v_o = A_2$$

which provides equation (4.4.17):

$$x(t) = v_o t e^{-ct} \qquad (4.4.17)$$

If the landing velocity, v_o; the mass, m; the damping coefficient, r; and the spring constant, k, are known, one can predict the movement, $x(t)$, of the particle (shot) during the contact with the heel. The same result can also be obtained experimentally, which may provide an opportunity to compare theory and experimental results.

The outlined procedure allows us to vary systematically one variable (e.g., the mass) for given additional parameter settings and, consequently, to understand the influence of this variable on the deformation in a functional sense. This procedure is called *simulation*. Simulation can, for instance, be used if the maximal compression of a material is known. If the material (k and r) and the velocity of touch down are known, the mass for which the material bottoms out can be determined. If the mass and the material are known, the velocity of touch down for which the material bottoms out can be determined. For a further discussion of simulation see chapter 4.8.

General comments on damping

Damping is a well known concept in mechanics. Expressions such as light, heavy, and critical damping are frequently used in mechanics and have their importance in biomechanical analysis. For the simplest case of a spring - damper system in parallel, critical damping is defined as:

$$r_{crit} = \sqrt{4mk} = 2m\Omega$$

where:

Ω = circular frequency

The three distinguished cases mentioned before are defined as:

Heavy damping: $r > r_{crit}$ Since c_1 and c_2 are negative, the position (x) approaches zero as t increases. The mass returns to its equilibrium position without any oscillations.

Critical damping: $r = r_{crit}$ A mass in a critically damped system regains its equilibrium position in the shortest possible time without any oscillations.

Light damping: $r < r_{crit}$ A mass oscillates around an equilibrium position with diminishing amplitude.

EXAMPLE 6 (a simple muscle model)

The ideas for this section derive from some considerations by Hörler (1972a). They deal with a simple muscle model. In the simplest possible (reasonable) case a muscle could be described with the help of a contractile element, a spring, and a damper. One may wonder whether such a simple mathematical model might predict the force-time characteristics of an isometric muscle contraction and, if yes, what mechanical characteristics of the system would best match the theoretical predictions of such a force-time curve.

Question:

Determine the mechanical characteristics of a simplistic muscle model, including a contractile element, a spring, and a damper.

Assumptions:

(1)   The muscle mass can be concentrated in one particle.

(2)   The mechanical properties of a muscle (which are typically distributed over the whole muscle) can be separated and modelled independently.

(3)   The elasticity of the considered muscle is assumed to be constant. Consequently, a simple linear spring can be used.

(4)   The damping behaviour of the considered muscle is assumed to be constant. Consequently, a simple damper can be used.

(5)   The muscle can be represented with a spring on one side of the mass and a damper in parallel with a contractile element on the other side of the mass.

(6)   The internal force produced by the contractile element will change from zero to its maximal value immediately at the time t = 0.

This means:

$$F(t) = 0 \quad \text{for} \quad t < 0$$
$$F(t) = F_{max} \quad \text{for} \quad t \geq 0$$

(7)   The symbols used in this example are:

t   =   time
m   =   muscle mass

k   =   spring constant
r   =   coefficient of viscous damping
x (t)   =   position of muscle mass (particle)
F (t)   =   internal force produced by the contractile element
K (t)   =   external force at muscle insertion, P(A)

(8)   The forces are drawn in the positive axis direction.

The FBD and the illustration of the situation are shown in Fig. 4.4.10.

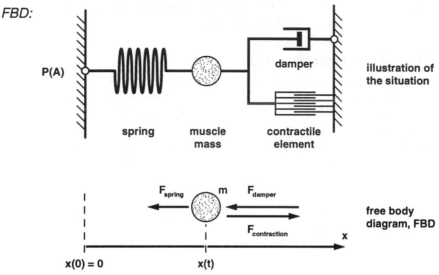

Figure 4.4.10   **Illustration of the simplistic muscle model (top) and the corresponding FBD (bottom) (from Hörler, 1972a, with permission).**

Equations of motion:
    The equation of motion for the particle is:

$$m\ddot{x}(t) = F(t) - kx(t) - r\dot{x}(t)$$

Initial conditions:

$$x(0) = 0$$

$$\dot{x}(0) = 0$$

Final conditions:

$$x(\infty) = x_{max}$$

$$\dot{x}(\infty) = 0$$

Solution:

The position before the time $t = 0$ is used as the zero point of the x-axis. The displacement of the mass, $[x(t) - x(0)]$ which corresponds to $x(t)$, also indicates the elongation of the spring. For the force, $K(t)$, acting at point P (A) one can write:

$$K(t) = kx(t)$$

Because the problem is similar to the problem discussed in the previous chapter, the various steps are not specifically presented. For non-critical damping $a \neq b$, the solution leads to the equation:

$$\frac{K(t)}{K_{max}} = 1 + \frac{b}{a-b} \cdot e^{-at} - \frac{a}{a-b} \cdot e^{-bt}$$

where:

$$a = \frac{r}{2m} + \sqrt{\frac{r^2}{m^2} - \frac{4k}{m}}$$

$$b = \frac{r}{2m} - \sqrt{\frac{r^2}{m^2} - \frac{4k}{m}}$$

For critical damping $a = b$, the solution leads to the equation:

$$\frac{K(t)}{K_{max}} = 1 - e^{-at} - te^{-at}$$

where:

$$a = b = \frac{r}{2m}$$

The result is a two parametric equation for the parameters a and b. The values of a and b can be determined with the help of an experiment. The experiment to determine a and b was performed for the biceps brachialis muscle group. The subject was strapped to a force measuring device which measured the force exerted by the lower arm in the direction of the line of action of the biceps (Fig. 4.4.11). At the time $t(0) = 0$, the subject had to perform a maximal isometric contraction, and the corresponding force, $K(t)$, was measured. The experiment was performed for four different angles between upper and lower arm (elbow angles φ). The experimentally determined result for $K(t)$ was simulated with the

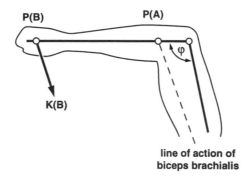

P(B)   P(A)

K(B)

φ

line of action of
biceps brachialis

Figure 4.4.11   **Schematic illustration of the experimental set-up for determining the coefficients a and b in the relative force equation.**

simple mechanical muscle model using a least square fit method. This procedure provided the values for the parameters a and b that are listed in Table 4.4.1.

Table 4.4.1   **Experimentally determined values for the parameters a and b for which the experimental and theoretical results for K(t) fit best.**

φ [°]	a [s^{-1}]	b [s^{-1}]
60	10.6	10.6
90	11.8	11.8
125	14.7	14.7
137	19.9	19.9

The results indicate that the experimental and theoretical K (t) results fit best when a = b. Based on the results of the previous section this corresponds to critical damping. The dimensions of the forces in the mechanical elements (spring and damper) are such that the displacement of the muscle mass occurs without any vibrations in the shortest possible time. The results from this experiment, therefore, suggest that (a) muscle properties are tuned for a certain outcome, and (b) the tuning of the "muscle" corresponds to the tuning of a critically damped movement. One may argue that this is the case in reality since muscle vibrations cannot be seen in such isometric contractions, at least not externally.

The calculations have also been performed with the assumption that the muscle force, F(t), does not change immediately but changes as a function of time.

$$F (t) \quad = \quad F_{max} (1 - e^{-ct}) \quad \text{for} \quad t \geq 0$$

The results for this more "realistic" approach support the same finding that muscle movement is tuned.

The idea of *muscle tuning* may, if further supported, be important for a number of reasons:

- Muscle wobbling should be minimal during locomotion which can be confirmed experimentally.
- Muscle tuning (and, therefore, muscle activity) should be influenced by the magnitude and/or frequency of external forces. It should be different for different surface/shoe combinations.
- Muscle tuning (and, therefore, muscle activity) should be influenced by actual muscle strength and, consequently, muscle fatigue.

These, of course, are only speculations. However, they illustrate how a simple one particle model may prompt a series of thoughts and sometimes findings.

EXAMPLE 7 (a model for "jumping ability")

The jumping ability of animals is rather surprising. One often tends to relate jumping ability to body size. However, this association is not without problems. First, "size" is not well defined. It is not clear, whether size corresponds to height, to weight, or to something else. Second, animals of similar "size" have completely different jumping abilities. A horse can jump rather high while a cow can barely jump, yet they have about a similar "size". Additionally, some small animals, such as fleas, are able to jump fantastic heights. It has been suggested, that the study of geometrically similar creatures may provide the answer to the question of how body "size" of body dimensions influence the jumping ability. The following model has been proposed earlier to solve this question (Hörler, 1973).

Question:

Develop a model to explain the relationship between body dimensions and the ability to jump.

Assumptions:

(1) Only geometrically similar creatures will be considered.

(2) The model is restricted to the vertical jump from standing position without wind-up movement.

(3) The joint angles of the starting position are the same for all "similar creatures".

(4) The "muscle force" responsible for the movement of the centre of mass can be described by a constant force, F.

(5) The human body considered in this context can be described by a particle with the mass, m.

The symbols used in the following are:

t	=	time
a	=	distance which the centre of mass moves upwards during contact with the ground
F	=	average muscle force responsible for movement
W	=	body weight = mg
H	=	height of flight
L	=	length (= height) of the body (body "size")
λ	=	similarity quotient

(6)   The comparison of a test body with a general body provides the similarity relations.

definition: $L = \lambda \cdot L_T$
length: $a = \lambda \cdot a_T$
because of the definition of the similarity quotient.

force: $F = \lambda^2 \cdot F_T$
because of the general rule that the muscular force increases proportionally with the physiological cross-sectional area of the muscle (which corresponds to the second power of the length).

weight: $W = \lambda^3 \cdot W_T$
because the weight increases proportionally with the volume (which corresponds to the third power of the length).

(7)   For the numerical calculations for a test subject (measured in an experiment):

a_T =   0.2 m
L_T =   1.6 m
H_T =   0.3 m

(8)   The problem can be solved using a one-dimensional approach.

An illustration of the situation and a free body diagram are shown in Fig. 4.4.12. The top part illustrates the position of the particle representing the jumper at different times; the bottom part shows the free body diagram.

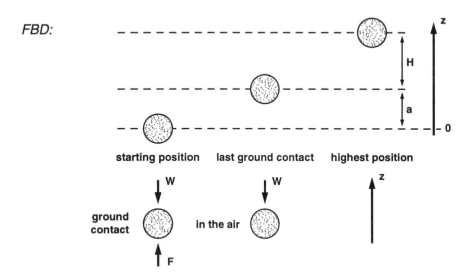

FBD:

Figure 4.4.12   Illustration of the general situation (top) and the free body diagram for the particle while in contact with the ground (bottom left) and while in the air (bottom right) (from Höerler, 1973, with permission).

Solution:

This problem uses energy conservation as one possible approach to reaching a solution. The reader may try to solve the problem in a different way.

The energy at take off corresponds to the work produced by the resultant force over the distance, a, that the particle is travelling vertically during "push-off".

$$(F - W) \cdot a = m \cdot g \cdot H$$

$$H = a\left[\frac{F}{W} - 1\right]$$

Using the test body:

$$H = \lambda \cdot a_T\left[\frac{\lambda^2 \cdot F_T}{\lambda^3 \cdot W_T} - 1\right]$$

$$H = a_T\left[\frac{F_T}{W_T} - \lambda\right]$$

$$H = a_T\left[\frac{F_T}{W_T} - \frac{L}{L_T}\right] \qquad (4.4.18)$$

Equation (4.4.18) describes the general relationship between the height of a jump, (H), and the "body size", (L), for a test subject, T, (Fig. 4.4.13).

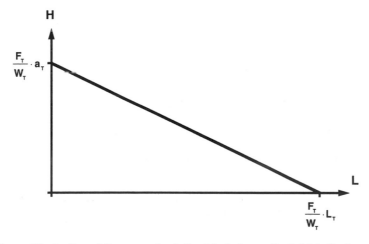

Figure 4.4.13   **Illustration of the general relationship between the height of a jump, H, and the length, L, of the body (body "size") for a test subject, T (from Höerler, 1973, with permission).**

This result shows that:

- The bigger the "body size" the less the height of jumping.
- There is a size limit. If the body exceeds this limit the subject cannot jump and once fallen it would not be able to raise itself to its legs.

Numerical considerations when using (4.4.18) for a test subject are:

$$H_T = a_T \left[\frac{F_T}{W_T} - \frac{L_T}{L_T} \right]$$

$$\frac{F_T}{W_T} = \frac{H_T}{a_T} + 1 = \frac{0.3}{0.2} + 1 = 2.5$$

Generally for a subject similar to the test subject:

$$H = 0.2 \cdot \left[2.5 - \frac{L}{1.6} \right]$$

$$H = \frac{1}{2} - \frac{L}{8}$$

where L and H are measured in metres.

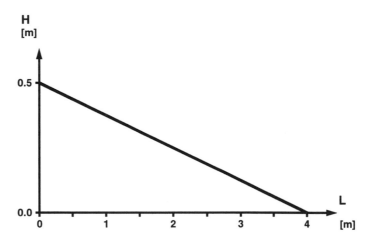

Figure 4.4.14   **Illustration of the relationship between body "size" and jumping height (from Höerler, 1973, with permission).**

The relationship between "body size", L, and "jumping height", H, is illustrated in Fig. 4.4.14. The graphical illustration for our test subject runs into problems for smaller sizes. One would, therefore, want to improve the model, which we will do in the next step.

EXAMPLE 8 (simple example for a system of particles)

This example is a continuation of the model for clumsiness (Example 7). It is an attempt to improve the initial model so that the results become more relevant for actual high jumping.

The relevance of this second step in this model is illustrated by a hypothetical discussion between athletes. The small athlete says to the tall one, "It is easy for you to jump high, because your centre of mass is already high at take-off". The tall athlete answers, "No, it is easy for you to jump high, because you need much less force to move your body mass, which is smaller than mine". Who is right?

Assumptions:

(1) The model of the human body consists of two particles. Half of the body mass is concentrated in the upper particle and the other half in the lower one (Fig. 4.4.15 and Fig. 4.4.16).

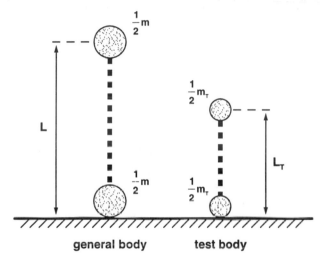

general body   **test body**

Figure 4.4.15   Schematic illustration of the two particle simulation (model) (from Höerler, 1972b, with permission).

(2) The movement analysed is the vertical jump without a wind-up movement.

(3) The "body" has the ability to pull together its two masses. This assumption is subdivided into two parts:

- One where the time for contraction is not a problem.
- One where the time for contraction is a problem.

(4) The length of the body ("size") in its starting position is 75% of its standing length.

(5) All the other assumptions that were made for the one particle model are used analogously in this second step.

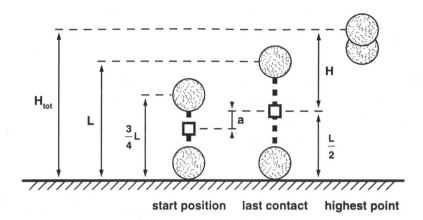

start position   last contact   highest point

Figure 4.4.16   Schematic illustration of the two body model at lowest position, at take-off, and at the highest point. The two masses are only connected by forces acting between them (from Höerler, 1972b, with permission).

Solution:

The solution will be determined in two steps, the first step where the subject has enough time to pull together and the second step where the subject does not have enough time to pull together.

The subject has enough time to pull together

The following calculations are made under the assumption that the subject has ample time to pull the two masses together. This may, of course, not correspond to reality.

Fig. 4.4.16 provides the equation:

$$H_{tot} = \frac{L}{2} + H$$

and using H from the previous section:

$$H = \frac{1}{2} - \frac{L}{8} \quad \text{(in metres)}$$

provides for H_{tot}:

$$H_{tot} = \frac{1}{2} + \frac{3}{8}L \tag{4.4.19}$$

This means that for the specific test-subject (L = 1.6 m) the total height of the jump would be 1.10 metres. The centre of mass would have its highest point 1.10 metres above ground (Fig. 4.4.17).

However, the human body usually does not have enough time to pull together during a jump. Consequently, the model will be developed in a further step with the attempt to incorporate this aspect too.

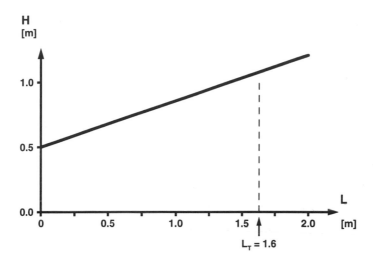

Figure 4.4.17   **Illustration of the relationship between H_{tot} and "body size" determined from the two mass model, if the subject has enough time to pull the two masses together (from Höerler, 1972, with permission).**

Calculations and considerations if there is not enough time to pull together

The assumption that the subject has enough time to pull together is not realistic. The "pulling together process" requires time, and it is more realistic to assume that this process depends on the acceleration of the masses and the available muscle forces. A possible model for this situation is illustrated in Fig. 4.4.18.

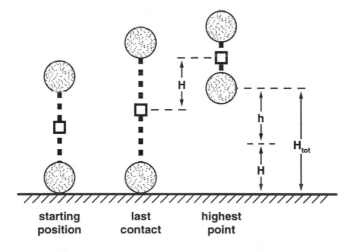

Figure 4.4.18   **Same situation as in 4.4.16 except that the "body" does not have enough time to contract (from Höerler, 1972b, with permission).**

The symbol h represents the distance the "body" is able to pull the lower mass towards the centre of mass during the time Δt of the upwards movement. This upwards movement is produced by a muscle force, and this muscle force is assumed to be the force F that was responsible for the upwards movement during ground contact.

$$H_{tot} \quad = \quad H + h$$

For the (vertical) displacement one can write the general relation:

$$x \quad = \quad \frac{1}{2}\ddot{x} \cdot \Delta t^2$$

if the force, F, is constant and the initial velocity $v_o = 0$.
    This relation can be used for:

(i)   The movement of the centre of mass during flight.
(ii) The movement of the lower mass during flight.

$$\text{(i)} \quad H \quad = \quad \frac{1}{2}g \cdot \Delta t^2 \tag{4.4.20}$$

because the only force acting during the flight is the gravitational force.

$$\text{(ii)} \quad h \quad = \quad \frac{1}{2}\ddot{x} \cdot \Delta t^2$$

$$\frac{1}{2}m\,\ddot{x}(t) \quad = \quad F - \frac{1}{2}W$$

$$\ddot{x}(t) \quad = \quad \frac{2}{m}\left[F - \frac{1}{2}W\right]$$

$$h \quad = \quad \frac{1}{2}\cdot\frac{2}{m}\left[F - \frac{1}{2}W\right]\Delta t^2$$

$$h \quad = \quad \frac{1}{m}\left[F - \frac{1}{2}W\right]\cdot\Delta t^2 \tag{4.4.21}$$

Elimination of Δt from (4.4.20) and (4.4.21) yields:

$$h \quad = \quad \frac{1}{m}\left[F - \frac{1}{2}W\right]\cdot 2\frac{H}{g} \quad = \quad 2H\left[\frac{F}{W} - \frac{1}{2}\right]$$

$$H_{tot} \quad = \quad H + 2H\cdot\frac{F}{W} - H \quad = \quad 2H\cdot\frac{F}{W}$$

H from (4.4.18) can be used:

$$\frac{F}{W} \quad = \quad \frac{F_T}{W_T}\cdot\frac{L_T}{L}$$

$$H_{tot} = 2 \cdot a_T \cdot \frac{F_T}{W_T}\left[\frac{F_T}{W_T} \cdot \frac{L_T}{L} - 1\right] \tag{4.4.22}$$

For the numerical example the total jumping height is determined by:

$$H_{tot} = \frac{4}{L} - 1 \tag{4.4.23}$$

The combination of (4.4.19) and (4.4.23) provides the relationship between the "body size", L, and the height of the jump, H_{tot}, that is illustrated in Fig. 4.4.19.

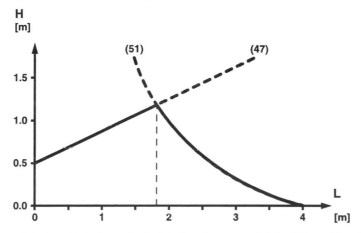

Figure 4.4.19   **Relationship between body "size" and jumping height if the subject lacks enough time to pull the two masses together (from Höerler, 1972b, with permission).**

The maximum of the curve corresponds to the ideal length of the body for this test subject if everything is changed geometrically similar.

Summary

The results of the two models using one and two particles for the description of the relationship between "body size" and jumping height can be summarized as follows:

(1)   Assuming geometrically similar body constructions, there would be an optimal body length ("size") for the vertical jump studied. In other words, the optical illusion in the Tarzan movies that show Tarzan as 3 metres tall but with the movement (jumping) abilities that he would have if his stature were normal is not realistic. Tarzan should be much clumsier.

(2)   This result is based on several assumptions. The most important is that the body weight increases faster than the muscle force for increasing body size (length).

(3)   The numerical considerations were made with a test-subject (a semi-sportive young woman) of 1.6 m body length. They showed that optimal body length

would be 1.83 m. The number, 1.83 m, is not important. This result suggests in a general sense that good high jumpers will be tall but not extremely tall.

(4)   The idea that the performance of a human or an animal is limited by mechanical constraints has been further studied for birds by Pennycuick (1993). He discussed the mechanical limitations on evolution and diversity for birds, and showed that there is a mass limit above which "birds" would not be able to fly. He found that there is no living species weighing more than 14 kg that flies, and that the diversity of birds dwindles as mass increases.

EXAMPLE 9 (elastic and viscoelastic shoe soles and surfaces)

The following model illustrates the influence of the elastic and viscous properties of heel pads, shoe soles and/or playing surfaces on the energy demands during locomotion (Anton and Nigg, 1990), and follows the publication of Nigg and Anton (in press) with the permission of the publisher.

Introductory comments

Three materials are used to "cushion" the landing of the heel during running: the running surface, the midsole material of the shoe, and the soft tissue of the heel. These materials have elastic and viscous components. In order to show the effects of these materials, a simple thought experiment is discussed (Alexander et al., 1986). It is illustrated in Fig. 4.4.20.

The first step in this thought experiment involves a mass and a spring that drops onto a rigid surface. The resultant force-time diagram (top of Fig. 4.4.20) shows the corresponding ground reaction force. The second step involves a second somewhat smaller mass, m, which is added at the lower end of the dropping structure. For this drop test, the force-time diagram looks different. It has a first force peak followed by a second force peak with a slower loading rate than the first peak (second group from the top in Fig. 4.4.20). This force time curve resembles a running force time curve. In the third step of the thought experiment, another spring is added to the dropping system. This spring could represent an additional elastic element - for instance a shoe sole. For this drop test, the force-time diagram resembles the first force-time diagram with the addition of a high frequency modulation of the original signal. In the last step (bottom of Fig. 4.4.20) the lower spring is replaced by a spring-dashpot combination. The force-time curve for this drop test shows an initial impact force peak with some subsequent dampened vibrations, followed by an active force peak.

The thought experiment may suggest that the combination of elastic and viscous elements in the surface-shoe-heel material may be of critical importance for economical running. Vibrations due to dominant elastic behaviour may be disadvantageous to running economy. It is speculated that muscular activity and work may be greater when high frequency vibrations are present than when there are no vibrations. Consequently, it is speculated that the combination of the elastic and viscous material properties of surface-shoe-heel may influence the running economy.

In order to examine these speculations, a theoretical model has been developed. The purpose of this model is to investigate how the work requirements in running are affected by different viscoelastic characteristics of the surface-shoe-heel interface.

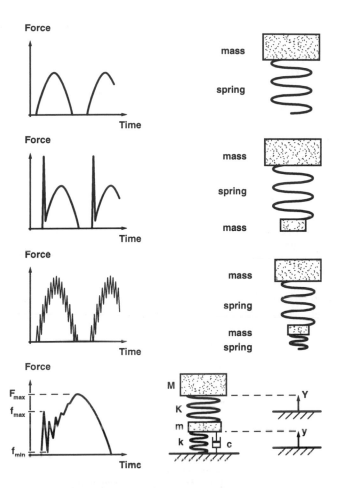

Figure 4.4.20 Schematic illustration of the effects of elastic and/or viscoelastic elements on the ground reaction forces during impact and active phases (from Alexander et al., 1986, with permission).

Models that have represented the combined effects of all the leg and hip muscles involved in running by passive elements such as springs and dampers (McMahon and Green, 1979; Blickhan, 1989) have been found to be inadequate for the purpose of this research. The very fact that serious running is strenuous seems to suggest that modelling muscles by energy conserving mechanical elements like springs may be inappropriate, because the total mechanical energy content in a system composed exclusively of masses and springs remains constant over time. Only the relative amounts of kinetic and potential energy change. In the human body, the question of how much work is performed in a mechanical system is, therefore, meaningless. Spring-mass systems are not suited to respond to the purpose of this investigation. Additionally, including damper elements in the system decreases the mechanical energy content over time. The lost energy cannot be regained for lack of active components in these models. As a result, models including exclusively masses, springs, and dampers are not ideal for describing the energy aspects of a sustained running motion.

The model described in this section represents an attempt to replace the passive mechanical elements (spring and dampers) between the foot and the rest of the body, with a strategic formulation of how a resultant force that represents the net effect of all the muscles between the foot and the rest of the body has to evolve over time in a running situation. The model derived is then used to study work requirements for various surface-shoe-heel characteristics.

The model

Question:

Determine the influence of the elastic and viscous elements that cushion the ground contact on the energy demands of a runner.

Assumptions:

(1)   The human body is subdivided into two masses. One mass, m_1, represents the foot of the support leg. The other mass, m, represents the rest of the body (Fig. 4.4.21). This

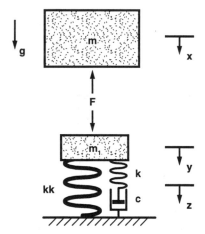

Figure 4.4.21   **Illustration of the mechanical model used in this example (from Nigg and Anton, in press, with permission).**

second mass, m, is called the "upper body" in the following.

(2)   The effects of all the extremities except the support leg on the upper body are neglected.

(3)   The horizontal velocity of the upper body is assumed to be constant. The model neglects its movement in the horizontal direction, considering only its vertical movement. The model is, therefore, one-dimensional.

(4)   A spring-damper combination, k, kk, and c, represents the combined material properties of the surface, the shoe midsole, and the human heel.

(5)   The surface is assumed to be rigid.

(6)   A force, F, acts between the upper body and the foot.

(7)   A force, F_G (ground reaction force), acts between the rigid ground and the lower end of the spring-damper combination.

(8)   During the flight phase the runner's body moves freely in the conservative gravitational force field. For that time interval the total mechanical energy (sum of kinetic and potential energy) is assumed (as a first approximation) to be constant.

(9)   The mathematical analysis is limited to the stance phase.

(10) During stance phase the upper body loses height and is simultaneously slowed down. The total mechanical energy of the body decreases from an initial value when it touches down to a minimal value and increases again towards take-off.

(11) The force, F, between the foot and the upper body develops so as to minimize the work it performs in bringing the upper body from touch-down to take-off. In its mathematical expression this assumption leads to an open loop optimal control problem.

(12) The muscles that generate the force, F, are assumed to be incapable of energy storage. Therefore, the work performed is counted positive for the upper body moving up and down.

(13) The problem can be solved one-dimensionally.

The free body diagram for this example is illustrated in Fig. 4.4.22.

FBD:

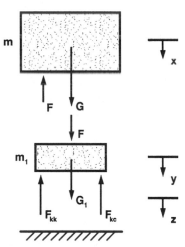

Figure 4.4.22   **FBD of the two "particle" model for the energy demand considerations during running.**

Equations of motion:

The equations of motion for the illustrated set-ups are:

$$m\ddot{x} \quad = \quad m \cdot g - F \tag{4.4.24}$$

$$m\ddot{y} \quad = \quad m_1 \cdot g + F - (kk + k) \cdot y + k \cdot z \tag{4.4.25}$$

$$\dot{z} \quad = \quad \frac{k}{c}(y - z) \tag{4.4.26}$$

where:

g   =   earth acceleration
x, y, z =   coordinates

Solution:

The question will be solved as an optimization problem. In this process the equations of motion act as constraints for the optimization problem.

$$\int_0^t \left[|F(\dot{x} - \dot{y})| + a \cdot F^2 + b \cdot f(\dot{F}) \right] dt \rightarrow \min \tag{4.4.27}$$

where:

$$f(\dot{F}) \quad = \quad (\dot{F} - \dot{F}_{max})^2 \quad \text{if} \quad \dot{F} > \dot{F}_{max}$$

$$f(\dot{F}) \quad = \quad 0 \quad \quad \text{if} \quad \dot{F}_{min} \le \dot{F} \le \dot{F}_{max}$$

$$f(\dot{F}) \quad = \quad (\dot{F} - \dot{F}_{min})^2 \quad \text{if} \quad \dot{F} < \dot{F}_{min}$$

The first term under the integral represents the work performed by the force, F. The second term under the integral represents the fact that the physiological cost to the system is higher at high force levels than at lower ones. The third term under the integral permits the limitation of the maximal rate of force increase and decrease, dF/dt. The second time derivative of F is the unknown function that will be determined by the optimization process. The second derivative of F was chosen as the unknown so that boundary conditions for F and dF/dt could be specified. The factors a and b permit the adjustment of the relative importance of the second and third terms under the integral with respect to the first term.

The ground reaction force, F_G, is given by the equation:

$$F_G \quad = \quad (kk + k) \cdot y - k \cdot z$$

Pontryagin's maximum principle is applied to solve the optimization problem. The first term under the integral supplies the work required for a one step cycle once the solution to the optimization process is substituted.

The model was initially tested using the following inputs:

Δt_1 =   0.1 s corresponding to running
Δt_2 =   0.6 s corresponding to walking
m   =   70 kg
m_1 =   7.5 kg
kk   = $2.5 \cdot 10^5$ N/m
k   = $2.5 \cdot 10^6$ N/m
c   = $8.4 \cdot 10^3$ kg/s

$$\dot{F}_{max} = +7.5 \cdot 10^4 \text{ N/s}$$
$$\dot{F}_{min} = -7.5 \cdot 10^4 \text{ N/s}$$

The stiffness and damping values chosen provide for a stiff and critically dampened foot-surface interface.

The initial and terminal boundary conditions chosen were:

for t = 0 s
x = 0 m
y = 0 m
z = 0 m
$\dot{x}$ = 0 m/s
$\dot{y}$ = 0 m/s
$\dot{z}$ = 0 m/s
F = 0 N
F_G = 0 N

for t = Δt_i (0.1 s or 0.6 s)

$\dot{x}$ = −0.6 m/s
$x - y$ = 0 m
F_G = 0 N

The second terminal boundary condition, $x - y = 0$ m, stipulates that the length of the leg at take-off is the same as at touch-down. However, take-off may occur with the masses m and m1 being located at different heights than at touch-down.

Results of the model and discussion

The output of the model consists of the ground reaction force, F_G, and the force, F, which acts between the upper body and the foot. The estimation of the ground reaction force is not needed since it is possible to measure the ground reaction force experimentally. However, the ground reaction force can be used to evaluate (validate) the model. If the ground reaction force predicted by the model and the actual ground reaction force are similar, confidence in the other results increases. Consequently, a first step compares the predicted ground reaction forces with the experimentally determined ground reaction forces. Fig. 4.4.23 shows the predicted ground reaction force for "running" (contact time = 0.1 s). The estimated ground reaction force shows some similarity to experimentally determined ground reaction forces (Cavanagh and Lafortune, 1980; Nigg and Lüthi, 1980). The estimate shows the initial impact force peak and the subsequent active force peak.

Fig. 4.4.24 shows the predicted ground reaction force for walking. The general characteristics for ground reaction forces for walking are present in this predicted force time curve. The curve has a camel-like shape as typical walking curves have.

The two comparisons between predicted and experimentally determined ground reaction forces show good agreement in shape as well as in magnitude. Based on this agreement, the credibility of the values of other variables determined by this model may increase. Note that only very general assumptions were made for the calculation with this

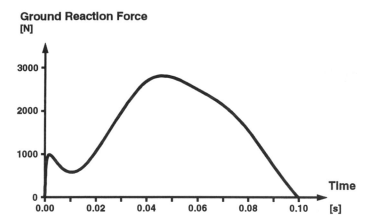

Figure 4.4.23   Predicted ground reaction force, F_G, for a contact time of 0.1 s (running) (from Nigg and Anton, in press, with permission).

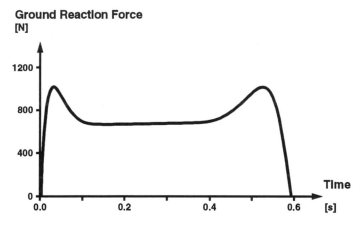

Figure 4.4.24   Predicted ground reaction force, F_G, for a contact time of 0.6 s (walking) (from Nigg and Anton, in press, with permission).

model. In addition to some geometrical assumptions (e.g., that landing and take-off speed are the same) the main assumption was that F is selected so that the mechanical work performed by F is minimal. This simple mechanical system with very basic assumptions produces ground reaction forces that correspond in magnitude and shape to the actual ground reaction forces measured experimentally. Furthermore, the shape of the ground reaction force changes from short to long contact times, the same way it changes from running to walking.

The second step in the model calculations estimated the forces, F, between "foot" and "upper body". In a first approximation, these forces can be considered as the forces in the "ankle joint". Fig. 4.4.25 illustrates the force-time diagram for the force F for the contact time of 0.1 s. Notice that the peak value of F is about 10% smaller than the peak value of the ground reaction force, F_G, for the same movement ($F_{max} = 2500$ N and $F_{Gmax} = 2800$

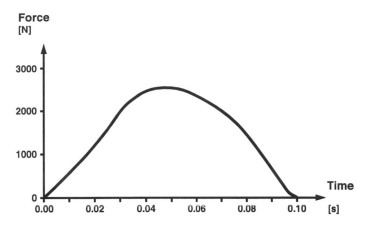

Figure 4.4.25   Illustration of the internal force, F, as a function of time for the short contact time, 0.1 s, corresponding to the movement running (from Nigg and Anton, in press, with permission).

N) due to dynamic effects at the foot level. Additionally, the impact peak in the ground reaction force, which is solely due to dynamic effects at the foot level, is absent in the force curve for F at the "ankle joint" level. The force-time curve, F(t), for the longer contact time of 0.6 s is nearly identical to the ground reaction force curve for the same contact time. Dynamic effects do not have a noticeable effect and no diagram for them is, therefore, shown.

In the third step the work performed by F was calculated. This calculation has been done for the following material constants of the elements between the foot and the surface:

$$
\begin{aligned}
kk &= 1.25 \cdot 10^5 \text{ N/m} \quad \text{(case 1)} \\
kk &= 2.5 \cdot 10^5 \text{ N/m} \quad \text{(case 2)} \\
k &= \text{variable from } 2.5 \cdot 10^5 \text{ N/m to } 6.25 \cdot 10^5 \text{ N/m} \\
c &= \text{variable from } 2.5 \cdot 10^3 \text{ kg/s to } 17.5 \cdot 10^3 \text{ kg/s}
\end{aligned}
$$

The material constant, kk, has been chosen in such a way that the maximal static deflection of the foot mass, m_1, under a load of 2500 N was 2 cm for case 1, and 1 cm for case 2. The values used correspond reasonably well to the actual forces and deflections in human movement.

The ranges for k and c have been chosen so that they extend from subcritical damping of a system composed solely of m_1, k, kk, and c to critical damping of a system including m, m_1, k, kk, and c. This range was assumed to cover the actual range of possibilities for running.

The estimations of the performed work during a step cycle (Fig. 4.4.26 and Fig. 4.4.27) indicate that the amount of work required is generally higher for case 1 (the softer spring constant for kk) than for case 2 (the harder spring constant for kk). For case 1, with the softer spring, kk, the work performed decreases steadily with increasing c and decreasing k. Higher values for c, the damper, make the foot-surface interface dynamically stiffer, which results in a lower damper deflection and consequently in lower damping.

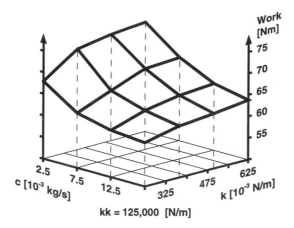

kk = 125,000 [N/m]

Figure 4.4.26   Work required per step cycle for kk = 1.25 · 10⁵ N/m and variable k and c (case 1) (from Nigg and Anton, in press, with permission).

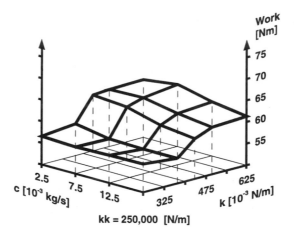

kk = 250,000 [N/m]

Figure 4.4.27   Work required per step cycle for kk = 2.50 · 105 N/m and variable k and c (case 2) (from Nigg and Anton, in press, with permission).

Increasing values of k communicate more force to the damper, which results in higher damper deflections and higher damping.

Case 2, in which the spring stiffness of kk is higher, presents a completely different relationship between the material properties of k and c and the work performed. The influence of the damping coefficient, c, is quite small, but the influence of k becomes more interesting. There is a critical range in k values over which the work requirements change rapidly, whereas, they remain fairly constant over the remaining intervals of k and c. From a practical standpoint, this means that there is a critical combination of material properties at which relatively small changes in material properties are associated with work increases of about 10%.

This example is only a first step in the attempt to determine the effect of various material properties and movement changes on the work requirements during running. The model does not allow us to provide specific details about the critical material properties for reduced or increased work requirements. As a matter of fact, one cannot really separate the material constants k and c. The required work depends on the combined effect of k and c. However, the model illustrates that specific combinations of material properties may be advantageous or disadvantageous from an energy point of view.

This simple model also shows that the work requirement is not exclusively dependent on how much energy is lost in the damper. Work is performed by the muscles in slowing down the upper body after touch-down and in accelerating it in anticipation of take-off. The amount of work performed is the sum of force multiplied by displacement increments. The functional dependencies of force and displacement on time are interdependent for the problem under investigation, and they depend on the visco-elastic foot-shoe-surface interface. They must assume a form that fulfils the task of bringing the mass m from touch-down to lift-off under the given boundary conditions. Note that even if c were set to zero, the resulting work requirement would still be unequal to zero and would assume different values for different spring constants kk.

There is evidence that part (but only part) of the kinetic energy at touch-down is stored in elastic tissue during the stance phase (Alexander and Vernon, 1975). This is not inconsistent with the approach taken here. Even if the muscles are required to perform only part of the work given by the first term in the integral equation, it would still make sense for this portion to be minimal. However, the muscle tendon units have been assumed in this approach to be unable to store energy. The fact that the theoretically estimated ground reaction force is close to the experimentally determined ground reaction force, may suggest that the idea of storage of elastic energy in the muscle-tendon units during running should be carefully reconsidered.

The modelling approach used here determined an integral force, F, which could be considered the result of all forces produced due to muscle activity, gravity, and inertia at the ankle joint level. The initial and terminal boundary conditions chosen were realistic for running. The resultant F_G-time curves as estimated from the model, and the F_G-time curves as determined from experiments, show good agreement in magnitude and shape. This may suggest that the presented optimal control model for running and its underlying postulate of minimum performed work is acceptable.

4.4.5 MECHANICAL MODELS USING RIGID BODIES

INTRODUCTORY COMMENTS

In the preceding sections it was assumed that each "body" or "system of interest" could be treated as a single particle or as a system of particles. Such an approach, however, is not always appropriate. In this section the systems of interest discussed are considered as rigid bodies.

Forces acting on a rigid body may be classified as either external or internal forces. External forces result from the action of other bodies. They include contact and remote forces and are responsible for the kinematic behaviour of the rigid body of interest.

The possible movements of a rigid body are translation and rotation.

The degree of freedom (DOF) of a rigid body is the number of independent variables necessary to describe its position in space.

The maximal number of variables necessary to describe translation is three. The maximal number of variables necessary to describe rotation is three. Consequently, in a general case, a rigid body has three translational and three rotational degrees of freedom.

Kinetics and kinematics of a rigid body may be described by using the two fundamental principles of linear and angular momentum, and the conservation of mechanical energy theorem:

- Conservation of linear momentum.
- Conservation of angular momentum.
- Conservation of energy.

These three sets of equations describe the movement of a rigid body. In most applications, however, not all of them are needed. Other ways to discuss movements of a rigid body have been discussed in chapter 1, however, in the following examples, the three laws of conservation are used. In a majority of such analyses in biomechanics the laws of conservation of linear and angular momentum are used.

The conservation of linear momentum can be written in its components as:

$$m\ddot{x} \quad = \quad F_{x1} + F_{x2} + ... + F_{xn}$$

$$m\ddot{y} \quad = \quad F_{y1} + F_{y2} + ... + F_{yn}$$

$$m\ddot{z} \quad = \quad F_{z1} + F_{z2} + ... + F_{zn}$$

where:

m = mass of the rigid body of interest
x, y, z = coordinates of the centre of mass of the rigid body
F_{ji} = force component of force i in j direction

The conservation of the angular momentum in the most general case can be written as:

$$\{M\} \quad = \quad [I]\,\{\alpha\} + [\omega]\,[I]\,\{\omega\} \tag{4.4.28}$$

where:

$\{\}$ = vector symbol
$[]$ = tensor symbol
$\{M\}$ = moment vector
$[I]$ = inertia tensor
$\{\alpha\}$ = angular acceleration vector
$[\omega]$ = angular velocity vector expressed as a skew symmetric second order tensor

$\{\omega\}$ = angular velocity vector

or in the form of equation (4.4.29):

$$\begin{bmatrix} M_x \\ M_y \\ M_z \end{bmatrix} = \begin{bmatrix} I_{xx} & -I_{xy} & -I_{xz} \\ -I_{yx} & I_{yy} & -I_{yz} \\ -I_{zx} & -I_{zy} & I_{zz} \end{bmatrix} \begin{bmatrix} \alpha_x \\ \alpha_y \\ \alpha_z \end{bmatrix} + \begin{bmatrix} 0 & -\omega_z & \omega_y \\ \omega_z & 0 & -\omega_x \\ -\omega_y & \omega_x & 0 \end{bmatrix} \begin{bmatrix} I_{xx} & -I_{xy} & -I_{yz} \\ -I_{yx} & I_{yy} & -I_{yz} \\ -I_{zx} & -I_{zy} & I_{zz} \end{bmatrix} \begin{bmatrix} \omega_x \\ \omega_y \\ \omega_z \end{bmatrix}$$

$\begin{array}{lll} I_{xx}, I_{yy}, I_{zz} & = & \text{moments of inertia with respect to, CM} \\ I_{ij} & = & \text{product of inertia with respect to the axes i and j} \\ \alpha_x, \alpha_y, \alpha_z & = & \text{angular acceleration components} \\ M_x, M_y, M_z & = & \text{moment components of moment, } \mathbf{M} \end{array}$

Consequently, a maximum of six equations are available (three for translation and three for rotation) for the mathematical description of the kinematics and/or kinetics of one rigid body. A two-dimensional example for one rigid body with four external forces acting on it is discussed below. Also given are the free body diagram (Fig. 4.4.28) and the corre-

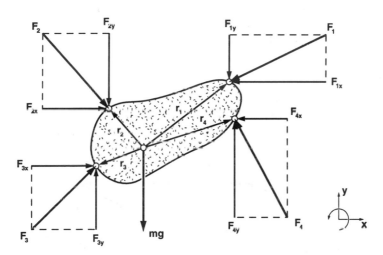

Figure 4.4.28   Rigid body with four forces acting externally.

sponding equations of motion.

$$m\ddot{x} \quad = \quad -F_{1x} + F_{2x} + F_{3x} - F_{4x}$$

$$m\ddot{y} \quad = \quad -F_{1y} - F_{2y} + F_{3y} + F_{4y} - mg$$

$$I_{zz}\ddot{\phi}_z \quad = \quad [(\mathbf{r}_1 \times \mathbf{F}_1) + (\mathbf{r}_2 \times \mathbf{F}_2) + (\mathbf{r}_3 \times \mathbf{F}_3) + (\mathbf{r}_4 \times \mathbf{F}_4)]_z$$

where:

×	=	sign for vector multiplication
r_i	=	vector from the origin to the point of application of the force F_i

Note that the right hand side of the moment equation is the component of the moment in z-axis direction, which is, of course, a scalar quantity. The subscript "z" beside the bracket indicates this fact.

Section 4.4.5 is again a section with many examples. The examples carry two messages. First, they attempt to illustrate different possibilities to model determinate systems of rigid bodies with simple mathematical tools. Second, they attempt to discuss the physical and biological interpretation which simple models of rigid bodies may include.

EXAMPLE 10 (stability in somersaulting)

Somersaults are performed in many sports activities, including gymnastics, diving, and ski acrobatics. Coaches and athletes know, based on practical experience, that some rotations are stable and others unstable. In other words, in some types of somersaults it is difficult to maintain the initial rotation around the same axis, while in others it is easy. Lay-out somersaults, for instance, are difficult, tucked somersaults easy to balance. Furthermore, it is known that the magnitude of the principal moments of inertia play an important role in the stability of a specific rotation. The purpose of this section is to provide two model considerations to shed light on somersaults.

Plausibility considerations

Consider a rectangular block of a homogeneous material with sides a, b, and c (Fig. 4.4.29) and principal axes 1, 2, and 3. This block is thrown into the air rotating

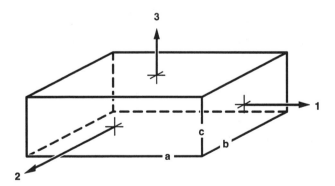

Figure 4.4.29   **Rigid body, with three principle axes. Axis 3 having the maximal and axis 1 having the minimal moment of inertia.**

around one of the three principle axes. During the airborne phase only gravity acts on the block, so the angular momentum, H, remains constant. One would, therefore, expect that the rotation, which at the beginning is around one particular principal axis of the rectangular block, should continue around the same principal axis during the whole airborne phase. These expectations can be experimentally verified for rotations around axes 1 and 3, the axes with the maximal (3) and minimal (1) values for the principal moments of inertia. However, rotations that start around axis 2, the axis with the median value for the principal moment of inertia, are unstable, and the block starts to twist and tilt (Nigg, 1974).

Question:

Illustrate with the conservation of energy the fact that some rotations are stable and others unstable.

Assumptions:

(1) Homogeneous rigid rectangular block.
(2) No air resistance.
(3) $a > b > c$.
(4) $I_{11} < I_{22} < I_{33}$.

Solution:

In this example, the conservation of energy is used.

Symbols:

I_{ii} = moment of inertia for a principal axis (central principal moments of inertia where central is referring to the centre of mass)

I_a = moment of inertia of the rectangular block with respect to a momentary axis of rotation "a"

E_{kr} = kinetic energy of rotation

$\mathbf{H}$ = angular momentum vector with the components (H_1, H_2, H_3)

ω = angular velocity about the principle axis i

For a principal-axis-system the kinetic energy of rotation is:

$$E_{kr} = \frac{1}{2} I_{11} \cdot \omega_1^2 + \frac{1}{2} I_{22} \cdot \omega_2^2 + \frac{1}{2} I_{33} \cdot \omega_3^2 \qquad (4.4.29)$$

where the axes 1, 2, and 3 are the principal axes of the body.

The components H_1, H_2, and H_3 of the angular momentum vector, $\mathbf{H}$, are given by:

$$H_i = I_{ii} \cdot \omega_i \quad (i = 1, 2, 3) \qquad (4.4.30)$$

Equation (4.4.29) may now be written in the form:

$$1 = \frac{H_1^2}{2E_{kr} \cdot I_{11}} + \frac{H_2^2}{2E_{kr} \cdot I_{22}} + \frac{H_3^2}{2E_{kr} \cdot I_{33}} \qquad (4.4.31)$$

Equation (4.4.31) represents the surface of an ellipsoid aligned with the body axes in the angular momentum space on which the tip of the angular momentum vector must lie.

The magnitude of the angular momentum vector is given by:

$$H^2 = H_1^2 + H_2^2 + H_3^2 \qquad (4.4.32)$$

This may be written in the form:

$$1 = \frac{H_1^2}{H^2} + \frac{H_2^2}{H^2} + \frac{H_3^2}{H^2} \qquad (4.4.33)$$

Equation (4.4.33) represents the surface of a sphere in the angular momentum space on which the tip of the angular momentum vector must lie.

Both equations (4.4.31) and (4.4.33) must be satisfied throughout a given motion. This means that the angular momentum vector will travel along the intersection of the surface of the ellipsoid and the surface of the sphere. The results will be discussed for special cases.

Special case 1:

The momentary axis of rotation coincides with principal axis 1.

$$I_{11} = I_a$$

Since I_{11} is the smallest moment of inertia, the length of the radius of the sphere is the same as the length of the smallest semi-axis (in direction 1) of the ellipsoid. Therefore, the ellipsoid and the sphere have two common points, or the sphere contacts the surface of the ellipsoid at the two points where they intersect with axis 1 (Fig. 4.4.30 left). A small vari-

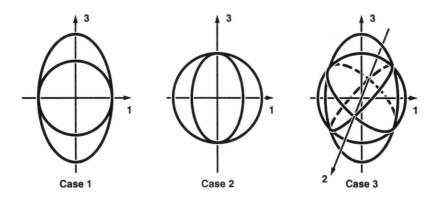

Case 1	Case 2	Case 3

Figure 4.4.30   **Illustration of stable and unstable mathematical solutions in the angular momentum space for the selected special cases 1, 2, and 3, which correspond to initial rotations around each of the principal axes of rotation.**

ation of the momentary angular velocity direction corresponds to a small variation of a stable state of equilibrium. No neighbouring point is a possible solution. Consequently, the rotation around the principal axis with the minimal value of moment of inertia is stable.

Special case 2:

The momentary axis of rotation coincides with principal axis 3.

$$I_{33} = I_a$$

Since $I_{33} > I_{22} > I_{11}$, the ellipsoid in this special case has two common points of contact with the sphere (Fig. 4.4.30 centre). All other points of the sphere surface fall outside the ellipsoid surface. A small variation of the momentary angular velocity direction signifies a small variation of a stable state of equilibrium. However, all neighbouring points of the solution point are not a possible solution. Consequently, the rotation around the principal axis with the maximal value of moment of inertia is a stable rotation.

Special case 3:

The momentary axis of rotation coincides with principal axis 2.

$$I_{22} = I_a$$

Since $I_{11} < I_{22} < I_{33}$, the sphere surface and the ellipsoid surface have two intersecting curves (Fig. 4.4.30 right side). All points of these two curves are possible solutions for the angular momentum vector A_a. The neighbouring points to the initial solution (axis 2 intersection) are possible solutions. A small variation of the momentary angular velocity direction signifies, therefore, a small variation in an unstable state of equilibrium, since all the neighbouring points of the initial solution are possible solutions too. Axis 2, therefore, may "move away" from the initial direction of the angular momentum. Consequently, rotation around the principal axis with the median value of moment of inertia is unstable.

Application to human movement:

In the standing position, the axis with the median value of moment of inertia is the transverse axis. The axis with the smallest moment of inertia is the longitudinal axis. The axis with the largest moment of inertia is the antero-posterior or sagittal axis. Estimated moments of inertia for the three "principal" axes of the human body in the standing position are summarized in Table 4.4.2.

Table 4.4.2   Estimated moments of inertia for the three "principal" axes, longitudinal, transverse, and antero-posterior.

AXIS	ESTIMATED MOMENT OF INERTIA [kgm²]
LONGITUDINAL	$\approx$ 1
TRANSVERSE	$\approx$ 12
ANTERO-POSTERIOR	$\approx$ 13

Rotation around the transverse axis occurs in the layout somersault. From the considerations above, we know that this rotation is unstable. The layout somersault is, therefore, not as simple as commonly assumed. On the other hand, initiating twisting during a layout somersault would seem to be much less difficult than generally assumed. To initiate a twist during the airborne somersault phase:

- There has to be an initial rotation around the axis with the median moment of inertia (for the layout somersault, the transverse axis).
- The body has to be in the layout position (Note: in the tucked position the moment of inertia around the transverse axis is the greatest).
- The initially symmetrical "body" must be made asymmetrical.

For the somersault in the tucked position the moments of inertia may be estimated using an ellipsoid approach with axis "a" as the transverse axis, axis "b" as the sagittal axis, and axis "c" as the longitudinal axis:

$$I_a = \frac{m}{5}(b^2 + c^2)$$

$$I_b = \frac{m}{5}(a^2 + c^2)$$

$$I_c = \frac{m}{5}(a^2 + b^2)$$

With the assumptions:

$$
\begin{aligned}
m &= 60 \text{ kg} \\
a &= 0.2 \text{ m} \\
b &= 0.3 \text{ m} \\
c &= 0.5 \text{ m}
\end{aligned}
$$

the moments of inertia in the tucked position are estimated to be:

transverse axis	$I_a =$	4.1 kgm^2
sagittal axis	$I_b =$	3.5 kgm^2
longitudinal axis	$I_c =$	1.6 kgm^2

Consequently, a rotation around axis "b", the sagittal axis, is unstable, while rotations around axis "a", the transverse axis, and axis "c", the longitudinal axis, are stable for somersaults in the tucked position. It is, therefore, more difficult to initiate a twisting movement from the tucked position.

The moments of inertia in the layout position for the transverse and sagittal axes are relatively close. A change in body configuration (e.g., backwards arching) may change which axis has the median moment of inertia, and make a previously unstable rotation stable.

The mechanics of twisting somersaults

The rotations of a rigid body during airborne movements in general, and the mechanics of twisting somersaults in particular, have been discussed extensively by Yeadon

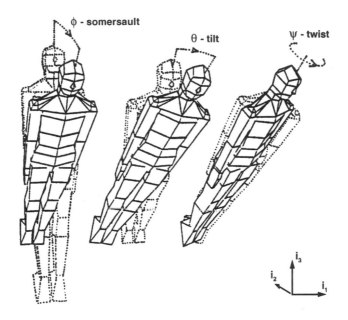

φ - somersault

θ - tilt

ψ - twist

Figure 4.4.31   **Illustration of the three rotational angles used to describe rotations of a rigid body during airborne motions (from Yeadon, 1984, with permission of Pergamon Press Ltd., Headington Hill Hall, Oxford 0X3 0BW, UK).**

(1993a, b, c, d, and 1984). Unstable rotations during somersaults have been discussed extensively in these publications, and the following paragraphs summarize their findings.

Rotations of a rigid body are typically described using three rotational angles, as illustrated in Fig. 4.4.31:

Φ   SOMERSAULT   Rotation about an axis fixed in space.
Ψ   TWIST   Rotation about a longitudinal body axis.
Θ   TILT   Rotation away from the somersault plane.

Airborne motions of a rigid body fall into the twisting and wobbling modes and the singular solution separating them. In the singular mode, the rotation occurs about the axis with the medial moment of inertia. The rotation is unstable. In the twisting mode, the twist increases monotonically. In the wobbling mode the twist oscillates about a mean value. The objective of this section is to derive the two modes and the singular solution and to discuss their implications for the airborne movement of humans.

Derivation of wobbling and twisting modes:

A rigid body with zero net torque conserves angular momentum, H:

$$\mathbf{H} \quad = \quad I_{ff} \quad \cdot \quad \omega_{fi} \tag{4.4.34}$$

where:

I_{ff} =   whole body inertia tensor

ω_{fi} =   angular velocity of the body relative to the inertial frame i.

The conservation of angular momentum can be expressed in terms of a local frame, f. For this purpose, I_{ff}, may be oriented with the principal axes of the body so that $(I_{ff})_f$ is a diagonal matrix:

$$(I_{ff})_f \quad = \quad \begin{bmatrix} A & 0 & 0 \\ 0 & B & 0 \\ 0 & 0 & C \end{bmatrix}$$

where:

A   =   principal moment of inertia for the principal axis $\mathbf{f}_1$,
B   =   principal moment of inertia for the principal axis $\mathbf{f}_2$, and
C   =   principal moment of inertia for the principal axis $\mathbf{f}_3$

In the frame f, the angular momentum, $(\mathbf{H})_f$, is:

$$(\mathbf{H})_f \quad = \quad R_3(\Psi) \cdot R_2(\Theta) \cdot R_1(\phi) \cdot (\mathbf{H})_i \tag{4.4.35}$$

where:

$R(\Psi)$ =   rotation matrix corresponding to the twist angle, Ψ,
$R(\Theta)$ =   rotation matrix corresponding to the tilt angle, Θ, and
$R(\Phi)$ =   rotation matrix corresponding to the somersault angle, Φ

The angular momentum, $(\mathbf{H})_i$, is set such that the tumbling motion takes place in a specific plane:

$$(\mathbf{H})_i \quad = \quad \begin{bmatrix} H \\ 0 \\ 0 \end{bmatrix}$$

which leads to:

$$(\mathbf{H})_f \quad = \quad \begin{bmatrix} H\cos\Theta\cos\phi \\ -H\cos\Theta\sin\phi \\ H\sin\Theta \end{bmatrix} \tag{4.4.36}$$

In the frame f, the angular velocity, ω_{fi}, is:

$$(\omega_{fi})_f \quad = \quad \begin{bmatrix} \cos\Theta\cos\Psi & \sin\Psi & 0 \\ -\cos\Theta\sin\Psi & \cos\Psi & 0 \\ \sin\Theta & 0 & 1 \end{bmatrix} \cdot \begin{bmatrix} \dot{\phi} \\ \dot{\Theta} \\ \dot{\Psi} \end{bmatrix} \tag{4.4.37}$$

The initial equation then takes the form:

$$
\begin{bmatrix} H\cos\Theta\cos\Psi \\ -H\cos\Theta\sin\Psi \\ H\sin\Theta \end{bmatrix} = \begin{bmatrix} A & 0 & 0 \\ 0 & B & 0 \\ 0 & 0 & C \end{bmatrix} \cdot \begin{bmatrix} \dot\phi\cos\Theta\cos\Psi + \dot\Theta\sin\Psi \\ -\dot\phi\cos\Theta\sin\Psi + \dot\Theta\cos\Psi \\ \dot\phi\sin\Theta + \dot\Psi \end{bmatrix} \qquad (4.4.38)
$$

In the absence of external torque, the conservation of rotational energy yields:

$$
E_{rot} = \frac{1}{2} \cdot (A\omega_1^2 + B\omega_2^2 + C\omega_3^2) \qquad (4.4.39)
$$

where:

E_{rot} = (constant) rotational energy
$\omega_1, \omega_2, \omega_3$ = components of the angular velocity, $(\omega_{fi})_f$

or in terms of Θ and Ψ:

$$
E_{rot} = \frac{1}{2} \cdot \left(\frac{H^2\cos^2\Theta\cos^2\Psi}{A} + \frac{H^2\cos^2\Theta\sin^2\Psi}{B} + \frac{H^2\sin^2\Theta}{C} \right) \qquad (4.4.40)
$$

This may alternately be written in either of the following two forms:

$$
\sin^2\Psi = (1 - c_\alpha^2\sec^2\Theta) \cdot \frac{B(A-C)}{C(A-B)} \qquad (4.4.41)
$$

or:

$$
\cos^2\Psi = (c_\alpha^2\sec^2\Theta - 1) \cdot \frac{A(B-C)}{C(A-B)} \qquad (4.4.42)
$$

where the coefficients c_α and c_β are defined as:

$$
c_\alpha^2 = \frac{\dfrac{H^2}{C} - 2E_{rot}}{\dfrac{H^2}{C} - \dfrac{H^2}{A}} \qquad\qquad c_\beta^2 = \frac{\dfrac{H^2}{C} - 2E_{rot}}{\dfrac{H^2}{C} - \dfrac{H^2}{B}}
$$

The *somersault rate* (angular velocity for somersault) may be expressed using the above equations as:

$$
\dot\phi = H\left[\frac{1}{C} - (\frac{1}{C} - \frac{1}{A}) c_\alpha^2\sec^2\Theta \right] \qquad (4.4.43)
$$

The *twist rate* (angular velocity for twist) may be expressed using the above equations as:

$$\dot{\Psi} \quad = \quad H\left[\frac{1}{C} - \frac{1}{A}\right] c_\alpha^2 \sec^2\Theta \sin\Theta \tag{4.4.44}$$

The constants c_α^2 and c_β^2 are positive if C is the minimal principal moment of inertia. For the following considerations we assume that A>B>C (note that the previous discussions are correct for A>B>C and for B>A>C).

The constant c_α^2 attains its minimal value of zero if:

$$E_{rot1} \quad = \quad \frac{H^2}{2C}$$

For a given angular momentum, the energy, E_{rot1}, corresponds to the largest possible energy of rotation, since A > B > C. In other words, if $c_\alpha^2 = 0$, the rotational energy of the rigid body is maximal.

The constant c_α^2 attains its maximal value of 1 if:

$$E_{rot2} \quad = \quad \frac{H^2}{2A}$$

For a given angular momentum, the energy, E_{rot2}, corresponds to the smallest possible energy of rotation, since A>B>C. In other words, if $c_\alpha^2 = 1$, the rotational energy of the rigid body is minimal.

Consequently, the coefficient c_α^2 is positive and between 0 and 1:

$$0 \quad \leq \quad c_\alpha^2 \quad \leq \quad 1$$

One may define a real value, α, by the relationship:

$$\cos^2\alpha \quad = \quad c_\alpha^2$$

The previous equations can be used to show that, for a twist angle $\Psi = 0$, the tilt angle, Θ, is equal to α:

$$\sin^2\Psi \quad = \quad (1 - c_\alpha^2 \sec^2\Theta) \cdot \frac{B\,(A-C)}{C\,(A-B)}$$

For $\Psi = 0$:

$$0 \quad = \quad (1 - c_\alpha^2 \sec^2\Theta)$$

$$c_\alpha^2 \quad = \quad \cos^2\alpha = \quad \cos^2\Theta$$

Therefore:

$$\Theta \quad = \quad \alpha \qquad \text{for } \Psi \ = \ 0 \qquad \text{(no twist)}$$

The same procedure will now be applied with c_α. The complication arises when $(c_\beta)^2$ is not restricted to be less than or equal to one. This case yields then, to the twist and wobble modes.

Using the expressions for c_α and c_β we may write:

$$\cos^2\alpha = \cos^2\alpha_o \cdot c_\beta^2$$

where:

$$\cos^2\alpha_o = \frac{A(B-C)}{B(A-C)}$$

This leads to:

$$\cos^2\alpha = \cos^2\Theta (1 - \sin^2\Psi \sin^2\alpha_o)$$

The case in which $\Psi = 90°$ corresponds to the case in which:

$$\cos^2\Theta = c_\beta^2$$

Now, there are three cases to consider: the case when $\alpha > \alpha_o$, the case when $\alpha < \alpha_o$, and the case when $\alpha = \alpha_o$.

Case 1 (the twisting mode):

This mode is also called the "high energy mode".

Let us consider a rod with A, B, C such that $A = B > C$. For this case, $\alpha_o = 0$ and, consequently:

$$\alpha \geq \alpha_o$$

In general, a rotating rigid body will be said to be in the rod mode for $\alpha > \alpha_o$.
For the rod mode:

$$0 < c_\beta^2 < 1$$

By analogy to the previous considerations, there exists an angle β for which:

$$\cos^2\beta = c_\beta^2$$

α and β are related by:

$$\left[\frac{1}{C} - \frac{1}{A}\right]\cos^2\alpha = \left[\frac{1}{C} - \frac{1}{A}\right]\cos^2\beta$$

which relates the values α and β of the tilt angle Θ to the values 0 and 90° for the twist angle Ψ.

The procedures of Whittaker (1937) and Whittaker and Watson (1962) may be used to obtain a general solution of the form:

$$y \quad = \quad dn\,(pt)$$

where:

$$y \quad = \quad \frac{\sin\Theta}{\sin\alpha}$$

$$dn \quad = \quad \text{a Jacobian elliptic function with modulus } k$$

$$p^2 \quad = \quad \frac{H^2\sin^2\alpha\,(A-C)\,(B-C)}{ABC^2}$$

$$t \quad = \quad time$$

The function $dn\,(pt)$ and the tilt angle Θ are found to be periodic with the time period $2K/p$ where K is the complete elliptic integral of the first kind (Bowman, 1953). This oscillation of the tilt angle is called nutation.

The equation for the twist rate shows that the twist rate has the sign of $H \cdot \sin\Theta$. The twist rate is monotonic, since $\sin\Theta$ is bounded between $\sin\alpha$ and $\sin\beta$.

$$\text{sign}\,(\dot\Psi) = \quad \text{sign}\,(H\sin\Theta)$$

The equations describing the connection between the tilt angle and the twist angle show that the situation $\Theta = \alpha$ corresponds to the zero and half twist, and, therefore, the time taken for a full twist is twice the period of the tilt angle, Θ, and the average twist rate is:

$$\dot\Psi_{av} \quad = \quad \frac{\pi p}{2K}$$

The somersault rate varies between H/A and H/B and the time period of the oscillations is the same as the time period of the tilt angle, Θ, namely $2K/p$. The average somersault rate is:

$$\dot\phi_{av} \quad = \quad H\left[\frac{1}{C} - \left(\frac{1}{C} - \frac{1}{A}\right)V\,(\alpha)\right]$$

where:

$$V\,(\alpha) \quad = \quad \left[\frac{\cos^2\alpha}{\cos^2\Theta}\right]_{av}$$

In summary, both the somersault angle, Φ, and the twist angle, Ψ, steadily increase, while the tilt angle, Θ, oscillates between α and β (assuming that H, α, $\beta > 0$).

The results for the twisting mode can be summarized as follows (Table 4.4.3) (Yeadon, 1984):

Table 4.4.3   **Summary of selected values for case 1, the twisting model.**

TIME [TWIST PERIODS]	TWIST ANGLE [RAD]	TILT ANGLE	TILT RATE [RAD/S]	SOMERSAULT RATE	TWIST RATE
QUARTER	$\dfrac{\pi}{2}$	β (min)	0	$\dfrac{H}{B}$ (max)	$\left[\dfrac{H}{C} - \dfrac{H}{B}\right]\sin\beta$
HALF	π	α (max)	0	$\dfrac{H}{A}$	$\left[\dfrac{H}{C} - \dfrac{H}{A}\right]\sin\alpha$ (max)
THREE QUARTER	$\dfrac{3\pi}{2}$	β (min)	0	$\dfrac{H}{B}$ (max)	$\left[\dfrac{H}{C} - \dfrac{H}{B}\right]\sin\beta$
FULL	2π	α (max)	0	$\dfrac{H}{A}$	$\left[\dfrac{H}{C} - \dfrac{H}{A}\right]\sin\alpha$ (max)

The rod movement may be described as a twisting somersault. This mode of motion will, henceforth, be referred to as the *twisting mode*.

Case 2 (the wobbling mode):

This case is also called the "low energy mode", and is analogous to the movement mode of a disc with the principal moments of inertia A, B, and C, where A>B=C. The derivations are not discussed in this section. They have been published by Yeadon (1993a, b, c, d).

The results can be summarized as:

- The somersault angle Φ steadily increases while,
- The tilt angle Θ oscillates between α and $-\alpha$, and
- The twist angle Ψ oscillates between Ψ_0 and $-\Psi_0$.

These results are summarized in Table 4.4.4 (Yeadon, 1984).

Table 4.4.4   **Summary of the values for the twist and tilt angles as functions of the period and of the tilt angle rate.**

TIME	TILT ANGLE	TWIST ANGLE	SOMERSAULT RATE
0	α	0	minimum
K/q	0	Ψ_0	maximum
2K/q	$-\alpha$	0	minimum
3K/q	0	$-\Psi_0$	maximum

The discussed movement of a disc may be described as a wobbling somersault. This mode will, henceforth, be referred to as the *wobbling mode*.

Case 3 (the unstable singular solution):

The sections on cases 1 and 2 have explored the motions associated with $\alpha > \alpha_0$ and $\alpha < \alpha_0$ respectively. This section discusses the singular solution that corresponds to $\alpha = \alpha_0$. For a rigid body with the principal moments of inertia $A > B > C$ the condition:

$$E_{rot} = \frac{1}{2} \cdot \frac{H^2}{B}$$

is satisfied by steady rotations about the axis corresponding to the intermediate moment of inertia, B. This solution is obvious. However, the condition is also satisfied by a motion in which the rotation about the intermediate axis in one direction leads to a half twist followed by a motion in the opposite direction. This latter set of solutions may be derived using the procedure of Whittaker (1937) and the fact that the condition:

$$E_{rot} = \frac{1}{2} \cdot \frac{H^2}{B}$$

implies both:

$$\sin^2\alpha_0 = \frac{C(A-B)}{B(A-C)}$$

$$\beta_0 = 0$$

where:

α_0 is the angle of tilt when the twist angle Ψ is zero, and
β_0 is the angle of tilt at the quarter twist position.

With this, the motion may now be quantitatively described (Table 4.4.5) (Yeadon, 1984). For ease of presentation, the time is defined as zero when the twist angle $\Psi = 0$.

The average twist rate is zero. The average somersault rate is H/B. This, of course is to be expected, since the motion has the same energy as the rotation (with rate of H/B) around the intermediate axis.

Both the motions that satisfy:

$$E_{rot} = \frac{1}{2} \cdot \frac{H^2}{B}$$

are unstable with respect to perturbations that are not in the rotation plane. The result of such a perturbation would be degeneration of the movement into either the twisting or the wobbling mode solution space.

- If the perturbation nudges the movement into the twisting mode, successive twists of 180° will appear at periodic intervals and the twists will be in the same direction.

- If the perturbation nudges the movement into the wobbling mode, the successive twists of nearly 180° will be in opposite directions. In between the 180° twists, little twisting action will be observed.

Table 4.4.5   Description of the motion for the unstable rotation case.

TIME	TWIST ANGLE [DEG]	SOMERSAULT RATE	TWIST RATE
- ∞	-90	$\dfrac{H}{B}$	0
0	0	$\dfrac{H}{A}$	$H\left[\dfrac{1}{C} - \dfrac{1}{A}\right]\sin\alpha$
+ ∞	-90	$\dfrac{H}{B}$	0

An arbitrarily small perturbation of the system can give rise to qualitatively disparate motions. The rigid body movement may, therefore, be described as an unstable somersault. This mode of motion will, henceforth, be described as an *unstable mode*.

Discussion:

Rotations about the axis with the minimal principal moment of inertia are stable because such motions are central in the twisting mode solution space, and a small perturbation merely moves the trajectory into a similar twisting movement.

Rotations about the axis with the maximal principal moment of inertia are similarly stable because they are central in the wobbling mode solution space, and a small perturbation moves the trajectory into a similar wobbling movement.

Rotations about the axis with the intermediate principal moment of inertia are unstable because they are an asymptotic limit of both the twisting and the wobbling mode solution spaces. A small perturbation applied outside the plane of rotation results in the adoption of either a twisting or a wobbling movement.

Torque-free rotational motions of a rigid body fall into two general classes (Fig. 4.4.32):

- The *twisting mode*.
- The *wobbling mode*.

In the twisting mode, the twist angle steadily increases, whereas, in the wobbling mode, the twist angle oscillates around a mean value. Which mode a body gets into depends on the relative magnitudes of the principal moments of inertia. This suggests that these movement modes change with changes in the body configuration that alter the relative magnitudes of the principal moments of inertia. One may use such changes in body configuration to change a non-twisting into a twisting somersault and vice-versa. Changes in body configurations can be used to initiate or terminate twisting movements in somer-

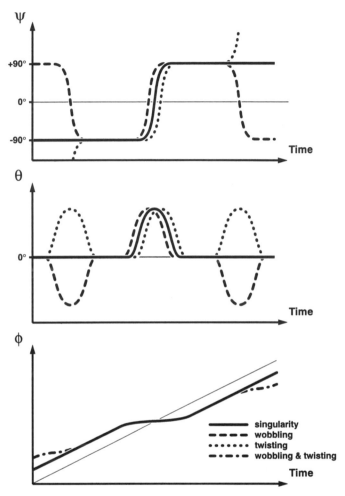

Figure 4.4.32   **Illustration for the twisting and the wobbling mode of rotations of a rigid body (from Yeadon, 1984, with permission).**

saults. As well, these considerations can be taken into account when determining the optimal time and position for someone to accelerate or decelerate the twisting motion (Yeadon, 1993a, b, c, d).

EXAMPLE 11   (arm movement with no muscles)

One simple approach to discussing movement of an extremity and "joint forces" is to use rigid body dynamics and not to consider any muscle involvement. For example, the movement of the upper and lower arm with respect to the shoulder joint in a horizontal plane may be described as the movement of a double pendulum. Certainly, such an approach can only approximate a very specific real life situation. Discussing this most simple example of multiple rigid body dynamics may, however, be helpful in showing a possible procedure for modelling human movement.

Question:

Determine the movement (position, velocity, and acceleration) of an idealized arm with no muscle influence in the horizontal plane.

Assumptions:

(1)   The elbow and shoulder are ideal hinge joints with no friction.

(2)   The shoulder joint is fixed in space.

(3)   The muscular influence is neglected.

(4)   The anthropometrical information with respect to length, moments of inertia, and masses are known.

(5)   The initial conditions for the angular position, $\varphi_i(0)$, and the angular velocity, $\dot{\varphi}_i(0)$, are known.

(6)   The problem can be solved two-dimensionally.

The illustration of the situation and the free body diagrams are given in Fig. 4.4.33.

FBD:

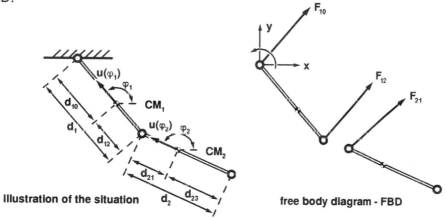

Figure 4.4.33   **Illustration of an idealized arm with no muscles, moving in a horizontal plane (left). Corresponding free body diagram (right).**

Solution:

The origin of the coordinate system is positioned in the shoulder joint.
For the following calculation the symbols used are:

$\mathbf{F}_{ik}$ =   force acting on body i from body k

$\mathbf{r}_i$ =   vector from the origin of the coordinate system to the centre of mass (CM) of segment i

$\mathbf{r}_{ik}$ =   vector from the CM of segment i to the joint between segments i and k

$\mathbf{u}\,(\varphi_i)$	=	unit vector from the CM of segment i to the joint between segment i and segment (i-1)
d_i	=	distance between joint (i -1) and joint i which corresponds to the length of segment i
d_{ik}	=	distance between the CM of segment i and joint k
I_i	=	moment of inertia of segment i with respect to the centre of mass of segment i

The system illustrated in Fig. 4.4.33 is a system with two DOF and with no friction in the joints. The corresponding equations of motion are:

Equations of motion:

Translation:

$$m_1\ddot{\mathbf{r}}_1 \;\;=\;\; \mathbf{F}_{10}+\mathbf{F}_{12} \tag{4.4.45}$$

$$m_2\ddot{\mathbf{r}}_2 \;\;=\;\; \mathbf{F}_{21} \tag{4.4.46}$$

Rotation:

$$I_1\ddot{\varphi}_1 \;\;=\;\; [\mathbf{r}_{10}\times\mathbf{F}_{10}]_z + [\mathbf{r}_{12}\times\mathbf{F}_{12}]_z$$

and with:

$$\mathbf{r}_{10} \;\;=\;\; d_{10}\cdot\mathbf{u}\,(\varphi_1)$$

$$\mathbf{r}_{12} \;\;=\;\; -d_{12}\cdot\mathbf{u}\,(\varphi_1)$$

The equations for rotation are:

$$I_1\ddot{\varphi}_1 \;\;=\;\; d_{10}\,[\mathbf{u}\,(\varphi_1)\times\mathbf{F}_{10}]_z - d_{12}\,[\mathbf{u}\,(\varphi_1)\times\mathbf{F}_{12}]_z \tag{4.4.47}$$

$$I_2\ddot{\varphi}_2 \;\;=\;\; d_{21}\,[\mathbf{u}\,(\varphi_2)\times\mathbf{F}_{21}]_z \tag{4.4.48}$$

Constraints:

$$\mathbf{r}_1 \;\;=\;\; -d_{10}\cdot\mathbf{u}\,(\varphi_1) \tag{4.4.49}$$

$$\mathbf{r}_2 \;\;=\;\; -d_1\cdot\mathbf{u}\,(\varphi_1) - d_{21}\cdot\mathbf{u}\,(\varphi_2) \tag{4.4.50}$$

where:

$$\mathbf{u}\,(\varphi_i) \;\;=\;\; (\cos\varphi_i,\ \sin\varphi_i,\ 0)$$

The general procedure for the solution has two steps:

- Step 1 - find $\mathbf{F}_{ik}$ in terms of $\mathbf{r}_i$ and their time derivatives, using the equations (4.4.45) and (4.4.46).

- Step 2 - find r_i and their time derivatives in terms of the angular coordinates, φ_i and their time derivatives, using the equations (4.4.49) and (4.4.50).

These steps will now be applied to this example:

Step 1:

$$-F_{12} \quad = \quad m_2 \ddot{r}_2 \quad = \quad +F_{21} \tag{4.4.51}$$

$$F_{10} \quad = \quad m_1 \ddot{r}_1 + m_2 \ddot{r}_2 \tag{4.4.52}$$

Equations (4.4.51) and (4.4.52) in equations (4.4.47) and (4.4.48):

$$I_1 \ddot{\varphi}_1 \quad = \quad d_{10} \left[u\,(\varphi_1) \times (m_1 \ddot{r}_1 + m_2 \ddot{r}_2) \right]_z + d_{12} \left[u\,(\varphi_1) \times m_2 \ddot{r}_2 \right]_z$$

$$I_1 \ddot{\varphi}_1 \quad = \quad d_{10} \left[u\,(\varphi_1) \times m_1 \ddot{r}_1 \right]_z + d_1 \left[u\,(\varphi_1) \times m_2 \ddot{r}_2 \right]_z$$

$$I_2 \ddot{\varphi}_2 \quad = \quad d_{21} \left[u\,(\varphi_2) \times m_2 \ddot{r}_2 \right]_z$$

$$I_1 \ddot{\varphi}_1 \quad = \quad m_1 d_{10} \left[u\,(\varphi_1) \times \ddot{r}_1 \right]_z + m_2 d_1 \left[u\,(\varphi_1) \times \ddot{r}_2 \right]_z \tag{4.4.53}$$

$$I_2 \ddot{\varphi}_2 \quad = \quad m_2 d_{21} \left[u\,(\varphi_2) \times \ddot{r}_2 \right]_z \tag{4.4.54}$$

Step 2 using (4.4.49) and (4.4.50):

$$\dot{r}_1 \quad = \quad -d_{10} \left[\frac{du\,(\varphi_1)}{d\varphi_1} \cdot \frac{d\varphi_1}{dt} \right] \quad = \quad -d_{10} \cdot u'\,(\varphi_1) \cdot \dot{\varphi}_1$$

$$\ddot{r}_1 \quad = \quad -d_{10} \left[u''\,(\varphi_1) \cdot \dot{\varphi}_1^2 + u'\,(\varphi_1) \cdot \ddot{\varphi}_1 \right] \tag{4.4.55}$$

Similar calculations lead to equation (4.4.56) for segment # 2:

$$\ddot{r}_2 \quad = \quad -d_1 \left[u''\,(\varphi_1) \cdot \dot{\varphi}_1^2 + u'\,(\varphi_1) \cdot \ddot{\varphi}_1 \right]$$

$$\quad -d_{21} \left[u''\,(\varphi_2) \cdot \dot{\varphi}_2^2 + u'\,(\varphi_2) \cdot \ddot{\varphi}_2 \right] \tag{4.4.56}$$

Rules for calculation:

$$u\,(\varphi_i) \times u'\,(\varphi_i) \quad = \quad (\cos\varphi_i,\ \sin\varphi_i,\ 0) \times (-\sin\varphi_i,\ \cos\varphi_i,\ 0)$$

$$\quad = \quad (0, 0,\ \cos^2\varphi_i + \sin^2\varphi_i)$$

$$\quad = \quad (0, 0, 1)$$

$$\mathbf{u}\,(\varphi_i) \times \mathbf{u}''\,(\varphi_i) \quad = (\cos\varphi_i, \ \sin\varphi_i, \ 0) \times (-\cos\varphi_i, \ -\sin\varphi_i, \ 0)$$

$$= (0, 0, 0)$$

$$\mathbf{u}'\,(\varphi_i) \times \mathbf{u}''\,(\varphi_i) \quad = (-\sin\varphi_i, \ \cos\varphi_i, \ 0) \times (-\cos\varphi_i, \ -\sin\varphi_i, \ 0)$$

$$= (0, 0, 1)$$

$$\mathbf{u}\,(\varphi_1) \times \mathbf{u}''\,(\varphi_2) \quad = (0, 0, \ \sin(\varphi_1 - \varphi_2))$$

$$\mathbf{u}\,(\varphi_1) \times \mathbf{u}'\,(\varphi_2) \quad = (0, 0, \ \cos(\varphi_1 - \varphi_2))$$

In the next step the equations (4.4.55) and (4.4.56) are substituted in (4.4.53) and (4.4.54) using the rules for calculating cross products:

$$I_1 \cdot \ddot{\varphi}_1 \quad = \quad m_1 \cdot d_{10} \{\mathbf{u}\,(\varphi_1) \times [-d_{10}\,[\mathbf{u}''\,(\varphi_1) \cdot \dot{\varphi}_1^2 + \mathbf{u}'\,(\varphi_1) \cdot (\ddot{\varphi}_1)]\,]\,\}$$

$$+ \quad m_2 \cdot d_1 \{\mathbf{u}\,(\varphi_1) \times [-d_1\,[\mathbf{u}''\,(\varphi_1) \cdot \dot{\varphi}_1^2 + \mathbf{u}'\,(\varphi_1) \cdot (\ddot{\varphi}_1)]\,]\,\}$$

$$+ \quad m_2 \cdot d_1 \{\mathbf{u}\,(\varphi_1) \times [-d_{21}\,[\mathbf{u}''\,(\varphi_2) \cdot \dot{\varphi}_2^2 + \mathbf{u}'\,(\varphi_2) \cdot (\ddot{\varphi}_2)]\,]\,\}$$

$$= \quad m_1 \cdot d_{10}\,(-d_{10} \cdot \ddot{\varphi}_1) + m_2 \cdot d_1\,(-d_1 \cdot \ddot{\varphi}_1)$$

$$- \quad m_2 \cdot d_1 \cdot d_{21}\,[\sin(\varphi_1 - \varphi_2) \cdot \dot{\varphi}_2^2 + \cos(\varphi_1 - \varphi_2) \cdot \ddot{\varphi}_2|$$

Which leads to equation (4.4.57):

$$[I_1 + m_1 d_{10}^2 + m_2 \cdot d_1^2] \cdot \ddot{\varphi}_1 + [m_2 \cdot d_1 \cdot d_{21} \cdot \cos(\varphi_1 - \varphi_2)] \cdot \ddot{\varphi}_2$$

$$= \quad -[m_2 d_1 d_{21} \cdot \sin(\varphi_1 - \varphi_2)] \cdot \dot{\varphi}_2^2 \qquad (4.4.57)$$

Equation (4.4.54) is treated the same way, which leads to equation (4.4.58):

$$m_2 d_1 \cdot d_{21} \cdot \cos(\varphi_2 - \varphi_1)\,]\,\ddot{\varphi}_1 + [I_2 + m_2 \cdot d_{21}^2]\,\ddot{\varphi}_2$$

$$= \quad -[m_2 d_1 d_{21} \sin(\varphi_2 - \varphi_1)] \cdot \dot{\varphi}_1^2 \qquad (4.4.58)$$

or, in a more general form:

$$A \cdot \ddot{\varphi}_1 + B \cdot \ddot{\varphi}_2 = \ E \cdot \dot{\varphi}_2^2$$

$$C \cdot \ddot{\varphi}_1 + D \cdot \ddot{\varphi}_2 = \ F \cdot \dot{\varphi}_1^2$$

This is a system of non-linear, second order differential equations. Such a system is usually solved using numerical methods. However, numerical solutions may well have more than one solution. Caution and intuition are important for an iteration approach.

Here, the calculations for an example with the following numerical values are presented:

$$
\begin{aligned}
m_1 &= 3 \text{ kg} \\
m_2 &= 2 \text{ kg} \\
d_{10} &= d_{12} = d_{21} = 0.15 \text{ m} \\
d_1 &= d_2 = 0.3 \text{ m} \\
I_1 &= I_2 = 0.01 \text{ kg m}^2
\end{aligned}
$$

and for the initial conditions, IC:

$$
\begin{aligned}
\varphi_1(0) &= 0 \\
\varphi_2(0) &= 0 \\
\dot{\varphi}_1(0) &= 2\pi/\text{sec} \\
\dot{\varphi}_2(0) &= \pi/\text{sec}
\end{aligned}
$$

The basic procedure for this numerical solution is to use the initial conditions for the time t to determine the angular position and velocity for the time t + Δt. Starting with the initial conditions at the time 0, which are known, a possible example for such an approach could be:

$$
\varphi(\Delta t) = \varphi(0) + \Delta t \cdot \dot{\varphi}(0)
$$

$$
\dot{\varphi}(\Delta t) = \dot{\varphi}(0) + \Delta t \cdot \ddot{\varphi}(0)
$$

or in a more general form:

$$
\varphi(t + \Delta t) = \varphi(t) + \Delta t \cdot \dot{\varphi}(t)
$$

$$
\dot{\varphi}(t + \Delta t) = \dot{\varphi}(t) + \Delta t \cdot \ddot{\varphi}(t)
$$

The angular accelerations are determined for each point in time using the equations:

$$
A \cdot \ddot{\varphi}_1 + B \cdot \ddot{\varphi}_2 = E \dot{\varphi}_2^2
$$

$$
C \cdot \ddot{\varphi}_1 + D \cdot \ddot{\varphi}_2 = F \dot{\varphi}_2^1
$$

which provide for the accelerations:

$$
\ddot{\varphi}_1 = \frac{DE\dot{\varphi}_2^2 - BF\dot{\varphi}_1^2}{AD - BC}
$$

$$
\ddot{\varphi}_2 = \frac{AF\dot{\varphi}_1^2 - CE\dot{\varphi}_2^2}{AD - BC}
$$

Using this approach, the angular position, velocity, and accelerations can easily be determined in steps of, for instance, 1 ms:

$$
\begin{array}{lll}
\text{for } t = 0 & \ddot{\varphi}_1 & = 0 \\
& \ddot{\varphi}_2 & = 0 \\
& \dot{\varphi}_1 & = 2\pi \\
& \dot{\varphi}_2 & = \pi \\
& \varphi_1 & = 0 \\
& \varphi_2 & = 0
\end{array}
$$

$$
\begin{array}{lll}
\text{for } t = 1 \text{ ms} & \ddot{\varphi}_1(1) & = -0.190919 \\
& \ddot{\varphi}_2(1) & = 0.51543 \\
& \dot{\varphi}_1(1) & = 6.28 \\
& \dot{\varphi}_2(1) & = 3.14 \\
& \varphi_1(1) & = 0.00628 \\
& \varphi_2(1) & = 0.00314
\end{array}
$$

$$
\begin{array}{lll}
\text{for } t = 2 \text{ ms} & \ddot{\varphi}_1(2) & = -0.382016 \\
& \ddot{\varphi}_2(2) & = 1.030977 \\
& \dot{\varphi}_1(2) & = 6.283 \\
& \dot{\varphi}_2(2) & = 3.142 \\
& \varphi_1(2) & = 0.01256 \\
& \varphi_2(2) & = 0.00628
\end{array}
$$

This iteration procedure can be executed for the time interval of interest. The results for this iteration procedure for the first 1000 ms (1 s) are illustrated in Fig. 4.4.34.

EXAMPLE 12   (a model for impact forces)

Impact forces have been defined (section 3.1) as forces that result from a collision of two objects that have a maximum force (impact force peak) earlier than 50 milliseconds after the first contact of the two objects. Impact forces occur during daily activities such as walking but are generally of more interest in activities with higher forces, such as running, jumping, and boxing.

The most commonly discussed and studied impact force variable is the maximal vertical impact force, F_{zi}, during landing in running. A simple mechanical model using two rigid bodies and two springs connecting these rigid bodies on each side was used (Nigg, 1986) to understand factors important in the development of an appropriate model for impact force analysis. The experiment was performed by dropping this mechanical structure from different heights onto a force plate that measured the vertical ground reaction force. Additionally, the vertical impact forces were simulated with a mathematical model that described this situation (Fig. 4.4.35).

The results of the experiment and the simulation (Nigg, 1986) indicate that the angle between the two rigid segments ("knee angle") and the landing velocity (of the "heel") are important factors influencing the magnitude of the measured or calculated maximal vertical impact forces, F_{zi}. However, the results suggest that the stiffness of and the force in the two springs do not influence the maximal vertical impact forces, F_{zi}, for impact forces

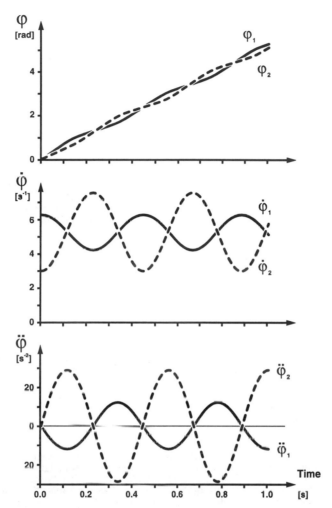

Figure 4.4.34   Graphical illustration of the numerical solution. Segment 1 and 2 pull and push each other. The amplitudes are, however, higher for the movement of segment 2 than for the movement of segment 1.

with a maximum earlier than 10 milliseconds after first ground contact. This result may be used to establish the assumptions for a model of the human body that can estimate impact forces. The results of this experiment suggest that, in a first approximation, muscle forces and muscle stiffness can be neglected in a model used to estimate impact forces in landing. This finding has been supported in an impact simulation study that included muscles with actual muscle properties (Gerritsen et al., 1993). Note: strength and stiffness of the springs do, however, influence the magnitude of the measured or calculated result for impact forces that reach their maximum later (see section 4.8).

Based on these considerations the following model was developed (Lemm, 1978; Denoth, 1986).

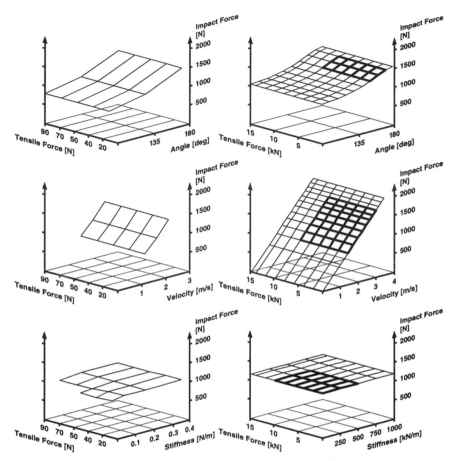

Figure 4.4.35   Maximum impact forces resulting from a drop of a two segment structure (left) and from a computer simulation of this same experiment (right). Top results: velocity = 2 m/s, stiffness of surface = 4000 N/m. Middle results: "knee angle" = 135°, stiffness of surface = 4000 N/m. Bottom results: velocity = 2 m/s, "knee angle" = 135°. The thicker lines indicate results for this physiological range (from Nigg, 1986, with permission).

Question:

Develop a model that allows for the determination of factors important for impact forces.

Assumptions:

(1)   The problem can be solved two-dimensionally.

(2)   The model consists of three rigid segments. Segment 1 corresponds to the leg. Segment 2 corresponds to the thigh. Segment 3 corresponds to the rest of the body.

(3)   The segments are assumed to have a length, L_i. The width and depth are assumed to be very small compared to the other geometrical measures.

(4)   The muscle influence is neglected.

(5) The movement of interest is heel-toe running.

(6) Running surface, heel pad, and ankle joint are considered one system with a linear spring constant. The spring constant is known.

(7) The masses, m_i, consist of a rigid and a non-rigid part. For this model the "rigid" masses are assumed to be:

$$m_1 \text{ (rigid)} = 0.5 \text{ m (leg)} \quad = \quad 50\% \text{ of the total leg mass}$$
$$m_2 \text{ (rigid)} = 0.5 \text{ m (thigh)} \quad = \quad 50\% \text{ of the total thigh mass}$$
$$m_3 \text{ (rigid)} = 0.2 \text{ m (rest)} \quad = \quad 20\% \text{ of the total rest mass}$$

(8) The moments of inertia, I_i, are known.

(9) The joints are assumed to be frictionless.

(10) The symbols used in this example are:

φ_i angle between the horizontal axis (x-axis) and segment i (long segment axis) measured at the posterior side

d_i length of segment i

$\mathbf{r}_i$ vector from the origin of the coordinate system to the centre of mass, CM_i, of segment i

where:

$$\mathbf{r}_1 = (x_1, y_1, m_1)$$
$$\mathbf{r}_2 = (x_2, y_2, m_1)$$
$$\mathbf{r}_3 = (x_3, y_3, m_1)$$

m_i mass of the element i

d_{ik} distance between the centre of mass of segment i, CM_i, and the end of segment i towards segment k

$m_i \mathbf{g}$ gravitational forces acting on segment i

where:

$$\mathbf{g} = (0, -g)$$

$\mathbf{F}_{ik}$ (actual) joint force in the joint between the segments i and k

where:

$$\mathbf{F}_{ik} = -\mathbf{F}_{ki}$$

$\mathbf{F}_{10}$ ground reaction force

$\mathbf{u}(\varphi_i)$ unit vector from the centre of mass of segment i, CM_i, directed towards segment (i + 1).

FBD:

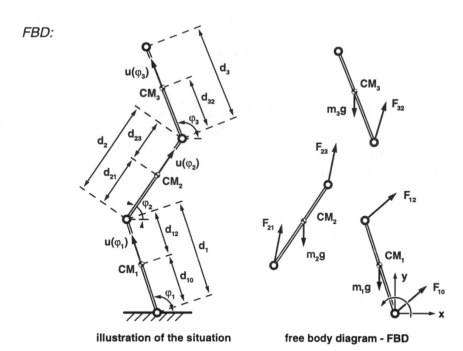

illustration of the situation   free body diagram - FBD

Figure 4.4.36   Free body diagram (right) and descriptive diagram (left) for the three segment model.

Equations of motion:

Translation:

$$m_1 \ddot{\mathbf{r}}_1 = m_1 \mathbf{g} + \mathbf{F}_{10} + \mathbf{F}_{12}$$

$$m_2 \ddot{\mathbf{r}}_2 = m_2 \mathbf{g} + \mathbf{F}_{21} + \mathbf{F}_{23}$$

$$m_3 \ddot{\mathbf{r}}_3 = m_3 \mathbf{g} + \mathbf{F}_{32}$$

Rotation:

Since the problem is two-dimensional the moment equations are written with respect to the centre of mass for the third component:

$$I_1 \ddot{\varphi}_1 = [\mathbf{M}(\mathbf{F}_{10}) + \mathbf{M}(\mathbf{F}_{12})]_z$$

$$= [(\mathbf{r}_{10} \times \mathbf{F}_{10}) + (\mathbf{r}_{12} \times \mathbf{F}_{12})]_z$$

$$I_1 \ddot{\varphi}_1 = [-d_{10}\mathbf{u}(\varphi_1) \times \mathbf{F}_{10} + d_{12}\mathbf{u}(\varphi_1) \times \mathbf{F}_{12}]_z$$

$$I_2 \ddot{\varphi}_2 = [-d_{21}\mathbf{u}(\varphi_2) \times \mathbf{F}_{21} + d_{23}\mathbf{u}(\varphi_2) \times \mathbf{F}_{23}]_z$$

$$I_3 \ddot{\varphi}_3 \quad = \quad [-d_{32} \mathbf{u}(\varphi_3) \times \mathbf{F}_{32}]_z$$

Constraints:

$$\mathbf{r}_2 \quad = \quad \mathbf{r}_1 + d_{12} \mathbf{u}(\varphi_1) + d_{21} \mathbf{u}(\varphi_2)$$

$$\mathbf{r}_3 \quad = \quad \mathbf{r}_2 + d_{23} \mathbf{u}(\varphi_2) + d_{32} \mathbf{u}(\varphi_3)$$

$$= \quad \mathbf{r}_1 + d_{12} \mathbf{u}(\varphi_1) + d_2 \mathbf{u}(\varphi_2) + d_{32} \mathbf{u}(\varphi_3)$$

and for the time derivatives of the constraint functions:

$$\dot{\mathbf{r}}_2 \quad = \quad \dot{\mathbf{r}}_1 + d_{12} \mathbf{u}'(\varphi_1) \dot{\varphi}_1 + d_2 \mathbf{u}'(\varphi_2) \dot{\varphi}_2$$

$$\ddot{\mathbf{r}}_2 \quad = \quad \ddot{\mathbf{r}}_1 + d_{12} [\mathbf{u}''(\varphi_1) \dot{\varphi}_1^2 + \mathbf{u}'(\varphi_1) \ddot{\varphi}_1]$$

$$+ \quad d_{21} [\mathbf{u}''(\varphi_2) \dot{\varphi}_2^2 + \mathbf{u}'(\varphi_2) \ddot{\varphi}_2]$$

$$\ddot{\mathbf{r}}_3 \quad = \quad \ddot{\mathbf{r}}_1 + d_{12} [\mathbf{u}''(\varphi_1) \dot{\varphi}_1^2 + \mathbf{u}'(\varphi_1) \ddot{\varphi}_1]$$

$$+ \quad d_2 [\mathbf{u}''(\varphi_2) \dot{\varphi}_2^2 + \mathbf{u}'(\varphi_2) \ddot{\varphi}_2]$$

$$+ \quad d_{32} [\mathbf{u}''(\varphi_3) \dot{\varphi}_3^2 + \mathbf{u}'(\varphi_3) \ddot{\varphi}_3]$$

where $\mathbf{u}'(\varphi_i)$ and $\mathbf{u}''(\varphi_i)$ are the first and second derivatives of the unit vectors with respect to φ_i, and where:

$$\mathbf{u}(\varphi_i) \quad = \quad (\cos\varphi_i, \sin\varphi_i, 0)$$

$$\mathbf{u}'(\varphi_i) \quad = \quad (-\sin\varphi_i, \cos\varphi_i, 0)$$

$$\mathbf{u}''(\varphi_i) \quad = \quad (-\cos\varphi_i, -\sin\varphi_i, 0) = \quad -\mathbf{u}(\varphi_i)$$

Elimination of $\mathbf{r}_2$, $\mathbf{r}_3$, $\mathbf{F}_{12}$, $\mathbf{F}_{23}$, and $\mathbf{F}_{10}$ provides:

$$c_{11} \ddot{x}_1 + c_{12} \ddot{y}_1 + c_{13} \ddot{\varphi}_1 + c_{14} \ddot{\varphi}_2 + c_{15} \ddot{\varphi}_3 \quad = \quad D_1$$

$$c_{21} \ddot{x}_1 + c_{22} \ddot{y}_1 + c_{23} \ddot{\varphi}_1 + c_{24} \ddot{\varphi}_2 + c_{25} \ddot{\varphi}_3 \quad = \quad D_2$$

$$c_{31} \ddot{x}_1 + c_{32} \ddot{y}_1 + c_{33} \ddot{\varphi}_1 + c_{34} \ddot{\varphi}_2 + c_{35} \ddot{\varphi}_3 \quad = \quad D_3$$

$$c_{41} \ddot{x}_1 + c_{42} \ddot{y}_1 + c_{43} \ddot{\varphi}_1 + c_{44} \ddot{\varphi}_2 + c_{45} \ddot{\varphi}_3 \quad = \quad D_4$$

$$c_{51} \ddot{x}_1 + c_{52} \ddot{y}_1 + c_{53} \ddot{\varphi}_1 + c_{54} \ddot{\varphi}_2 + c_{55} \ddot{\varphi}_3 \quad = \quad D_5$$

or, in a more general form:

$$C\ddot{X} = D$$

where:

$$\ddot{X} = (\ddot{x}_1, \ddot{y}_1, \ddot{\varphi}_1, \ddot{\varphi}_2, \ddot{\varphi}_3)$$

and where the matrix C is symmetric with $c_{ik} = c_{ki}$. The corresponding coefficients are described by Lemm (1978).

The solution of such a system of differential equations can be performed using numerical techniques. In the case of this specified example, it was solved using the 4th order Runge-Kutta approach. Lemm (1978) and Denoth (1986) used this model to develop the concept of an effective mass.

EXAMPLE 13 (a wobbling mass model)

Rigid body models together with inverse dynamics methods have typically been used to estimate the internal forces acting on the musculo-skeletal system during human movement. However, in reality, the body is not composed of a set of linked rigid bodies. Rather, each body segment consists of a rigid part (bone), and a non-rigid part (skin, muscle, ligament, tendon, connective tissue, and other soft tissue structures). During an impact, such as heel-strike in running or landing from a vertical jump, the skeletal structures of the body experience sudden, high accelerations, whereas, the soft tissue's movement is delayed, initiating damped vibrations of the soft tissue relative to the bone. In estimating internal forces proximal to the point of contact using inverse dynamics, the externally applied force and the acceleration of the segments of interest are typically measured. The measured ground reaction force (assuming impact with the ground) contains inertial components of both rigid and soft tissue structures and is not influenced by the rigid body assumption. However, acceleration measurements may introduce errors in the estimation of internal forces because accelerations of the bone lead to overestimation and accelerations of the soft tissue to underestimation of the internal forces.

The potential errors associated with rigid body models led to the conclusion that the approximation of the human body with rigid segments is justified only for movements that are not too rapid and is irrelevant for high impacts (Denoth et al., 1984). In order to account for the effect of relative displacements between the rigid and soft tissues on the joint forces during impact of the foot with the ground, a two-dimensional model of the body was developed, in which each segment was a combination of a skeletal part and a soft tissue part (Denoth et al., 1984). This model was further refined by Gruber (1987), who used it to illustrate the potential errors in estimating the forces and moments at the knee and hip joints using a rigid body model.

Question:

Determine the resultant forces and moments at the knee and hip joints when landing on the heel of one foot from a vertical jump (Fig. 4.4.37).

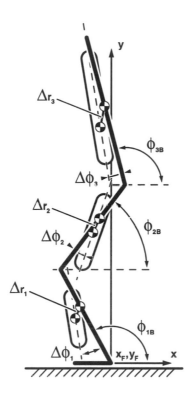

Figure 4.4.37   Illustration of the three-link model with skeletal and soft tissue masses and a mass-less foot (from Gruber, 1987, with permission).

Assumptions:

(1)   All motion occurs in the sagittal plane.

(2)   The human body is composed of three segments: the shank, the thigh, and the trunk, numbered segments 1, 2, and 3, respectively.

(3)   The segments are joined together by frictionless hinge joints.

(4)   Movements of the upper extremities are neglected, and their masses are included in the mass of the trunk.

(5)   All skeletal structures within a segment are modelled as a single rigid body with the geometry of a cylinder with length L_{ir}, diameter d_{ir}, and mass m_{ir}, for $i=1,2,3$.

(6)   The soft tissues, or wobbling mass, of each segment are modelled as a rigid body with a geometry of a cylinder of length L_{iw}, diameter d_{iw}, and mass m_{iw}, for $i=1,2,3$.

(7)   The moment of inertia of each skeletal mass and each soft tissue mass in the plane of motion can be approximated from the length and radius of each cylinder.

(8) Relative motion between the soft tissues and the skeletal mass of each segment has three degrees of freedom, two translational and one rotational.

(9) Quasi-elastic, strongly damped forces and moments are associated with relative movement between the soft tissue and skeletal mass of each segment. These forces and moments act at the centres of mass of the soft and skeletal tissues.

Symbols:

The symbols used for this example are (all vectors are defined in an inertial reference frame):

m_{ir} = mass of the rigid part of segment i (corresponds to mass of the bone)

m_{iw} = mass of the soft tissue of segment i (corresponds to the "wobbling mass"). Note that the soft tissue will be called "wobbling mass" in the following

$\mathbf{g}$ = acceleration due to gravity

$\mathbf{r}_{ik}$ = position vector from the centre of mass of the rigid part of segment i to the joint between segments i and k

$\mathbf{r}_{ir}$ = position of the centre of mass of the rigid part of segment i in the inertial reference frame

$\mathbf{r}_{iw}$ = position of the centre of mass of the wobbling mass of segment i in the inertial reference frame

φ_{ir} = angle that each rigid segment makes with the horizontal axis of the inertial reference frame in the sagittal plane

φ_{iw} = angle that each wobbling mass segment makes with the horizontal axis of the inertial reference frame in the sagittal plane

$\mathbf{F}_{ik}$ = force acting on segment i from segment k

$\mathbf{F}_{ie}$ = elastic force associated with the translation of the wobbling mass

$\mathbf{F}_{id}$ = damping force associated with the translation of the wobbling mass

$\mathbf{M}_{ik}$ = resultant moment acting on segment i at the joint between segments i and k

$\mathbf{M}_{ie}$ = elastic moment associated with the rotation of the wobbling mass

$\mathbf{M}_{id}$ = damping moment associated with the rotation of the wobbling mass

The free body diagram for the skeletal segments of the model is shown in Fig. 4.4.38.

FBD:

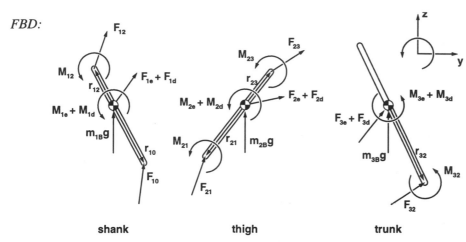

| shank | thigh | trunk |

Figure 4.4.38   FBD of the skeletal parts of the wobbling mass model.

The free body diagram for the soft tissue segments of the model is shown in Fig. 4.4.39.

FBD:

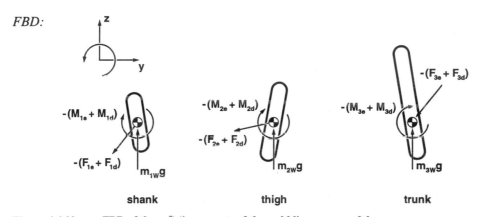

| shank | thigh | trunk |

Figure 4.4.39   FBD of the soft tissue parts of the wobbling mass model.

Equations of motion:

The equations of motion for each of the rigid segments of the model are:

Translation:

$$m_{1r}\ddot{\mathbf{r}}_{1r} = \mathbf{F}_{10} + \mathbf{F}_{12} + \mathbf{F}_{1e} + \mathbf{F}_{1d} + m_{1r}\mathbf{g}$$

$$m_{2r}\ddot{\mathbf{r}}_{2r} = \mathbf{F}_{21} + \mathbf{F}_{23} + \mathbf{F}_{2e} + \mathbf{F}_{2d} + m_{2r}\mathbf{g}$$

$$m_{3r}\ddot{\mathbf{r}}_{3r} = \mathbf{F}_{32} + \mathbf{F}_{3e} + \mathbf{F}_{3d} + m_{3r}\mathbf{g}$$

Rotation:

$$I_{1r}\ddot{\varphi}_{1r} = \left[(\mathbf{r}_{10} \times \mathbf{F}_{10}) + (\mathbf{r}_{12} \times \mathbf{F}_{12}) + \mathbf{M}_{12} + \mathbf{M}_{1e} + \mathbf{M}_{1d}\right]_z$$

$$I_{2r}\ddot{\varphi}_{2r} = \left[(\mathbf{r}_{21} \times \mathbf{F}_{21}) + (\mathbf{r}_{23} \times \mathbf{F}_{23}) + \mathbf{M}_{21} + \mathbf{M}_{23} + \mathbf{M}_{2e} + \mathbf{M}_{2d}\right]_z$$

$$I_{3r}\ddot{\varphi}_{3r} = \left[(\mathbf{r}_{32} \times \mathbf{F}_{32}) + \mathbf{M}_{32} + \mathbf{M}_{3e} + \mathbf{M}_{3d}\right]_z$$

The equations of motion for each of the soft tissue segments (wobbling masses) are:

$$m_{iw} \cdot \ddot{\mathbf{r}}_{iw} = -\mathbf{F}_{1e} - \mathbf{F}_{1d} + m_{iw} \cdot \mathbf{g}$$

$$m_{2w} \cdot \ddot{\mathbf{r}}_{2w} = -\mathbf{F}_{2e} - \mathbf{F}_{2d} + m_{2w} \cdot \mathbf{g}$$

$$m_{3w} \cdot \ddot{\mathbf{r}}_{3w} = -\mathbf{F}_{3e} - \mathbf{F}_{3d} + m_{3w} \cdot \mathbf{g}$$

Rotation:

$$I_{1w} \cdot \ddot{\varphi}_{1w} = \left[-\mathbf{M}_{1e} - \mathbf{M}_{1d}\right]_z$$

$$I_{2w} \cdot \ddot{\varphi}_{2w} = \left[-\mathbf{M}_{2e} - \mathbf{M}_{2d}\right]_z$$

$$I_{3w} \cdot \ddot{\varphi}_{3w} = \left[-\mathbf{M}_{3e} - \mathbf{M}_{3d}\right]_z$$

The model has 14 degrees of freedom with the following inputs:

- The initial conditions for segmental orientations, $\varphi_{1r}(0)$ were determined from high speed film of a subject performing the required movement.
- Foot acceleration was measured as a function of time using an accelerometer placed on the lateral malleolus of the subject.
- The vertical and antero-posterior components of the ground reaction force were measured throughout the movement.
- The length of each segment was determined based on measurements taken of the subject.
- The distribution of the subject's mass between the three segments was determined based on the published cadaver data of Clauser et al. (1969).
- The distribution of the segmental masses to the skeletal and soft tissue masses was done based on empirical measurements and by trial and error.

- Moments of inertia were estimated based on the length of each segment and an assumption for the radius of each cylindrical segment.
- The stiffness and damping coefficients associated with relative translations between skeletal and soft tissue masses were determined experimentally.
- The stiffness and damping coefficients associated with the relative rotation between skeletal and soft tissue masses were determined by trial and error.

Solution:

The set of coupled equations of motion were solved using standard integration techniques. The initial conditions for the rigid segments were determined from film analysis. The initial conditions for each wobbling segment were assumed to be identical to the initial conditions of the corresponding rigid segment 1.

The differences between the results predicted for the wobbling mass model and the rigid body mass model are summarized in Table 4.4.6. The values were estimated from the force-time and moment-time graphs of Gruber's (1987) publication.

Table 4.4.6   Comparison of selected maximal results during impact from the rigid body and the wobbling mass model. Approximate values were taken from the presented force/moment-time curves based on Gruber (1987), with permission.

SEGMENT	VARIABLE	UNIT	PREDICTED PEAK VALUES	
			WOBBLING MASS MODEL	RIGID BODY MODEL
KNEE	F_{ant}	N	200	800
	F_{post}	N	no peak	1000
	M_{12} (flexion)	Nm	no peak	200
HIP	F_{ant}	N	200	1100
	F_{post}	N	300	2500
	M_{23} (flexion)	Nm	no peak	1200
	M_{23} (extension)	Nm	no peak	800

Furthermore, the graphs for the anterior-posterior forces in the knee and hip joint (Fig. 4.4.40) as well as the flexion/extension moments of these two joints (Fig. 4.4.41) are shown as a function of time.

The results can be summarized as follows:

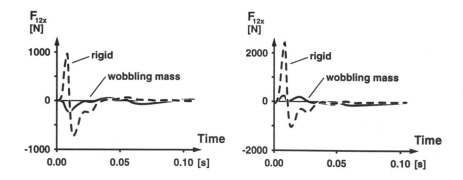

Figure 4.4.40   Horizontal components of the resultant forces at the knee joint (left), and at the hip joint (right), estimated by the rigid body and wobbling mass models for a drop jump movement with landing at the heel (from Gruber, 1987, with permission).

(1)   The anterior-posterior forces in the knee and hip joint were predicted differently by the rigid body and wobbling mass models. The rigid body model predicted high impact peaks in the anterior-posterior direction, while the wobbling mass model predicted only small anterior-posterior forces during the impact phase.

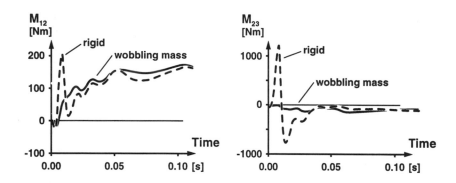

Figure 4.4.41   Resultant moments at the knee joint (left), and at the hip joint (right), estimated by the rigid body and wobbling mass models for a drop jump movement with landing at the heel (from Gruber, 1987, with permission).

(2)   The flexion/extension moments at the knee and hip joint were predicted differently by the rigid body and wobbling mass models. The rigid body model predicted high impact peaks, while the wobbling mass model predicted no impact peaks.

(3)   In general, the wobbling mass model predicted smaller internal forces and moments than did the rigid body model.

Comments

This example brings to light two important considerations: the magnitude of the absolute error for the results predicted with the two models and errors that may have contributed to the differences in the results of the wobbling mass and rigid body models.

Errors in the results of both the wobbling mass model and the rigid body model arise from a number of sources, including the various assumptions that have been made and the experimentally measured inputs that have been used. The accelerations measured using the skin mounted accelerometer at the malleolus seemed unusually high for the move being performed. LaFortune (1991) has shown that skin mounted accelerometers can overestimate peak accelerations of the tibia by as much as 50% in comparison to bone mounted accelerometers.

The differences in the results between the two models are of greater interest, however, in the context of this book. One would like to know whether the wobbling mass model exerts a corrective influence over the results or, in other words, whether or not the reduction of internal loading in the knee and hip joint predicted by the wobbling mass model corresponds to reality. Possible errors that may affect the results are that muscles are attached to one specific segment (for instance, gastrocnemius) and that the soft tissue is allowed to rotate. However, Gruber (1987) does not elaborate on these questions, and they cannot be discussed further in this context.

In summary, the wobbling mass model provides a more appropriate conceptual representation of the mechanics of the body during an impact than a rigid body model does. The accuracy of the two models, however, cannot be determined since they both rely on numerous simplifications and assumptions, and the results predicted by the models may be influenced by errors in the experimentally measured input parameters.

4.4.6 COMMENTS FOR SECTION 4.4

Section 4.4 discussed various mathematical models that were determinate, which means that the number of unknowns and equations were equal. This is a very specific situation for biomechanical modelling and, as discussed in the next section, is not at all common. However, in summary:

(1)   It may be possible in biomechanics to discuss specific questions using simple deterministic models. The complexity of a model is not a priority related to the power and potential of a model to answer a question or to increase our understanding of a situation.

(2)   Even the very simple two-dimensional models that use two or three rigid bodies and no muscles provide results that must be solved with the help of numerical techniques. These numerical techniques, however, usually give more than one solution, so caution is required in using them.

4.4.7 REFERENCES

Alexander, R.M. (1992) Simple Models of Walking and Jumping. *Human Movement Science*. **11,** pp. 3-9.

Alexander, R.M., Bennett, M.B., and Ker, R.F. (1986) Mechanical Properties and Functions of the Paw Pads of Some Mammals. *J. Zoology*. **209,** pp. 405-419.

Alexander, R.M. and Vernon, A. (1975) The Mechanics of Hopping by Kangaroos. (Macropodidae). *J. Zool.* **177,** pp. 265-303.

Anton, M.G. and Nigg, B.M. (1990) An Optimal Control Model for Running. *Proc. 6th Biennial Conf. of the Can. Soc. for Biomechanics.* pp. 61-62.

Blickhan, R. (1989) The Spring-mass Model for Running and Hopping. *J. Biomechanics.* **22 (11-12),** pp. 1217-1227.

Bowman, F. (1953) *Introduction to Elliptic Functions with Applications.* Dover, New York.

Cavanagh, P.R. and Lafortune, M.A. (1980) Ground Reaction Forces in Distance Running. *J. Biomechanics.* **13,** pp. 397-406.

Clauser, C.E., McConville, J.T., and Young, J.W. (1969) Weight, Volume, and Centre of Mass Segments of the Human Body. *AMRL Technical Report (TR-69-70).* Wright Patterson Air Force Base, Ohio. pp. 69-70.

Denoth, J., Gruber, K., Ruder, H., and Keppler, M. (1984) Forces and Torques During Sports Activities with High Accelerations. *Biomechanics Current Interdisciplinary Research* (eds. Perren, S.M. and Schneider, E.). Martinuis Nijhoff Pub., Dordrecht, Netherlands. pp. 663-668.

Denoth, J. (1985) The Dynamic Behaviour of a Three Link Model of the Human Body During Impact with the Ground. *Biomechanics IX-A* (eds. Winter, D., Norman, R.W., Wells. R.P., Hayes, K.C., and Patta, A.E.). Human Kinetics, Champaign, IL. pp. 102-106.

Denoth, J. (1986) Load on the Locomotor System and Modelling. *Biomechanics of Running Shoes* (ed. Nigg, B.M.). Human Kinetics, Champaign, IL. pp. 63-116.

Gerritsen, K.G.M. and van den Bogert, A.J. (1993) Direct Dynamics Simulation of the Impact Phase in Heel-toe Running. *Proc. 4th International Symposium on Computer Simulation in Biomechanics.* BML1-6 - BML1-13.

Gruber, K. (1987) Entwicklung eines Modells zur Berechnung der Kräfte im Knie- und Hüftgelenk bei sportlichen Bewegungsabläufen mit hohen Beschleunigungen. Ph.D. Thesis, Universitat Tübingen, Germany.

Hörler, E. (1972a) Mechanisches Modell zur Beschreibung des isometrischen und des dynamischen Muskelkraftverlaufs. *Jungend und Sport, Nov.* pp. 271-273.

Hörler, E. (1972b) Hochsprungmodell. *Jungend und Sport, Aug.* pp. 381-383.

Hörler, E. (1973) Clumsiness and Stature: A Study of Similarity. *Biomechanics III* (eds. Cerquiglini, S., Venerando, A., and Wartenweiler, J.). University Park Press, Baltimore. S. Karger AG, Basel. **8,** pp. 146-150.

LaFortune, M.A. (1991) Three-dimensional Acceleration of the Tibia During Walking and Running. *J. Biomechanics.* **24,** pp. 877-886.

Lemm, R. (1978) *Passive Kräfte bei sportlichen Bewegungen.* Unpublished Diplomarbeit, ETH Zürich.

McMahon, T.A. and Green, P.R. (1979) The Influence of Track Compliance on Running. *J. Biomechanics.* **12 (12),** pp. 893-904.

Misevich, K.W. and Cavanagh, P.R. (1984) Material Aspects of Modelling Shoe/foot Interaction. *Sport Shoes and Playing Surfaces* (ed. Frederick, E.C.). Human Kinetics, Champaign, IL. pp. 47-75.

Nigg, B.M. (1974) Analysis of Twisting and Turning Movements. *Biomechanics IV* (eds. Nelson, R.C. and Morehouse, C.A.). University Park Press, Baltimore. pp. 279-283

Nigg, B.M. and Denoth, J. (1980) *Sportplatzbeläge.* Juris Verlag, Zürich.

Nigg, B.M. and Lüthi, S.M. (1980). Bewegungsanalysen beim Laufschuh. *Sportwissenschaft.* **3,** pp. 309-320.

Nigg, B.M. (1986) Biomechanical Aspects of Running. *Biomechanics of Running Shoes* (ed. Nigg, B.M.). Human Kinetics Pub. Inc., Champaign, IL. pp. 1-25.

Nigg, B.M. and Anton, M. (In press) Energy Aspects for Elastic and Viscoelastic Shoe Soles and Surfaces. *Med. Sc. Sports and Exercise.*

Pennycuick, C.J. (1993) Mechanical Limits to Evolution and Diversity. *Proc. 14th Biomechanics Congress.* pp. 20-21.

Seireg, A. and Avikar, A.J. (1973) A Mathematical Model for Evaluation of Forces in Lower Extremities of the Musculo-skeletal System. *J. Biomech.* **6 (3),** pp. 313-326.

Whittaker, E.T. (1937) *A Treatise on the Analytical Dynamics of Particles and Rigid Bodies.* Cambridge University Press, Cambridge.

Whittaker, E.T. and Watson, G.N. (1962) *A Course of Modern Analysis* (5th Ed.). Cambridge University Press, Cambridge.

Yeadon, M.R. (1984) *The Mechanics of Twisting Somersaults*. (Unpublished Ph.D. Thesis) Loughborough University of Technology, UK.

Yeadon, M.R. (1993a) The Biomechanics of Twisting Somersaults, Part 1: Rigid Body Motions. *J. Sports Sciences*. **11 (3)**, pp. 187-198.

Yeadon, M.R. (1993b) The Biomechanics of Twisting Somersaults, Part 2: Contact Twist. *J. Sports Sciences*. **11 (3)**, pp. 199-208.

Yeadon, M.R. (1993c) The Biomechanics of Twisting Somersaults, Part 3: Aerial Twist. *J. Sports Sciences*. **11 (3)**, pp. 209-218.

Yeadon, M.R. (1993d) The Biomechanics of Twisting Somersaults, Parts 4: Partitioning Performances Using the Tilt Angle. *J. Sports Sciences*. **11 (3)**, pp. 219-225.

4.5 MATHEMATICALLY INDETERMINATE SYSTEMS

HERZOG, W.
BINDING, P.

4.5.1 DEFINITIONS AND COMMENTS

Distribution problem:	Calculation of internal forces acting on the musculo-skeletal system using the known resultant joint forces and moments. The distribution problem is usually underdetermined.
Mathematical system:	A set of mathematical equations and inequality relationships.
• Determinate system:	A mathematical system is called determinate if the number of system equations equals the number of unknowns.
• Overdetermined system:	A mathematical system is called overdetermined if it has more system equations than unknowns.
• Underdetermined system:	A mathematical system is called underdetermined if it has fewer system equations than unknowns.

4.5.2 SELECTED HISTORICAL HIGHLIGHTS

1973	Seireg & Arvikar	Were the first to use mathematical optimization to solve the distribution problem.
1978	Pedotti et al.	Were the first to use a non-linear optimization approach to solve the distribution problem.
1981	Crowninshield & Brand	Were the first to use a non-linear optimization approach based on a physiological criterion to solve the distribution problem.
1984	Dul et al.	Were the first to attempt a validation of the results of their distribution problems using direct muscle force measurements.

4.5.3 INTRODUCTION

If a mathematical system contains more equations than unknowns, it is said to be over-determined, and in general, there will be no solution. For example, the system shown below contains three equations (4.5.1) to (4.5.3), and two unknowns, x and y.

$$x + y \;=\; 7 \tag{4.5.1}$$

$$x - y \;=\; 3 \tag{4.5.2}$$

$$x + 2y \;=\; 12 \tag{4.5.3}$$

From equations (4.5.1) and (4.5.2) one obtains a unique solution for x (x=5) and y (y=2); however, this solution does not satisfy equation (4.5.3). If, for example, equations (4.5.1) and (4.5.3) are used to solve for x and y, the solution will be different (i.e., x=2, y=5), and it will not satisfy equation (4.5.2). Solutions will exist if the right hand sides obey special conditions (e.g., if 12 is replaced by 9 in equation (4.5.3)).

If a system contains more unknowns than equations, it is said to be underdetermined, and in general, there will be an infinite number of possible solutions. For example, the system shown below contains one equation, (4.5.4), and two unknowns, x and y.

$$x \cdot y \;=\; 12 \tag{4.5.4}$$

Possible solutions for this equation include:

    (x=1, y=12),   or

    (x=2, y=6),   or

    (x=3, y=4),   or

    (x=120, y=0.1),   etc.

In biomechanics, there is a specific underdetermined problem, the so-called distribution problem (Crowninshield and Brand, 1981a), which has received overwhelming attention in the past two decades. The distribution problem is used to solve for internal forces acting on the musculoskeletal system by using the known resultant joint forces and moments. In the calculation of internal forces, the number of unknowns typically exceeds the number of available equations describing the mechanics of the system. In order to find a unique solution for the distribution problem in biomechanics, the underdetermined system may be made determinate either by increasing the number of system equations (e.g., Pierrynowski and Morrison, 1985) or by decreasing the number of unknowns (e.g., Paul, 1965) until the number of equations and unknowns is the same. However, when increasing the number of system equations, some non-trivial assumptions must be made, and when decreasing the number of unknowns, some relevant information about the specific system behaviour may get lost. For these and some well-supported physiological reasons, researchers have attempted to solve the distribution problem in biomechanics by using mathematical optimization approaches. In this chapter, we will focus on solving the distribution problem in biomechanics using optimization theory.

4.5.4 BASIC CONCEPTS

When attempting to determine internal (= ligamentous, muscular, bony contact) forces in and around a biological joint from the known resultant joint forces and moments, certain modelling assumptions must be made. These assumptions include how a joint is defined and how forces are transmitted across the joint by internal structures. A brief discussion of these assumptions follows.

A biological joint is typically defined as a point that may be associated with an anatomical landmark (e.g., the lateral malleolus for the ankle joint), or that may be defined mathematically and may move relative to the bones that make up the joint (e.g., the instantaneous centre of zero velocity concept). In mechanics, we typically think of a joint as a point that is "contained" in both segments that make up the joint. When performing an analysis of internal forces, a fictitious surface, which is not necessarily planar, is passed through the anatomical joint space, and this surface severs all tissues that transverse the joint. It is typically assumed that bony contact regions, muscles, and ligaments are the only structures that transmit non-negligible forces across a joint. Furthermore, for each structure that transmits force across a joint, the point of application of that force, Q, is chosen in such a way that the moment produced by that structure about point Q is zero (Fig. 4.5.1). Of course, each structure will produce a moment about points other than point

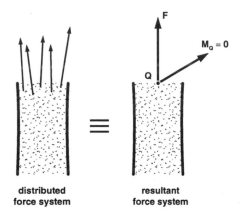

Figure 4.5.1   **Equipollent replacement of a distributed force system (for example, in a ligament attaching to a bone) by a resultant force and moment. In biomechanics, we tend to associate the point of application of the resultant force with a point, Q, where the resultant moment of the distributed force system is zero.**

Q, in particular about the joint centre, O.

4.5.5 JOINT EQUIPOLLENCE EQUATIONS

The joint equipollence equations relate the muscular, ligamentous, and bony contact forces to the resultant joint force and moment. Using the assumptions made above, the force and moment equipollence equations are as follows (Fig. 4.5.2):

$$\mathbf{F} = \sum_{i=1}^{N} (\mathbf{F}_i^m) + \sum_{j=1}^{P} (\mathbf{F}_j^l) + \sum_{k=1}^{Q} (\mathbf{F}_k^c) \tag{4.5.5}$$

$$\mathbf{M}_o = \sum_{i=1}^{N} (\mathbf{r}_{i/0} \times \mathbf{F}_i^m) + \sum_{j=1}^{P} (\mathbf{r}_{j/0} \times \mathbf{F}_j^l) + \sum_{k=1}^{Q} (\mathbf{r}_{k/0} \times \mathbf{F}_k^c) \tag{4.5.6}$$

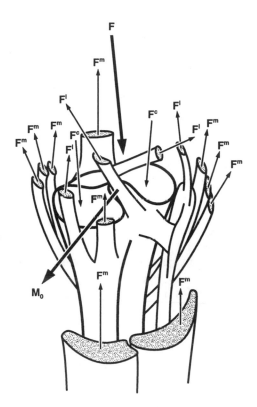

Figure 4.5.2 Force distribution in a joint and its equipollent replacement by a resultant external force (F) and a resultant external joint moment (M₀) (from Crowninshield and Brand, 1981a, with permission).

where:

F	=	variable resultant external joint force
M$_o$	=	variable resultant external joint moment
F^m	=	internal muscular forces
F^l	=	internal ligamentous forces
F^c	=	internal bony contact forces
r$_{i/0}$	=	location vector for muscular force i
r$_{j/0}$	=	location vector for ligamentous force j
r$_{k/0}$	=	location vector for bony contact force k
N	=	integer indicating the number of muscular forces
P	=	integer indicating the number of ligamentous forces
Q	=	integer indicating the number of bony contact forces

The resultant joint force, $\mathbf{F}$, and joint moment, $\mathbf{M}_o$, may be obtained using the inverse dynamics approach (e.g., Andrews, 1974). Since equations (4.5.5) and (4.5.6) are two vector equations, they yield six scalar equations in a three-dimensional system. The unknowns include all muscular, ligamentous, and bony contact force vectors, as well as the corresponding location vectors. Therefore, the number of unknowns exceeds the number of system equations. Using anatomical information from cadaver or imaging studies, the unknown location vectors, as well as the direction of the muscular and ligamentous force vectors, may be determined or estimated. This anatomical information reduces the number of unknowns substantially, leaving just the magnitudes of all internal force vectors and the direction of the bony contact force vectors as unknowns. Therefore, the number of scalar unknowns (SU) in the system represented by equations (4.5.5) and (4.5.6) is equal to:

$$SU = N + P + 3Q \tag{4.5.7}$$

Where N, P, and Q represent the number of muscles, ligaments, and bony contact areas of the joint under consideration. In general, the number of unknowns will exceed the number of available system equations (i.e., six scalar equations in the three-dimensional case), or:

$$N + P + 3Q > 6 \tag{4.5.8}$$

Therefore, the problem is mathematically underdetermined.

4.5.6 SOLVING MATHEMATICALLY UNDERDETERMINED SYSTEMS USING OPTIMIZATION THEORY

Any mathematically underdetermined system may be made determinate by decreasing the number of unknowns and/or increasing the number of system equations until the number of unknowns and system equations match. These approaches have been used to solve the underdetermined distribution problem in biomechanics (Paul, 1965; Morrison, 1968; Pierrynowski and Morrison, 1985). However, the approach used most often to solve the distribution problem is mathematical optimization. Optimization procedures are not only an elegant way of solving this type of mathematical problem, but are also believed to be good indicators of the physiology underlying force-sharing among internal structures. This belief goes as far back as Weber and Weber (1836), who stated that locomotion is performed in such a way as to optimize (i.e., minimize) metabolic cost.

Optimization problems, in general, are defined by three quantities: the cost function, the design variables, and the constraint functions. The cost function is the function to be optimized. For the distribution problem in biomechanics, cost functions have been defined as:

Minimize ϕ where:

$$\phi = \sum_{i=1}^{N} F_i^m \qquad \text{(e.g., Seireg \& Arkivar, 1973)} \tag{4.5.9}$$

or:

$$\phi = \sum_{i=1}^{N} (F_i^m / pcsa_i)^3 \quad \text{(e.g., Crowninshield \& Brand, 1981b)} \qquad (4.5.10)$$

or:

$$\phi = \sum_{i=1}^{N} (F_i^m / M_{maxi})^3 \quad \text{(e.g., Herzog, 1987)} \qquad (4.5.11)$$

where:

F_i^m = force magnitude of the ith-muscle
$pcsa_i$ = physiological cross-sectional area of the ith-muscle
M_{maxi} = variable maximal moment that the ith-muscle can produce as a function
         of its instantaneous contractile conditions
N   = total number of muscles considered

Design variables are the variables that are systematically changed until the cost function is optimized and all constraint functions are satisfied. The design variables must be contained in the cost function, and for the distribution problem, they typically are the magnitudes of the individual (muscle) forces.

The constraint functions restrict the solution of the optimization approach to certain boundary conditions. For example, in the distribution problem, typical inequality constraints are:

$$F_i^m \geq 0, \quad \text{for } i = 1, \, , N \qquad (4.5.12)$$

and typical equality constraints are:

$$M_o = \sum_{i=1}^{N} (r_{i/o} \times F_i^m) \qquad (4.5.13)$$

which indicate that muscular forces must always be zero or positive (tensile), and resultant joint moments, M_o, are assumed to be satisfied by the vector sum of all moments produced by the muscular forces. An optimization problem may not have constraint functions, and then is referred to as an unconstrained problem.

UNCONSTRAINED PROBLEM: ONE DESIGN VARIABLE

When trying to find optimal solutions to a problem, the difficulty of solving the problem analytically is directly related to the number of design variables and constraints. An unconstrained problem with one design variable can typically be solved in a straightforward way.

Let us assume that we are trying to find local minima of the following cost function:

$$f(x) \;=\; x^2 + 2x + 5 \tag{4.5.14}$$

The design variable in this case is x. The necessary and sufficient conditions for local minima of equation (4.5.14) are given by equations (4.5.15) and (4.5.16), respectively:

$$\frac{df(x)}{dx} \;=\; 0 \tag{4.5.15}$$

$$\frac{d^2f(x)}{dx^2} \;>\; 0 \tag{4.5.16}$$

Solving our problem for the necessary condition gives:

$$\frac{df(x)}{dx} \;=\; 0 \;=\; 2x + 2 \tag{4.5.17}$$

or:

$$x \;=\; -1$$

The point x = -1 is called a stationary point. If $d^2f(x)/dx^2$ is positive for x = -1, then the stationary point represents a minimum; if it is negative, the stationary point is a maximum

$$\frac{d^2f(x)}{dx^2} \;=\; 2 \tag{4.5.18}$$

(for any value of x).

Therefore, the solution x = -1 is a minimum. Since the solution, x = -1, is the only stationary point, we also know that it is a global minimum (Fig. 4.5.3).

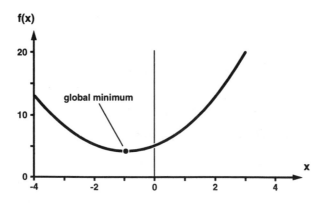

Figure 4.5.3   **Illustration of the equation $f(x) \;=\; x^2 + 2x + 5$.**

As a second example, let us assume that we would like to find local minima of the function:

$$f(x) = x^3 - 7.5x^2 + 18x - 10 \qquad (4.5.19)$$

Taking first and second derivatives with respect to the design variable (x), and testing the necessary and sufficient conditions (equations (4.5.15) and (4.5.16)) gives:

$$\frac{df(x)}{dx} = 0 = 3x^2 - 15x + 18 \qquad (4.5.20)$$

$$\frac{d^2f(x)}{dx^2} = 6x - 15 \qquad (4.5.21)$$

Stationary points may be obtained from equation (4.5.20). They are $x_1 = 2$ and $x_2 = 3$. Replacing the two solutions in equation (4.5.21), we realize that solution one, x_1, yields -3, and solution two, x_2, yields +3. Therefore, solutions one and two represent a local maximum ($x_1 = 2$) and a local minimum ($x_2 = 3$), respectively (Fig. 4.5.4).

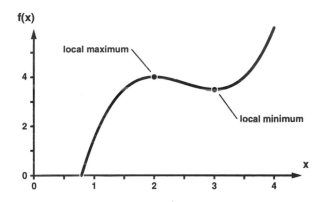

Figure 4.5.4   **Illustration of the equation** $f(x) = x^3 - 7.5x^2 + 18x - 10$.

UNCONSTRAINED PROBLEM: MORE THAN ONE DESIGN VARIABLE

When trying to find a solution to an unconstrained, multi-design variable problem, a necessary condition for a minimum solution is that the first partial derivatives of the cost function, ($f(x)$), with respect to each design variable, x_i, are equal to zero:

$$\frac{\partial f(x)}{\partial x_i} = 0 \quad \text{for } i = 1,..., N \qquad (4.5.22)$$

where N is the total number of design variables. A sufficient condition for x to give a local minimum of this type of problem is that the Hessian matrix, $H(x)$, given by:

$$
H(x) = \begin{bmatrix} \dfrac{\partial^2 f(x)}{\partial x_1^2} & \cdots & \dfrac{\partial^2 f(x)}{\partial x_1 \partial x_n} \\ \cdots & \cdots & \cdots \\ \dfrac{\partial^2 f(x)}{\partial x_n \partial x_1} & \cdots & \dfrac{\partial^2 f(x)}{\partial x_n^2} \end{bmatrix}, \tag{4.5.23}
$$

is positive definite (if -H(x) is positive definite, then x gives a local maximum). There are various ways to test this definiteness (for example, all eigenvalues of H must be positive) and we shall illustrate an easily computed test in the following.

EXAMPLE 1

Question:
   Find a minimum solution of the function $f(x_1, x_2)$, where:

$$
f(x_1, x_2) = 5x_1 - \frac{x_1^2 \cdot x_2}{16} + \frac{x_2^2}{4x_1} \tag{4.5.24}
$$

The necessary conditions for stationary points (i.e., equation (4.5.22)) for this problem become:

$$
\frac{\partial f(x_1, x_2)}{\partial x_1} = 0 = 5 - \frac{2x_1 x_2}{16} - \frac{x_2^2}{4x_1^2} \tag{4.5.25}
$$

$$
\frac{\partial f(x_1, x_2)}{\partial x_2} = 0 = -\frac{x_1^2}{16} + \frac{2x_2}{4x_1} \tag{4.5.26}
$$

Solving equations (4.5.25) and (4.5.26) gives two stationary points, P_1 and P_2, where:

$$
P_1(x_1 = 4, x_2 = 8)
$$

and:

$$
P_2(x_1 = -4, x_2 = -8)
$$

It suffices, for a minimum solution, that the Hessian matrix be positive definite. The second partial derivatives of equation (4.5.24) are:

$$
\frac{\partial^2 f(x_1, x_2)}{\partial x_1^2} = -\frac{x_2}{8} + \frac{x_2^2}{2x_1^3} \tag{4.5.27}
$$

$$\frac{\partial^2 f(x_1,x_2)}{\partial x_2^2} = \frac{1}{2x_1} \tag{4.5.28}$$

$$\frac{\partial^2 f(x_1,x_2)}{\partial x_1 \partial x_2} = -\frac{x_1}{8} - \frac{x_2}{2x_1^2} \tag{4.5.29}$$

Therefore, the Hessian matrix for the first stationary point, P_1 ($x_1 = 4, x_2 = 8$), becomes:

$$H(4,8) \quad = \quad \begin{bmatrix} -\dfrac{1}{2} & -\dfrac{3}{4} \\ -\dfrac{3}{4}+\dfrac{1}{8} \end{bmatrix} \tag{4.5.30}$$

We calculate the two "principal" determinants $d = \det\begin{bmatrix} -\frac{1}{2} \end{bmatrix}$ and $D = \det H(4,8)$. Then, $H(4,8)$ is positive definite if d and D are positive.

Since $d = \begin{bmatrix} -\frac{1}{2} \end{bmatrix}$ and $D = \begin{bmatrix} -\frac{5}{8} \end{bmatrix}$, $H(4,8)$ is not positive definite. In fact, since d and D are negative, the stationary point P_1 does not represent a local minimum. (P_1 is actually a saddle point.) The above procedure may be repeated with the second stationary point, P_2 ($x_1 = -4, x_2 = -8$); again, one finds that P_2 does not represent a local minimum but another saddle point.

There is an important class of problems for which equation (4.5.22) is sufficient on its own for x to give a minimum, which is even global. This is the class of convex problems, i.e., problems for which f is a convex function. There are several tests for convexity: for example, it suffices if H(y) is non-negative definite for all $y = (y_1, y_2)$.

GENERAL CONSTRAINED PROBLEM

The distribution problem in biomechanics is typically solved by minimizing the cost function of a general constrained problem. Specifically, suppose we want to minimize f(x) subject to the constraints h_i (x) = 0 (i = 1, 2,..., s) where f and h_i are differentiable and $x = (x_1, ..., x_N)$ is a vector of design variables which are assumed to be non-negative. These non-negativity constraints $x_j \geq 0$ (j = 1, 2, ..., N) make the problem of the "mathematical programming" type, and in certain cases, it can be solved via the KarushKuhn-Tucker (KKT) conditions.

One such case, which we shall consider, is where f is convex (see above) and each h_i is affine, i.e., $h_i(x) = a_i^T x + b_i$ for some (row) vector a_i and constant b_i. In this case, the KKT conditions are necessary and sufficient for a global minimum, and they admit the following interpretation.

First select the design variables which are zero, and label them x_k (the other design variables are positive). Then solve the *equality* constrained problem of minimizing f(x) subject to:

$$\text{all} \quad h_i(x) = 0 \quad \text{and all} \quad x_k = 0 \tag{4.5.31}$$

This problem can be attacked by Lagrange multipliers λ_i and μ_k. Let:

$$L(x) \quad = \quad f(x) + \sum_{i=1}^{s} \lambda_i h_i(x) + \sum_k \mu_k x_k$$

and solve the mathematical system consisting of the stationary point condition:

$$0 = \frac{\partial L}{\partial x_j}(x) \quad (j = 1, 2, ..., N) \tag{4.5.32}$$

together with the equality constraints (4.5.31).

If N is large, the task of selecting which design variables should be zero can be formidable. For small N, however, the problem may be tackled as follows: first try the case where no x_k is zero (so all x_j are positive and the final summation is absent from $L(x)$). If the system (4.5.31 and 4.5.32) is soluble with all $x_j \geq 0$, then we have found the global minimum. If not, then we try the case where just one x_k is zero. There are N such possibilities, and if any of the resulting systems has a solution with all $x_j \geq 0$, then again we have found the minimum. If not, then we try those cases where exactly two of the x_k are zero, and so on.

In the following subsection, we shall illustrate this procedure for a case with N = 2.

EXAMPLE 2

The following example is based on the work by Crowninshield and Brand (1981b). These authors derived their particular optimization approach based on experimentally determined stress versus endurance-time relations of skeletal muscles. The intent of their approach was to solve the distribution problem by assigning forces to the muscles involved in a particular task, such that endurance time of the task (i.e., the time the task can be maintained) was maximized. It can be shown that this intent was not achieved by Crowninshield and Brand (1981b), however, a formal proof of this statement goes beyond the scope of this book.

In the approach proposed by Crowninshield and Brand (1981b) to solve the distribution problem (i.e., equations (4.5.5) and (4.5.6)), the force equation (4.5.5) was not considered, but the moment equation (4.5.6) was used, assuming that ligamentous and bony contact forces do not contribute significantly to the resultant external joint moment. The cost function (equation (4.5.10)) requires the minimization of the sum of the cubed muscular stresses, which was said to be equivalent to maximizing muscular endurance. The optimization approach of Crowninshield and Brand (1981b) may thus be formulated as:

Minimize the cost function ϕ, where:

$$\phi \quad = \quad \sum_{i=1}^{N} (F_i^m / pcsa_i)^3 \tag{4.5.33}$$

and:

F_i^m,   for i = 1,..., N are the design variables

subject to the constraints:

$$F_i^m \geq 0 , \qquad i = 1,..., N \tag{4.5.34}$$

$$\mathbf{M}_o = \sum_{i=1}^{N} (\mathbf{r}_{i/0} \times \mathbf{F}_i^m) \tag{4.5.35}$$

where all symbols have been defined before.

Let us assume, for the sake of simplicity, that the system of interest is a one-joint, planar system in which the joint of interest is crossed by two agonistic muscles. Let us further assume that we are interested in calculating the force-sharing between these two muscles; that is, we would like to express the force in muscle 1, F_1^m, as a function of the force in muscle 2, F_2^m. Therefore, we minimize:

$$\phi = \sum_{i=1}^{2} (F_i^m/pcsa_i)^3 \tag{4.5.36}$$

subject to:

$$M_o = \sum_{i=1}^{2} (r_{i/0} \cdot F_i^m) \tag{4.5.37}$$

or:

$$0 = \sum_{i=1}^{2} (r_{i/0} \cdot F_i^m) - M_o \tag{4.5.38}$$

and we admit only solutions for muscular forces that satisfy $F_i^m \geq 0$. Strictly speaking, ϕ is not convex for all values of F_i^m . Nevertheless, this problem can be converted to a convex one of the type considered in section 4.5.6 (Herzog and Binding, 1993), and as a result the necessary and sufficient conditions for a minimum solution are:

$$\frac{\partial L(F)}{\partial F_i^m} = 0 \tag{4.5.39}$$

where L depends on which F_i^m are selected to be zero. As in section 4.5.6.3, we first try the case where both F_i^m are positive. Then:

$$L = \sum_{i=1}^{2} (F_i^m/pcsa_i)^3 + \lambda \left[\sum_{i=1}^{2} (r_{i/0} \cdot F_i^m) - M_o \right] \tag{4.5.40}$$

The design variables are F_i^m; therefore:

$$\frac{\partial L\,(F)}{\partial F_1^m} = \left[3\,(F_1^m)^2/pcsa_1^3\right] + \lambda r_{1/0} = 0 \tag{4.5.41}$$

$$\frac{\partial L\,(F)}{\partial F_2^m} = \left[3\,(F_2^m)^2/pcsa_2^3\right] + \lambda r_{2/0} = 0 \tag{4.5.42}$$

Solving equations (4.5.41) and (4.5.42) for λ, and setting them equal to one another, yields:

$$\frac{3\,(F_1^m)^2}{(pcsa_1^3 \cdot r_{1/0})} = \frac{3\,(F_2^m)^2}{(pcsa_2^3 \cdot r_{2/0})} \tag{4.5.43}$$

and simplifying equation (4.5.43), and solving it for F_1^m, results in:

$$F_1^m = \left(\frac{r_{1/0}}{r_{2/0}}\right)^{1/2} \cdot \left(\frac{pcsa_1}{pcsa_2}\right)^{3/2} \cdot F_2^m \tag{4.5.44}$$

Equation (4.5.44) shows that the force in muscle 1, F_1^m, is a linear function of the force in muscle 2, F_2^m. The relation of force-sharing between the two muscles is influenced by the ratio of their moment arms $(r_{1/0}/r_{2/0})^{1/2}$ and their physiological cross-sectional areas $(pcsa_1/pcsa_2)^{3/2}$.

Let us now solve the force-sharing between these two muscles using a numerical example, where:

$$M_o = 10$$
$$pcsa_1 = pcsa_2 = 1$$
$$r_{1/0} = 1 \text{ and } r_{2/0} = 2$$

All values are in arbitrary (but, we assume, consistent) units. Therefore, the force-sharing equation for this problem (4.5.44) becomes:

$$F_1^m = (\tfrac{1}{2})^{1/2} \cdot (\tfrac{1}{1})^{3/2} \cdot F_2^m \tag{4.5.45}$$

or:

$$F_1^m = 0.71 F_2^m \quad \text{and} \quad F_2^m = 1.41 F_1^m \tag{4.5.46}$$

and the constraint equation (4.5.37) becomes:

$$10 \quad = \quad 1 \cdot F_1^m + 2 \cdot F_2^m \tag{4.5.47}$$

Solving equations (4.5.46) and (4.5.47) for F_1^m and F_2^m gives $F_1 = 2.61$ and $F_2 = 3.69$ with an associated cost (equation (4.5.36)) of 68.0. Both F_i^m are indeed positive, so we have found the global minimum.

The graphical solution of this problem is illustrated in Fig. 4.5.5. It has F_1^m on the horizontal axis and F_2^m on the vertical axis.

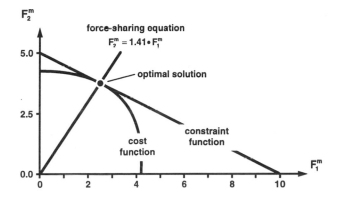

Figure 4.5.5   **Graphic illustration of the optimal solution of force sharing between two muscles in a one degree of freedom system (for details see text).**

4.5.7 REFERENCES

Andrews, J.G. (1974) Biomechanical Analysis of Human Motion. *Kinesiology.* Amer. Assoc. for Health, Phys. Ed., & Rec., Washington, D.C. **IV,** pp. 32-42.

Crowninshield, R.D. and Brand, R.A. (1981a) The Prediction of Forces in Joint Structures: Distribution of Intersegmental Resultants. *Exerc. Sport Sci. Reviews* (ed. Miller, D.I.). **9,** pp. 159-181.

Crowninshield R.D. and Brand R.A. (1981b) A Physiologically Based Criterion of Muscle Force Prediction in Locomotion. *J. Biomechanics.* **14 (11),** pp. 793-802.

Dul, J., Johnson, G.E., Shiavi, R., and Townsend, M.A. (1984) Muscular Synergism - II: A Minimum-fatigue Criterion for Load Sharing Between Synergistic Muscles. *J. Biomechanics.* **17 (9),** pp. 675-684.

Herzog, W. (1987) Individual Muscle Force Estimations using a Non-linear Optimal Design. *J. Neuroscience Methods.* **21,** pp. 167-179.

Herzog, W. and Binding, P. (1993) Co-contraction of Pairs of Antagonistic Muscles: Analytical Solution for Planar Static Non-linear Optimization Approaches. *Math. Biosciences.* **118,** pp. 83-95.

Morrison, J.B. (1968) Bioengineering Analysis of Force Actions Transmitted by the Knee Joint. *Bio-Medical Eng.* **3,** pp. 164-170.

Paul, J.P. (1965) Bio-engineering Studies of the Forces Transmitted by Joints - II: Engineering Analysis. *Biomechanics and Related Bioengineering Topics* (ed. Kenedi, R.M.). Pergamon Press, Oxford. pp. 369-380.

Pedotti, A., Krishnan, V.V., and Starke, L. (1978) Optimization of Muscle-force Sequencing in Human Locomotion. *Math. Biosciences.* **38,** pp. 57-76.

Pierrynowski, M.R. and Morrison, J.B. (1985) Estimating the Muscle Forces Generated in the Human Lower Extremity when Walking: A Physiological Solution. *Math. Biosciences.* **75,** pp. 43-68.

Seireg, A. and Arvikar, R.J. (1973) A Mathematical Model for Evaluation of Force in Lower Extremities of the Musculo-skeletal System. *J. Biomechanics.* **6 (3),** pp. 313-326.

Weber, W. and Weber, E. (1836) *Mechanik der menschlichen Gehwerkzeuge.* W. Fischer Verlag, Goettingen.

4.6 GENERAL CONSIDERATIONS ON DETERMINATE AND INDETERMINATE SYSTEMS

NIGG, B.M.

4.6.1 INTRODUCTION

Chapters 4.4 and 4.5 presented information on mathematically determinate and indeterminate systems, providing in-depth descriptions of selected topics and examples. This section will summarize and synthesize solutions to these systems.

There are two classes of mathematical systems: determinate and indeterminate. Indeterminate systems themselves can be subclassified into overdetermined and underdetermined systems (Fig. 4.6.1).

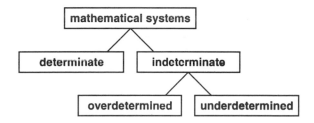

Figure 4.6.1   **Classification of mathematical systems.**

The characteristics of these systems are:

• Determinate	$n = u$
• Indeterminate	
◆ overdetermined	$n > u$
◆ underdetermined	$n < u$

where:

n = number of equations
u = number of unknowns

Note that mathematically determinate systems ($n = u$) have typically a unique solution in biomechanical applications. Furthermore, mathematically indeterminate systems typically do not have a unique solution in biomechanical applications. However, this statement is not true for the general mathematical cases of determinate and indeterminate systems.

Mathematically determinate and indeterminate systems are solved in different ways in biomechanical applications. The possible approaches are outlined, grouped, and discussed in the following pages.

4.6.2 DETERMINATE SYSTEMS

Mathematically determinate systems have the same number of equations and unknowns (n = u). In biomechanical applications they can be uniquely solved using analytical or numerical methods. Chapter 4.4 demonstrated how analytical solutions for very simple models could be devised. However, for a slightly more complex model (e.g., two rigid bodies with no muscles), analytical solutions become cumbersome, and numerical methods are preferred. Software for numerical solutions of a determinate system is readily available.

4.6.3 INDETERMINATE SYSTEMS

OVERDETERMINED SYSTEMS

An overdetermined mathematical system is defined as a system containing more equations than unknowns (n > u). There are at least two major strategies for solving such systems:

- Reduce constraints, and
- Soften constraints.

The strategy of reducing constraints consists of reducing the number of equations. This may be done by using physiological considerations or accuracy concerns - it may even be done arbitrarily. This approach is seldom, if ever, used in biomechanical modelling projects.

The strategy of softening constraints consists of changing the expectations for the solution. Instead of requiring an exact solution, satisfying all equations, one searches for a set of values where the error is minimal. A "least squares fit" technique is often used in finding the set of values for the unknowns best fitting the constraints. Overdetermined models are rather unusual in biomechanics research because most models attempt to describe the complexity of the locomotor system and, therefore, have more unknowns than equations.

The softening constraints approach to solving overdetermined systems has been used in biomechanics. Morlock and Nigg (1991) for example, presented four different three-dimensional models of the human foot, having one, six, twelve, and eighteen degrees of freedom (DOF). The models with twelve and eighteen DOF were overdetermined and were solved by minimizing the residuals.

UNDERDETERMINED SYSTEMS

A mathematically underdetermined system is defined as a system containing more unknowns than equations (n < u). Three different types of strategies can be used to solve this type of system (Nigg, 1989):

- *Reduction* of the number of unknowns,
- *Addition* of system equations:
  - ◆physiological
  - ◆neurophysiological
  - ◆additional criteria (e.g., optimization), and
- *Discussion* of possible solution spaces.

Reduction

The "reduction method" is used to reduce the number of unknowns until they equal the number of system equations. This is usually done by combining different muscles into one "muscle group" (Fig. 4.6.2).

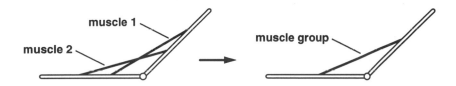

Figure 4.6.2   Schematic illustration of the reduction method for a two segment model with (originally) two muscles. The two muscles (left) are combined into one muscle group (right).

Examples of this method for walking are found in the work of Paul (1965) for the hip joint, Morrison (1970) for the knee joint, and Procter (1980) for the ankle joint complex. Baumann and Stucke (1980) used the reduction approach to estimate forces at the ankle joint, the Achilles tendon, and the knee joint during sports activities.

The reduction model provides a mathematically unique solution for the bone-to-bone forces in joints and for the forces in muscle groups. This method does not, however, provide a solution for individual muscle forces.

Addition

The "addition method" is used to increase the number of system equations until they equal the number of unknowns. This approach provides actual values for bone-to-bone forces as well as for individual muscle forces.

Additional constraints may be based on physiological principles and/or mathematical techniques, such as optimization methods (Seireg and Arvikar, 1973; Pedotti et al., 1978; Crowninshield and Brand, 1981; Herzog, 1987). Models with more physiological content seem, at first glance, to be more reasonable than models that use mathematical techniques, which may appear more constructed and, one may argue, are less in tune with reality. This view, however, is rather naive because if all the physiological and neurological rules were known, it would be easy to add them in the form of equations and constraints, but these rules are unknown, opening approaches to discussion and criticism.

Whenever muscles are involved in a model, one may (and probably should) argue that their neurophysiological aspect (control system) should be included. Such inclusion may add equations for solving the underdetermined mathematical problem. Hatze (1981) used additional neurophysiological constraints in a model with the control system. Pierrynowski (1982) used a pattern generator that purportedly mimicked the neurophysiological nature of the musculoskeletal system. In reality, additional mechanical constraints dictated by the geometry of the muscle moments arms, were imposed. Consequently, Pierrynowski's approach is not truly neurophysiological. Denoth (1985) used the electro mechanical delay of muscles (a neurophysiological aspect) to solve the indeterminacy problem for impact forces.

The control system approach to modelling the human locomotor system incorporates physiological as well as neurological aspects into the model. This approach, however, usually results in complicated mathematical systems that have only become solvable in recent years thanks to the increasing power of computers. As a result, only a few researchers in biomechanics have presented models featuring neurophysiological elements.

Discussion

The "discussion method" changes the mathematical problem into a philosophical one. Instead of solving the indeterminacy problem mathematically, various solution spaces are discussed. For example, the forces that would act in a joint if only one muscle worked, or if two muscles were to work at similar magnitudes could be discussed. This approach might yield a comprehensive understanding of different, especially clinical, possibilities (Denoth, 1985). The discussion approach, however, will be unable to provide a unique solution as to when actual forces in the locomotor system should be estimated. This approach does not provide actual values for bone-to-bone forces nor for individual muscle forces.

4.6.4 GENERAL COMMENTS

It might be thought that some ranking of the methods should be possible. However, ranking seems inappropriate since not enough knowledge and insight are available to solve the problems mathematically. The two methods, reduction and addition, differ primarily in where they locate their assumptions and uncertainty. The "reduction method" concentrates them at the onset of model development, whereas, the "addition method" adds assumptions and uncertainty along the way.

As a general rule, one may follow the guideline that the approach is dictated by the problem at hand. For example, it is not necessary to solve the distribution problem if one is "only" interested in bone-to-bone contact forces.

4.6.5 REFERENCES

Baumann, W. and Stucke, H. (1980) Sportspezifische Belastungen aus der Sicht der Biomechanik. *Die Belastungstoleranz des Bewegungapparates* (eds. Cotta, H., Krahl, H., and Steinbrueck, K.). Thieme, Stuttgart. pp. 55-64.

Crowninshield, R.D. and Brand, R.A. (1981) A Physiologically Based Criterion of Muscle Force Prediction in Locomotion. *J. Biomechanics.* **14 (11)**, pp. 793-801.

Denoth, J. (1985) Load on the Locomotor System in Modelling. *Biomechanics of Running Shoes* (ed. Nigg, B.M.). Human Kinetics, Champaign, IL. pp. 63-116.

Hatze, H. (1981) *Myocybernetic Control Models of Skeletal Muscle.* Characteristics & Applications. University of South Africa Press, Pretoria.

Herzog, W. (1987) Considerations for Predicting Individual Muscle Forces in Athletic Movements. *International Journal of Sport Biomechanics, University of Calgary.* Human Kinetics, Champaign, IL. **3 (2)**, pp. 128-141.

Morlock, M. and Nigg, B.M. (1991) Theoretical Considerations and Practical Results on the Influence of the Representation of the Foot for the Estimation of Internal Forces with Models. *Clinical Biomechanics.* **6**, pp. 3-13.

Morrison, J.B. (1970) The Mechanics of the Knee Joint in Relation to Normal Walking. *J. Biomechanics.* **3**, pp. 51-71.

Nigg, B.M. (1989) Assessment of Load Effects in the Reduction and Treatment of Injuries. *Future Directions in Exercise and Sport Science Research* (eds. Skinner, J.S., Corbin, C.B., Landers, D.M., Martin, P.E., and Wells, C.L.). Human Kinetics, Champaign, IL. pp. 181-193.

Paul, J.P. (1965) Bioengineering Studies of the Forces Transmitted by Joints. *Engineering Analysis, Biomechanics and Related Bioengineering Topics* (ed. Kennedy, R.M.). Pergamon Press, Oxford. pp. 369-380.

Pedotti, A., Krishnan, V.V., and Starke, L. (1978) Optimization of Muscle-force Sequencing in Human Locomotion. *Mathematical Biosciences.* **38**, pp. 57-76.

Pierrynowski, M.R. (1982) *A Physiological Model for the Solution of Individual Muscle Force During Normal Human Walking.* (Unpublished Doctoral Dissertation) Simon Fraser University, B.C., Canada.

Procter, P. (1980) *Ankle Joint Biomechanics.* (Unpublished Doctoral Dissertation) University of Strathclyde, Glasgow, Scotland.

Seireg, A. and Arvikar, R.J. (1973) A Mathematical Model for Evaluation of Forces in Lower Extremities of the Musculo-skeletal System. *J. Biomechanics.* **6 (3)**, pp. 313-326.

4.7 ENERGY CONSIDERATIONS

EPSTEIN, M.

4.7.1 DEFINITIONS AND COMMENTS

Absolute temperature: θ: a positive temperature measure consistent with the ideal gas law.

Clausius-Duhem inequality: The form of the second law of thermodynamics for a continuous non-homogeneous system.

Conservative force field: A force-field derived from a potential, V.

Degrees of freedom: Independent geometric parameters completely specifying the configuration of a system.

Endergonic reaction: A chemical reaction for which the Gibbs free-energy of the reactants is smaller than that of the products.

Endothermic reaction: A chemical reaction for which the enthalpy of the reactants is smaller than that of the products.

Enthalpy: $H = U - pV$

A function of state for systems that can be characterized by pressure and volume alone.

Entropy: S: a thermodynamical function of state postulated in the second law of thermodynamics.

Exergonic reaction: A chemical reaction for which the Gibbs free-energy of the reactants is larger than that of the products.

Exothermic reaction: A chemical reaction for which the enthalpy of the reactants is larger than that of the products.

First law of thermodynamics: The time rate of change of the sum of kinetic plus internal energy is equal to the power of the external forces plus the heating input.

Gibbs free-energy: $G = H - \theta S$

A thermodynamical function of state.

Helmholtz free-energy: $F = U - \theta S$

A thermodynamical function of state.

Homogeneous process: A thermodynamical process for which all points of a system are at the same state for any given instant.

Internal energy: U: a thermodynamical function of state postulated in the first law of thermodynamics.

Kinetic energy of a particle: $E_{kin} = \frac{1}{2}mv^2$

Lagrange multipliers:	λ: additional variables used to introduce geometrical constraints in the field equations.
Mechanical power (or working) of a force:	$P = \mathbf{F} \cdot \mathbf{v}$
Metabolic cycle:	The process of dissociation of high energy compounds, such as glucose, to produce the free-energy needed to sustain animal life.
Potential energy:	V: a scalar time-independent field such that the force field is (minus) its spatial gradient.
Principle of conservation of mechanical energy:	When all forces acting on a system of particles derive from a potential, the sum of kinetic plus potential energy is a constant of the motion of the system.
Principle of virtual work:	The equilibrium of a mechanical system is equivalent to the identical vanishing of the total virtual work performed by all external and internal forces through all virtual displacement of the system.
Thermoelastic effect:	The emission (or absorption) of heat by a thermoelastic material subjected to an isothermal contraction (or expansion).
Virtual displacement:	A set of small changes in the values of the degrees of freedom of a system.
[]	Represents end of examples.

4.7.2 SELECTED HISTORICAL HIGHLIGHTS

4th Century, B.C.	Aristotle used the term "energy" to denote something "in action".
1695 Leibniz	Formulated explicitly the principle of conservation of mechanical energy for a particle, in terms of "live force" (kinetic energy) and "dead force" (potential energy).
1744 Euler	Following a suggestion of his teacher, D. Bernoulli (1700-1782), successfully used a definition of the "potential force" (strain energy) of an elastic bar to solve the buckling problem by means of variational calculus.
1788 Lagrange	Adopted the principle of virtual work as the basis for his analytical mechanics.
1798/1799 Rumford/ Davy	Count Rumford (Benjamin Thompson) and Sir Humphry Davy proved experimentally that heat is not a substance but a form of energy.

1824	Carnot	Published his *Réflexions sur la Puissance Motrice du feu*, in which he laid down the foundation for the second law of thermodynamics.
1850	Clausius	Introduced the term "entropy" for what essentially was Carnot's "caloric".
1851	Kelvin	Reconciled the work of Carnot with that of Joule, to give a precise formulation of the first and second laws of thermodynamics.
1857	Pasteur	Established the catalytic role of microorganisms in the process of fermentation.
1864	Heidenhain	Established that the work potential and the heat production of a muscle increased with externally applied loads, up to a maximum.
1878	Gibbs	Developed the fundamentals of chemical thermodynamics.
1913	Hill	Started his famous experiments on heat measurements of contracting frog skeletal muscle.
1923	Fenn	While working in Hill's laboratory, demonstrated that energy output increased with work done against an opposing force in stimulated frog muscle, thus ruling out the viscoelastic theory of Hill and others.
1938	Hill	Found an experimental relation between muscular force and velocity, and attempted a thermodynamic interpretation.
1941	Lipmann	Discovered the role of ATP as a kind of common currency of metabolic energy.
1957	A.F. Huxley	Proposed the "Cross-bridge Theory" of muscular contraction which allowed a consistent view of mechanical and energetic properties of skeletal muscle.

4.7.3 INTRODUCTION

The literature of biomechanics seems to deal with energy considerations haphazardly, leaving the reader at a loss as to whether or not a valid physical law has been used and how. A typical example is the calculation of the so-called efficiency in running or other repetitive motions. In most treatments, assumptions are made along the way whose validity is never properly discussed and, in the end, energy balance is seldom enforced. For this reason, this chapter will take the reader slightly farther afield into the theoretical realm, in the hope that the extra effort expended will eventually pay off in terms of the realization that the energetics of the animal body is an extremely complex phenomenon. It involves chemical, electrical, and thermomechanical processes that cannot be brushed aside with simple-minded "experiments" at the global level.

Disappointingly perhaps, no "solutions" are presented, because they simply do not exist at this stage. How is metabolic energy spent? The answer, most probably is, not in directly producing externally observable mechanical work but rather in a delicate process unleashed by neural activation, which controls muscular contraction at the microscopic level. Thus, two tasks that produce the same mechanical work may require vastly different amounts of metabolic energy, and tasks that produce no work at all, such as holding a weight, require amounts of metabolic energy dictated by the activation of the muscles involved. It appears that any reliable answer to problems of energy consumption must include at least a model of muscular involvement and activation. Simple ideas such as "joint moments" will not do. Models of muscular architecture do already exist and others are in the process of development. They are quite involved and require the use of high-speed electronic computers. But so do models of buildings and dams and other inanimate objects; and muscles are so much more complex! It is into such models that thermodynamic considerations will eventually be added and this chapter has been conceived with that end in mind.

4.7.4 THE "PURELY MECHANICAL" CASE

Let us start with the familiar situation of a material particle of mass, m, which is acted upon, at a certain instant, by a force, **F**. According to Newton's second law, the particle must move with an instantaneous acceleration, **a**, given by:

$$\mathbf{a} = \frac{\mathbf{F}}{m} \tag{4.7.1}$$

The kinetic energy, E_{kin}, of the particle is defined as the scalar quantity:

$$E_{kin} = \frac{1}{2} m \mathbf{v} \cdot \mathbf{v} = \frac{1}{2} m v^2 \tag{4.7.2}$$

where, **v**, is the velocity vector whose magnitude is the speed, v, and where "·" denotes the "dot product" of vectors.

As the particle moves, the kinetic energy will, in general, vary. Its rate of variation can be obtained by evaluating its time derivative:

$$\frac{dE_{kin}}{dt} = \frac{\partial E_{kin}}{\partial \mathbf{v}} \cdot \frac{d\mathbf{v}}{dt} = m\mathbf{v} \cdot \mathbf{a} = (m\mathbf{a}) \cdot \mathbf{v} = \mathbf{F} \cdot \mathbf{v} \tag{4.7.3}$$

In obtaining the final result, equation (4.7.1) was used. If the dot product $\mathbf{F} \cdot \mathbf{v}$ is identified as the "working" (or "mechanical power"), P, of the force, **F**, equation (4.7.3) can be rewritten as:

$$\frac{dE_{kin}}{dt} = P \tag{4.7.4}$$

In words: the rate of change of the kinetic energy of a particle is equal to the working of the total force acting on it. This is the statement of the balance of mechanical energy for a single particle. If we look upon the working as a cause, the balance principle asserts that its effect is to produce a change of kinetic energy. Note, however, that since the principle was derived from Newton's second law it contains essentially no new information.

EXAMPLE 1

Question:

A mass, m, attached to a fixed point, A, by an inextensible string, is acted upon by its own weight. If the mass moves on a horizontal circle, show that its speed cannot change.

Solution:

A free-body diagram of the mass in an arbitrary position along the circle reveals that both forces acting on it (the tension, **T**, of the string and the weight, **P**) are perpendicular to the velocity vector, **v**. The total working is, therefore, zero. It follows from equation (4.7.4) that:

$$\frac{dE_{kin}}{dt} = 0$$

whence:

$$E_{kin} = \frac{1}{2}mv^2 = \text{constant}$$

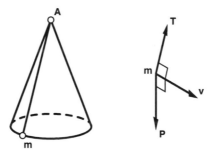

Figure 4.7.1   **Mass, m, attached to a fixed point, A, by an inextensible string.**

[]

An important particular case may arise when the force, **F**, depends only on the position presently occupied by the particle (and not on, say, its velocity or the time variable). In such a case it may happen that the value of the force at a point is given by the gradient of a scalar function, -V, of position, namely:

$$F_i = -\frac{\partial V}{\partial x_i}, \quad (i = 1, 2, 3) \tag{4.7.5}$$

where the subscript denotes Cartesian components in a fixed Cartesian coordinate system x_1, x_2, x_3. The function, V, is called a potential (or potential energy), and the force field,

F, is said to derive from the potential, V, or to be a conservative force field with potential, V.

If the position vector, **r**, with components x_1, x_2, x_3, is introduced, equation (4.7.5) can also be written as:

$$\mathbf{F} = -\frac{\partial V}{\partial \mathbf{r}} \tag{4.7.6}$$

As the particle moves in space, it traverses points of varying potential, so it makes sense to calculate the rate of change of the potential:

$$\frac{dV}{dt} = \frac{\partial V}{\partial \mathbf{r}} \cdot \frac{d\mathbf{r}}{dt} = -\mathbf{F} \cdot \mathbf{v} = -P \tag{4.7.7}$$

where equation (4.7.6) and the definition of velocity:

$$\mathbf{v} = \frac{d\mathbf{r}}{dt} \tag{4.7.8}$$

have been used. We conclude that in the case of a conservative force field the working of the force is equal to minus the rate of change of the potential energy. Combining this result with the energy balance, equation (4.7.4), yields:

$$\frac{d(E_{kin} + V)}{dt} = 0 \tag{4.7.9}$$

or:

$$E_{kin} + V = \text{constant} \tag{4.7.10}$$

In words: when the forces acting on a particle derive from a potential, the sum of the kinetic and potential energies is a constant of the motion. This is the so-called *"principle of conservation of mechanical energy"*.

EXAMPLE 2

Question:

Show that the field of gravity near the earth's surface derives from a potential.

Solution:

If a Cartesian coordinate system with x_3 pointing in the upward vertical direction is used, the force of gravity on a mass, m, is given by:

$$F_1 = F_2 = 0$$

$$F_3 = -mg$$

where:

$$g \quad = \quad 9.81 \text{ m/s}^2$$

Therefore, the function:

$$V \quad = \quad mgx_3$$

is a potential, by equation (4.7.5).

[]

EXAMPLE 3

Question:

Show that the gravitational force field of a heavy body derives from a potential.

Solution:

According to Newton's law of gravitation (the "inverse square" law), the force exerted by a body of mass, M, upon a particle of mass, m, is given by:

$$\mathbf{F} \quad = \quad -\frac{Mm}{r^3}\mathbf{r}$$

or:

$$F_i \quad = \quad -\frac{Mm}{r^3}x_i \quad (i = 1, 2, 3)$$

where a Cartesian system of coordinates with origin at the centre of mass, **M**, is assumed. The magnitude of the position vector, **r**, is:

$$r \quad = \quad \sqrt{x_1^2 + x_2^2 + x_3^2}$$

Defining:

$$V \quad = \quad -\frac{Mm}{r}$$

we obtain by direct calculation:

$$\frac{\partial V}{\partial x_i} \quad = \quad \frac{Mm}{r^2}\frac{\partial r}{\partial x_i} \quad = \quad (\frac{(Mm)}{r^2})\frac{x_i}{\sqrt{x_1^2 + x_2^2 + x_3^2}} \quad = \quad \frac{Mmx_i}{r^3} \quad = \quad -F_i$$

which shows that, V, is a potential.

[]

EXAMPLE 4

Question:

Construct a force field that does not derive from a potential.

Solution:

Since for a twice differentiable function the second derivatives are symmetrical, when a force field derives from a potential the components of the force must satisfy the condition:

$$\frac{\partial F_i}{\partial x_j} = -\frac{\partial}{\partial x_j}\left(\frac{\partial V}{\partial x_i}\right) = -\frac{\partial}{\partial x_i}\left(\frac{\partial V}{\partial x_j}\right) = \frac{\partial F_j}{\partial x_i}$$

To construct a force field without potential it is, therefore, enough to specify components that violate that condition. For example:

$$F_1 = x_2$$
$$F_2 = 0$$
$$F_3 = 0$$

[]

EXAMPLE 5

Question:

Show that the ordinary frictionless pendulum satisfies the principle of conservation of mechanical energy.

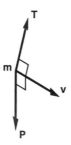

Figure 4.7.2   **Forces and velocity vectors for a frictionless pendulum.**

Solution:

The free-body diagram shows that there are two forces acting on the mass: the gravitational pull, **P**, and the tension, **T**, in the string. But if the string is inextensible, its tension is always perpendicular to the velocity vector. It, therefore, does no work. All the work is done by the gravitational force, which, as shown in Example 2, derives from a potential.

[]

Many geometrical constraints, such as the inextensibility of the string in the example above, satisfy the condition of doing no work on any possible motion of the system. The forces associated with such "workless" constraints can, therefore, be excluded from energy considerations.

EXAMPLE 6

Question:

Show that when air friction is included in the model, the preceding example does not abide by the law of conservation of mechanical energy.

Solution:

The drag force exerted by the air is a function, among other factors, of the speed of the moving element.

[]

When a *system of particles* is considered, it is obvious that as long as each particle is treated individually all the previous considerations apply. In particular, the total change of kinetic energy of the system is equal to the sum of the workings of all the forces acting on each particle.

In many applications, however, it is important to consider the system as a whole and to distinguish between external forces (resulting from interactions between the system and the rest of the universe) and internal forces (representing the interactions among the particles). The internal forces are usually assumed to abide by Newton's third law (action and reaction).

EXAMPLE 7

If the solar system is considered to be a system of particles (largely isolated from the rest of the universe), it is not acted upon by any external forces. The internal forces are those prescribed by Newton's universal gravitation law (or "inverse square" law). If the earth and the moon alone are considered a system, then their mutual attraction is an internal force, and the sun's gravitational pull (and to a much lesser extent, that of other planets) provides the external forces.

[]

An important particular case is that in which the internal force between each pair of particles of the system is a function only of the distance between the particles. Then it follows that the internal forces derive from a potential. Indeed, let, $\mathbf{r}_I$ and $\mathbf{r}_J$, denote the position vectors of particles I and J, respectively. Then the internal force described is of the type:

$$\mathbf{F}_{IJ} \;=\; -\mathbf{F}_{JI} \;=\; \frac{\mathbf{r}_J - \mathbf{r}_I}{s} f(s) \tag{4.7.11}$$

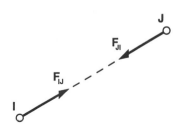

Figure 4.7.3   Internal, mutual attraction forces between two particles.

where:

$f(s)$ = some scalar function of the distance

s = $|\mathbf{r}_J - \mathbf{r}_I|$ = $\sqrt{(\mathbf{r}_J - \mathbf{r}_I) \cdot (\mathbf{r}_J - \mathbf{r}_I)}$

Define:

$$E(s) \;=\; \int_{s_o}^{s} f(\tau)\, d\tau \tag{4.7.12}$$

or, in other words, let $E(s)$ be a primitive function of, f, in the sense of calculus (assuming, f, to be integrable). Then:

$$\frac{\partial E}{\partial \mathbf{r}_J} \;=\; \frac{dE}{ds}\frac{\partial s}{\partial \mathbf{r}_J} \;=\; f(s)\frac{(\mathbf{r}_J - \mathbf{r}_I)}{\sqrt{(\mathbf{r}_J - \mathbf{r}_I) \cdot (\mathbf{r}_J - \mathbf{r}_I)}} \;=\; -\mathbf{F}_{JI} \tag{4.7.13}$$

The working of the pair $\mathbf{F}_{IJ}$ and $\mathbf{F}_{JI}$ is evaluated as:

$$\mathbf{F}_{IJ} \cdot \mathbf{v}_I + \mathbf{F}_{JI} \cdot \mathbf{v}_J \;=\; -\frac{\partial E}{\partial \mathbf{r}_I} \cdot \mathbf{v}_I - \frac{\partial E}{\partial \mathbf{r}_J} \cdot \mathbf{v}_J$$

$$= \; \frac{\partial E}{\partial (\mathbf{r}_J - \mathbf{r}_I)} \cdot \mathbf{v}_I - \frac{\partial E}{\partial (\mathbf{r}_J - \mathbf{r}_I)} \cdot \mathbf{v}_J$$

$$= \quad -\frac{\partial E}{\partial (r_J - r_I)} \cdot (v_J - v_I)$$

$$= \quad -\frac{\partial E}{\partial (r_J - r_I)} \cdot \frac{d(r_J - r_I)}{dt} = -\frac{dE}{dt} \qquad (4.7.14)$$

What has been done for one pair can be done for all, and redefining, E, as the sum of all the potentials we obtain:

$$\frac{d(E_{kin} + E)}{dt} \quad = \quad P_{ext} \qquad (4.7.15)$$

where P_{ext} is the working of the external forces. The potential, E, is sometimes called the "elastic energy" of the system. It is nothing but the potential of the internal forces.

[]

EXAMPLE 8

Deformable solids are sometimes idealized as point masses interconnected by springs and/or other elements to form regular lattices. From the preceding discussion it follows that this model will have an elastic energy if, and only if, the interconnecting elements are elastic (linear or non-linear) springs. If other elements, such as dampers or friction elements, are introduced, the working of the internal forces cannot be represented as the time derivative of a function of position alone. Also, if the internal force at one instant depends on the past history of the distance ("materials with memory"), an elastic potential does not exist.

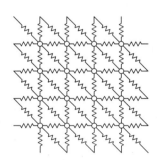

Figure 4.7.4   **Illustration of point masses interconnected by springs.**

[]

EXAMPLE 9 (advanced)

At the other extreme, a deformable solid can be represented as a continuum. The point masses are replaced by a mass density, and the internal forces become the (negative of the) stress tensor σ. The working of the internal forces per unit volume is now obtained by contracting the stress tensor with the velocity gradient, so that:

$$P_{int} \quad = \quad -\int_V (\sigma : \nabla v) \, dV \tag{4.7.16}$$

where:

$V \quad = \quad$ volume occupied by the solid

The external working consists of two terms:

$$P_{ext} \quad = \quad \int_V f \cdot v \, dV + \int_S t \cdot v \, dS \tag{4.7.17}$$

where:

$f \quad = \quad$ external force per unit volume
$t \quad = \quad$ force per unit area of the boundary S

This surface traction is related to the stress at the boundary through the exterior unit normal, n, by:

$$t \quad = \quad \sigma \cdot n \tag{4.7.18}$$

The kinetic energy is given by:

$$E_{kin} \quad = \quad \int_V \frac{1}{2} \rho v \cdot v \, dV \tag{4.7.19}$$

where:

$\rho \quad = \quad$ mass density

[]

EXAMPLE 10

If the interconnecting elements from Example 8 were replaced by rigid (massless) links, then the working of the internal forces would be zero, since - by the definition of rigidity - there can be no relative velocity between the particles. Thus, a rigid system abides by the law:

$$\frac{dE_{kin}}{dt} \quad = \quad P_{ext} \tag{4.7.20}$$

Since rigidity reduces the number of degrees of freedom of the system to six (three translations and three rotations, rather than the larger three translations per particle), relatively simple expressions can be derived for P_{ext} and E_{kin} that involve angular velocities and the so-called tensor of inertia. A particular case, in which these expressions simplify even more is that of a lamina (of constant thickness, say) constrained to move on its plane. Denoting by, C, its centre of mass, the kinetic energy is given by a translational and a rotational component, as follows:

$$E_{kin} = \frac{1}{2}m\mathbf{v}_c \cdot \mathbf{v}_c + \frac{1}{2}I_c\omega^2 \qquad (4.7.21)$$

where:

$\mathbf{v}_c$ = velocity of the centre of mass

m   = total mass

I_c = moment of inertia with respect to an axis perpendicular to the lamina at C

ω = angular speed

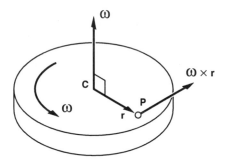

Figure 4.7.5   **Illustration of a rotating lamina of constant thickness.**

The velocity, $\mathbf{v}$, of an arbitrary point, P, whose position vector with respect to, C, is, $\mathbf{r}$, is:

$$\mathbf{v} = \mathbf{v}_c + \boldsymbol{\omega} \times \mathbf{r} \qquad (4.7.22)$$

where:

$\boldsymbol{\omega}$ = angular velocity vector, perpendicular to the lamina, whose magnitude is ω and whose direction abides by the "corkscrew rule"

[]

In the case of a system of interacting rigid bodies, similar considerations as those for systems of particles apply.

4.7.5 THE PRINCIPLE OF VIRTUAL WORK AND ITS APPLICATIONS

It should be clear from the foregoing theory and examples that the law of energy balance is a direct consequence of Newton's law of motion. One may ask whether the process can be reversed and the law of motion derived from the assumption that energy balance is satisfied. Except for systems with just one degree of freedom, the answer must be no, since, if nothing else, the statement of energy balance can be expressed with a scalar equation, while the law of motion is vectorial. One of the main advantages of properly formulated scalar expressions is that their form remains invariant under coordinate transformations.

In mechanics, it is possible to replace the laws of equilibrium (and motion) with a single scalar identity[1]. This is known as the *principle of virtual work* (d'Alembert's principle). Its usefulness lies in the fact that it facilitates the systematic derivation of exact or approximate equations of balance in any coordinate system, even in the presence of arbitrary geometrical constraints. Since biomechanical models tend to be constrained systems, a rather detailed treatment of the principle of virtual work is presented here, as well as several illustrative examples.

Any of the possible geometric configurations of a given mechanical system can be specified using a set of independent parameters called the *degrees of freedom* of the system.

EXAMPLE 11

Question:

For each of the following systems, determine a set of degrees of freedom:

(a)   A particle in space.
(b)   A plane pendulum.
(c)   A point on a surface.
(d)   A rigid body in space.
(e)   A beam.
(f)   A general continuum.

Solution:

(a)   The position of a particle in space can be described by, for instance, specifying its Cartesian coordinates x, y, and z in a fixed frame. Cylindrical or spherical coordinates may also be used.

---

1. It is always possible to replace a given vector equation with a single scalar identity - rather than an equation. Every solution of the vector equation will satisfy the identity, but there will generally be solutions of the scalar identity that are not smooth enough to satisfy the original vector equations. Such solutions are called "weak" and are beyond the scope of this chapter.

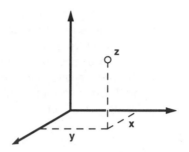

Figure 4.7.6   **Cartesian coordinate system.**

(b)   A plane pendulum that is inextensible (and incontractible) can be characterized by its angular deviation from the vertical. Alternatively, the x coordinate shown below can be used as long as the domain is restricted to $-\frac{\pi}{2} \le \theta \le \frac{\pi}{2}$.

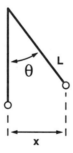

Figure 4.7.7   **Plane pendulum, characterized by its angular deviation from the vertical.**

(c)   Any system of curvilinear coordinates on a surface furnishes an admissible set of degrees of freedom for a material particle constrained to move on it. If the surface is finite in extent and the physical situation is such that the particle may fall off, thereby, acquiring a third degree of freedom, the nature of the mechanical system undergoes a sudden change, which requires special treatment. Normally, attention is restricted to mechanical systems whose degrees of freedom can be varied smoothly. Technically, they constitute a differentiable manifold (the "configuration space") whose dimension is equal to the number of degrees of freedom.

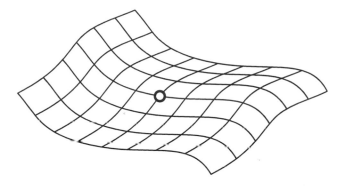

Figure 4.7.8   **Material particle constrained to move on surface.**

(d)   For a rigid body in space, it is enough to specify the position of one of its points,
O′, and the orientation of a rigid orthogonal triad attached to the body at O′. This
necessitates a total of six parameters, three for O′ and three for the angular orientations (such as Euler's angles θ, ϕ, and Ψ). Alternatively (and less appealingly)
one may choose three non-collinear points, A, B, and C, in the body, and provide
x_A, y_A, z_A, x_B, y_B, and x_C. The missing coordinates can be obtained from the
known (invariable) distances between the points.

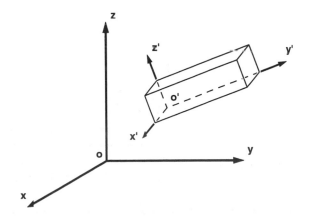

Figure 4.7.9   **Rigid body in space.**

(e)   A beam on two simple supports can achieve any shape given by a twice differentiable function $y = f(x)$, such that $f(0) = f(L) = 0$. Thus, the configuration space is infinite dimensional.

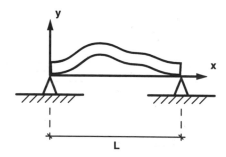

Figure 4.7.10   Beam on two simple supports.

(f)   In the case of a general continuum, three functions of position, representing, say, the three Cartesian components of the displacement vector field, need to be given, viz:

$$u = u(x,y,z)$$

$$v = v(x,y,z)$$

$$w = w(x,y,z)$$

These functions may be subject to constraints if part or all of the boundary of the body is fixed.

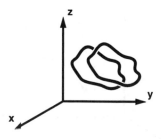

Figure 4.7.11   Representation for the description of the position of a general continuum.

[]

A *virtual displacement* is, loosely speaking, a set of small changes, δx_i, imposed on given values, x_i, of the degrees of freedom of a system. By definition, virtual displacements do not violate the geometrical constraints of the system.

EXAMPLE 12

Question:

For each of the cases of Example 11 show a virtual displacement.

Solution:

(a)   A particle in space:

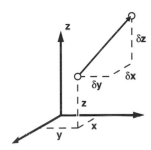

Figure 4.7.12   **Illustration of a particle in space.**

(b)   A plane pendulum:

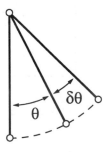

Figure 4.7.13   **Illustration of a plane pendulum.**

(c)   A point on a surface:

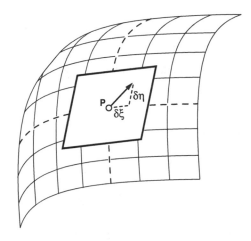

Figure 4.7.14   **Representation of a point on a surface.**

This example shows that virtual displacements live in the so-called *"tangent manifold"* of the configuration space.

(d)   A rigid body in space:

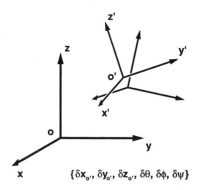

$$\{\delta x_{o'}, \delta y_{o'}, \delta z_{o'}, \delta\theta, \delta\phi, \delta\psi\}$$

Figure 4.7.15   **Representation of a rigid body in space.**

(e)   A beam:

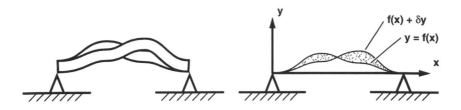

Figure 4.7.16   **Representation of a beam.**

Notice that the virtual displacement function δy (shaded) satisfies the geometric constraints (at the supports).

(f)   A general continuum:

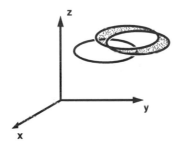

Figure 4.7.17   **Illustration of a general continuum.**

A virtual displacement consists of a vector field $\{\delta u,\ \delta v,\ \delta w\}$ schematically shown by the shaded volume of the figure.

[]

Forces acting on a mechanical system perform *virtual work* on any given virtual displacement, according to the definition of mechanical work.

The principle of virtual work asserts that for a system to be in an equilibrium[1] configuration, it is necessary and sufficient that the total virtual work performed by the external and internal forces vanish identically for all virtual displacements that can be imposed on that configuration.[2]

---

1. For simplicity we consider only statics, but the principle can be extended to dynamics.

EXAMPLE 13

Question:

Using the principle of virtual work, derive the equations of equilibrium of a particle in space.

Solution:

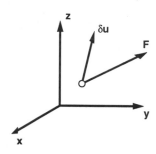

Figure 4.7.18   Particle in space.

If the coordinates of x, y, and z of the particle are chosen as degrees of freedom, the virtual work of the total force, **F**, acting on the particle is given by:

$$\mathbf{F} \cdot \delta\mathbf{u} = F_x\delta x + F_y\delta y + F_z\delta z$$

where:

F_x, F_y, and F_z =   Cartesian components of the force, F

δx, δy, and δz =   Cartesian components of the virtual displacement, $\delta\mathbf{u}$

The principle of virtual work demands that, for equilibrium:

$$F_x\delta x + F_y\delta y + F_z\delta z \;\equiv\; 0$$

Since the quantities δx, δy, and δz can be chosen arbitrarily and independently of each other, the only way that the identity can hold true is if the coefficients vanish, i.e.:

$$Fx \;=\; 0$$
$$Fy \;=\; 0$$
$$Fz \;=\; 0$$

[]

---

2. Strictly speaking there may be "weak" solutions, i.e., solutions of the virtual work identity that are not smooth enough to qualify as admissible static solutions.

EXAMPLE 14

Question:

Derive the equations of equilibrium for a rigid lamina in the plane, subjected to the action of n forces, F_i, i = 1 ,..., n, acting at n given points, P_i, of the lamina.

Solution:

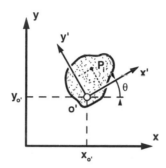

Figure 4.7.19   **Rigid lamina in a plane.**

If the x, y coordinates of a point $0'$ attached to the lamina, and the orientation θ of a pair of ("moving") axes x´, y´ attached to the lamina, are chosen as degrees of freedom, it is not difficult to see that a point, P_i, of the lamina with coordinates x_i', y_i' will have the x, y coordinates:

$$x_i = x_{o'} + x_i' \cos\theta - y_i' \sin\theta$$

$$y_i = y_{o'} + x_i' \sin\theta + y_i' \cos\theta$$

A virtual displacement $\{\delta x_{o'}, \delta y_{o'}, \delta\theta\}$ results in the small changes[1]:

$$\delta x_i = \delta x_{o'} - x_i' \sin\theta\delta\theta - y_i' \cos\theta\delta\theta$$

$$\delta y_i = \delta y_{o'} + x_i' \cos\theta\delta\theta - y_i' \sin\theta\delta\theta$$

The virtual work of all the forces is, therefore, given by:

$$VW = \sum_{i=1}^{n} (F_{x_i}\delta_x^i + F_{y_i}\delta_{y_i}) = \sum_{i=1}^{n} (F_{x_i}\delta x_{o'} + F_{y_i}\delta y_{o'}$$

---

1. These small changes, or "variations", can be calculated as differentials.

$$+ \quad (-F_{x_i} \sin\theta + F_{y_i} \cos\theta) \, x'_i \, \delta\theta$$

$$- \quad (F_{x_i} \cos\theta + F_{y_i} \sin\theta) \, y'_i \, \delta\theta \,)$$

Noting that:

$$-F_{x_i} \sin\theta + F_{y_i} \cos\theta \quad = \quad F_{y'_i}$$

$$F_{x_i} \cos\theta + F_{y_i} \sin\theta \quad = \quad F_{x'_i}$$

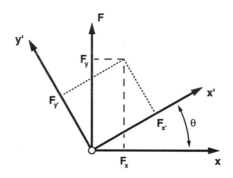

Figure 4.7.20   **Moving axes, x' and y', attached to the lamina.**

we can write:

$$\text{VW} \quad = \quad \delta x_{o'} \left(\sum_{i=1}^{n} F_{x_i} \right) + \delta y_{o'} \left(\sum_{i=1}^{n} F_{y_i} \right)$$

$$+ \quad \delta\theta \left(\sum_{i=1}^{n} (F_{y'_i} x'_i - F_{x'_i} y'_i) \right)$$

For this expression to vanish identically we must have:

$$\sum_{i=1}^{n} F_{x_i} \quad = \quad 0$$

$$\sum_{i=1}^{n} F_{y_i} = 0$$

$$\sum_{i=1}^{n} F_{y_i'} x_i' - F_{x_i'} y_i' = 0$$

The last equation expresses the vanishing of the sum of the moments of the forces with respect to the (arbitrary) point 0′.

[]

It is to be noted that whatever internal forces develop to maintain the rigidity of the lamina of Example 14, they play no role in the virtual work expression, since, by definition of rigidity, the points on which they are applied undergo no relative motion. This automatic elimination of workless constraint forces is one of the most powerful and useful features of the principle of virtual work.

EXAMPLE 15

Question:

Derive the law of equilibrium of a simple lever.

Solution:

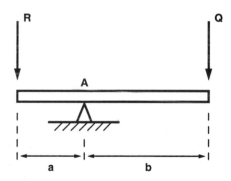

Figure 4.7.21   Simple lever.

Although a lever is a particular case of a rigid lamina, the existence of the fixed support, A, results in a significant simplification by reducing the number of degrees of freedom from three to just one.

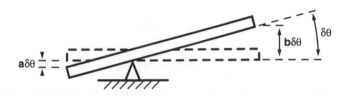

Figure 4.7.22   Lever with angle of rotation, θ.

Indeed, adopting the angle of rotation θ as the single degree of freedom, and considering the horizontal position as a possible equilibrium configuration, we evaluate the virtual work as:

$$VW = Ra\delta\theta - Qb\delta\theta \equiv 0$$

whence:

$$Ra = Qb$$

which is the well known condition of equilibrium of a lever.

[]

It is important to notice in Example 15 that the support reaction does not appear in the virtual work expression because, by definition, the support cannot move. Again, the power of the virtual work method in excluding workless forces of constraint is exhibited.

EXAMPLE 16

Question:

This example is closer in spirit to what might appear in a muscle model. Consider the deformable hinged triangle formed by three deformable bars, AB, BC, and AC, subjected to a vertical force, P, at the apex, C. No limitation is imposed on the magnitude of the deflections of C ("geometrical non-linearity"). It is required to find the equilibrium configuration(s) when point A is fixed and point B is prevented from moving vertically. The initial (unstretched) length of all three bars is L.

Note: in many cases, the stiffness of the structure permits the assumption of very small deflections ("geometrical linearity"). Although this simplification is common in many engineering applications, the flexibility of biological tissues, as well as the mechanism-like interconnection of the elements, makes it necessary to retain the utmost geometrical generality in biological modelling.

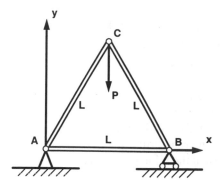

Figure 4.7.23 Deformable hinged triangle formed by three deformable bars.

Solution:

If the coordinates x_c, y_c, and x_B are adopted as degrees of freedom, the virtual work of the external forces is simply:

$$EVW = -P\delta y_c$$

In addition to this work, as each of the bars deforms, it exerts internal forces at the hinges, as shown. The work of these internal forces will be calculated as:

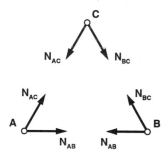

Figure 4.7.24 Internal forces at the hinges of a deformable hinged triangle.

$$IVW = -N_{AB}\delta L_{AB} - N_{BC}\delta L_{BC} - N_{AC}\delta L_{AC}$$

where:

$\delta L_{AB}=$ small change in length L_{AB} of bar AB
$\delta L_{BC}=$ small change in length L_{BC} of bar BC
$\delta L_{AC}=$ small change in length L_{AC} of bar AC

resulting from the infinitesimal virtual displacements, δx_C, δy_C, and δx_B.

The minus signs stem from the fact that if the forces are assumed to be in tension (as shown), they perform negative work on the nodes as they tend to drift farther apart. (It is customary to define the internal virtual work as that done by the forces acting on the bars rather than the nodes. Then the principle of virtual work reads EVW ≡ IVW).

We have:

$$L_{AB} \quad = \quad x_B$$

$$L_{BC} \quad = \quad \sqrt{(x_C - x_B)^2 + y_C^2}$$

$$L_{AC} \quad = \quad \sqrt{x_C^2 + y_C^2}$$

Whence:

$$\delta L_{AB} \quad = \quad \delta x_B$$

$$\delta L_{BC} \quad = \quad \frac{(x_C - x_B)(\delta x_C - \delta x_B) + y_C \delta y_C}{\sqrt{(x_C - x_B)^2 + y_C^2}}$$

$$\delta L_{AC} \quad = \quad \frac{x_C \delta x_C + y_C \delta y_C}{\sqrt{x_C^2 + y_C^2}}$$

The identity:

$$EVW + IVW \equiv 0$$

implies the three equilibrium equations:

$$\frac{N_{BC}(x_C - x_B)}{\sqrt{(x_C - x_B)^2 + y_C^2}} + \frac{N_{AC} x_C}{\sqrt{x_C^2 + y_C^2}} \quad = \quad 0$$

$$P + \frac{N_{BC} y_C}{\sqrt{(x_C - x_B)^2 + y_C^2}} + \frac{N_{AC} y_C}{\sqrt{x_C^2 + y_C^2}} = 0$$

$$-N_{AB} + \frac{N_{BC} (x_C - x_B)}{\sqrt{(x_C - x_B)^2 + y_C^2}} = 0$$

These equations become a system in the three unknowns x_C, y_C, and x_B once constitutive equations for the internal forces are provided. In the case of elasticity, for example, these forces are functions of the length changes only:

$$N_{AB} = f(L_{AB} - L) \quad \text{etc.}$$

But any other law can be accommodated (viscoelasticity, contractility, etc.). The resulting system of equations is highly non-linear and it is unlikely that, in the most general cases, a direct derivation could have proceeded so simply and in a virtually foolproof manner. In this relatively simple example, it is not difficult to see that the multipliers of the internal forces in the final equations represent the cosines of the angles of the deformed bars with the horizontal and vertical directions, so that the equations obtained are simply statements of the equilibrium of joints C and B in their free directions.

[]

Sometimes the constraints are so involved that it is difficult, if not impossible, to implement them directly by choosing appropriate degrees of freedom. In such cases the principle of virtual work can be easily modified by means of so-called *"Lagrange multipliers"*. Let a system originally having n degrees of freedom, u_i (i = 1, ..., n), be subjected to a geometrical constraint of the form:

$$g(u_1, u_2, ..., u_n) = 0$$

where:

g = a differentiable function of n variables whose variation can be calculated as:

$$\delta g = \frac{\partial g}{\partial u_1} \delta u_1 + ... + \frac{\partial g}{\partial u_n} \delta u_n$$

Instead of expressing one of the variables, u_n say, in terms of the others by algebraic elimination, thus, reducing the number of variables to n-1, it can be shown that the modified virtual work statement:

$$EVW + IVW - \lambda \delta g - g \delta \lambda \equiv 0$$

in the independent n+1 variables, u_1, u_2,...,u_n, and λ, leads to a system of equations of equilibrium. The new variable, λ, called a Lagrange multiplier, often has a definite physical meaning (such as the force necessary to maintain the given constraint). Any number of constraints can be handled in a similar manner by introducing one new Lagrange multiplier per constraint.

EXAMPLE 17

Question:

In some skeletal muscle models, extra stiffness is attributed to the supposed incompressibility of the muscle. By analogy, in the two-dimensional model of Example 16, assume that the area of the triangle ABC is preserved and derive an appropriate system of equilibrium equations.

Solution:

Although the number of true degrees of freedom is now only two, we choose to express equilibrium in terms of four variables. The area of the deforming triangle ABC is given by:

$$a \quad = \quad \frac{1}{2} x_B y_C$$

and, in view of the initial equilateral geometry, the constraint of area conservation reads:

$$g(x_C, y_C, x_B) \quad = \quad \frac{1}{2} x_B y_C - \frac{\sqrt{3}}{4} L^2 \quad = \quad 0$$

Supplementing the principle of virtual work in the manner described above leads straightforwardly to the system of four equations:

$$\frac{N_{BC}(x_C - x_B)}{\sqrt{(x_C - x_B)^2 + y_C^2}} + \frac{N_{AC} x_C}{\sqrt{x_C^2 + y_C^2}} \quad = \quad 0$$

$$P + \frac{N_{BC} y_C}{\sqrt{(x_C - x_B)^2 + y_C^2}} + \frac{N_{AC} y_C}{\sqrt{x_C^2 + y_C^2}} - \frac{1}{2} \lambda x_B \quad = \quad 0$$

$$-N_{AB} + \frac{N_{BC}(x_C - x_B)}{\sqrt{(x_C - x_B)^2 + y_C^2}} - \frac{1}{2} \lambda y_C \quad = \quad 0$$

$$\frac{1}{2} x_B y_C - \frac{\sqrt{3}}{4} L^2 \quad = \quad 0$$

[]

As well as serving rather impressively for the derivation of equilibrium equations of constrained systems in an accurate, error-free manner, the principle of virtual work is widely used in the construction of approximate theories of structures (beams, plates, shells) and in the all-important generation of approximate equilibrium equations for numerical solution by digital computers (the so-called "Finite Element Method"). These topics fall outside the scope of this presentation.

4.7.6 THE THERMOMECHANICAL CASE

The treatment of the purely mechanical case shows that the energy balance equation:

$$\frac{dE_{kin}}{dt} = P_{ext} + P_{int} \tag{4.7.23}$$

is a direct result of the mechanical laws of motion, and, as such, contains essentially no new information. For a system of particles (or, in the limit, a continuum) to fall into the "purely mechanical" category, the internal forces must depend on kinematic variables alone: the total internal force at a point can, at most, depend on the past history of the motions of all the particles in the system. Otherwise, the system has to be recognized as not purely mechanical, or mechanically indeterminate. Such is the case, for instance, of the electrodynamics of deformable media, in which the mechanical equations are coupled with Maxwell's equations for the electromagnetic field. But since our interest is mainly the laws of thermodynamics, we shall confine our attention to the thermomechanical case. A thermomechanical system consists of an underlying mechanical system (such as a deformable body) endowed with an added non-mechanical variable, T, called "empirical temperature". For physical reasons the system must be idealized as a continuum and the temperature may vary from point to point. The internal forces (stresses) may depend on the temperature (and/or its derivatives, and/or its history) in addition to the dependence on mechanical variables (strains, strain rates, etc.). The collection of the variables on which the internal stresses may depend characterize the "state" of the system.

In addition to the mechanical working of the external forces, a new power input is recognized as "heating". This non-mechanical power is denoted, Q, and is measured in units of energy per unit time. It is determined experimentally by means of the so-called "calorimetric experiments".

The *first law of thermodynamics* asserts that for every thermomechanical system there exists a characteristic function U of the state of the system (mechanical and thermal variables) such that:

$$\frac{d(U + E_{kin})}{dt} = P_{ext} + Q \tag{4.7.24}$$

for every process that the system can undergo. This function, U, is called the internal energy of the system. Since the balance of mechanical energy always applies, equation (4.7.24) can also be written as:

$$\frac{dU}{dt} = Q - P_{int} \tag{4.7.25}$$

The mechanical equations and the first law of thermodynamics, when supplemented with "constitutive equations" characterizing the material at hand, and with appropriate boundary and initial conditions, form a complete system of equations that allows us to determine the evolution of the system in terms of positions and temperatures for any prescribed external forces and heatings.

EXAMPLE 18   (thermostatics of a gas)

Consider a gas in a cylinder with a movable piston to which a force, F, can be applied. Assuming that whatever motion is impressed to the piston will be slow enough to permit the gas to evolve from one state of dynamic equilibrium to another, we see that the state of the gas can be characterized by just two variables: its volume, V, and its (uniform) temperature, T. Since the movements are very slow (no "inertia" effects) at each instant of time, the force, F, is balanced by the internal pressure, p, in the gas:

$$F \quad = \quad pA \tag{4.7.26}$$

where:

$$A \quad = \quad \text{area of the piston}$$

The working of the external force is:

$$Fv \quad = \quad pA\frac{dx}{dt} \quad = \quad -p\frac{dV}{dt} \tag{4.7.27}$$

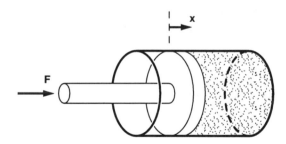

Figure 4.7.25   **Gas in a cylinder with a movable piston.**

The gas must now be characterized by two "constitutive equations". The first one, sometimes called the equation of state, is of the form:

$$p \quad = \quad p(V,T) \tag{4.7.28}$$

In the so-called ideal gas, for example, this equation reads:

$$p \quad = \quad k\frac{\theta}{V} \tag{4.7.29}$$

where k is a constant and θ is the so-called absolute temperature.

The second constitutive equation is:

$$U = U(V,T) \tag{4.7.30}$$

It can be shown that U for an ideal gas depends on T alone. These two constitutive equations characterize the gas involved and must be determined experimentally. The first law of thermodynamics establishes that:

$$\frac{dU}{dt} = -p\frac{dV}{dt} + Q \tag{4.7.31}$$

with kinetic energy being neglected.

A problem in gas thermostatics is specified by giving the external heating supply, Q, and the external force, both as functions of time. Since giving the force is, by equation (4.7.26), tantamount to giving the pressure, equation (4.7.28) effectively allows us to determine the volume as a function of temperature and time:

$$V = f(T, t) \tag{4.7.32}$$

so that:

$$\frac{dV}{dt} = \frac{\partial f}{\partial T}\frac{dT}{dt} + \frac{\partial f}{\partial t} \tag{4.7.33}$$

Similarly:

$$\frac{dU}{dT} = \frac{\partial U}{\partial V}\frac{dV}{dt} + \frac{\partial U}{\partial T}\frac{dT}{dt} \tag{1.7.34}$$

$$= \left(\frac{\partial U}{\partial V}\frac{df}{dT} + \frac{\partial U}{\partial T}\right)\frac{dT}{dt} + \frac{\partial U}{\partial V}\frac{\partial f}{\partial t} \tag{4.7.35}$$

Introducing (4.7.33) and (4.7.34) in (4.7.35) yields:

$$\left[(p + \frac{\partial U}{\partial V})\frac{\partial f}{\partial T} + \frac{\partial U}{\partial T}\right]\frac{dT}{dt} + (p + \frac{\partial U}{\partial V})\frac{df}{dt} = Q \tag{4.7.36}$$

which is an ordinary differential equation in T. This shows that in the case of thermostatics of a gas, the first law of thermodynamics provides the evolution equation for the temperature for any given external force and heating input as functions of time.

[]

EXAMPLE 19 (a rigid heat conductor)

We assume that a material has been given which, in first approximation, may be considered rigid, and we consider the case in which the material is fixed in the inertial frame of the observer, so that the kinetic energy is zero, such as in a laboratory test to measure the thermal conductivity of a solid specimen. For simplicity, we consider the one-dimensional situation of a wire thermally insulated on its lateral surface.

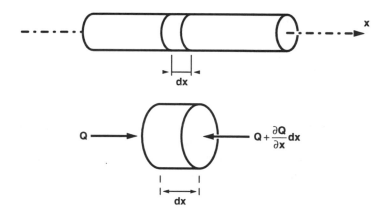

Figure 4.7.26   **Wire, thermally insulated on its lateral surface.**

Considering an elementary length dx of the wire and calling, Q, the heating supply through the left cross-section, we conclude that the net heating supply is:

$$\frac{\partial Q}{\partial x} dx$$

Denoting by, u, the internal energy per unit volume and by, A, the area of the cross-section, the rate of change of internal energy in the elementary slice is given by:

$$\frac{\partial U}{\partial t} = \frac{\partial u}{\partial t} A dx$$

(The partial derivative is intended to distinguish the time dependence from the material particle dependence). In the absence of mechanical contributions the first law of thermodynamics implies:

$$\frac{\partial u}{\partial t} = \frac{1}{A} \frac{\partial Q}{\partial x}$$

The constitutive laws that characterize the material give Q and u as functionals of the temperature. For many materials:

$$Q = -Ak \frac{\partial T}{\partial x} \qquad \text{(Fourier's law)}$$

and:

$$u = cT$$

with k and c as constants. It follows that:

$$c\frac{\partial T}{\partial t} + k\frac{\partial^2 T}{\partial x^2} \quad = \quad 0$$

which is the well known heat conduction equation.

[]

EXAMPLE 20 (a general continuum - advanced)

In the same spirit as with other quantities, we introduce an internal energy per unit mass, u. The heating consists of two terms: a supply, r, per unit mass (such as in a microwave oven) and a supply, h, per unit area of the boundary, S (such as in a conventional oven). Thus:

$$Q \quad = \quad \int_V \rho r\, dV + \int_S h\, dS \qquad (4.7.37)$$

The heating h through the boundary can be shown to be the dot product of a vector, $-\mathbf{q}$ (called the heat flux vector) and the exterior unit normal:

$$h \quad = \quad -\mathbf{q} \cdot \mathbf{n} \qquad (4.7.38)$$

This vector, $-\mathbf{q}$, is defined on any area element (not just the actual boundary of the body), since the laws are assumed valid for any part of the body as well as for the whole. The heat flux vector is the thermal counterpart of the stress tensor. Its dependence on the temperature, temperature gradient, strains and/or other thermomechanical variables is to be given by a constitutive law based on experiments.

Returning to the equations (4.7.37) and (4.7.38) and using the divergence theorem of vector calculus, we obtain:

$$Q \quad = \quad \int_V (\rho r - \mathrm{div}\,\mathbf{q})\, dV \qquad (4.7.39)$$

Using equations (4.7.25) and (4.7.16) yields:

$$\frac{d}{dt}\int_V \rho u\, dV \quad = \quad \int_V (\rho r - \mathrm{div}\,\mathbf{q} + \sigma : \nabla\mathbf{v})\, dV \qquad (4.7.40)$$

Because of mass conservation:

$$\frac{d}{dt}(\rho\, dV) \quad = \quad 0 \qquad (4.7.41)$$

so that the left-hand side becomes:

$$\frac{d}{dt}\int_V \rho u\, dV \quad = \quad \int_V \rho\frac{du}{dt}\, dV \qquad (4.7.42)$$

Technically, d/dt stands here for the so-called material time derivative (derivative following the particle's motion). It is given by:

$$\frac{du}{dt} = \frac{\partial u}{\partial t} + \frac{\partial u}{\partial \mathbf{r}} \cdot \mathbf{v} \tag{4.7.43}$$

when u is conceived as a function of the spatial position vector and of time (Eulerian description).

Since equation (4.7.42) is assumed to hold for arbitrary volumes, it follows that:

$$\rho \frac{du}{dt} = \rho r - \mathrm{div}\, \mathbf{q} + \sigma : \nabla \mathbf{v} \tag{4.7.44}$$

This is the differential version of the first law of thermodynamics in a continuum.

[]

As has been noted above, the mechanical equations (mass conservation and balance of linear and angular momentum) together with the first law of thermodynamics provide, when combined with constitutive equations that characterize the material properties, a complete system of equations for the evolution of the motion and the temperature of the system. The addition of any new physical principle appears, therefore, to be superfluous. The fact, however, that certain processes do not seem to occur spontaneously (such as a gas collecting in one region of a container) and that some material properties never occur (such as a fluid with negative viscosity) indicates the existence of an underlying principle of directionality in nature, formally known as the second law of thermodynamics.

The second law of thermodynamics asserts the existence of a constitutive quantity called entropy whose time rate of change is never less than a quantity constructed out of the heating and a special temperature measure. Unlike the empirical temperature this "absolute temperature", as it is called, is always a positive number, and its scale is proportional to the volumetric changes of an ideal gas at constant pressure.

For a homogeneous process, namely one for which all points of the system are at the same state in every given instant, the second law reads:

$$\frac{dS}{dt} \geq \frac{Q}{\theta} \tag{4.7.45}$$

where:

S = entropy
θ = absolute temperature

A process is called reversible if the equality in (4.7.45) holds at all times. Otherwise it is irreversible.

EXAMPLE 21   (advanced)

In the non-homogeneous case, such as an arbitrarily deforming continuum, an entropy density per unit mass is assumed to exist, and the second law reads:

$$\frac{du}{dt}\int_V \rho s dV \quad \geq \quad \int_S \frac{h}{\theta} dS + \int_V \frac{\rho r}{\theta} dV$$

Using the divergence theorem with $h = -\mathbf{q} \cdot \mathbf{n}$ (as in Example 20) yields the differential version:

$$\rho\frac{ds}{dt} \quad \geq \quad -\operatorname{div}\left(\frac{\mathbf{q}}{\theta}\right) + \frac{\rho r}{\theta}$$

Combining this result with the differential statement of the first law (equation (4.7.44) of Example 20), so as to eliminate the term containing the volumetric heating r, yields:

$$\rho\theta\frac{ds}{dt} \quad \geq \quad \rho\frac{du}{dt} + \frac{1}{\theta}\mathbf{q}\cdot\nabla\theta - \sigma{:}\nabla\mathbf{v}$$

This is the differential version of the second law of thermodynamics in a continuum. It is known as the Clausius-Duhem inequality.

[]

For homogeneous processes of systems whose state is completely defined by the present values of the volume and the temperature, some specific results can be obtained. We now turn our attention to such systems. Many of these results can be generalized for arbitrary systems.

Our functions of state, so far, are the pressure, the internal energy, and the entropy, viz:

$$
\begin{aligned}
p &= p(V, \theta) \\
U &= U(V, \theta) \\
S &= S(V, \theta)
\end{aligned}
\tag{4.7.46}
$$

Other functions of state can be defined as combinations of these. The most commonly used are the enthalpy:

$$H = U + pV \tag{4.7.47}$$

the free energy (or Helmholtz's function):

$$F = U - \theta S \tag{4.7.48}$$

and the free enthalpy (or Gibbs' free-energy function):

$$G = H - \theta S \tag{4.7.49}$$

The usefulness of the enthalpy can be explored by plugging its definition into the formulation of the first law of thermodynamics (equation (4.7.24)) to obtain:

$$\frac{d(H + E_{kin})}{dt} = Q + P_{ext} + p\frac{dV}{dt} + V\frac{dp}{dt} \tag{4.7.50}$$

which can be written as:

$$\frac{dH}{dt} = Q - P_{int} + p\frac{dV}{dt} + V\frac{dp}{dt} \qquad (4.7.51)$$

Strictly speaking, for a homogeneous process we must have:

$$P_{int} = p\frac{dV}{dt} \qquad (4.7.52)$$

so that equation (4.7.51) simplifies to:

$$\frac{dH}{dt} = Q + V\frac{dp}{dt} \qquad (4.7.53)$$

which shows that the heat input in an isobaric (i.e., constant-pressure) process is measured by the enthalpy increase:

$$\int_{t_1}^{t_2} Qdt = H_2 - H_1 \qquad (p = constant) \qquad (4.7.54)$$

Thermodynamicists use, in practice, the volume-temperature state space even for some processes that are not strictly homogeneous, such as a fluid agitated by a stirrer. In such cases the external working input is divided into that part which is due to the volumetric change and that which is not, and equation (4.7.54) may not directly apply.

Many chemical reactions of interest take place under atmospheric conditions and can, therefore, be considered as isobaric processes. If the enthalpy of the reactants is larger than the enthalpy of the products, the right-hand side of equation (4.7.54) is negative, implying that heat has been emitted to the atmosphere. Such reactions are called exothermic. Conversely, when the reaction results in an increase of enthalpy, it is called endothermic.

Choosing a reference temperature and pressure (usually 25°C and 1 atm), the "heat of formation" (or enthalpy change) for the reaction that produces one mol of a given substance, at the reference temperature and pressure, out of its component elements can be tabulated. These values can be used to calculate heats of reaction at constant pressure for more involved chemical reactions.

EXAMPLE 22

Question:

Calculate the heat of reaction of:

$$2H_2S(g) + 3O_2(g) \rightarrow 2H_2O(l) + 2SO_2(g)$$

at standard conditions of temperature and pressure.

Solution:

We note first that the given equation has already been balanced from the point of view of mass, yielding the "stoichiometric" coefficients shown. Secondly, the parenthetical letter after each substance indicates whether it is in the gaseous (g), liquid (l), or solid (s) state. This is important, since the heats of formation are highly dependent on the phase, the difference being the latent heat for the substance at the given temperature and pressure.

From a table of standard heats of formation we find the following values:

Table 4.7.1   Selected standard heats of formation.

SUBSTANCE	Δh (kJ/mol)
H_2S (g)	-20.6
O_2 (g)	0
H_2O (l)	-285.8
SO_2 (g)	-296.8

The elements themselves are assigned a zero value in their natural state. We obtain:

$$\Delta H_{reactants} = 2 \text{ mol } (-20.6\frac{kJ}{mol}) + 3 \text{ mol } (0) = -41.2 \text{ kJ}$$

$$\Delta H_{products} = 2 \text{ mol } (-285.8\frac{kJ}{mol}) + 2 \text{ mol } (-296.8\frac{kJ}{mol}) = -1165.2 \text{ kJ}$$

$$\text{Heat input} = \Delta H_{products} - \Delta H_{reactants} = -1165.2 \text{ kJ} + 41.2 \text{ kJ} = -1124 \text{ kJ}$$

The negative sign shows that heat is flowing out (exothermic reaction). Note that the value obtained corresponds to the reaction of 2 mols of hydrogen sulphide. For any given mass, the heat of reaction can be easily calculated by reading from the periodic table the atomic molar masses for sulphur (32.06) and hydrogen (1.01). For hydrogen sulphide (H_2S) we obtain, therefore:

$$32.06 + 2 (1.01) = 34.08 \text{ g/mol}$$

[]

To reveal the physical significance of Helmholtz's free-energy we start by using its definition, equation (4.7.48), to write the second law, equation (4.7.45), as:

$$\frac{dV}{dt} - \frac{dF}{dt} - \frac{d\theta}{dt}S \geq Q \tag{4.7.55}$$

Combining this result with the first law, equation (4.7.24), we obtain:

$$-P_{ext} \quad \leq \quad -\frac{dF}{dt} - \frac{d\theta}{dt}S - \frac{dE_{kin}}{dt} \tag{4.7.56}$$

EXAMPLE 23

Question:

For a substance undergoing homogeneous processes only and abiding by constitutive equations of the form:

$$p \quad = \quad p\,(V, \theta),$$

$$U \quad = \quad U\,(V, \theta), \text{ and}$$

$$S \quad = \quad S\,(V, \theta),$$

show that all processes are reversible and that the free-energy function of Helmholtz acts as a potential. Obtain particular expressions for an ideal gas.

Solution:

In terms of the free-energy the second law can be written as equation (4.7.56), which is equivalent to:

$$\frac{dF}{dt} + S\frac{d\theta}{dt} + P_{int} \quad \leq \quad 0 \tag{4.7.57}$$

or:

$$\frac{dF}{dt} + S\frac{d\theta}{dt} + p\frac{dV}{dt} \quad \leq \quad 0 \tag{4.7.58}$$

where F, S, and p are, in our case, functions of V and θ alone. By the chain rule of differentiation we get:

$$(\frac{\partial F}{\partial V} + p)\frac{dV}{dt} + (\frac{\partial F}{\partial \theta} + S)\frac{d\theta}{dt} \quad \leq \quad 0 \tag{4.7.59}$$

Since the coefficients of dV/dt and $d\theta/dt$ are independent of these rates, we conclude that the left-hand side of (4.7.59) is a linear function of dV/dt and $d\theta/dt$. A linear function cannot be non-positive for all values of the variables. Therefore, it must vanish identically. It follows that all processes are reversible and that:

$$p \quad = \quad -\frac{\partial F}{\partial V} \tag{4.7.60}$$

and:

$$S \quad = \quad -\frac{\partial F}{\partial \theta} \tag{4.7.61}$$

so, F, acts as a potential.

In particular, since mixed partial derivatives are symmetrical, it follows that:

$$\frac{\partial p}{\partial \theta} = \frac{\partial S}{\partial V} \tag{4.7.62}$$

which is known as Maxwell's law for a gas.

An ideal gas abides by:

$$p = NR\frac{\theta}{V} \tag{4.7.63}$$

where:

N = number of mols in the system
R = universal gas constant = $8.3144 \ \ J \cdot {}^{\circ}K^{-1} \cdot mol^{-1}$

A simple integration now yields:

$$F = N(\hat{f}(\theta) - R\theta \ln\frac{V}{V_o}) \tag{4.7.64}$$

where:

V_o = a reference volume
$\hat{f}(\theta)$ = a function of temperature only, characteristic of the gas

Noting that, for an ideal gas:

$$\ln\frac{V}{V_o} = \ln(\frac{\theta}{\theta_o}) - \ln\frac{p}{p_o} \tag{4.7.65}$$

where:

p_o, θ_o = reference values associated with V_o

the free-energy function can also be written as:

$$F = N(f(\theta) - R\theta \ln\frac{p}{p_o}) \tag{4.7.66}$$

where:

$f(\theta)$ = a function that characterizes the gas

[]

EXAMPLE 24

Question:

Discuss the consequences of the second law for a gas with internal friction (or viscosity).

Solution:

We start with the general form:

$$p = p(V, \theta, \dot{V})$$

$$U = U(V, \theta, \dot{V})$$

$$S = S(V, \theta, \dot{V})$$

where:

$$\dot{V} = dV/dt$$

(The a priori inclusion of the same list of independent variables in all constitutive laws is known as the *principle of equipresence*).

In terms of the free-energy:

$$F = U - S\theta = F(V, \theta, \dot{V}) \tag{4.7.67}$$

the second law states:

$$\frac{dF}{dt} + S\frac{d\theta}{dt} + p\frac{dV}{dt} \leq 0 \tag{4.7.68}$$

By the chain rule we obtain:

$$\left(\frac{\partial F}{\partial V} + p\right)\frac{dV}{dt} + \left(\frac{\partial F}{\partial \theta} + S\right)\frac{d\theta}{dt} + \frac{\partial F}{\partial \dot{V}}\frac{d\dot{V}}{dt} \leq 0 \tag{4.7.69}$$

It immediately follows, as in Example 23, that the linear terms must vanish, viz:

$$\frac{\partial F}{\partial \dot{V}} = 0 \tag{4.7.70}$$

and:

$$\frac{\partial F}{\partial \theta} = -S \tag{4.7.71}$$

so that neither the free-energy nor the entropy (not the internal energy) can depend on, $\dot{V}$.

We are left with the residual inequality:

$$\left(\frac{\partial F}{\partial V} + p\right)\frac{dV}{dt} \leq 0 \tag{4.7.72}$$

Let us assume, for example, that the viscosity effect is linear, i.e.:

$$p = p_o(V,\theta) - v\dot{V} \tag{4.7.73}$$

where:

p_o = a constitutive function
v = a material constant

It follows that:

$$\frac{\partial F}{\partial V} = p_o \tag{4.7.74}$$

and:

$$-v\left(\frac{dV}{dt}\right)^2 \leq 0 \tag{4.7.75}$$

In conclusion: reversible processes are possible if and only if:

v = 0 (no viscosity; all homogeneous processes are reversible)

or:

v > 0 (only "static" processes are reversible: $dV/dt = 0$)

Notice that v cannot be negative without violating the second law (for non-static processes).

[]

EXAMPLE 25 (the thermoelastic effect)

To account for some observed facts in muscle energetics, such as the release or absorption of extra heat by an active muscle during a small and rapid shortening or lengthening, the so-called "thermoelastic effect" has been invoked (Woledge et al., 1985, p. 235). This effect is also of some importance in metals (Kestin, 1979, p. 367) and particularly in rubber and other elastomers, which exhibit a somewhat anomalous behaviour (Treloar, 1975, p. 24).

We present here a rigorous treatment of the thermoelastic effect and extend it to include velocity dependent material behaviour. In the process, some interesting thermodynamic restrictions are obtained and shown to be consistent with Hill's velocity law (Hill, 1938). Finally, a general expression for the rate of heat emitted in an isothermal process is derived which shows that the effect is magnified during contraction of a velocity-dependent material as compared with its purely thermoelastic counterpart.

For clarity, we consider a one-dimensional situation, represented by a wire made of a thermoelastic material subjected to uniform strains and uniform temperature fields. The wire is assumed to carry no force at a reference length, L_o, and a reference temperature, θ_o, and to behave in accordance with thermoelastic constitutive laws of the form:

$$f = f(e, \tau) \tag{4.7.76}$$

$$U = U(e, \tau) \tag{4.7.77}$$

and:

$$S = S(e, \tau) \tag{4.7.78}$$

where the meaning of the symbols is as follows:

> e = elongation
> τ = temperature change
> f = force, positive in tension
> U = internal energy
> S = entropy

The second law of thermodynamics asserts that for every process:

$$Q \leq \theta \dot{S} \tag{4.7.79}$$

where, Q, is the heating input (in Joule/sec, say) and, θ, denotes absolute temperature. A superimposed dot stands for the time derivative. Introducing the free-energy of Helmholtz:

$$F = U - \theta S \tag{4.7.80}$$

and making use of the first law of thermodynamics in the form:

$$\dot{U} = Q + f\dot{e} \tag{4.7.81}$$

the second law can be rewritten as:

$$\dot{F} + S\dot{\theta} - f\dot{e} \leq 0 \tag{4.7.82}$$

Using the chain rule and collecting terms, yields:

$$\left(\frac{\partial F}{\partial e} - f\right)\dot{e} + \left(\frac{\partial F}{\partial \tau} + S\right)\dot{\tau} \leq 0 \tag{4.7.83}$$

Since this inequality is to be satisfied identically for all processes, and since the bracketed quantities are independent of $\dot{e}$ and $\dot{\tau}$, it follows that:

$$\frac{\partial F}{\partial e} = f \tag{4.7.84}$$

and:

$$\frac{\partial F}{\partial \tau} = -S \tag{4.7.85}$$

In the purely thermoelastic case, therefore, the free-energy acts as a potential for both the force and the entropy. By the symmetry of mixed partial derivatives, equations (4.7.84) and (4.7.85) imply the integrability condition:

$$\frac{\partial p}{\partial \tau} = -\frac{\partial S}{\partial e} \tag{4.7.86}$$

also known as Maxwell's law.

Returning briefly to equation (4.7.83), one observes that all uniform (non-heat-conducting) processes in a purely thermoelastic material turn out to be reversible. Indeed, the equal sign will always be enforced in equation (4.7.83) and, a fortiori, in equation (4.7.79). Thus:

$$Q = \theta \dot{S} = \theta \left(\frac{\partial S}{\partial e} \dot{e} + \frac{\partial S}{\partial \tau} \dot{\tau} \right) \tag{4.7.87}$$

For an isothermal process:

$$\dot{\theta} = \dot{\tau} = 0 \tag{4.7.88}$$

and from equations (4.7.86) and (4.7.87), we finally obtain:

$$Q = -\theta \frac{\partial f}{\partial \tau} \dot{e} \tag{4.7.89}$$

Now, the material property:

$$\alpha = -\frac{\partial f}{\partial \tau} \tag{4.7.90}$$

represents the change in force wrought by a change in temperature when the length is kept fixed. Most real materials have a tendency to elongate when heated, so that preventing elongation while raising the temperature will result in a compression, thereby, yielding a positive value for, α (some materials, such as rubber, exhibit the opposite behaviour). It follows that an isothermal contraction ($\dot{e} < 0$) is accompanied by a heat emission ($Q < 0$), and, conversely, that an isothermal expansion entails heat absorption according to the formula:

$$Q = \alpha \theta \dot{e} \tag{4.7.91}$$

In the case of skeletal muscle, the constitutive laws should be modified, according to overwhelming experimental evidence, to include at least a velocity dependence, namely:

$$f = f(e, \tau, \dot{e}) \tag{4.7.92}$$

$$U = U(e, \tau, \dot{e}) \tag{4.7.93}$$

and:

$$S = S(e, \tau, \dot{e}) \tag{4.7.94}$$

Proceeding exactly as before, the second law boils down to the inequality:

$$(\frac{\partial F}{\partial e} - f)\dot{e} + (\frac{\partial F}{\partial \tau} + S)\dot{\tau} + \frac{\partial F}{\partial \dot{e}}\ddot{e} \leq 0 \tag{4.7.95}$$

which is linear in the rates $\dot{\tau}$ and $\ddot{e}$, but non-linear in $\dot{e}$. It follows, then, that the identical satisfaction of this inequality implies:

$$\frac{\partial F}{\partial \dot{e}} = 0 \tag{4.7.96}$$

$$\frac{\partial F}{\partial \tau} = -S \tag{4.7.97}$$

and the residual inequality:

$$(\frac{\partial F}{\partial e} - f)\dot{e} \leq 0 \tag{4.7.98}$$

Note, in passing, that equation (4.7.96) together with equation (4.7.97) prescribe that not only the free-energy, but also the entropy and, consequently, the internal energy, are independent of the velocity of deformation, leaving the force, f, as the only velocity-dependent quantity.

As for Maxwell's law, equation (4.7.98) clearly shows that it no longer holds. Nevertheless, it is still possible to evaluate the extent of the "thermoelastic" effect. Indeed, equation (4.7.98) implies that, as far as the variable $\dot{e}$ is concerned, the function:

$$(\frac{\partial F}{\partial e} - f)\dot{e}$$

has a maximum at $\dot{e} = 0$. Assuming differentiability it follows that:

$$\frac{\partial F}{\partial e} = f_o \tag{4.7.99}$$

where we have denoted:

$$f_o = f_o(e, \tau) = f(e, \tau, 0) \tag{4.7.100}$$

We have, therefore, a modified version of Maxwell's law, namely:

$$\frac{\partial S}{\partial e} = -\frac{\partial f_o}{\partial \tau} \tag{4.7.101}$$

or, one might say, Maxwell's law is still valid for "static" processes. The property:

$$\alpha = -\frac{\partial f_o}{\partial \tau} \tag{4.7.102}$$

has the same meaning as before and it should be measured in a "static" (reversible) process. To obtain the heat rate associated with an isothermal $(\dot{\tau} = 0)$ process, we use the first law:

$$Q = \dot{U} - f\dot{e} = \dot{F} + \dot{\tau}S + \theta\dot{S} - f\dot{e}$$

$$= (\frac{\partial F}{\partial e} + \theta\frac{\partial S}{\partial e} - f)\dot{e}$$

$$= [\alpha\theta - (f - f_o)]\dot{e} \tag{4.7.103}$$

For a force-velocity relation which is monotonically increasing, the extra term $(f - f_o)\dot{e}$ is always positive, so that in a contraction *more* heat than before is emitted and, conversely, less heat than before is absorbed in expansion.

A further consequence of the maximal condition is that the second partial derivative of $(\frac{\partial F}{\partial e} - f)\dot{e}$ with respect to $\dot{e}$ at $\dot{e} = 0$, is non-positive, which implies, after some analysis, that at $\dot{e} = 0$:

$$\frac{\partial f}{\partial \dot{e}} \geq 0 \tag{4.7.104}$$

which is consistent with Hill's force-velocity law described elsewhere in this book. It should be emphasized, however, that in discussing the thermoelastic effect in velocity-dependent materials we have tacitly assumed that no chemical reactions are triggered by the shortening. For this to be the case, the change in length has to be so rapid that the only energy released is that stored by the elasticity of the cross-bridges. That the heat released in slower processes turns out, in experiments, to be qualitatively consistent with the thermoelastic effect in a non-reacting velocity-dependent material may have been a contributing factor to some reluctance to abandon Gasser and Hill's (1924) original hypothesis, later discarded, that skeletal muscle is essentially viscoelastic.

[]

Returning to equation (4.7.56) for an isothermic process with no appreciable kinetic energy, we obtain:

$$-P_{ext} \quad \leq \quad -\frac{dF}{dt} \tag{4.7.105}$$

so that:

$$\int_{t_1}^{t_2} -P_{ext}\,dt \quad \leq \quad F_1 - F_2 \tag{4.7.106}$$

In words: the mechanical work that can be extracted from a (closed) system in an iso-thermal process, is less than or equal to the decrease in Helmholtz's free-energy. The equal sign corresponds to the reversible process.

A stronger result can be proven by observing that heat cannot spontaneously flow against the temperature fall. Even if the process is not isothermic, but provided that it starts and ends at the temperature of the surroundings (e.g., the atmosphere) and that it exchanges heat only with the surroundings, the upper limit for the work that can be extracted from the system is the same (equation (4.7.106)).

A similar result can be obtained for Gibbs' function. Indeed, from its definition and equation (4.7.106), it follows that:

$$\int_{t_1}^{t_2} -P_{ext} \, dt \quad \leq \quad G_1 - G_2 - p_1 V_1 + p_2 V_2 \tag{4.7.107}$$

If the pressures at the initial and final states are equal to the atmospheric pressure p_o, we can write:

$$\int_{t_1}^{t_2} -P_{ext} \, dt \quad - \quad p_o (V_2 - V_1) \quad \leq \quad G_1 - G_2 \tag{4.7.108}$$

The left-hand side is easily recognized as the work done by the system in addition to compensating the work of the atmospheric pressure against the system. This extra work is sometimes called "useful work" and, according to equation (4.7.108), it is limited by the decrease in free enthalpy.

The importance of Gibbs' function resides also in the fact that it can be used as a rigorous criterion for spontaneity of a process and equilibrium of a system. Roughly speaking, a process can be initiated at a constant temperature and pressure only if it leads to a decrease in Gibbs' function. Correspondingly, if a system is in equilibrium, its Gibbs' function must be at a relative minimum. To see this we note that for a process at constant pressure the left-hand side of equation (4.7.108) vanishes identically, so we must have:

$$G_2 \quad \leq \quad G_1 \tag{4.7.109}$$

Equation (4.7.109) shows that a process at constant temperature and pressure can take place spontaneously only if the outcome of the process has a lower value of Gibbs' function than the start. For equilibrium, no spontaneous processes can occur, so that Gibbs' function must be at a minimum.

EXAMPLE 26

Question:

Determine whether the following chemical reaction will tend to occur spontaneously at standard conditions:

$$2HNO_3 \, (l) + Ag(s) \rightarrow Ag \, NO_3 \, (s) + NO_2 \, (g) + H_2O \, (l)$$

Solution:

From a table of free-energies (Gibbs) of formation, we obtain:

Table 4.7.2   Selected free-energies (Gibbs) of formation.

SUBSTANCE	ΔG (kJ/mol)
HNO_3 (l)	-79.9
Ag (s)	0
Ag NO_3 (s)	-32
NO_2 (g)	51.9
H_2O (l)	-237.2

$$\Delta G° = \Delta G_{products} - \Delta G_{reactants}$$

$$= (1\ mol)\ (-32\frac{kJ}{mol}) + (1\ mol)\ (51.9\frac{kJ}{mol})$$

$$+ (1\ mol)\ (-237.2\frac{kJ}{mol}) - \left[(2\ mol)\ (-79.9\frac{kJ}{mol}) + (1\ mol) \cdot (0) \right]$$

$$= -57.5\ kJ\ <\ 0$$

The change in Gibbs' free-energy is negative, so the reaction tends to occur spontaneously.

[]

Although in the last example the difference $\Delta G°$ between the free-energies of the reactants and the products was taken as an indication of spontaneity, a more careful analysis shows that, in a closed system, as the reaction progresses the proportions of reactants and products change continuously, and so does the free-energy. Eventually a minimum is reached, corresponding to (dynamic) equilibrium, before the reactants are exhausted. Actually, the reaction could have also started from the products and proceeded in the reverse direction to equilibrium. The value of $\Delta G°$ is thus only an indication of whether the equilibrium is "closer" to the reactants ($\Delta G° > 0$) or to the products ($\Delta G° < 0$).

To see this in detail, let us consider first a mixture of n ideal gases. Denoting by m_i and M_i, the mass of the i-th gas present in the mixture and its molar mass, respectively, the corresponding number of moles, N_i, will be:

$$N_i = \frac{m_i}{M_i} \tag{4.7.110}$$

The number of moles in the mixture is:

$$N \quad = \quad \sum_{i=1}^{n} N_i \tag{4.7.111}$$

and the molar fractions are defined as:

$$v_i \quad = \quad \frac{N_i}{N}, \quad (i = 1, ..., n) \tag{4.7.112}$$

so that:

$$\sum_{i=1}^{n} v_i = 1 \tag{4.7.113}$$

Defining the partial pressures by:

$$p_i \quad = \quad v_i p, \quad (i = 1, ..., n) \tag{4.7.114}$$

it follows that the pressure of the mixture satisfies:

$$p \quad = \quad \sum_{i=1}^{n} p_i \tag{4.7.115}$$

Assuming, moreover, that the mixture itself behaves like an ideal gas we obtain:

$$p_i \quad = \quad v_i p \quad = \quad (\frac{N_i}{N}) \frac{NR\theta}{V} \quad = \quad N_i \frac{R\theta}{V} \tag{4.7.116}$$

which shows that an ideal gas in a mixture behaves as though it occupied the whole volume at the temperature of the mixture (Dalton's law). In particular, from Example 24, the free-energy of each component is given by:

$$F_i \quad = \quad N_i \left(f_i(\theta) + R\theta \ln \frac{p_i}{p_o} \right) \tag{4.7.117}$$

and, consequently, Gibbs' function for the i-th component is:

$$G_i \quad = \quad N_i \left(g_i(\theta) + R\theta \ln \frac{p_i}{p_o} \right) \tag{4.7.118}$$

with:

$$g_i \quad = \quad f_i + R\theta$$

The total free-energy of the mixture is:

$$G = \sum_{i=1}^{n} G_i \qquad (4.7.119)$$

and its differential at constant mixture pressure and temperature is given by:

$$dG = \sum_{i=1}^{n} \left(g_i(\theta) + R\theta \ln \frac{p_i}{p_o} \right) dN_i \qquad (4.7.120)$$

In deriving this formula, the fact that:

$$d(\ln p_i) = d(\ln v_i p) = \frac{dv_i}{v_i} \qquad (4.7.121)$$

in conjunction with:

$$\sum_{i=1}^{n} dv_i = 0 \qquad (4.7.122)$$

which follows from equation (4.7.113), was used to cancel a whole term. The reader should supply the missing steps.

Consider now a stoichiometrically balanced chemical reaction of the type:

$$n_1 C_1(g) + n_2 C_2(g) \rightarrow n_3 C_3(g) + n_4 C_4(g) \qquad (4.7.123)$$

where:

$$C_1, ..., C_4 = \text{ideal gas compounds}$$

Although the molar fractions v_1, v_2, v_3, v_4 are not subject to any direct constraints, their changes are. Indeed, stoichiometry (mass conservation), demands that:

$$-\frac{dN_1}{n_1} = -\frac{dN_2}{n_2} = \frac{dN_3}{n_3} = \frac{dN_4}{n_4} = dx \qquad (4.7.124)$$

Combining this result with equation (4.7.120) we get:

$$dG = dx [-n_1 g_1(\theta) - n_2 g_2(\theta) + n_3 g_3(\theta) + n_4 g_4(\theta)$$

$$+ R\theta \ln \frac{\left(\dfrac{p_3}{p_o}\right)^{n_3} \left(\dfrac{p_4}{p_o}\right)^{n_4}}{\left(\dfrac{p_1}{p_o}\right)^{n_1} \left(\dfrac{p_2}{p_o}\right)^{n_2}} \quad] \qquad (4.7.125)$$

For a minimum of, G, this expression must vanish, so that at equilibrium the partial pressures satisfy the equation:

$$\frac{\left(\dfrac{P_3}{P_o}\right)^{n_3}\left(\dfrac{P_4}{P_o}\right)^{n_4}}{\left(\dfrac{P_1}{P_o}\right)^{n_1}\left(\dfrac{P_2}{P_o}\right)^{n_2}} \quad = \quad \exp\left(\frac{n_1 g_1 + n_2 g_2 - n_3 g_3 - n_4 g_4}{R\theta}\right) = K(\theta) \quad (4.7.126)$$

The function $K(\theta)$ depends on temperature alone, and for a given temperature it is called the equilibrium constant of the reaction. A similar result is obtained for any number of reactants and products. Although this result was obtained for ideal gases, its validity can be justified in a more general context.

Assuming that the $\Delta G°$ for an equation such as (4.7.123) would be calculated for all reactants and products at the standard pressure, p_o, we obtain from equation (4.7.118):

$$\Delta G°(\theta) \quad = \quad -n_1 g_1(\theta) - n_2 g_2(\theta) + n_3 g_3(\theta) + n_4 g_4(\theta) \qquad (4.7.127)$$

which, comparing with equation (4.7.126), yields the useful relation:

$$\Delta G° \quad = \quad -R\theta \ln K(\theta) \qquad (4.7.128)$$

EXAMPLE 27

Question:

$N_1°$ and $N_2°$ mols of ideal gases C_1 and C_2, respectively, are permitted to react at a given pressure, p, and temperature, θ, to produce ideal gases, C_3 and C_4, according to the stoichiometrically balanced equation:

$$n_1 C_1 + n_2 C_2 \rightleftharpoons n_3 C_3 + n_4 C_4$$

Knowing the value of $K(\theta)$, determine the number of mols of reactants and products at equilibrium.

Solution:

Let μ be an unknown quantity representing the advance of the reaction. The present molar quantities of each of the gases are given by:

$$N_1 \quad = \quad N_1° - n_1 \mu \qquad (4.7.129)$$

$$N_2 \quad = \quad N_2° - n_2 \mu \qquad (4.7.130)$$

$$N_3 \quad = \quad n_3 \mu \qquad (4.7.131)$$

$$N_4 \quad = \quad n_4 \mu \qquad (4.7.132)$$

The total number of mols is:

$$N = \sum_{i=1}^{4} N_i = N_1^\circ + N_2^\circ + (\Delta n)\mu$$

where $\Delta n = n_3 + n_4 - (n_1 + n_2)$, so that the partial pressures are given by:

$$p_1 = \left(\frac{N_1^\circ - n_1\mu}{N_1^\circ + N_2^\circ + \Delta n\mu} \right) p$$

$$p_2 = \left(\frac{N_2^\circ - n_2\mu}{N_1^\circ + N_2^\circ + \Delta n\mu} \right) p$$

$$p_3 = \left(\frac{n_3\mu}{N_1^\circ + N_2^\circ + \Delta n\mu} \right) p$$

$$p_4 = \left(\frac{n_4\mu}{N_1^\circ + N_2^\circ + \Delta n\mu} \right) p$$

For equilibrium we must have, according to equation (4.7.126):

$$\frac{\mu^{n_3 + n_4}}{(N_1^\circ - n_1\mu)^{n_1}(N_2^\circ - n_2\mu)^{n_2}} = \frac{K(\theta)}{n_3^{n3} n_4^{n4}} \left[\frac{p}{(N_1^\circ + N_2^\circ + \Delta n\mu) p_o} \right]^{n_1 + n_2 - n_3 - n_4}$$

which provides a non-linear equation for μ at equilibrium. Plugging this value of μ into (4.7.129), (4.7.130), (4.7.131), and (4.7.132) gives the desired solution.

[]

EXAMPLE 28

Question:
Express the equilibrium constant in terms of molar concentrations rather than partial pressures.

Solution:
The molar concentration, c_i, is defined as:

$$c_i = \frac{N_i}{V}$$

where:

V = volume of the mixture, which is assumed to behave like an ideal gas

Therefore:

$$c_i = \frac{N_i p}{N R \theta} = \frac{p_i}{R \theta}$$

From equation (4.7.126) we then obtain:

$$K(\theta) = \left(\frac{R\theta}{p_o}\right)^{\Delta n} \frac{c_3^{n3} c_4^{n4}}{c_1^{n1} c_2^{n2}}$$

The value:

$$K_c(\theta) = \frac{c_3^{n3} c_4^{n4}}{c_1^{n1} c_2^{n2}}$$

is a function of temperature alone and is called the equilibrium constant in terms of concentrations. Its usefulness can be extended beyond gases to the calculation of chemical equilibrium of solutions.

[]

EXAMPLE 29

Question:

N_1°, N_2°, N_3°, and N_4° mols of ideal gases C_1, C_2, C_3, and C_4, respectively, are permitted to react at a given pressure, p, and temperature, θ, according to the stoichiometrically balanced equation:

$$n_1 C_1 + n_2 C_2 \rightleftharpoons n_3 C_3 + n_4 C_4$$

Knowing the value of $K(\theta)$, determine the direction in which the reaction will proceed spontaneously.

Solution:

Letting, μ, be a molar parameter representing the advance of the reaction, as in Example 27, we have at any instant:

$$N_1 = N_1^\circ - n_1 \mu \tag{4.7.133}$$

$$N_2 = N_2^\circ - n_2 \mu \tag{4.7.134}$$

$$N_3 \;=\; N_3^{\circ} + n_3 \mu \tag{4.7.135}$$

$$N_4 \;=\; N_4^{\circ} + n_4 \mu \tag{4.7.136}$$

Using these values in equation (4.7.120) yields:

$$\frac{dG}{d\mu} \;=\; -R\theta \ln K(\theta) + R\theta \ln \left[\frac{(N_3^{\circ})^{n3} (N_4^{\circ})^{n4}}{(N_2^{\circ})^{n2} (N_1^{\circ})^{n1}} \left(\frac{p}{N^{\circ} p_o}\right)^{\Delta n} \right] \tag{4.7.137}$$

where:

$$N^{\circ} \;=\; \sum_{i=1}^{4} N_i^{\circ}$$

Since a reaction proceeds spontaneously in the direction of decreasing free-energy, if the value of the slope, as given by equation (4.7.137) is negative, the reaction will proceed forward ($C_1 + C_2 \to C_3 + C_4$), and vice-versa. Note that the first term on the right-hand side of (4.7.137) is nothing but ΔG°, which shows that the direction of a spontaneous reaction is governed by a combination of ΔG° and the initial concentrations of reactants of products.

[]

EXAMPLE 30 (the metabolic cycle)

The ultimate source of energy for living organisms is solar power. By the process of photosynthesis, plants convert solar energy into chemical energy according to endergonic (free-energy absorbing) reactions such as:

$$6CO_2 + 6H_2O \qquad \to \quad C_6H_{12}O_6 + 6O_2$$
$$\text{(glucose)}$$

Partial or total reversal of this reaction in animals, in a process known as catabolism, results in the release of the free-energy needed to sustain life. But in order for this energy not to be totally dissipated as heat it must first be stored in highly mobile high-energy molecules, the most important of which is adenosine triphosphate (ATP). The production of ATP during catabolism is made possible by chemical coupling: an enzyme permits an endergonic $(\Delta G^{\circ} > 0)$ reaction to be driven by an exergonic $(\Delta G^{\circ} < 0)$ reaction so that the overall reaction is still exergonic and, therefore, spontaneous.

The free-energy carried by ATP to the body cells is released in the hydrolysis reaction:

$$ATP + H_2O \qquad \to \quad ADP + phosphate$$
$$\text{(adenosine}$$
$$\text{diphosphate)}$$

with:

$$\Delta G^o = -30kJ/mol$$
(at 1 atm and 25°C)

The free-energy made available by this reaction is used to produce mechanical and transport work as well as to biosynthesize other molecules (again, through coupling). The ADP produced is eventually resynthesized into ATP during catabolism, thus, closing the metabolic cycle.

[]

EXAMPLE 31

Question:

Calculate the value of the equilibrium constant for the hydrolysis of ATP at 25°C, and discuss its implications.

Solution:

From equation (4.7.128):

$$\Delta G^o + R\theta \ln K \, (\theta) = 0$$

whence:

$$K = \exp\left(-\frac{\Delta G^o}{R\theta}\right)$$

with:

$$\Delta G^o = \quad -30,000 \text{J mol}^{-1}$$

$$R = \quad -8.3144 \text{J °K}^{-1} \text{mol}^{-1}$$

$$\theta = \quad 273 + 25 = 298 \text{ °K}$$

We obtain:

$$K = \quad 1.8 \times 10^5$$

The implication of this very high value is that at equilibrium the ratio of the concentration of ADP to ATP would be so high that practically no ATP would be left. Under normal circumstances, however, the metabolic cycle guarantees a continual supply of ATP that keeps its concentration constant and the reaction highly spontaneous.

[]

4.7.7 COMMENTS ON SKELETAL MUSCLE MODELS

Apart from its importance in having established an experimentally based force-velocity relation in skeletal muscle (see section 2.5.4), A.V. Hill's famous 1938 paper was crucial in legitimizing a line of approach to muscular mechanics based on energetics. Indeed, Hill's paper concerned itself mainly with measurements of heat production, and it was through hypothesizing (as it turned out, incorrectly) certain relations between measured heat rates and velocities, that Hill's essentially correct mechanical model was originally "derived".

Hill's experiment is the so-called isotonic quick-release, in which the muscle is first maximally stimulated under isometric (i.e., fixed length) conditions until it attains its maximal force f_0, and then suddenly released at a constant load, $f < f_0$. The observed events are summarized in Fig. 4.7.27.

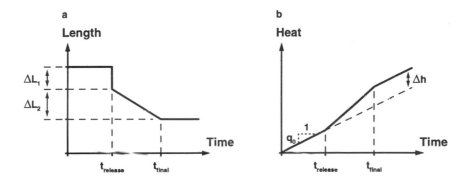

Figure 4.7.27   Length-time and heat-time relationships for Hill's experiment.

It appears from Fig. 4.7.27a that the shortening consists of two successive parts: an instantaneous shortening, ΔL_1, and a constant velocity shortening, ΔL_2. The instantaneous component increases with the load differential $f_0 - f$, and was thus attributed to an elastic element (EE) connected in series with another element, called the contractile element (CE) (Fig. 4.7.28) which encompasses all time-dependent characteristics.

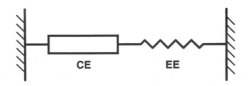

Figure 4.7.28   Elastic and contractile elements.

To characterize the contractile element (in the absence of detailed knowledge about the internal microscopic structure of muscle), Hill observed after many experiments that there seemed to be a definite relation between the heat released during shortening and the shortening itself. Observing the general behaviour of the heat released versus time curve (Fig. 4.7.27b), it follows that heat is released at a constant rate during the isometric totally stimulated phase. This heat rate was called "maintenance heat" by Hill. Upon isotonic release an excess heat, Δh, is liberated at a higher rate until the shortening stops, at which point the maintenance heat rate is resumed. Hill suggested that the extra heat, Δh, was proportional to the constant speed of shortening, $\Delta \dot{L}_2$:

$$\Delta h = a \Delta \dot{L}_2$$

where the proportionality factor "a" is a characteristic of a given muscle (here, ΔL_2 is considered a positive quantity in shortening).

Next, Hill proceeded to evaluate the extra energy, ΔE (in excess of maintenance heat) released during the constant speed phase. This consists of the extra heat, Δh, plus the work done by, f, throughout the shortening, ΔL_2. Denoting by v the shortening velocity, the rate of extra energy release is seen to be:

$$\Delta \dot{E} = \frac{\Delta E}{\Delta t} = \frac{\Delta h}{\Delta t} + f \frac{\Delta L_2}{\Delta t} = a \frac{\Delta L_2}{\Delta t} + f \frac{\Delta L_2}{\Delta t} = (a + f) v$$

On the basis of his experimental evidence, Hill proceeded to postulate that this rate is proportional to the force differential, viz:

$$\Delta \dot{E} = b (f_o - f) \tag{4.7.138}$$

where b is another constant characteristic of the muscle. Equating the last two expressions, it follows that:

$$(a + f) v = b (f_o - f) \tag{4.7.139}$$

which is a purely mechanical force-velocity dependence characterizing the contractile element. Hill himself observed that this relation could have been obtained directly by purely mechanical measurements. In retrospect, it is fortunate that this direct route was not followed. For, although the assumptions embodied in equations (4.7.138) and (4.7.139) were later proven to be inadequate, they opened the door to much speculation on the nature of the chemical reactions and mechanisms, whereby, muscular contraction is generated, culminating with the now generally accepted concepts of the Cross-bridge Theory.

In using the first law of thermodynamics for a system, such as muscle, in which chemical reactions take place, it is necessary to include a term reflecting the energy input arising from the reaction, namely:

$$\dot{E}_{kinn} + \dot{U} + \sum_i \Delta H_i \mu_i = Q + P_{ext}$$

where ΔH_i is the enthalpy change associated with the i-th reaction, and μ_i is a parameter measuring the rate of advance of the reaction (e.g., in, mol-sec^{-1}). Assuming on the one hand, that the only significant reactions taking place are ATP hydrolysis and resynthesis (reactions for which ΔH is known from in-vitro experiments), and that the energy released is not stored temporarily in some internal-energy reservoir, and, on the other hand, measuring heat and work production, it is possible to compare the chemical input ("explained energy") with the heat and work output ("observed energy"). It has been found (Woledge et al., 1985, p. 250) that in the first few seconds of the development of an isometric contraction, a significant positive difference exists between observed and explained energy. This "unexplained energy" is usually attributed to a n reaction that takes place only at the outset of the tetanus. But the satisfaction of energy balance after the initial phase points to the fact that ATP splitting is indeed the main source of energy in muscular contraction. It should be noted that only the free-energy of a chemical reaction can be converted into mechanical work, so that any efficiency considerations should be based upon this free energy. Retracing the multiple steps involved in muscle contraction, it should be possible, in principle, to derive an expression for chemical power (i.e., free-energy rate requirements) in terms of muscle force, length, and velocity of deformation. Such an attempt has been made by Hardt (1978), who provides an explicit formula based on the Cross-bridge Theory and a number of simplifying assumptions. An attractive feature of Hardt's formula is that it assigns a non-zero value of free-energy requirement to an isometric state (zero velocity of deformation). This treatment of the isometric case is consistent with the concept of maintenance heat mentioned above, and permits the evaluation of energy costs of activities which produce no externally observable mechanical work, such as holding a weight.

An extra variable to be incorporated in some form or other in a comprehensive muscle model is the so-called *"activation"*. For a group of muscle fibres, activation represents either the proportion of fibres which are fully activated, or the uniform partial activation of each fibre, or a combination of both. The result can be expressed as a coefficient, between 0 and 1, by which the force function is called. The difficulties associated with the determination of activation include the impracticability of localized measurements (as opposed to a global EMG), and the determination of a control criterion to predict the order and extent to which fibre groups are activated.

A comprehensive muscle model can be envisaged to encompass:

(1)   An accurate geometrical representation of the muscle. This element of the model is one of the easiest to satisfy, although some difficulties arise with an exact representation of, for instance, angles of pinnation of internal fibres.

(2)   An accurate formulation of the non-linear equations of motion. This step is best achieved with the help of the principle of virtual work, which allows the incorporation of any kind of geometrical constraints, as described in this chapter.

(3)   An experimentally consistent mathematical representation of the constitutive behaviour of each component. For the passive components, such as tendinous tissue, this is a relatively simple task. For the fibres themselves, one has to resort to a model based on Hill's equation (extended to include expansion as well as con-

traction) composed with the force length dependence and with an activation scaling.

(4)   A control model of activation. This is perhaps the most challenging open question in any model. One way to handle it is by requiring a criterion of optimality (such as minimum consumption of metabolic energy), but it is not clear that such a criterion is supported by experimental evidence. Another approach is to establish a recruiting criterion, but again, experimental evidence is inconclusive in this respect.

(5)   A metabolic energy requirement equation permitting the calculation of chemical energy consumption as a function of the evolution of mechanical events at the level of each fibre.

Only when such a satisfactory model is constructed for a muscle or group of muscles, can an activity be fully modelled, analysed, and assessed in a rational and consistent manner.

4.7.8 REFERENCES

Eisenberg, D. and Crothers, D. (1979) *Physical Chemistry with Applications to the Life Sciences*. The Benjamin/Cummings Publishing Co., Menlo Park.

Gasser, H.S. and Hill, A.V. (1924) The Dynamics of Muscular Contraction. *Proc. Roy. Soc.* **96 (B)**, pp. 398-437.

Hardt, D.E. (1978) A Minimum Energy Solution for Muscle Force Control During Walking. Ph.D. Thesis, Dept. of Mech. Eng., Massachusetts Institute of Technology.

Hill, A.V. (1938) The Heat of Shortening and the Dynamic Constants of Muscle. *Proc. R. Soc.* **126 (B)**, pp. 136-195.

Jones, J.B. and Hawkins, G.A. (1986) *Engineering Thermodynamics* (2nd Ed.). John Wiley & Sons, New York.

Kestin, J. (1979) *A Course in Thermodynamics*. Hemisphere Publishing Co., Washington.

Stryer, L. (1988) *Biochemistry* (3rd Ed.). W.H. Freeman and Co., New York.

Treloar, L.R.G. (1975) *The Physics of Rubber Elasticity*. Clarendon Press, Oxford.

Truesdell, C.A. (1969) *Rational Thermodynamics*. McGraw-Hill Book Co., New York.

Woledge, R.C., Curtin, N.A., and Homsher, E. (1985) *Energetic Aspects of Muscle Contraction*. Academic Press, London.

4.8 SIMULATION

NIGG, B.M.
VAN DEN BOGERT, A.J.

4.8.1 DEFINITIONS AND COMMENTS

Analytical solution:
General solution of a mathematical problem using symbolic manipulation (e.g., paper and pencil).

Model parameter:
A numerical value that represents a certain mechanical property of the system.
Comment: a model parameter is constant in time (e.g., spring constant).

Model variable:
A numerical value that represents a certain aspect of the mechanical state of the system.
Comment: a model variable changes its value with time (e.g., the bone to-bone force in a joint).

Numerical solution:
Solution of a mathematical problem for specific values of model parameters (e.g., using a computer).

Simulation:
Predicting the behaviour of a complex system by repeatedly applying simple principles.
Comment: simulation can be done with paper and pencil. However, the repeated execution of a simple task is best done using a computer program. Thus, in practice, the expression "simulation" is commonly understood as a synonym for "computer simulation".

4.8.2 SELECTED HISTORICAL HIGHLIGHTS

Simulation is a rather young tool, as applied to biomechanical problems. It seems, therefore, difficult to speak of "historical highlights". Nevertheless, some important facts from the relatively young "history" of simulation should be listed.

1968 Riddle & Kane
Used a three-segment model to simulate the effect of arm movement for reorienting astronauts in space.

1970 Miller
Simulated somersaults with a four-segment model.

1971 Passerello & Huston
Developed a ten-segment model of the human body to simulate movement in space.

1976 Hatze
Presented a simulation model for the optimization of human movement.

1979 van Gheluwe
Simulated back somersaults with full twists.

1981	Hatze	Presented a comprehensive model for simulating human motion and applied it to the take-off phase of the long jump.
1984	Yeadon	Presented a simulation model for the analysis of twisting somersaults.
1987		First International Symposium on Computer Simulation in biomechanics.
1987	Winters & Stark	Discussed the advantages and disadvantages of changing the complexity of muscle models.
1989	Haug	Developed software for computer simulation of mechanical problems (DADS).
1991	Pandy & Zajac	Determined optimal muscle coordination for vertical jumping, using a four-segment musculo-skeletal model.
1992	van Soest	Developed a software subroutine package (SPACAR) for simulating the behaviour of biomechanical systems.

4.8.3 INTRODUCTION

Scientific investigations traditionally include experimental and theoretical work. Factual results from experimental work (i.e., observations) and theory are (or should be) constantly interacting. Science must start with facts and end with facts, no matter what theoretical structures it builds in between (Einstein, from Kemeny, 1959). Theoretical work develops and uses principles (laws) that may result in predictions about the behaviour of an actual system. Experimental work provides the necessary factual background for the development of theories and is additionally applied to support or reject theoretical predictions (see chapter 4.2).

Mathematical simulation is one component of the theoretical research approach.

> **Simulation predicts the behaviour of complex systems by repeatedly applying simple principles.**

In physics, for instance, simulation may be used to predict non-equilibrium thermodynamic properties of gases. A model is developed to describe the many particles in the gas of interest. In this model, each particle moves according to the (simple) equation of a point mass, and the force between two particles is a simple function of the distance between them. The system as a whole will show a complex behaviour due to the large number of particles, despite the simple principles governing the interaction. The simulation of such a complex system is possible today due to the substantial increase in processing speed of computers.

Two types of simulation approaches are commonly used: discrete and continuous simulation.

DISCRETE SIMULATION

Discrete simulation uses variables that can assume a finite (limited) number of values. The governing functions of discrete simulations are probabilities. The following are examples:

(1)   Simulate a roulette game to improve playing strategies.
(2)   Optimize the allocation of personnel and/or equipment in a production process with varying work load.

CONTINUOUS SIMULATION

Continuous simulation uses real number variables. The governing functions are typically ordinary or partial differential equations, which, generally, cannot be solved analytically. Mechanical models for simulation always fall into this category. The following are examples:

(1)   Constructions of steel or concrete structures are commonly tested in a computer simulation before they are actually built. Finite element methods are often applied in this process to describe the structures.
(2)   Mechanical stresses in bones are simulated in an analogous way (Huiskes and Chao, 1983).
(3)   Blood flow can be simulated with the variables, velocity and pressure, depending on both time and location (which makes the simulation computationally intensive).

This chapter discusses models describing human or animal movement that use the assumption of rigid body mechanics. A rigid body model, represented by a set of equations of motion, can be used either for simulation or for the analysis of movement. When *simulating* movement, the forces are used as input to the equations of motion. From these equations, the linear and angular accelerations are determined and integrated twice to obtain all the variables that describe the movement of each segment (Fig. 4.8.1, top). This mechanical process is called *forward dynamics* or *direct dynamics*.

When *analyzing* movement in order to determine internal forces, for instance, movement (positional and time information) is used as input into the system. This input is differentiated twice to obtain the linear and angular accelerations for each segment (Fig. 4.8.1, bottom). The determination of internal forces, however, is usually not a deterministic procedure, since there are more unknowns than equations (see chapters 4.4 and 3.5). This mechanical process is called *inverse dynamics*.

The terms *forward* and *inverse dynamics* reflect the causality of the modelling process. Forces are considered the cause of movement. Consequently, a direct dynamic analysis starts with the cause (the force), while an inverse dynamic analysis starts with the result (the movement). One and only one movement of a system of interest can be the result of a known force acting on it. However, a specific movement can be the result of an infinite number of possible force-time functions. Thus, the forward dynamic approach is deterministic while the inverse dynamic approach has an infinite number of possible solutions (see chapter 4.2).

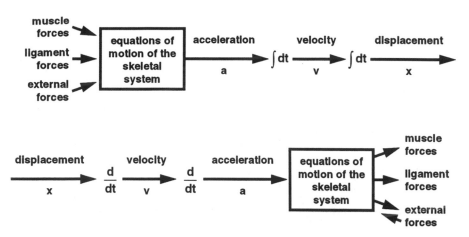

Figure 4.8.1   Schematic illustration of the similarities between the processes of simulation and in inverse dynamics analysis. Note: As shown in the diagram for inverse dynamics, external forces may be known or unknown, depending on the experimental set-up (e.g., with or without force plates).

4.8.4 SIMULATION OF MOVEMENT

Aspects of movement of a human or animal body can be studied at different levels of complexity:

Skeletal system:	A system of rigid bodies.
Musculo-skeletal system:	A system that includes rigid bodies, muscle-tendon units, and ligaments.
Neuro-musculo-skeletal systems:	A system that includes rigid bodies, muscle-tendon units, ligaments, and muscular control.

EXAMPLE OF A NEURO-MUSCULO-SKELETAL SYSTEM

A schematic description of the components and the function of a neuro-musculo-skeletal model is illustrated in Fig. 4.8.2.

Muscles are represented by mathematical equations that produce force as a function of activation, muscle length, and muscle lengthening velocity. The muscle length is the distance between the two points of attachment and depends on the relative position of the bones (rigid structures). The lengthening velocity depends on the relative velocity of these attachment points. This representation allows for the inclusion of such muscle properties as "force-length" and "force-velocity" relationships (chapter 2.5) into the model.

Ligaments are represented as elastic elements, and their viscous properties are neglected. In this representation, the force in a ligament depends on its length, which is determined by bone positions.

External forces are modelled depending on positions and velocities of body segments. Sometimes the contact between foot and ground is modelled as a rigid constraint (Hatze, 1981).

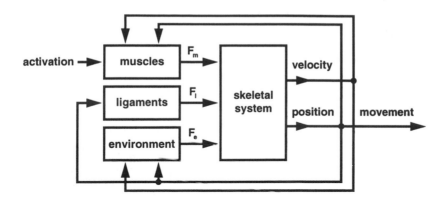

Figure 4.8.2   Schematic illustration of the components and function of a neuro-musculo-skeletal system.

The *input variables* for this type of model comprise the activation pattern of the muscle as a function of time. Thus, the model only allows pre-programmed control and does not account for changes in muscle activity that result from a movement that has been sensed by reflexes.

The *output variable* of this type of model is the movement. However, several other variables may be deduced from the model, including internal forces.

CONSTRUCTION OF A SIMULATION MODEL

Several steps are necessary in the construction of a simulation model. They are discussed for the model from Fig. 4.8.2 in the following paragraphs:

Components:

Select the level of complexity of the model, i.e., what should be included and what should not.
Comment: unnecessary components may reduce the reliability of the model output.

Segment connections:

Determine the type of segment connections appropriate for each connection. Segment connections include:

- Hinge joints, and
- Ball-and-socket joints.

Other assumptions may be chosen for specific joint connections if necessary.

Segment parameters:

Determine the parameters that describe the particular body segments. Segment parameters include:

- Distance between joints,
- Segment mass,

- Moment of inertia/inertia matrix, and
- Location of centre of mass.

The inertial properties (mass, moment of inertia, centre of mass) are usually estimated from external measurements such as length and perimeter of the body segment, using mathematical regression models based on cadaver data (e.g., Yeadon and Morlock, 1989). Some parameters may not be important for a given simulation; model and rough estimations may be sufficient (e.g., head in running).

External parameters: Determine parameters that govern the magnitude and/or direction of external forces.
Comment: material properties of ground, shoe, and heel, for instance, influence the ground reaction forces in running.

Line of action: Define the rules governing the determination of the line of action of muscle-tendon units. Lines of action have been defined as straight lines (Seireg and Arvikar, 1973; Crowninshield and Brand, 1981) or with the help of mathematical regression models (Spoor et al., 1990).

Mechanical properties: Determine the mechanical properties of passive elastic structures. Passive elastic structures include ligaments and capsules.

Muscle properties: Determine the muscle properties, so that muscle force can be calculated from the activation level, the muscle length, and muscle lengthening velocity. Practically all simulation models have used Hill's muscle model (Hill, 1938) or a variation thereof, to represent muscle properties. Important muscle-specific parameters include:

- Maximal isometric force,
- Optimal length of the contractile element,
- Width of the force-length relationship,
- Maximal contraction velocity,
- Shape factor of the force-velocity relationship,
- Stiffness of the parallel elastic element,
- Resting length of the series elastic element, and
- Stiffness of the series elastic element.

(An overview of muscle models suitable for musculoskeletal simulation can be found in Winters and Stark, 1987).

NUMERICAL METHODS

Simulation models require numerical solutions. Two options are possible for the implementation of a computer simulation:

(1)   Using standard numerical methods:

The equations of motion are formulated as a system of ordinary linear differential equations. This system is solved by using standard numerical methods (Shampine and Gordon, 1975). This process has the advantages that the researcher has complete control over the process and that the program can be tailored to the needs of the user. The process has the disadvantages that much effort is required for programming, and that it is difficult to develop the software to make it easily adaptable for new situations.

(2)   Using dedicated computer software:

The simulation is solved using dedicated computer software such as DADS, ADAMS, and SPACAR (Nikravesh, 1984; Haug, 1989; van Soest, 1992). These programmes accept a high level description of the model (a list of segments with their connections). From this description, the equations of motion are generated and solved. This approach is recommended when the mechanical model has many components, making the equations of motion and their solution rather complex. An important advantage of this approach is that the researcher does not have to cope with typical problems in software development, which do not relate directly to the actual question of interest. The process has the disadvantage that the researcher loses direct control over the simulating process (many "black boxes"). As well, the inclusion of non-standard model components (muscles, shoes, etc.) requires that the program be modified, which necessitates knowledge of part of the source code.

4.8.5 RELEVANCE OF SIMULATION MODELS

Simulation models can be used for several different purposes:

(1)   Simulation is a powerful tool for improving our *understanding* of a system. Successive modifications of a simulation model will reveal which model property is responsible for a particular aspect of the model's behaviour.

(2)   Simulation may be applied to *optimize* the design of a system. Muscle activation patterns, for instance, may be modified until a certain performance criterion is fulfilled.

Examples: A minimum time kicking movement (Hatze, 1976). A maximum height vertical jump (Pandy et al., 1990; Pandy and Zajac, 1991).

(3)   Simulation may be applied to study *internal forces* and their dependence on model parameters.

Examples: estimation of tendon and ligament forces in the forelimb of a horse, and their dependence on ground hardness and horseshoe shape (van den Bogert and Schamhardt, 1992).

4.8.6 EXAMPLE OF A MUSCULO-SKELETAL SIMULATION MODEL

Question:

As an example of the application of simulation, the task of dynamic knee extension is discussed. The question of interest is which properties of the system determine the time required to move the knee joint from 90° flexion to full extension while the subject is in a sitting position. (Fig. 4.8.3). This problem is similar to, but much simpler than the two-

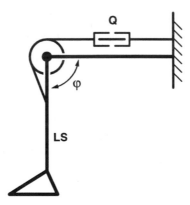

Figure 4.8.3   **Illustration of the example used for illustration of the simulation process. Q = quadriceps muscle; LS = leg segment.**

segment, five-muscle model of Hatze (1976). The influences of anatomy and muscle architecture are assessed.

THE RIGID-BODY MODEL

Fig. 4.8.3 shows the model used. The foot and shank are combined into one rigid leg segment, and the knee extensors are combined into one muscle.

Assumptions:
(1)   The problem can be treated two-dimensionally.
(2)   The line of action of the quadriceps muscle is assumed to run along a circular pulley, which represents the patella.

EM:

The equation of motion for this system is:

$$\ddot{\varphi} = \frac{M}{I}$$

Symbols:

φ = angle between thigh and shank (zero at full flexion)

$\ddot{\varphi}$ = angular acceleration at the knee joint

M = total moment on the leg segment with respect to the knee joint axis: M is the sum of two terms, one due to the extensor muscle force, F, and one in the opposite direction due to gravity which only exists if φ is not equal to $\pi/2$:

$$M = r_F \cdot F - mg \cdot r_{CM} \cdot \sin{(\varphi - \frac{\pi}{2})}$$

I = moment of inertia of the leg segment with respect to the knee joint axis: inertial properties for a leg segment can be used from literature (Winter, 1979). For a male subject with a total body mass of 80 kg and height of 1.80 m, the moment of inertia is:

$$I = 0.1832 \text{ kg.m}^2.$$

m = segment mass of the lower segment: the segment mass, m, of the lower leg plus foot is assumed to be 4.88 kg (Winter, 1979)

r_{CM} = distance from the joint axis to the segment centre of mass: this distance is 0.264 m (Winter, 1979)

g = acceleration of gravity (9.81 m/s^2)

r_F = moment arm (radius of the pulley) of the quadriceps muscle with respect to a joint centre: the distance r_F can be measured from a radiograph, assuming a joint centre. However, a more accurate alternative is to measure tendon displacement as a function of joint angle in a cadaver (Visser et al., 1990). With a circular pulley, tendon travel should be a linear function of the joint angle with slope r_F (if φ is expressed in radians). When a straight line is fitted to the experimental data (Visser et al., 1990), a moment arm equal to 8% of the length of the thigh is obtained. The "typical subject" used in this example has a thigh length of 0.41 m (Winter, 1979); therefore, $r_F = 0.033$ m

THE MUSCLE MODEL

Assumptions:

(1) The muscle force, F, will be generated by a Hill muscle model.
(2) For this maximum-effort activity the muscle is maximally activated throughout the movement.
(3) The tendon is inextensible.

(4) The force-velocity relationship for a shortening muscle fibre has the hyperbolic shape described by Hill (1938).

With these assumptions, the muscle force only depends on the fibre length, L, and the fibre contraction velocity, v, which is the first time derivative of the muscle fibre length, L.

$$F = F_{max} \cdot f_1 (l) \cdot f_2 (v)$$

The functions $f_1 (L)$ and $f_2 (v)$ are equal to one for an isometric (v=0) contraction at optimal fibre length $(L = L_{opt})$.

Symbols:

L   = muscle fibre length: assuming an inextensible tendon and a constant moment arm, the fibre length L depends on φ according to:

$$L = L_0 - r_F \cdot \varphi$$

L_0 = fibre length at full flexion $(\varphi=0)$: L_0 is the distance between origin and insertion at the angle $\varphi=0$, minus the length of the tendon. This parameter is difficult to determine experimentally, and is important because it determines in which part of the force-length relationship the muscle will operate. For this example, it is assumed that the fibres are at optimal length when the movement starts $(\varphi(t=0) = \pi/2,$ $\dot{\varphi} (t=0)=0)$, which gives:

$$L_0 = L_{opt} + (\pi/2) \cdot r_F$$

L_{opt} = optimal fibre length: the optimal fibre length, L_{opt}, is typically determined by counting the number of sarcomeres in a muscle fibre under a microscope. The results from several fibres from the same muscle should be averaged to get a reliable value. This number is multiplied by the optimal length of a sarcomere (about 2 μm, Gordon et al., 1966) to obtain the optimal fibre length. This procedure typically results in values between 0.08 and 0.10 m for the vasti muscles and for the rectus femoris muscle. In this example, a mean value of 0.09 m is used

v   = fibre contraction velocity with:

$$v = -\frac{dL}{dt}$$

The fibre contraction velocity, v, depends on the joint angular velocity according to the first derivative of the equation for L:

$$v = r_F \cdot \dot{\varphi}$$

v_{max} = maximal fibre contraction velocity

$f_1(L)$ = normalized force length-relationship which is, in this example, approximated by the parabolic function:

$$f_1(L) = 1 - \left(\frac{L - L_{opt}}{w \cdot L_{opt}}\right)^2$$

$f_2(v)$ = normalized force-velocity relationship: the relationship is written as:

$$f_2(v) = \frac{v_{max} - v}{v_{max} + cv}$$

The function $f_2(v)$ is characterized by two parameters, the maximum contraction velocity v_{max} and a shape parameter c. They are related to the traditional parameters a and b (see section 2.5) as follows:

$$c = \frac{F_{max}}{a}$$

$$v_{max} = \frac{b \cdot F_{max}}{a}$$

Literature sources suggest that $v_{max} \approx 5.L_{opt}s^{-1}$, depending slightly on the fibre type distribution (fast muscles have a larger v_{max}). This means that an unloaded muscle fibre contracts at a speed of 5 fibre lengths per second. In other words, the maximum possible active contraction of 0.5 fibre lengths requires about 100 milliseconds when the muscle is not loaded. The parameter c determines the curvature of the force-velocity relationship. The curve is linear when c = 0, and has an infinitely sharp curvature when c = ∞. The value 2.5 is typically found in the literature

F_{max} = maximal force the muscle can produce isometrically: F_{max} is estimated from the physiological cross-sectional area (PCSA). In this example, F_{max} is assumed to be approximately 12,000 N for the total quadriceps group

w = relative width of the force-length relationship: literature data suggests a typical value of w = 0.5, which means that muscle fibres can contract or lengthen 50% before they lose their force-generating capabilities, starting from optimal length (Gordon et al., 1966)

THE STATE EQUATION OF THE ENTIRE MODEL

After substituting the muscle model and gravitational moment into the equation of motion, the model is represented by the following second-order differential equation:

$$\ddot{\varphi} = \frac{r_F \cdot F_{max}}{I} \left(1 - \left(\frac{r_F\left(\frac{\pi}{2} - \varphi\right)}{w \cdot L_{opt}}\right)^2\right) \left(\frac{v_{max} - r_F \cdot \dot{\varphi}}{v_{max} + c \cdot r_F \cdot \dot{\varphi}}\right) - \frac{mg \cdot r_{CM}}{I} \sin\left(\varphi - \frac{\pi}{2}\right)$$

The differential equation can be solved using a computer program (program KICK.C in C in Appendix A), the initial conditions, and numerical values for the parameters:

Initial conditions:

$$\varphi(0) = \pi/2$$
$$\dot{\varphi}(0) = 0$$

Parameters:

m	=	4.88 kg
g	=	9.81 m/s^{-2}
r_{CM}	=	0.264 m
I	=	0.1832 kgm^2
r_F	=	0.033 m
F_{max}	=	12,000 N
L_{opt}	=	0.09 m
w	=	0.5
v_{max}	=	0.45 m/s^{-1}
c	=	2.5

Comments:

Even though a large number of parameters is required for this model, it has been demonstrated that it is possible to obtain good estimates for the parameter values from the literature. A potential problem when combining parameters from different sources is, that they come from different individuals. Whenever possible, the model parameters should be made consistent by appropriate scaling, as was done for the moment arm, r_F, in the present model. The properties of the contractile elements may seem complex, but in fact only two muscle parameters are muscle-specific: the maximal isometric force F_{max}, and the optimal fibre length L_{opt}. Both are easily determined from cadaver measurements, and representative values can also be found in the literature for various muscles. The other muscle parameters are properties of 'muscle tissue' in general, and, therefore, have identical values for different muscles. A possible exception is the ratio v_{max}/L_{opt}, which may vary slightly, depending on the fibre type distribution.

RESULTS

Fig. 4.8.5 shows results of the simulation, with the nominal parameter values. Full knee extension is reached after 0.140 seconds. When simulation is used as a research tool, it is important to validate the results. In this case, it would be relatively simple to measure knee angle as a function of time in an experimental situation. When the experiment and simulation produce drastically different results, the errors in the model must be identified. Often, adjustment of one or more parameter values will result in realistic simulation results. Parameter adjustment to 'fit' the model is, however, not without risk; there are many possible combinations of parameter values which produce similar angle-time patterns. It is, therefore, always best to estimate model parameters using independent methods. Confidence in the model may be increased even further by changing the conditions in the experiment, and performing the corresponding simulation. In this example, attaching different weights to the foot would be a good test for the model.

The simulation may be used to answer specific questions such as "why has evolution resulted in a pulley radius of 3.3 cm?". A larger patella would increase the moment arm of the quadriceps, making the limb stronger. It is easy to increase the moment arm in the model by 50%, and the results of the simulation are shown in Fig. 4.8.4. As expected, the

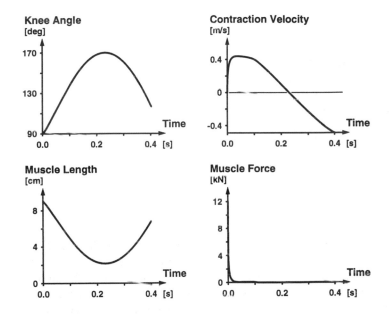

Figure 4.8.4   **Results of the simulation example with a 50% larger moment arm of the patella. Variables as in previous figure.**

initial acceleration is increased. But with the same movement, the muscle fibres have to shorten more and faster, when the pulley is larger. This results in so much loss of muscle force that full extension is not even reached. The maximum joint angle of 169.12 degrees is reached after 0.229 seconds. It can be concluded from this simple simulation that a larger moment arm makes small movements faster but large movements impossible. Possibly, the moment arms of the muscles are optimized for the range of motion of the joints during typical daily activities.

The foregoing simulation has demonstrated that the shortening capabilities of the muscle fibres are limiting the performance. One may try to increase performance by changing the muscle architecture. The muscle in the example had a fibre length, L_{opt}, of only 9 cm, which means that the muscle has a pennate architecture.

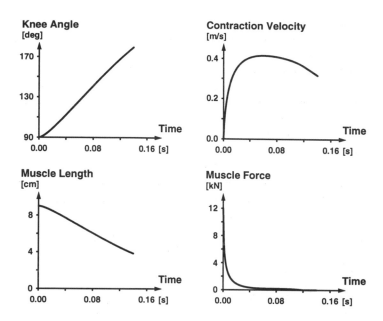

Figure 4.8.5   **Results of the simulation of the selected simple example. Four variables are shown as a function of time: joint angle, fibre length, fibre contraction velocity, and muscle force.**

The length of the thigh is sufficient to increase the fibre length by 50%, to a value of 13.5 cm, resulting in a more parallel-fibred architecture. This also increases the maximum

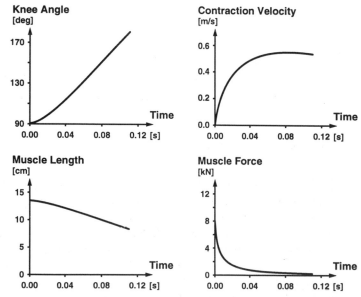

Figure 4.8.6   **Results for the simulation of the selected example after changing the muscle architecture. The number of sarcomeres along the length of each fibre was increased by 50%, and, consequently, the PCSA was reduced to 2/3 of its original value to keep a constant volume.**

contraction velocity, v_{max}, by 50% (because $v_{max} = 5\ L_{opt}$) to a value of 0.675 m/s. However, when the fibres are longer, there are also fewer fibres if the volume of the muscle is kept constant. Thus, only muscle architecture is changed, not size. This means that the PCSA is reduced to 2/3 of its original value, and, therefore, F_{max} is now 8000 N. This simulation is shown in Fig. 4.8.6. Even though the initial acceleration is reduced by 33%, the movement is still executed faster (0.110 seconds) because the muscle force can be maintained for a longer time. The modified muscle architecture is obviously more optimal for this particular task. For other, more important activities, such as walking or running, short fibres and a large PCSA could be better. Presumably humans are not optimized for kicking movements.

Modifications to the anatomy and muscle architecture may be interesting for those interested in functional anatomy, but have hardly any practical relevance. In this respect, models with more segments and more muscles are much more useful. For the knee extension task, it was immediately clear that maximal activation would give the best performance. This is no longer true for more complex models. For example, an optimal muscle activation (coordination) pattern has been found for vertical jumping, using a four-segment model (Pandy and Zajac, 1991). These simulation and optimization techniques are potentially useful in improving movement technique in many sports.

4.8.7 REFERENCES

Crowninshield, R.D. and Brand, R.A. (1981) A Physiologically Based Criterion of Muscle Force Prediction in Locomotion. *J. Biomechanics.* **14 (11),** pp. 793-801.

Gordon, A.M., Huxley, A.F., and Julian, F.J. (1966) The Variation in Isometric Tension with Sarcomere Length in Vertebrate Muscle Fibres. *J. Physiol.* **184,** pp. 170-192.

Hatze, H. (1976) The Complete Optimization of a Human Motion. *Math. Biosci.* **28,** pp. 99-135.

Hatze, H. (1981) A Comprehensive Model for Human Motion Simulation and its Application to the Take-off Phase of the Long Jump. *J. Biomechanics.* **14 (3),** pp. 135-142.

Haug, E.J. (1989) *Computer-aided Kinematics and Dynamics of Mechanical Systems. Basic Methods.* Allyn & Bacon, Boston.

Hill, A.V. (1938) The Heat of Shortening and the Dynamic Constants of Muscle. *Proc. Roy. Soc.* **126 (B),** pp. 136-195.

Huiskes, R. and Chao, E.Y.S. (1983) A Survey of Finite Element Analysis in Orthopedic Biomechanics: The First Decade. *J. Biomechanics.* **16 (6),** pp. 385-409.

Kemeni, J.G. (1959) *A Philosopher Looks at Science.* D. van Nostrand Comp., Princeton, NJ.

Miller, D.I. (1970) *A Computer Simulation Model of the Airborne Phase of Diving.* (Ph.D. Dissertation) Pennsylvania State University, University Park, PA.

Nikravesh, P.E. (1984) Some Methods for Dynamic Analysis of Constrained Mechanical Systems: A Survey. *Computer-aided Analysis and Optimization of Mechanical System Dynamics (NATO ASI-F9)* (ed. Haug, E.J.). Springer Verlag, Berlin.

Pandy, M.G., Zajac, F.E., Sim, E., and Levine, W.S. (1990) An Optimal Control Model for Maximum-height Human Jumping. *J. Biomechanics.* **23 (12),** pp. 1185-1198.

Pandy, M.G. and Zajac, F.E. (1991) Optimal Muscular Coordination Strategies for Jumping. *J. Biomechanics.* **24,** pp. 1-10.

Passerello, C.E. and Huston, R.L. (1971) Human Attitude Control. *J. Biomechanics.* **4 (2),** pp. 95-102.

Riddle, C. and Kane, T.R. (1968) Reorientation of the Human Body by Means of Arm Motions. *Technical Report 182, 62 (NASA CR-95362).* Division of Engineering Mechanics, Stanford University, Palo Alto, CA.

Seireg, A. and Arvikar, R.J. (1973) A Mathematical Model for Evaluation of Forces in Lower Extremities of the Musculo-skeletal System. *J. Biomechanics.* **6 (3),** pp. 313-326.

Shampine, L.F. and Gordon, M.K. (1975) *Computer Solution of Ordinary Differential Equations: The Initial Value Problem*. W.H. Freeman, San Francisco, CA.

Spoor, C.W., van Leeuwen, J.L., Meskers, C.G.M., Titulaer, A.F., and Huson, A. (1990) Estimation of Instantaneous Moment Arms of Lower-leg Muscles. *J. Biomechanics*. **23 (12)**, pp. 1247-1259.

van den Bogert, A.J. and Schamhardt, H.C. (In press) Modelling and Simulation of Animal Locomotion. *Acta. Anatomica*.

van Gheluwe, B. (1974) A New Three-dimensional Filming Technique Involving Simplified Alignment and Measurement Procedures. *Biomechanics IV* (eds. Nelson, R.C. and Morehouse, C.A.). University Park Press, Baltimore. pp. 476-482.

van Soest, A.J., Schwab, A.L., Bobbert, M.F., and van Ingen Schenau, G.J. (1992) SPACAR: A Software Subroutine Package for Simulation of the Behaviour of Biomechanical Systems. *J. Biomechanics*. **25**, pp. 1241-1246.

Visser, J.J., Hoogkamer, J.E., Bobbert, M.F., and Huijing, P.A. (1990) Length and Moment Arm of Human Leg Muscles as a Function of Knee and Hip-joint Angles. *Eur. J. Appl. Physiol.* **61 (5-6)**, pp. 453-460.

Winter, D.A. (1979) *Biomechanics of Human Movement*. Appendix A. John Wiley & Sons, New York.

Winters, J.M. and Stark, L. (1987) Muscle Models: What is Gained and What is Lost by Varying Model Complexity. *Biological Cybernetics*. **55**, pp. 403-420.

Yeadon, M.R. and Morlock, M. (1989) The Appropriate use of Regression Equations for the Estimation of Segmental Inertia Parameters. *J. Biomechanics*. **22 (6-7)**, pp. 683-689.

APPENDIX A: KICK.C

See section 4.8.6 regarding the usage of this program.

```c
#include <stdio.h>
#include <math.h>
main ()
{
/* Declare and initialize model parameters: */
float    m = 4.88,/* segment mass                                          */
         I = 0.1832,/* segment moment of inertia                           */
         rcm = 0.264,/* distance between CM and joint                       */
         g = 9.81, /* acceleration of gravity                              */
         rf = 0.033,/* moment arm of quadriceps                            */
         Fmax = 12000,/* maximum isometric force                           */
         Lopt = 0.09,/* optimal fibre length                               */
         w = O.5, /* width (relative to Lopt) of F-L curve*/
         vmax = 0.45,/* maximal contraction velocity                       */
         c = 2.5,   /* shape parameter of F-V curve                        */

/* Declare model variables: */
         t,        /* time                                                 */
         alfa, alfadot,/* joint angle and angular velocity                 */
         alfadotdot,/* joint angular acceleration                          */
         Fmus,    /* muscle force                                          */
         Mmus, Mgra,/* moments due to muscle and gravity*/

/* Constants */
         pi = 3.141592653589793,halfpi = pi/2.0,
         dt = 0 001;/* step size for integration */

/* Set initial values */
         t = 0.0;
         alfa = halfpi;
         alfadot = 0.0;

/* Simulate motion until full extension, or .4 seconds elapsed */
         while (alfa < pi && t < 0.4) {

/*       force length relationship */
                  Fmus = Fmax * (1.0 - pow(rf*(halfpi-alfa)/w/Lopt, 2.0));
                  if (Fmus < 0.0) Fmus = 0.0;

/*       multiply by force-velocity relationship */
                  Fmus = Fmus * (vmax - rf*alfadot)/(vmax + c*rf*alfadot);
                  if (Fmus < 0.0) Fmus = 0.0;

/*       calculate muscle moment and gravitational moment */
                  Mmus = Fmus * rf;
                  Mgra = - m * g * rcm * sin(alfa - halfpi);

/*       calculate angular acceleration */
                  alfadotdot = (Mmus + Mgra)/I;

/*       print output:
         time, angle, contraction velocity, fibre length, muscle force */
                  printf("%8.3f %8.2f %8.4f %8.4f %8.2f\n", t, alfa*180/pi,
                                    rf*alfadot, Lopt+rf*(halfpi-alfa), Fmus);
/*       integrate from t to t+dt */
                  alfadot                  = alfadot + alfadotdot * dt;
                  alfa                     = alfa + alfadot *dt;
                  t                        = t + dt;
         }
}
```

INDEX

Index compiled by Geoffrey C. Jones